ENGINEERING HYDROLOGY

Fourth Edition

About the Author

K Subramanya is a retired Professor of Civil Engineering at the Indian Institute of Technology, Kanpur. He obtained his bachelor's degree in Civil Engineering from Mysore University and a master's degree from the University of Madras. Further, he obtained another master's degree and PhD degree from the University of Alberta, Edmonton, Canada. He has taught at IIT Kanpur for over 30 years and has extensive teaching and research experience in the area of Fluid Mechanics, Hydrology and Hydraulic Engineering. During his tenure at IIT Kanpur, Prof. Subramanya worked as Visiting Faculty at the Asian Institute of Technology, Bangkok, and at Concordia University, Montreal, Canada, for a short while.

He has authored several successful books for McGraw Hill Education (India). Besides the current book, his other books include *Flow in Open Channels*; *Fluid Mechanics* and *Hydraulic Machines–Problems and Solutions*; and *Hydraulic Machines*.

Dr Subramanya has published over eighty technical papers in national and international journals and conferences. He is a Fellow of the Institution of Engineers (India), Fellow of the Indian Society for Hydraulics, Member of Indian Society for Technical Education and Member, Indian Water Resources Association.

Currently, Prof. Subramanya resides in Bangalore and is active as a practicing consultant in Water Resources Engineering. He can be contacted at ***subramanyak1@gmail.com***

ENGINEERING HYDROLOGY

Fourth Edition

K Subramanya
Former Professor of Civil Engineering Department
Indian Institute of Technology
Kanpur

McGraw Hill Education (India) Private Limited
NEW DELHI

McGraw Hill Education Offices
New Delhi New York St Louis San Francisco Auckland Bogotá Caracas
Kuala Lumpur Lisbon London Madrid Mexico City Milan Montreal
San Juan Santiago Singapore Sydney Tokyo Toronto

McGraw Hill Education (India) Private Limited

Published by McGraw Hill Education (India) Private Limited
P-24, Green Park Extension, New Delhi 110016

Engineering Hydrology, 4e

ISBN (13): 978-9-38-328653-9
ISBN (10): 9-38-328653-9

Vice President and Managing Director: *Ajay Shukla*

Head—Higher Education (Publishing and Marketing): *Vibha Mahajan*
Publishing Manager (SEM & Tech. Ed.): *Shalini Jha*

Senior Editorial Researcher: *Harsha Singh*
Manager—Production Systems: *Satinder S Baveja*
Assistant Manager—Editorial Services: *Sohini Mukherjee*
Production Executive: *Anuj K Shriwastava*

Assistant General Manager—Higher Education (Marketing): *Vijay Sarathi*
Product Specialist: *Sachin Tripathi*
Senior Graphic Designer—Cover: *Meenu Raghav*

General Manager—Production: *Rajender P Ghansela*
Manager—Production: *Reji Kumar*

Typeset at Print-O-World, 2579, Mandir Lane, Shadipur, New Delhi 110 008, and printed at Sai Printo Pack, A 102/4, Okhla Industrial Area, Phase-II, New Delhi - 110020

Cover Printer : Sai Printo Pack

DRZCRRXORABAR

Dedicated
to
My Mother

Content

Preface *vii*

1. Introduction **1**

1.1 Introduction *1*
1.2 Hydrologic Cycle *1*
1.3 Water-Budget Equation *4*
1.4 World Water Balance *7*
1.5 Global Freshwater Resources *10*
1.6 History of Hydrology *11*
1.7 Applications In Engineering *12*
1.8 Sources of Data *13*
References 14
Revision Questions 14
Problems 14
Objective Questions 15

2. Precipitation **17**

2.1 Introduction *17*
2.2 Forms of Precipitation *18*
2.3 Weather Systems for Precipitation *18*
2.4 Characteristics of Precipitation in India *20*
2.5 Measurement of Precipitation *25*
2.6 Raingauge Network *29*
2.7 Preparation of Data *31*
2.8 Presentation of Rainfall Data *37*
2.9 Mean Precipitation Over An Area *40*
2.10 Depth-Area-Duration Relationships *45*
2.11 Frequency of Point Rainfall *49*
2.12 Maximum Intensity/Depth-Duration-Frequency Relationship *53*
2.13 Probable Maximum Precipitation (PMP) *60*
2.14 World's Greatest Observed Rainfall *61*
2.15 Monsoon Studies in India *61*
2.16 Rainfall Data In India *63*
References 64
Revision Questions 64
Problems 65
Objective Questions 70

3. Abstractions from Precipitation **73**
3.1 Introduction *73*
3.2 Evaporation Process *73*
3.3 Evaporimeters *75*
3.4 Empirical Evaporation Equations *77*
3.5 Analytical Methods of Evaporation Estimation *79*
3.6 Reservoir Evaporation and Methods for its Reduction *82*
3.7 Transpiration *84*
3.8 Evapotranspiration *84*
3.9 Measurement of Evapotranspiration *86*
3.10 Evapotranspiration Equations *86*
3.11 Potential Evapotranspiration Over India *93*
3.12 Actual Evapotranspiration (AET) *94*
3.13 Interception *95*
3.14 Depression Storage *96*
3.15 Infiltration *97*
3.16 Infiltration Capacity *98*
3.17 Measurement of Infiltration *100*
3.18 Modelling Infiltration Capacity *102*
3.19 Classification of Infiltration Capacities *109*
3.20 Infiltration Indices *110*
References 115
Revision Questions 116
Problems 116
Objective Questions 119

4. Streamflow Measurement **122**
4.1 Introduction *122*
4.2 Measurement of Stage *123*
4.3 Measurement of Velocity *126*
4.4 Area-Velocity Method *131*
4.5 Dilution Technique of Streamflow Measurement *135*
4.6 Electromagnetic Method *138*
4.7 Ultrasonic Method *139*
4.8 Indirect Methods *140*
4.9 Stage-Discharge Relationship *145*
4.10 Extrapolation of Rating Curve *153*
4.11 Hydrometry Stations *155*
References 157
Revision Questions 157
Problems 158
Objective Questions 161

5. Runoff **164**
5.1 Introduction *164*

5.2 Hydrograph *167*
5.3 Runoff Characteristics of Streams *168*
5.4 Catchment Characteristics *169*
5.5 Runoff Volume (Yield) *173*
5.6 SCS-CN Method of Estimating Runoff Volume *185*
5.7 Flow-Duration Curve *194*
5.8 Flow-Mass Curve *197*
5.9 Sequent Peak Algorithm *206*
5.10 Droughts *210*
5.11 Surface-Water Resources of India *218*
5.12 Glacier Resources of Indian Himalayas *224*
5.13 Environmental Flows *225*
References 231
Revision Questions 232
Problems 233
Objective Questions 238

6. Hydrographs **240**
6.1 Introduction *240*
6.2 Factors Affecting Runoff Hydrograph *241*
6.3 Components of a Hydrograph *245*
6.4 Base Flow Separation *248*
6.5 Effective Rainfall (ER) *250*
6.6 Unit Hydrograph *252*
6.7 Derivation of Unit Hydrograph *259*
6.8 Unit Hydrographs of Different Durations *263*
6.9 Use and Limitations of Unit Hydrograph *270*
6.10 Duration of the Unit Hydrograph *270*
6.11 Distribution Graph *271*
6.12 Synthetic Unit Hydrograph *272*
6.13 Instantaneous Unit Hydrograph (IUH) *282*
References 287
Revision Questions 287
Problems 287
Objective Questions 292

7. Floods **296**
7.1 Introduction *296*
7.2 Rational Method *296*
7.3 Empirical Formulae *302*
7.4 Unit Hydrograph Method *304*
7.5 Flood Frequency Studies *305*
7.6 Gumbel's Method *308*
7.7 Log-Pearson Type III Distribution *317*

7.8 Partial Duration Series *321*
7.9 Regional Flood-Frequency Analysis *322*
7.10 Extremes of Extremes-Envelope Curve of Highest Floods in the World *322*
7.11 Data for Frequency Studies *324*
7.12 Design Flood *324*
7.13 Design Storm *326*
7.14 Risk, Reliability and Safety Factor *329*
References 331
Revision Questions 332
Problems 332
Objective Questions 337

8. Flood Routing **340**
8.1 Introduction *340*
8.2 Basic Equations *341*
8.3 Hydrologic Storage Routing (Level Pool Routing) *342*
8.4 Attenuation *351*
8.5 Hydrologic Channel Routing *352*
8.6 Hydraulic Method of Flood Routing *359*
8.7 Routing in Conceptual Hydrograph Development *360*
8.8 Clark's Method for IUH *360*
8.9 Nash's Conceptual Model *364*
8.10 Flood Control *373*
8.11 Flood Control in India *377*
8.12 Global Warming and its Impact on Water Resources of India *378*
References 381
Revision Questions 382
Problems 382
Objective Questions 386

9. Groundwater **389**
9.1 Introduction *389*
9.2 Forms of Subsurface Water *389*
9.3 Saturated Formation *391*
9.4 Aquifer Properties *393*
9.5 Geologic Formations as Aquifers *400*
9.6 Compressibility of Aquifers *401*
9.7 Equation of Motion *404*
9.8 Wells *415*
9.9 Steady Flow into a Well *416*
9.10 Open Wells *421*
9.11 Unsteady Flow in a Confined Aquifer *424*
9.12 Well Loss *430*
9.13 Specific Capacity *431*

9.14 Sea-Water Intrusion *431*
9.15 Recharge *432*
9.16 Groundwater Resource *437*
9.17 Groundwater Monitoring Network in India *441*
References 442
Revision Questions 442
Problems 443
Objective Questions 448

10. Erosion and Reservoir Sedimentation **452**
10.1 Introduction *452*
10.2 Erosion Processes *452*
10.3 Estimation of Sheet Erosion *454*
10.4 Channel Erosion *457*
10.5 Movement of Sediment from Watersheds *459*
10.6 Sediment Yield from Watersheds *460*
10.7 Trap Efficiency *465*
10.8 Density of Sediment Deposits *467*
10.9 Distribution of Sediment in the Reservoir *472*
10.10 Life of a Reservoir *483*
10.11 Reservoir Sedimentation Control *484*
10.12 Erosion and Reservoir Sedimentation Problems in India *485*
References 489
Revision Questions 490
Problems 490
Objective Questions 493

Appendix A – Multiple Choice Questions **495**

Appendix B – Abbreviations **509**

Appendix C – Conversion Factors **512**

Appendix D – Answers to Objective Questions **513**

Appendix E – Answers to Problems **514**

Index **526**

Preface

Overview

Water is vital to life and development in all parts of the world. In third-world countries, where the agricultural sector plays a key role in their economic growth, the management of water resources is an item of high priority in their developmental activities. The basic inputs in the evaluation of water resources are from hydrological parameters and the subject of hydrology forms the core in the evaluation and development of water resources. In the civil engineering curriculum, this subject occupies an important position.

During my long teaching experience, I have felt a strong need for a textbook oriented to the Indian environment and written in a simple and lucid style. The present book is a response to the same. This book is intended to serve as a text for a first course in engineering hydrology at the undergraduate level in the civil engineering discipline. Students specialising in various aspects of water-resources engineering, such as water-power engineering and agricultural engineering will find this book useful. This book also serves as a source of useful information to professional engineers working in the area of water resources evaluation and development.

Engineering hydrology encompasses a wide spectrum of topics and a book like the present one meant for the first course must necessarily maintain a balance in the blend of topics. The subject matter has been developed in a logical and coherent manner and covers the prescribed syllabi of various Indian universities. The mathematical part is kept to the minimum and emphasis is placed on the applicability to field situations relevant to Indian conditions. SI units are used throughout the book.

Further, the book is unique in being India-centric in its use of examples, applications and description of water-resources information and practices and this aspect may be of use to many professionals and NGOs working in the water sector.

Objective of the Fourth Edition

This book has been receiving very encouraging response from users: students, teachers and professionals alike. In response to a feedback by some users who felt the need for inclusion of a few more topics in the book, a thorough review of the book was undertaken. The fourth edition is the outcome of that exercise.

The main aim of this edition is to make the book meet practically all the needs of a textbook for the first course in Engineering Hydrology at Indian technical universities and engineering institutions. Towards this end, the redundant material of the third edition has been thoroughly pruned and appropriated new material has been added to make the book up to date and useful for a wider range of users.

While the book is designed essentially to meet the requirements of the undergraduate textbook in Engineering Hydrology, it will also be of help and use to students taking competitive examinations conducted by the UPSC, like the Central Engineering Services, Central Civil Services and Indian Forestry Services Examinations.

New to this Edition

It is believed that the new subject matter that has been added gives an introduction to important emerging areas of study and professional interest and would help users keep themselves well informed.

The significant changes to the present edition are the following:

- Comprehensive coverage and clear explanations of concepts presented to field situations relevant to Indian conditions
- Emphasis on current water-resources scenario in India with authentic and updated statistical data
- Contains latest technologies and applications like GPS, GIS and Remote Sensing in various aspects of applied hydrology
- Addition of new topics:
    - Global warming and its impact on the water resources of India
    - Environmental flows
    - Glacier resources of India
    - Global freshwater resources
    - Envelope curves of highest floods in the world
    - Catchment characteristics
    - Effect of urbanisation on runoff from a catchment
    - Sea-water intrusion in coastal aquifers
- Refreshed pedagogy:
    - Solved Examples: 107
    - Numerical Problems: 220
    - Review Questions: 139
    - Objective Type Questions: 245
    - Figures: 176
    - Tables: 105 (updated wherever necessary)

Chapter Organisation

Designed essentially for a one-semester course, the material in the book is presented in ten chapters. The hydrologic cycle and world water balance are covered in *Chapter 1*. Aspects of precipitation, essentially rainfall, are dealt in sufficient detail in *Chapter 2*. Hydrologic abstractions including evapotranspiration and infiltration are presented in *Chapter 3*. Streamflow-measurement techniques and assessment of surface-flow yield of a catchment form the subject matter of *Chapters 4 and 5*.

The characteristics of flood hydrographs and the unit hydrograph theory together with an introduction to instantaneous unit hydrograph are covered in sufficient detail with numerous worked examples in *Chapter 6*. Floods, a topic of considerable importance, constitute the subject matter of *Chapters 7 and 8*. While in *Chapter 7*, the flood-peak estimation and frequency studies are described in detail, *Chapter 8* delas with the aspects of flood routing, flood control and forecasting. Basic information on the hydrological aspects of groundwater has been covered in *Chapter 9*. Finally, erosion and reservoir sedimentation are presented in *Chapter 10*.

Numerous worked examples, a set of problems and a set of multiple-choice questions are provided at the end of each chapter to enable students gain good comprehension of the subject. Questions and problems included in the book are largely original and are designed to enhance the capabilities of comprehension, analysis and application of the student.

Online Learning Centre

The online Learning Centre of this book can be accessed at ***http://www.mhhe.com/subramanya/eh4*** and contains the following material:

For Instructors: Solution Manual and PowerPoint slides

For Students: Model Question Papers, List of Abbreviations, Additional References and Web links for further reading

Acnowledgements

I am grateful to UNESCO for permission to reproduce several figures from their publication, *Natural Resources of Humid Tropical Asia—Natural resources Research XII* ** UNESCO, 1974; the Director General of Metrology, India Metrological Department, Government of India for permission to reproduce several maps; M/s Leupold and Stevens Inc., Beaverton, Oregon, USA, for photographs of hydrometeorological instruments; M/s Alsthom-Atlantique, Neyrtec, Grenoble France, for photographs of several Neyrtec Instruments, M/s Lawrence and Mayo (India) Pvt. Ltd., New Delhi, for the photograph of a current meter.

Thanks are due to Professor K V G K Gokhale for his valuable suggestions and to Sri Suresh Kumar for his help. I wish to thank my student friends who helped in this endeavour in many ways. The financial support received under the Quality Improvement Programme (QIP), Government of India, through the Indian Institute of Technology, Kanpur, for the preparation of the manuscript is gratefully acknowledged.

The following reviewers of the typescript have provided valuable inputs to the contents of this edition. I would like to express my thanks to these reviewers and to all those who have directly or indirectly helped me in bringing out this book.

Vijay Kumar Dwivedi	*National Institute of Technology (NIT)* *Durgapur, West Bengal*
K K Khatua	*National Institute of Technology (NIT) Rourkela, Odisha*
Mujib Ahmad Ansari	*Aligarh Muslim University, Aligarh, Uttar Pradesh*
Amit Vishwakarma	*University Institute of Technology,* *Rajiv Gandhi Proudyogiki Vishwavidyalaya (RGPV),* *Bhopal, Madhya Pradesh*
Tuhin Subhra Konar	*Heritage Institute of Technology, Kolkata, West Bengal*

P L Patel	*Sardar Vallabhbhai National Institute of Technology (SVNIT), Surat, Gujarat*
P V Timbadiya	*Sardar Vallabhbhai National Institute of Technology (SVNIT), Surat, Gujarat*
Gayathri Palatarthi	*G S Moze College of Engineering, Pune, Maharashtra*
M R Swaminathan	*College of Engineering, Guindy, Chennai*
K R Suresh	*BMS College of Engineering, Bangalore, Karnataka*
Ramalinga Reddy	*Reva Institute of Technology, Bangalore, Karnataka*
B L Shivkumar	*R V College of Engineering, Bangalore, Karnataka*
Mahesh Chandra	*Dayanand Sagar College of Engineering, Bangalore, Karnataka*

Feedback

Comments and suggestions for further improvement of the book would be greatly appreciated. I can be contacted at the following e-mail address: ***subramanyak1@gmail.com***.

K Subramanya

Publisher's Note

Remember to write to us! We look forward to receiving your feedback, comments and ideas to enhance the quality of this book.

You can reach us at ***tmh.civilfeedback@gmail.com***. Kindly mention the title and author's name as the subject. In case you spot piracy of this book, please do let us know.

CHAPTER

1 INTRODUCTION

1.1 INTRODUCTION

Hydrology means the science of water. It is the science that deals with the occurrence, circulation and distribution of water of the earth and the earth's atmosphere. As a branch of earth science, it is concerned with the water in streams and lakes, rainfall and snowfall, snow and ice on the land and water occurring below the earth's surface in the pores of the soil and rocks. In a general sense, hydrology is a very broad subject of an inter-disciplinary nature drawing support from allied sciences, such as meteorology, geology, statistics, chemistry, physics and fluid mechanics.

Hydrology is basically an applied science. To further emphasise the degree of applicability, the subject is sometimes classified as

1. **Scientific hydrology**—the study which is concerned chiefly with academic aspects.
2. **Engineering or applied hydrology**—a study concerned with engineering applications.

In a general sense, engineering hydrology deals with (i) estimation of water resources, (ii) the study of processes such as precipitation, runoff, evapotranspiration and their interaction, and (iii) the study of problems such as floods and droughts, and strategies to combat them.

This book is an elementary treatment of engineering hydrology with descriptions that aid in a qualitative appreciation and techniques which enable a quantitative evaluation of the hydrologic processes that are of importance to a civil engineer.

1.2 *Hydrologic Cycle*

Water occurs on the earth in all its three states, viz. liquid, solid and gaseous, and in various degrees of motion. Evaporation of water from water bodies such as oceans and lakes, formation and movement of clouds, rain and snowfall, streamflow and groundwater movement are some examples of the dynamic aspects of water. The various aspects of water related to the earth can be explained in terms of a cycle known as the *hydrologic cycle.*

Figure 1.1 is a schematic representation of the hydrologic cycle. A convenient starting point to describe the cycle is in the oceans. Water in the oceans evaporates

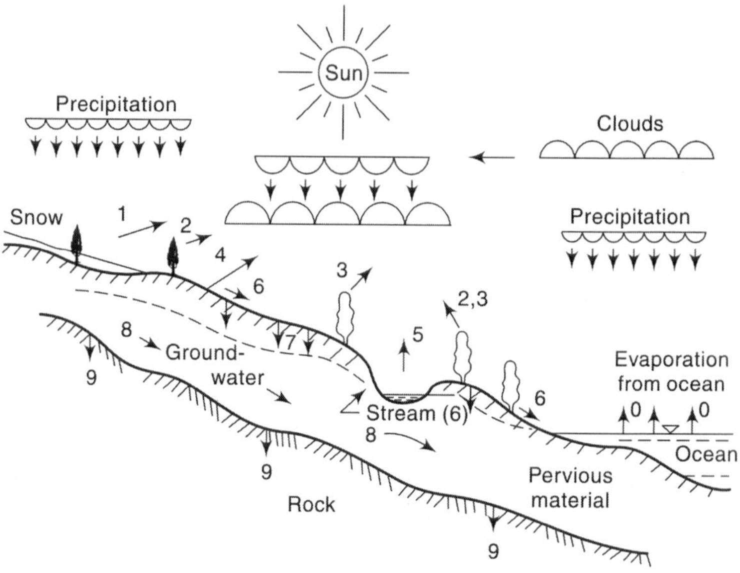

0 = Evaporation from ocean
1 = Raindrop evaporation
2 = Interception
3 = Transpiration
4 = Evaporation from land
5 = Evaporation from water bodies
6 = Surface runoff
7 = Infiltration
8 = Groundwater
9 = Deep percolation

Fig. 1.1 ***The Hydrologic Cycle***

due to the heat energy provided by solar radiation. The water vapour moves upwards and forms clouds. While much of the clouds condense and fall back to the oceans as rain, a part of the clouds is driven to the land areas by winds. There they condense and *precipitate* onto the landmass as rain, snow, hail, sleet, etc. A part of the precipitation may *evaporate* back to the atmosphere even while falling. Another part may be *intercepted* by vegetation, structures and other such surface modifications from which it may be either evaporated back to the atmosphere or move down to the ground surface.

A portion of the water that reaches the ground enters the earth's surface through *infiltration,* enhances the moisture content of the soil and reaches the groundwater body. Vegetation sends a portion of the water from under the ground surface back to the atmosphere through the process of *transpiration.* The precipitation reaching the ground surface after meeting the needs of infiltration and evaporation moves down the natural slope over the surface and through a network of gullies, streams and rivers to reach the ocean. The groundwater may come to the surface through springs and other outlets after spending a considerably longer time than the surface flow. The portion of the precipitation which by a variety of paths above and below the surface of the earth reaches the stream channel is called *runoff.* Once it enters a stream channel, runoff becomes *stream flow.*

The sequence of events as above is a simplistic picture of a very complex cycle that has been taking place since the formation of the earth. It is seen that the hydrologic cycle is a very vast and complicated cycle in which there are a large number of paths of varying time scales. Further, it is a continuous recirculating cycle in the sense that there is neither a beginning nor an end or a pause. Each path of the hydrologic cycle involves one or more of the following aspects: (i) transportation of water, (ii) temporary storage, and (iii) change of state. For example, (a) the process of rainfall has the change of state and transportation, and (b) the groundwater path has storage and transportation aspects.

The main components of the hydrologic cycle can be broadly classified as *transportation (flow) components* and *storage components* as below:

Transportation components	Storage components
• Precipitation	• Storage on the land surface (Depression storage, Ponds, Lakes, Reservoirs, etc)
• Evaporation	• Soil moisture storage
• Transpiration	• Groundwater storage
• Infiltration	
• Runoff	

Numerous diagrams have been designed to illustrate the salient aspects of the hydrologic cycle. Figure 1.1, described earlier, is a typical one of the descriptive type. A qualitative representation of the hydrologic cycle first introduced by Horton is shown in Fig. 1.2. This diagram, popularly known as Horton's representation of

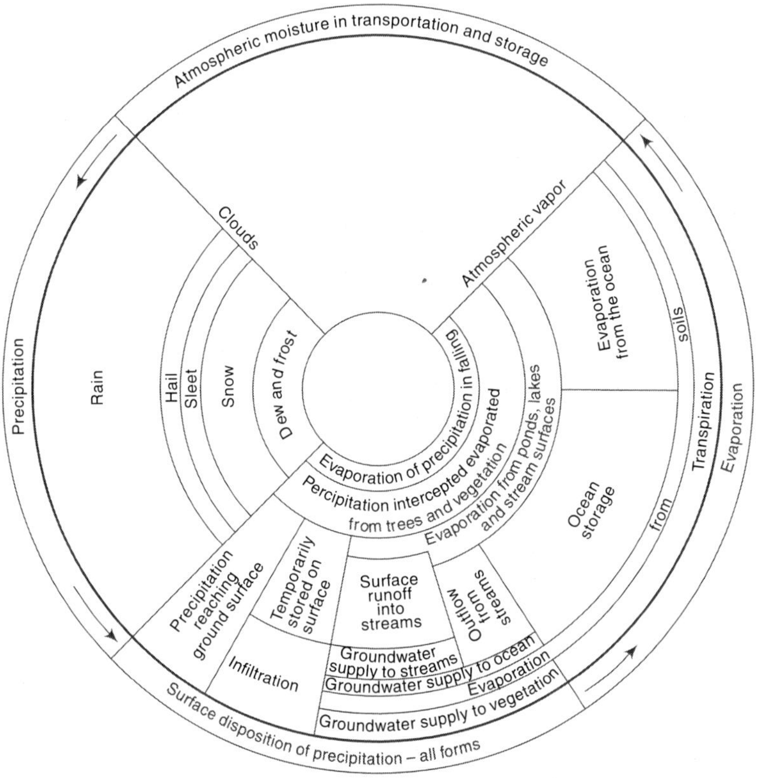

Fig. 1.2 ***Horton's Representation of the Hydrological Cycle (Ref: 1).***

the hydrologic cycle, illustrates very clearly the storage and transportation components and their relative positions in the cycle. The figure is read in the counter-clockwise manner and is self-explanatory.

Schematically, the interdependency of the transportation components can be represented as in Fig. 1.3. The quantities of water going through various paths of the hydrological cycle in a given system can be described by the continuity principle known as *water-budget equation* or *hydrologic equation.*

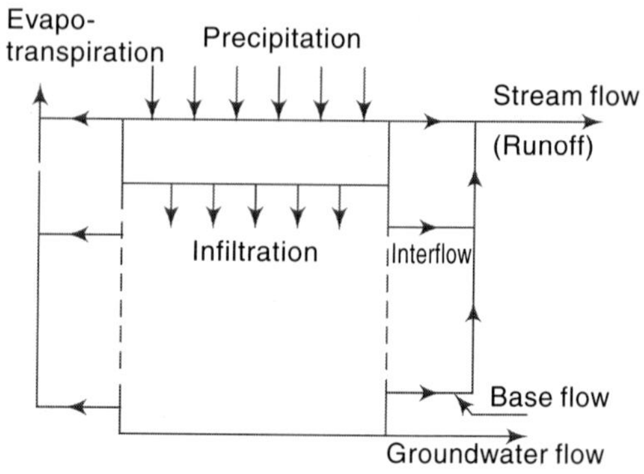

Fig. 1.3 ***Transportation Components of the Hydrologic Cycle***

It is important to note that the total water resources of the earth is constant and the sun is the source of energy for the hydrologic cycle. A recognition of the various processes such as evaporation, precipitation and groundwater flow helps one to study the science of hydrology in a systematic way. Also, one realises that humans can interfere with virtually any part of the hydrologic cycle, e.g. through artificial rain, evaporation suppression, change of vegetal cover and land use, extraction of groundwater, etc. Interference at one stage can cause serious repercussions at some other stage of the cycle.

The hydrological cycle has important influences in a variety of fields including agriculture, forestry, geography, economics, sociology and politics. Engineering applications of the knowledge of the hydrologic cycle, and hence of the subjects of hydrology, are found in the design and operation of projects dealing with water supply, irrigation and drainage, water power, flood control, navigation, coastal works, salinity control and recreational uses of water.

1.3 *Water-Budget Equation*

1.3.1 Catchment Area

The area of land draining into a stream or a water course at a given location is known as the *catchment area.* It is also called *drainage area* or *drainage basin.* In the *USA*, it is known as *watershed.* A catchment area is separated form its neighbouring areas by a ridge called *divide* in the *USA* and *watershed* in *UK* (Fig. 1.4). The areal extent of the catchment is obtained by tracing the ridge on a topographic map to delineate the catchment and measuring the area by

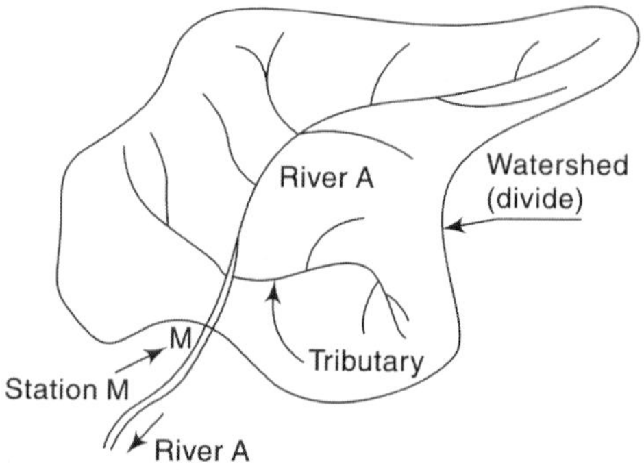

Fig. 1.4 ***Schematic Sketch of Catchment of River A at Station M***

a *planimeter.* It is obvious that for a river while mentioning the catchment area, the station to which it pertains (Fig. 1.4) must also be mentioned. It is normal to assume the groundwater divide to coincide with the surface divide. Thus, the catchment area affords a logical and convenient unit to study various aspects relating to the hydrology and water resources of a region.

Further, it is probably the singlemost important drainage characteristic used in hydrological analysis and design. Some of the geometrical parameters of the catchment as related to runoff process are described in Sec. 5.4, Chapter 5.

1.3.2 Water-Budget Equation

For a given problem area, say a catchment, in an interval of time Δt, the continuity equation for water in its various phases is written as

Mass inflow – Mass outflow = Change in mass storage

If the density of the inflow, outflow and storage volumes are the same,

$$V_i - V_0 = \Delta S \tag{1.1}$$

where V_i = inflow volume of water into the problem area during the time period, V_0 = outflow volume of water from the problem area during the time period, and ΔS = change in the storage of the water volume over and under the given area during the given period. In applying this continuity equation [Eq. (1.1)] to the paths of the hydrologic cycle involving change of state, the volumes considered are the equivalent volumes of water at a reference temperature. In hydrologic calculations, the volumes are often expressed as average depths over the catchment area. Thus, for example, if the annual stream flow from a 10 km^2 catchment is 10^7 m^3, it corresponds to a depth of $\left(\dfrac{10^7}{10 \times 10^6}\right)$ = 1 m = 100 cm. Rainfall, evaporation and often runoff volumes are expressed in units of depth over the catchment.

While realising that all the terms in a hydrological water budget may not be known to the same degree of accuracy, an expression for the water budget of a catchment for a time interval Δt is written as

$$P - R - G - E - T = \Delta S \tag{1.2-a}$$

In this, P = precipitation, R = surface runoff, G = net groundwater flow out of the catchment, E = evaporation, T = transpiration and ΔS = change in storage.

The storage S consists of three components as

$$S = S_s + S_{sm} + S_g$$

where S_s = surface water storage,

S_{sm} = water in storage as soil moisture, and

S_g = water in storage as groundwater.

Thus in Eq. (1.2-a), $\Delta S = \Delta S_s + \Delta S_{sm} + \Delta S_g$

All terms in Eq. (1.2-a) have the dimensions of volume. Note that all these terms can be expressed as depth over the catchment area (e.g. in centimetres), and in fact this is a very common unit.

In terms of rainfall–runoff relationship, Eq. (1.2-a) can be represented as

$$R = P - L \tag{1.2-b}$$

where L = losses = water not available to runoff due to infiltration (causing addition to soil moisture and groundwater storage), evaporation, transpiration and surface storage.

Details of various components of the water-budget equation are discussed in subsequent chapters. Note that in Eqs (1.2-a and b), the net import of water into the catchment, from sources outside the catchment, by action of humans is assumed to be zero.

Example 1.1 A lake had a water surface elevation of 103.200 m above datum at the beginning of a certain month. In that month, the lake received an average inflow of 6.0 m^3/s from surface runoff sources. In the same period, the outflow from the lake had an average value of 6.5 m^3/s. Further, in that month, the lake received a rainfall of 145 mm and the evaporation from the lake surface was estimated as 6.10 cm. Write the water-budget equation for the lake and calculate the water surface elevation of the lake at the end of the month. The average lake-surface area can be taken as 5000 ha. Assume that there is no contribution to or from the groundwater storage.

Solution

In a time interval Δt, the water budget for the lake can be written as

Input volume – Output volume = Change in storage of the lake

$$(\bar{I}\,\Delta t + PA) - (\bar{Q}\Delta t + EA) = \Delta S$$

where $\bar{I}$ = average rate of inflow of water into the lake, $\bar{Q}$ = average rate of outflow from the lake, P = precipitation, E = evaporation, A = average surface area of the lake and ΔS = change in storage volume of the lake.

Here $\Delta t = 1$ month $= 30 \times 24 \times 60 \times 60 = 2.592 \times 10^6$ s = 2.592 Ms

In one month:

$$\text{Inflow volume} = \bar{I}\,\Delta t = 6.0 \times 2.592 = 15.552 \text{ M m}^3$$

$$\text{Outflow volume} = \bar{Q}\Delta t = 6.5 \times 2.592 = 16.848 \text{ M m}^3$$

$$\text{Input due to precipitation} = PA = \frac{145 \times 5000 \times 100 \times 100}{1000 \times 10^6} \text{ M m}^3 = 7.25 \text{ M m}^3$$

$$\text{Outflow due to evaporation} = EA = \frac{6.10}{100} \times \frac{5000 \times 100 \times 100}{10^6} = 3.05 \text{ M m}^3$$

Hence $\quad \Delta S = 15.552 + 7.25 - 16.848 - 3.05 = 2.904 \text{ M m}^3$

Change inelevation $\quad \Delta z = \dfrac{\Delta S}{A} = \dfrac{2.904 \times 10^6}{5000 \times 100 \times 100} = 0.058$ m

New water surface elevation at the end of the month = 103.200 + 0.058

= 103.258 m above the datum.

Example 1.2 A small catchment of 150 ha area received a rainfall of 10.5 cm in 90 minutes due to a storm. At the outlet of the catchment, the stream draining the catchment was dry before the storm and experienced a runoff lasting for 10 hours with an average discharge of 1.5 m^3/s. The stream was again dry after the runoff event. (a) What is the amount of water which was not available to runoff due to combined effect of infiltration, evaporation and transpiration? What is the ratio of runoff to precipitation?

Solution

The water-budget equation for the catchment in a time Δt is

$$R = P - L \qquad (1.2\text{-b})$$

where L = losses = water not available to runoff due to infiltration (causing addition to soil moisture and groundwater storage), evaporation, transpiration and surface storage. In the present case, Δt = duration of the runoff = 10 hours.

Note that the rainfall occurred in the first 90 minutes and for the rest 8.5 hours, the precipitation was zero.

(a) P = Input due to precipitation in 10 hours

$= 150 \times 100 \times 100 \times (10.5/100) = 157{,}500 \text{ m}^3$

R = Runoff volume = outflow volume at the catchment outlet in 10 hours

$= 1.5 \times 10 \times 60 \times 60 = 54{,}000 \text{ m}^3$

Hence, losses $L = 157{,}500 - 54{,}000 = 103{,}500 \text{ m}^3$

(b) Runoff/rainfall = 54,000/157,500 = 0.343

(This ratio is known as *runoff coefficient* and is discussed in Chapter 5)

Example 1.3 A catchment area of 140 km^2 received 120 cm of rainfall in a year. At the outlet of the catchment, the flow in the stream draining the catchment was found to have an average rate of (i) 1.5 m^3/s for the first 3 months, (ii) 2.0 m^3/s for 6 months and (iii) 3.5 m^3/s for the remaining 3 months. (a) What is the runoff coefficient of the catchment? (ii) If the afforestation of the catchment reduces the runoff coefficient to 0.35, what is the increase in the abstraction from precipitation due to infiltration, evaporation and transpiration for the same annual rainfall of 120 cm?

Solution

(i) *Before Afforestation*

Consider a period $= \Delta t = 1$ year

Input volume to the catchment through precipitation =

$$\mathcal{V}_i = 140 \times 10^6 \times \left(\frac{120}{100}\right) = 168 \text{ Mm}^3$$

Runoff = Output volume = $\mathcal{V}_o = (1.5 \times 3) + (2 \times 6) + (3.5 \times 3) = 27 \text{ Mm}^3\text{.month}$

$$= 27\left(\frac{365}{12}\right) \times 24 \times 60 \times 60 = 70.956 \times 10^6 \text{ m}^3 = 70.956 \text{ Mm}^3$$

$$\text{Runoff coefficient} = \frac{70.956}{168.0} = 0.4224$$

Abstraction volume = $168.0 - 70.956 = 97.044 \text{ Mm}^3$

(ii) *After Afforestation*

Runoff = $0.35 \times 168 = 58.8 \text{ Mm}^3$

New abstraction volume = $168.0 - 58.8 = 109.2 \text{ Mm}^3$

Increase in abstraction = $109.20 - 97.044 = 12.156 \text{ Mm}^3$

1.4 *World Water Balance*

The total quantity of water in the world is estimated to be about 1386 million cubic kilometres (M km^3). About 96.5% of this water is contained in the oceans as saline water. Some of the water on the land amounting to about 1% of the total water is also saline. Thus, only about 35.0 M km^3 of fresh water is available. Out of this about 10.6 M km^3 is both liquid and fresh and the remaining 24.4 M km^3 is contained in frozen state as ice in the polar regions and on mountain tops and glaciers. An estimated distribution of water on the earth is given in Table 1.1.

Table 1.1 ***Estimated World Water Quantities***

Item	Area (M km^2)	Volume (M km^3)	Percent total water	Percent fresh water
1. Oceans	361.3	1338.0	96.5	—
2. Groundwater				
(a) fresh	134.8	10.530	0.76	30.1
(b) saline	134.8	12.870	0.93	—
3. Soil moisture	82.0	0.0165	0.0012	0.05
4. Ice caps and glaciers	16.0	24.0235	1.7	68.6
5. Other ice and snow	0.3	0.3406	0.025	1.0
6. Lakes				
(a) fresh	1.2	0.0910	0.007	0.26
(b) saline	0.8	0.0854	0.006	—
7. Wetlands	2.7	0.01147	0.0008	0.03
8. Rivers	148.8	0.00212	0.0002	0.006
9. Biological water	510.0	0.00112	0.0001	0.003
10. Atmospheric water	510.0	0.01290	0.001	0.04
Total: (a) All kinds of water	510.0	1386.0	100.0	
(b) Fresh water	148.8	35.0	2.5	100.0

The global annual water balance is shown in Table 1.2.

Table 1.2 ***Global Annual Water Balance***

Item	Ocean	Land
1. Area (M km^2)	361.30	148.8
2. Precipitation (km^3/year)	458,000	119,000
(mm/year)	1270	800
3. Evaporation (km^3/year)	505,000	72,000
(mm/year)	1400	484
4. Runoff to ocean		
(i) Rivers (km^3/year)		44,700
(ii) Groundwater (km^3/year)		2,200
Total runoff (km^3/year)		47,000
(mm/year)		316

It is seen from Table 1.2 that the annual evaporation from the world's oceans and inland areas are 0.505 and 0.072 M km^3 respectively. Thus, over the oceans about

9% more water evaporates than that falls back as precipitation. Correspondingly, there will be excess precipitation over evaporation on the land mass. The differential, which is estimated to be about 0.047 M km^3 is the runoff from land mass to oceans and groundwater outflow to oceans. It is interesting to know that less than 4% of this total river flow is used for irrigation and the rest flows down to sea.

These estimates are only approximate and the results from different studies vary; the chief cause being the difficulty in obtaining adequate and reliable data on a global scale.

The volumes in various phases of the hydrologic cycle (Table 1.1), as also the rate of flow in that phase (Table 1.2), do vary considerably. The average duration of a particle of water to pass through a phase of the hydrologic cycle is known as the *residence time* of that phase. It could be calculated by dividing the volume of water in the phase by the average flow rate in that phase. For example, by assuming that all the surface runoff to the oceans comes from the rivers,

From Table 1.1, the volume of water in the rivers of the world = 0.00212 M km^3

From Table 1.2, the average flow rate of water in global rivers = 44700 km^3/year

Hence, residence time of global rivers, T_r = 2120/44700 = 0.0474 year = 17.3 days.

Similarly, the residence time for other phases of the hydrological cycle can be calculated (Prob. 1.6). It will be found that the value of T_r varies from phase to phase. In a general sense, the shorter the residence time, the greater is the difficulty in predicting the behaviour of that phase of the hydrologic cycle.

Annual water balance studies of the sub-areas of the world indicate interesting facts. The water balance of the continental land mass is shown in Table 1.3(a). It is interesting to see from this table that Africa, in spite of its equatorial forest zones, is the driest continent in the world with only 20% of the precipitation going as runoff. On the other hand, North America and Europe emerge as continents with highest runoff. Extending this type of analysis to a smaller land mass, viz. the Indian subcontinent, the long-term average runoff for India is found to be 46%.

Table 1.3(a) ***Water Balance of Continents in mm/year*** (Ref. 2)

Continent	Area (M km^2)	Precipitation	Total runoff	Runoff as % of precipitation	Evapo-transpiration
Africa	30.3	686	139	20	547
Asia	45.0	726	293	40	433
Australia	8.7	736	226	30	510
Europe	9.8	734	319	43	415
N. America	20.7	670	287	43	383
S. America	17.8	1648	583	35	1065

Water-balance studies on the oceans indicate that there is considerable transfer of water between the oceans and the evaporation and precipitation values vary from one ocean to another (Table 1.3(b)).

Table 1.3(b) ***Water Balance of Oceans mm/year*** (Ref. 2)

Ocean	Area (M km^2)	Precipitation	Inflow from adjacent continents	Evaporation	Water exchange with other oceans
Atlantic	107	780	200	1040	–60
Arctic	12	240	230	120	350
Indian	75	1010	70	1380	–300
Pacific	167	1210	60	1140	130

Each year, the rivers of the world discharge about 44,700 km^3 of water into the oceans. This amounts to an annual average flow of 1.417 Mm^3/s. The world's largest river, the Amazon, has an annual average discharge of 200,000 m^3/s, i.e. one-seventh of the world's annual average value. India's largest river, the Brahmaputra, and the second largest, the Ganga, flow into the Bay of Bengal with mean annual average discharges of 16,200 m^3/s and 15,600 m^3/s respectively.

1.5 *Global Freshwater Resources*

Table 1.1 lists the components of global freshwater resources in Col.5. In this section, the important components of fresh water resources in the world are described. Figure 1.5 illustrates schematically the distribution of global freshwater resources.

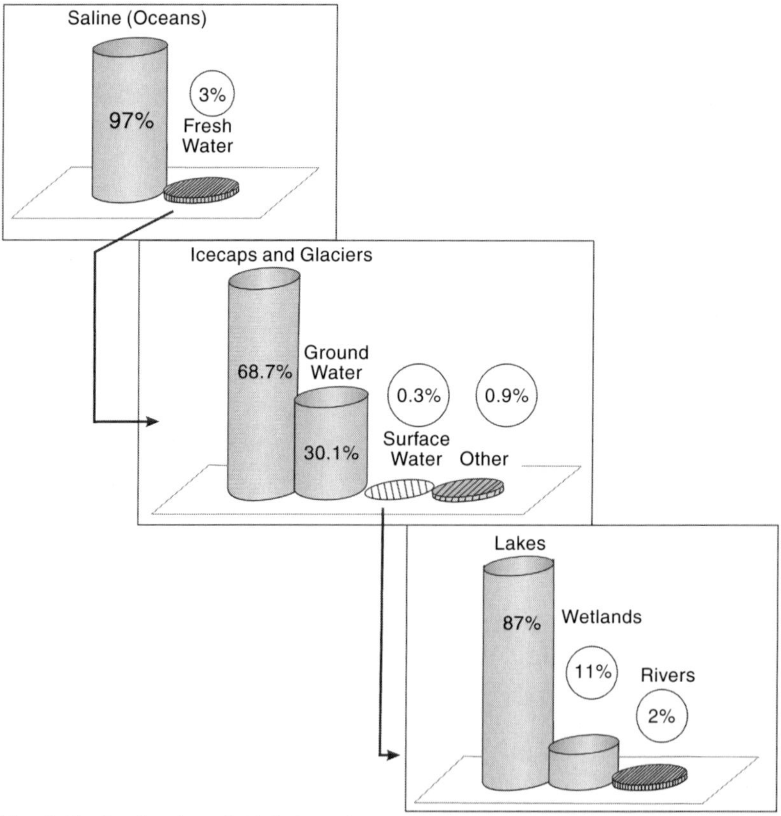

Fig. 1.5 ***Distribution of Global Freshwater***

1. Glaciers and Permanent Ice Caps

This form of freshwater accounts for about 69% of the total freshwater resources of the globe. They are located in about 10% of the world's land mass. It is estimated that Greenland and Antarctica account for most of this category of water. Obviously, being located in most inaccessible and uninhabited places of the world, this freshwater resource is not readily available for human use.

2. Groundwater

Ground water represents the second-most abundantly available freshwater resource and constitutes about 30% of fresh water resources of the globe. More than 1.5 billion people in the world are known to depends on groundwater for their drinking water requirements.

Groundwater forms an important component of hydrological studies and some basic aspects of groundwater are dealt with in Chapter 9 of this book

3. Freshwater Lakes

Natural freshwater lakes account for about 0.26% of the freshwater resources. More than 50% of these lakes are found in Canada.

4. Reservoirs

Reservoirs are artificial lakes created by humans through construction of dams across rivers. Most of the water in these reservoirs , estimated to be of the order of 4300 km^3, are used for beneficial purposes such as irrigation, drinking water, hydropower generation and industrial use.

5. Wetlands

A part of the freshwater resource (amounting to about 0.04%) is distributed in the globe as wetland, marshes, lagoons, swamps, bogs and mires. These water-bearing bodies play a very important role in maintaining the freshwater ecology as well as in the recharge of groundwater.

6. Rivers

Flowing water in rivers form one of the most important part of freshwater resource sustaining human activity and ecology in the world. Even though this component forms a tiny fraction (0.006%) of freshwater resource, it forms the core of human activity related to natural water use. A substantial part of the subject of engineering hydrology deals with river flow. Some important aspects related to river flow are discussed in chapters 4, 5, 6, 7 and 8 of this book.

1.6 *History of Hydrology*

Water is the prime requirement for the existence of life and thus it has been humankind's endeavour from time immemorial to utilise the available water resources. History has instances of civilisations that flourished with the availability of dependable water supplies and then collapsed when the water supply failed. Numerous references exist in Vedic literature to groundwater availability and its utility. During 3000 BC, groundwater development through wells was known to people of the Indus Valley civilisation as revealed by archaeological excavations at Mohenjodaro. Quotations in ancient Hindu scriptures indicate the existence of knowledge of the hydrologic cycle even as far back as the Vedic period. The first description of the raingauge and its use is contained in the *Arthashastra* by Chanakya (300 BC). Varahamihira's (AD 505–587) *Brihatsamhita* contains descriptions of the raingauge, wind vane and prediction procedures for rainfall. Egyptians knew the importance of the stage measurement of rivers, and records of the stages of the Nile dating back to 1800 BC have been located. The knowledge of the hydrologic cycle came to be known to Europe much later, around AD 1500.

Chow[1] classifies the history of hydrology into eight periods:

1. Period of speculation—prior to AD 1400
2. Period of observation—1400–1600
3. Period of measurement—1600–1700
4. Period of experimentation—1700–1800
5. Period of modernisation—1800–1900
6. Period of empiricism—1900–1930
7. Period of rationalisation—1930–1950
8. Period of theorisation—1950–to–date

Most of the present-day science of hydrology has been developed since 1930, thus giving hydrology the status of a young science. The worldwide activities in water-resources development since the last few decades by both developed and developing countries aided by rapid advances in instrumentation for data acquisition and in the computer facilities for data analysis have contributed towards the rapid growth rate of this young science.

1.7 *Applications In Engineering*

Hydrology finds its greatest application in the design and operation of water-resources engineering projects, such as those for (i) irrigation, (ii) water supply, (iii) flood control, (iv) water power, and (v) navigation. In all these projects, hydrological investigations for the proper assessment of the following factors are necessary:

1. The capacity of storage structures such as reservoirs.
2. The magnitude of flood flows to enable safe disposal of the excess flow.
3. The minimum flow and quantity of flow available at various seasons.
4. The interaction of the flood wave and hydraulic structures, such as levees, reservoirs, barrages and bridges.

The hydrological study of a project should necessarily precede structural and other detailed design studies. It involves the collection of relevant data and analysis of the data by applying the principles and theories of hydrology to seek solutions to practical problems.

Many important projects in the past have failed due to improper assessment of hydrological factors. Some typical failures of hydraulic structures are (i) overtopping and consequent failure of an earthen dam due to an inadequate spillway capacity, (ii) failure of bridges and culverts due to excess flood flow, and (iii) inability of a large reservoir to fill up with water due to overestimation of the stream flow. Such failure, often called *hydrologic failures,* underscore the uncertainty aspect inherent in hydrological studies.

Various phases of the hydrological cycle, such as rainfall, runoff, evaporation and transpiration, are all nonuniformly distributed both in time and space. Further, practically, all hydrologic phenomena are complex and at the present level of knowledge, they can at best be interpreted with the aid of probability concepts. Hydrological events are treated as random processes and the historical data relating to the event are analysed by statistical methods to obtain information on probabilities of occurrence of various events. The probability analysis of hydrologic data is an important component of present-day hydrological studies and enables the engineer to take suitable design decisions consistent with economic and other criteria to be taken in a given project.

1.8 *Sources of Data*

Depending upon the problem at hand, a hydrologist would require data relating to the various relevant phases of the hydrological cycle playing on the problem catchment. The data normally required in the studies are

- Weather records—temperature, humidity and wind velocity
- Precipitation data
- Stream-flow records
- Evaporation and evapotranspiration data
- Infiltration characteristics of the study area
- Soils of the area
- Land use and land cover
- Groundwater characteristics
- Physical and geological characteristics of the area
- Water-quality data

In India, hydro-meteorological data are collected by the India Meteorological Department (*IMD*) and by some state government agencies. The Central Water Commission (*CWC*) monitors flow in major rivers of the country. Stream-flow data of various rivers and streams are usually available from the State Water Resources/Irrigation Department. Groundwater data will normally be available with Central Groundwater Board (*CGWB*) and state government groundwater development agencies. Data relating to evapotranspiration and infiltration characteristics of soils will be available with state government organisations such as Department of Agriculture, Department of Watershed Development and Irrigation Department. The physical features of the study area have to be obtained from a study of topographical maps available with the Survey of India. The information relating to geological characteristics of the basin under study will be available with the Geological Survey of India and the State Geology Directorate. Information relating to soils at an area are available from relevant maps of National Bureau of Soil Survey and Land Use Planning (*NBSS&LUP*), 1996. Further additional or specific data can be obtained from the State Agriculture Department and the State Watershed Development Department. Land use and land-cover data would generally be available from State Remote Sensing Agencies. Specific details will have to be derived through interpretation of multispectral multiseason satellite images available from National Remote Sensing Agency (*NRSA*) of Government of India. Central and State Pollution Control Boards, *CWC* and *CGWB* collect water-quality data.

REFERENCES

1. Chow, V.T., (Ed), *Handbook of Applied Hydrology,* McGraw-Hill, New York, NY, 1964.
2. Schendel, V., "The world's water resources and water balance", *Natural Resources and Development,* Inst. for Sci. Coop, Hannover, Germany, pp. 8–14, Vol. 1, 1975.
3. *UNESCO*, "World Water Balance and Water Resources of the Earth", *Studies and Reports in Hydrology,* 25, *UNESCO,* Paris, France, 1978.
4. Van der Leeden, *Water Resources of the World,* Water Information Center, Port Wasghington, N.Y., USA, 1975.

REVISION QUESTIONS

1.1 Describe the hydrologic cycle. Explain briefly humankind's interference in various parts of this cycle.

1.2 Discuss the hydrological water budget with the aid of examples.

1.3 What are the significant features of global water-balance studies?

1.4 Explain briefly the distribution of fresh water resources in the world.

1.5 List the major activities in which hydrological studies are important.

1.6 Describe briefly the sources of hydrological data in India.

PROBLEMS

1.1 Two and half centimetres of rain per day over an area of 200 km^2 is equivalent to average rate of input of how many cubic metres per second of water to that area?

1.2 The watershed at a site on a stream has 77000 ha area. The mean annual precipitation at this place is 950 mm. (i) If 25% of rainfall reaches the basin outlet as stream flow, estimate the mean annual flow rate of the stream in m^3/s. (ii) If watershed treatment, including afforestation, causes the runoff rate to reduce to 22%, what is the increase in the volume of abstraction from all causes for the same mean annual rainfall of 950 mm?

1.3 Estimate the constant rate of withdrawal from a 1375 ha reservoir in a month of 30 days during which the reservoir level dropped by 0.75 m in spite of an average inflow into the reservoir of 0.5 Mm^3/day. During the month, the average seepage loss from the reservoir was 2.5 cm, total precipitation on the reservoir was 18.5 cm and the total evaporation was 9.5 cm.

1.4 A river reach had a flood wave passing through it. At a given instant, the storage of water in the reach was estimated as 15.5 ha.m. What would be the storage in the reach after an interval of 3 hours if the average inflow and outflow during the time period are 14.2 m^3/s and 10.6 m^3/s respectively?

1.5 A catchment has four sub-areas. The annual precipitation and evaporation from each of the sub-areas are given below.

Assume that there is no change in the groundwater storage on an annual basis and calculate for the whole catchment the values of annual average (i) precipitation, and (ii) evaporation. What are the annual runoff coefficients for the sub-areas and for the total catchment taken as a whole?

Sub-area	Area Mm^2	Annual precipitation mm	Annual evaporation mm
A	10.7	1030	530
B	3.0	830	438
C	8.2	900	430
D	17.0	1300	600

1.6 Estimate the residence time of the following by using Tables 1.1 and 1.2.
(a) Global atmospheric moisture
(b) Global groundwater by assuming that only the fresh groundwater runs off to the oceans
(c) Ocean water

OBJECTIVE QUESTIONS

1.1 The percentage of earth covered by oceans is about
(a) 31% (b) 51% (c) 71% (d) 97%

1.2 The percentage of total quantity of water in the world that is saline is about
(a) 71% (b) 33% (c) 67% (d) 97%

1.3 The percentage of total quantity of fresh water in the world available in the liquid form is about
(a) 30% (b) 70% (c) 11% (d) 51%

1.4 If the average annual rainfall and evaporation over land masses and oceans of the earth are considered, it would be found that
(a) over the land mass the annual evaporation is the same as the annual precipitation
(b) about 9% more water evaporates from the oceans than what falls back on them as precipitation
(c) over the ocean about 19% more rain falls than what is evaporated
(d) over the oceans about 19% more water evaporates than what falls back on them as precipitation

1.5 Considering the ratio of annual precipitation to runoff = r_0 for all the continents on the earth,
(a) Asia has the largest value of the ratio r_0
(b) Europe has the smallest value of r_0
(c) Africa has the smallest value of r_0
(d) Australia has the smallest value of r_0

1.6 In the hydrological cycle, the average residence time of water in the global
(a) atmospheric moisture is larger than that in the global rivers
(b) oceans is smaller than that of the global groundwater
(c) rivers is larger than that of the global groundwater
(d) oceans is larger than that of the global groundwater

1.7 A watershed has an area of 300 ha. Due to a 10 cm rainfall event over the watershed, a stream flow is generated and at the outlet of the watersheds it lasts for 10 hours. Assuming a runoff/rainfall ratio of 0.20 for this event, the average stream-flow rate at the outlet in this period of 10 hours is
(a) 1.33 m^3/s (b) 16.7 m^3/s (c) 100 m^3/minute (d) 60,000 m^3/h

1.8 Rainfall of 20 mm/h intensity occurred over a watershed of 100 ha area for a duration of 6 h. Measured direct runoff volume in the stream draining the watershed was found to be 30,000 m^3. The precipitation not available to runoff in this case is

(a) 9 cm (b) 3 cm (c) 17.5 mm (d) 5 mm

1.9 A catchment of area 120 km^2 has three distinct zones as below:

Zone	Area (km^2)	Annual runoff (cm)
A	61	52
B	39	42
C	20	32

The annual runoff from the catchment is

(a) 126.0 cm (b) 42.0 cm (c) 45.4 cm (d) 47.3 cm

CHAPTER 2

PRECIPITATION

2.1 INTRODUCTION

The term *precipitation* denotes all forms of water that reach the earth from the atmosphere. The usual forms are rainfall, snowfall, hail, frost and dew. Of all these, only the first two contribute significant amounts of water. Rainfall being the predominant form of precipitation causing stream flow, especially the flood flow in a majority of rivers in India, unless otherwise stated, the term *rainfall* is used in this book synonymously with precipitation. The magnitude of precipitation varies with time and space. Differences in the magnitude of rainfall in various parts of a country at a given time and variations of rainfall at a place in various seasons of the year are obvious and need no elaboration. It is this variation that is responsible for many hydrological problems, such as floods and droughts. The study of precipitation forms a major portion of the subject of hydrometeorology. In this chapter, a brief introduction is given to familiarise the engineer with important aspects of rainfall, and, in particular, with the collection and analysis of rainfall data.

For precipitation to form: (i) the atmosphere must have moisture, (ii) there must be sufficient nuclei present to aid condensation, (iii) weather conditions must be good for condensation of water vapour to take place, and (iv) the products of condensation must reach the earth. Under proper weather conditions, the water vapour condenses over nuclei to form tiny water droplets of sizes less than 0.1 mm in diameter. The nuclei are usually salt particles or products of combustion and are normally available in plenty. Wind speed facilitates the movement of clouds while its turbulence retains the water droplets in suspension. Water droplets in a cloud are somewhat similar to the particles in a colloidal suspension. Precipitation results when water droplets come together and coalesce to form larger drops that can drop down. A considerable part of this precipitation gets evaporated back to the atmosphere. The net precipitation at a place and its form depend upon a number of meteorological factors, such as the weather elements like wind, temperature, humidity and pressure in the volume region enclosing the clouds and the ground surface at the given place.

2.2 Forms of Precipitation

Some of the common forms of precipitation are rain, snow, drizzle, glaze, sleet and hail.

1. Rain

It is the principal form of precipitation in India. The term *rainfall* is used to describe precipitations in the form of water drops of sizes larger than 0.5 mm. The maximum size of a raindrop is about 6 mm. Any drop larger in size than this tends to break up into drops of smaller sizes during its fall from the clouds. On the basis of its intensity, rainfall is classified as follows:

Type	Intensity
1. Light rain	Trace to 2.5 mm/h
2. Moderate rain	2.5 mm/h to 7.5 mm/h
3. Heavy rain	> 7.5 mm/h

2. Snow

Snow is another important form of precipitation. Snow consists of ice crystals which usually combine to form flakes. When fresh, snow has an initial density varying from 0.06 to 0.15 g/cm^3 and it is usual to assume an average density of 0.1 g/cm^3. In India, snow occurs only in the Himalayan regions.

3. Drizzle

A fine sprinkle of numerous water droplets of size less than 0.5 mm and intensity less than 1 mm/h is known as drizzle. In this, the drops are so small that they appear to float in the air.

4. Glaze

When rain or drizzle comes in contact with cold ground at around 0°C, the water drops freeze to form an ice coating called *glaze* or *freezing rain.*

5. Sleet

It is frozen raindrops of transparent grains which form when rain falls through air at subfreezing temperature. In Britain, *sleet* denotes precipitation of snow and rain simultaneously.

6. Hail

It is a showery precipitation in the form of irregular pellets or lumps of ice of size more than 8 mm. Hails occur in violent thunderstorms in which vertical currents are very strong.

2.3 Weather Systems for Precipitation

For the formation of clouds and subsequent precipitation, it is necessary that the moist air masses cool to form condensation. This is normally accomplished by adiabatic cooling of moist air through a process of being lifted to higher altitudes. Some of the terms and processes connected with weather systems associated with precipitation are given below.

1. Front

A *front* is the interface between two distinct air masses. Under certain favourable conditions, when a warm air mass and cold air mass meet, the warmer air mass is lifted over the colder one with the formation of a front. The ascending warmer air cools adiabatically with the consequent formation of clouds and precipitation.

2. Cyclone

A *cyclone* is a large low-pressure region with circular wind motion. Two types of cyclones are recognised: tropical cyclones and extratropical cyclones.

(a) Tropical Cyclone A tropical cyclone, also called *cyclone* in India, *hurricane* in USA and *typhoon* in South-East Asia, is a wind system with an intensely strong depression with *MSL* pressures sometimes below 915 mbars. The normal areal extent of a cyclone is about 100–200 km in diameter. The isobars are closely spaced and the winds are anticlockwise in the northern hemisphere. The centre of the storm, called the *eye,* which may extend to about 10–50 km in diameter, will be relatively quiet. However, right outside the eye, very strong winds/reaching to as much as 200 kmph exist. The wind speed gradually decreases towards the outer edge. The pressure also increases outwards (Fig. 2.1). The rainfall will normally be heavy in the entire area occupied by the cyclone.

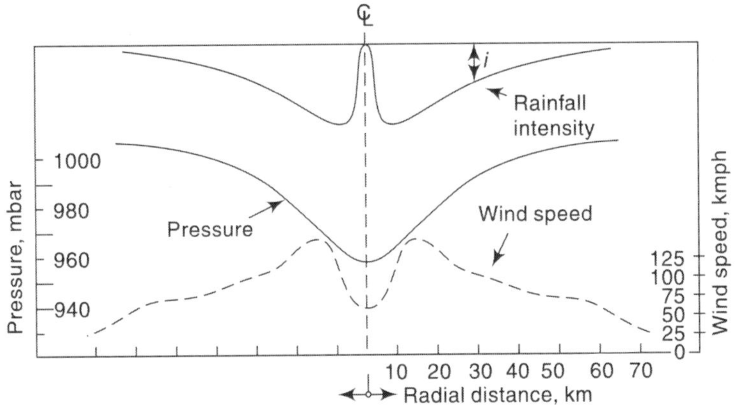

Fig. 2.1 ***Schematic Section of a Tropical Cyclone***

During summer months, tropical cyclones originate in the open ocean at around 5–10° latitude and move at speeds of about 10–30 kmph to higher latitudes in an irregular path. They derive their energy from the latent heat of condensation of ocean water vapour and increase in size as they move on oceans. When they move on land, the source of energy is cut off and the cyclone dissipates its energy very fast. Hence, the intensity of the storm decreases rapidly. Tropical cyclones cause heavy damage to life and property on their land path and intense rainfall and heavy floods in streams are its usual consequences. Tropical cyclones give moderate to excessive precipitation over very large areas, of the order of 10^3 km^2, for several days.

(b) Extratropical Cyclone These are cyclones formed in locations outside the tropical zone. Associated with a frontal system, they possess a strong counter-clockwise wind circulation in the northern hemisphere. The magnitude of precipitation and wind velocities are relatively lower than those of a tropical cyclone. However, the duration of precipitation is usually longer and the areal extent also is larger.

3. Anticyclones

These are regions of high pressure, usually of large areal extent. The weather is usually calm at the centre. Anticyclones cause clockwise wind circulations in the northern hemisphere. Winds are of moderate speed, and at the outer edges, cloudy and precipitation conditions exist.

4. Convective Precipitation

In this type of precipitation, a packet of air which is warmer than the surrounding air due to localised heating rises because of its lesser density. Air from cooler surroundings flows to take up its place, thus setting up a convective cell. The warm air continues to rise, undergoes cooling and results in precipitation. Depending upon the moisture, thermal and other conditions, light showers to thunderstorms can be expected in convective precipitation. Usually, the areal extent of such rains is small, being limited to a diameter of about 10 km.

5. Orographic Precipitation

The moist air masses may get lifted up to higher altitudes due to the presence of mountain barriers and consequently undergo cooling, condensation and precipitation. Such a precipitation is known as *orographic precipitation.* Thus, in mountain ranges, the windward slopes have heavy precipitation and the leeward slopes have light rainfall.

2.4 *Characteristics of Precipitation in India*

From the point of view of climate, the Indian subcontinent can be considered to have two major seasons and two transitional periods as

- South-west monsoon (June–September) (Summer Monsoon)
- Transition-I, post-monsoon (October–November)
- Winter season (December–February)
- Transition-II, Summer (Pre-monsoon). (March–May)

The chief precipitation characteristics of these seasons are given below.

2.4.1 South-West Monsoon (June–September) (Summer Monsoon)

The south-west monsoon (popularly known as *monsoon*) is the principal rainy season of India when over 75% of the annual rainfall is received over a major portion of the country. Excepting the south-eastern part of the peninsula and Jammu and Kashmir, for the rest of the country, the south-west monsoon is the principal source of rain with July as the month which has maximum rain. The monsoon originates in the Indian ocean and heralds its appearance in the southern part of Kerala by the end of May. The onset of monsoon is accompanied by high south-westerly winds at speeds of 30–70 kmph and low-pressure regions at the advancing edge. The monsoon winds advance across the country in two branches: (i) the Arabian sea branch, and (ii) the Bay of Bengal branch. The former sets in at the extreme southern part of Kerala and the latter at Assam, almost simultaneously in the first week of June. The Bay branch first covers the north-eastern regions of the country and turns westwards to advance into Bihar and UP. The Arabian sea branch moves northwards over Karnataka, Maharashtra and Gujarat. Both the branches reach Delhi around the same time by about the fourth week of June. A low-pressure region, known as *monsoon trough,* is formed between the two branches. The trough extends from the Bay of Bengal to Rajasthan and the precipitation pattern over the country is generally determined by its position. The monsoon winds increase from June to July and begin to weaken in September. The withdrawal of the monsoon, marked by a substantial rainfall activity, starts in September in the northern part of the country. The onset and withdrawal of the monsoon at various parts of the country are shown in Fig. 2.2(a) and Fig. 2.2(b).

The monsoon is not a period of continuous rainfall. The weather is generally cloudy with frequent spells of rainfall. Heavy rainfall activity in various parts of the country owing to the passage of low-pressure regions is common. Depressions formed in the Bay of Bengal at a frequency of 2–3 per month move along the trough causing excessive precipitation of about 100–200 mm per day. Breaks of about a week in which the rainfall activity is the least is another feature of the monsoon. The south-west monsoon rainfall over the country is indicated in Fig. 2.3. As seen from this figure, the heavy rainfall areas are Assam and the north-eastern region with 200–400 cm, west coast and Western Ghats with 200–300 cm, West Bengal with 120–160 cm, UP, Haryana and the Punjab with 100–120 cm. The long-term average monsoon rainfall over the country is estimated as 87.7 cm.

2.4.2 Post-Monsoon (October-November)

As the south-west monsoon retreats, low-pressure areas form in the Bay of Bengal and a north-easterly flow of air that picks up moisture in the Bay of Bengal is formed. This air mass strikes the east coast of the southern peninsula (Tamil Nadu) and causes rainfall. Also, in this period, especially in November, severe tropical cyclones form in the Bay of Bengal and the Arabian sea. The cyclones formed in the Bay of Bengal are about twice as many as in the Arabian Sea. These cyclones strike the coastal areas and cause intense rainfall and heavy damage to life and property.

2.4.3 Winter Season (December-February)

By about mid-December, disturbances of extra tropical origin travel eastwards across Afghanistan and Pakistan. Known as *western disturbances,* they cause moderate to heavy rain and snowfall (about 25 cm) in the Himalayas, and Jammu and Kashmir. Some light rainfall also occurs in the northern plains. Low-pressure areas in the Bay of Bengal formed in these months cause 10–12 cm of rainfall in the southern parts of Tamil Nadu.

2.4.4 Summer (Pre-monsoon) (March-May)

There is very little rainfall in India in this season. Convective cells cause some thunderstorms mainly in Kerala, West Bengal and Assam. Some cyclone activity, dominantly on the east coast, also occurs.

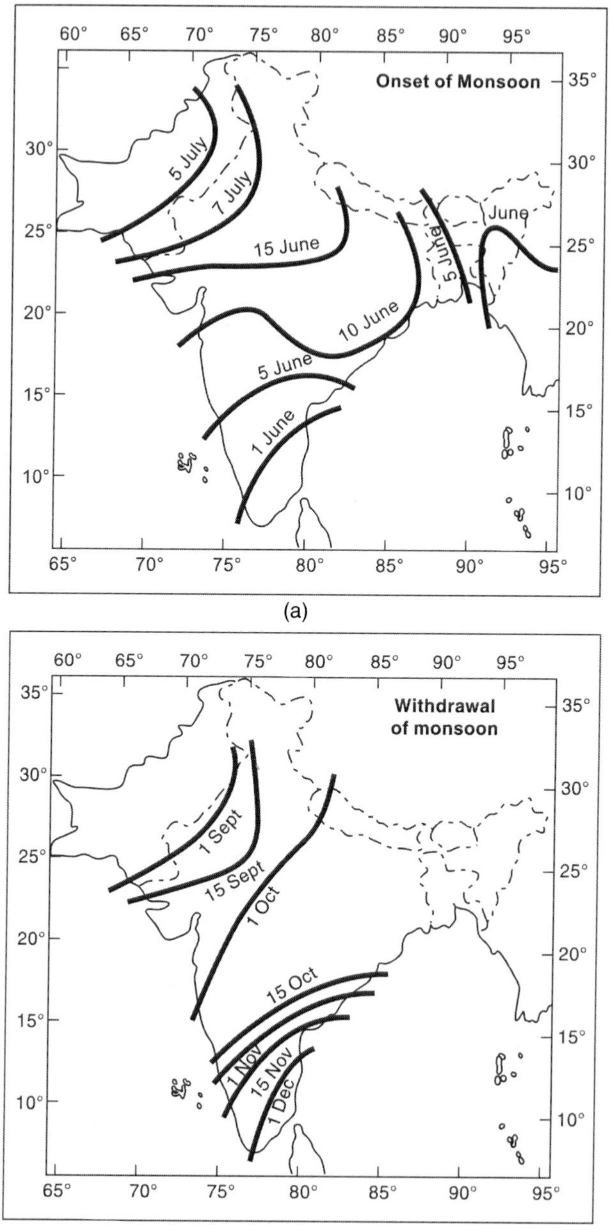

Fig. 2.2 ***(a) Normal Dates of Onset of Monsoon (b) Normal Dates of Withdrawal of Monsoon***

(Reproduced from Natural Resources of Humid Tropical Asia—Natural Resources Research, XII. © UNESCO, 1974, with permission of UNESCO)

The territorial waters of India extend into the sea to a distance of 12 nautical miles measured from the appropriate baseline

Responsibility for the correctness of the internal details on the map rests with the publisher.

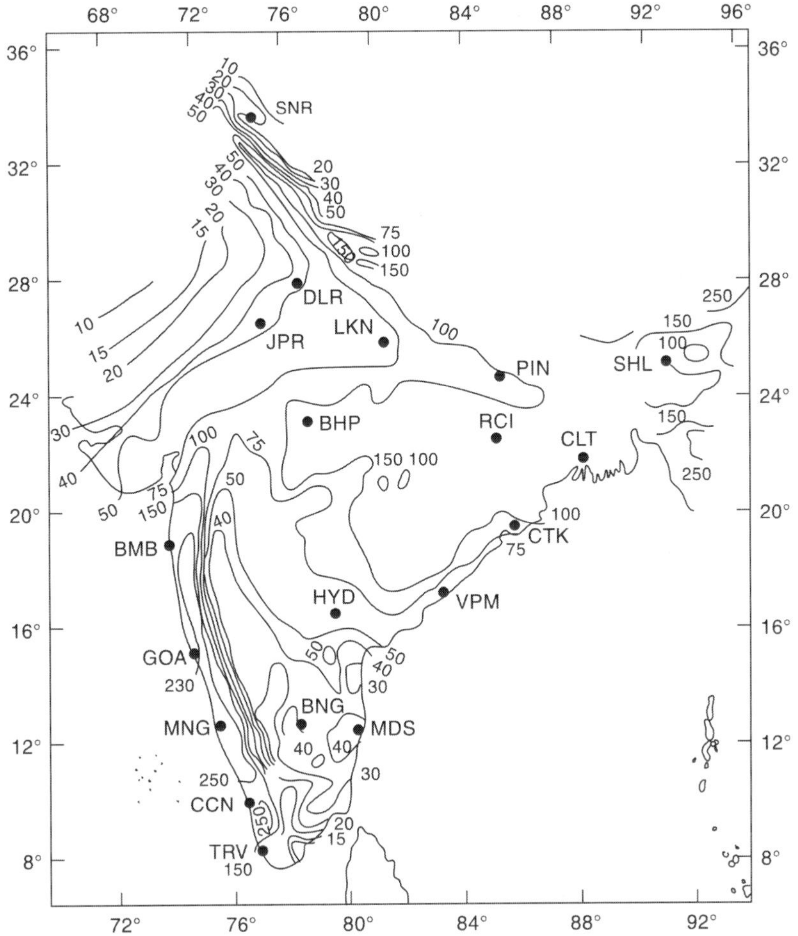

Fig. 2.3 ***South-west Monsoon Rainfall (cm) over India and Neighbourhood***

(Reproduced with permission from India Meteorological Department)

Based upon Survey of India map with the permission of the Surveyor General of India © Government of India Copyright 1984

The territorial waters of India extend into the sea to a distance of 12 nautical miles measured from the appropriate baseline

Responsibility for the correctness of the internal details on the map rests with the publisher.

2.4.5 Annual Rainfall

The annual rainfall over the country is shown in Fig. 2.4. Considerable areal variation exists for the annual rainfall in India with high rainfall of the magnitude of 200 cm in Assam and north-eastern parts and the Western Ghats, and scanty rainfall in eastern Rajasthan and parts of Gujarat, Maharashtra and Karnataka. The average annual rainfall for the entire country is estimated as 118.3 cm.

It is well known that there is considerable variation of annual rainfall in time at a place. The coefficient of variation,

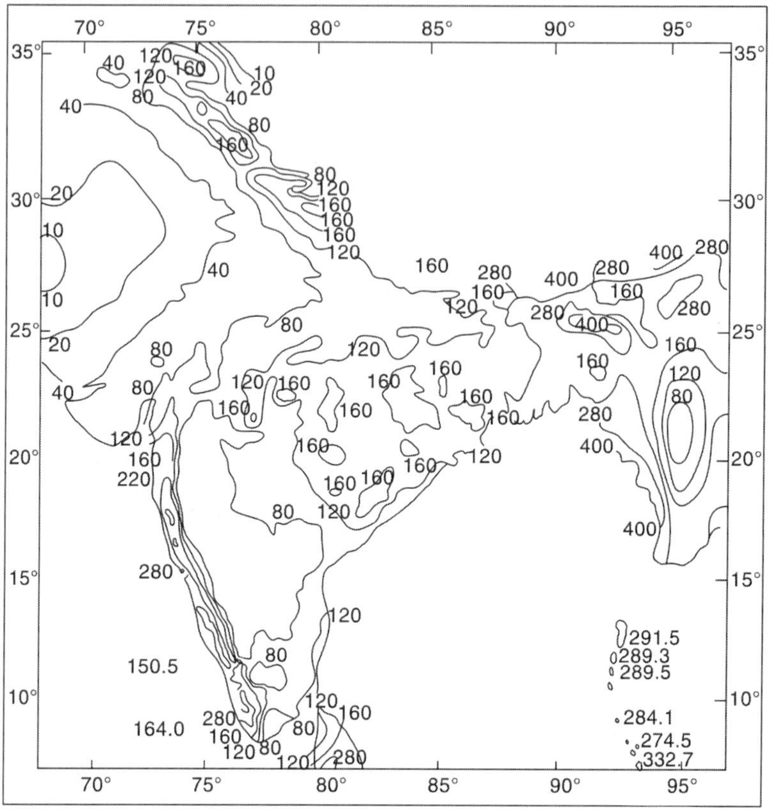

Fig. 2.4 ***Annual Rainfall (cm) over India and Neighbourhood***

(Reproduced from Natural Resources of Humid Tropical Asia—Natural Resources Research, XII. © UNESCO, 1974, with permission of UNESCO)

Based upon Survey of India map with the permission of the Surveyor General of India © Government of India Copyright 1984

The territorial waters of India extend into the sea to a distance of 12 nautical miles measured from the appropriate baseline

Responsibility for the correctness of the internal details on the map rests with the publisher.

$$C_v = \frac{100 \times \text{Standard deviation}}{\text{Mean}}$$

of the annual rainfall varies between 15 and 70, from place to place with an average value of about 30. Variability is least in regions of high rainfall and largest in regions of scanty rainfall. Gujarat, Haryana, Punjab and Rajasthan have large variability of rainfall.

Some of the interesting statistics relating to the variability of the seasonal and annual rainfall of India are as follows:

- A few heavy spells of rain contribute nearly 90% of total rainfall.
- While the average annual rainfall of the country is 118 cm, average annual rainfall varies from 10 cm in the western desert to 1100 cm in the north-east region.
- More than 50% rain occurs within 15 days and less than 100 hours in a year.

- More than 80% of seasonal rainfall is produced in 10–20% rain events, each lasting 1–3 days.

2.5 *Measurement of Precipitation*

2.5.1 Rainfall

Precipitation is expressed in terms of the depth to which rainfall water would stand on an area if all the rain were collected on it. Thus, 1 cm of rainfall over a catchment area of 1 km^2 represents a volume of water equal to 10^4 m^3. In the case of snowfall, an equivalent depth of water is used as the depth of precipitation. The precipitation is collected and measured in a *raingauge.* Terms such as *pluviometer, ombrometer* and *hyetometer* are also sometimes used to designate a raingauge.

A raingauge essentially consists of a cylindrical-vessel assembly kept in the open to collect rain. The rainfall catch of the raingauge is affected by its exposure conditions. To enable the catch of raingauge to accurately represent the rainfall in the area surrounding the raingauge, standard settings are adopted. For setting up a raingauge, the following considerations are important:

- The ground must be level and in the open and the instrument must present a horizontal catch surface.
- The gauge must be set as near the ground as possible to reduce wind effects but it must be sufficiently high to prevent splashing, flooding, etc.
- The instrument must be surrounded by an open fenced area of at least 5.5 m × 5.5 m. No object should be nearer to the instrument than 30 m or twice the height of the obstruction.

Raingauges can be broadly classified into two categories as (i) nonrecording raingauges, and (ii) recording gauges.

1. Nonrecording Gauges

The nonrecording gauge extensively used in India is the *Symons' gauge.* It essentially consists of a circular collecting area of 12.7 cm (5.0 inch) diameter connected to a funnel. The rim of the collector is set in a horizontal plane at a height of 30.5 cm above the ground level. The funnel discharges the rainfall catch into a receiving vessel. The funnel and receiving vessel are housed in a metallic container. Figure 2.5 shows the details of the installation. Water contained in the receiving

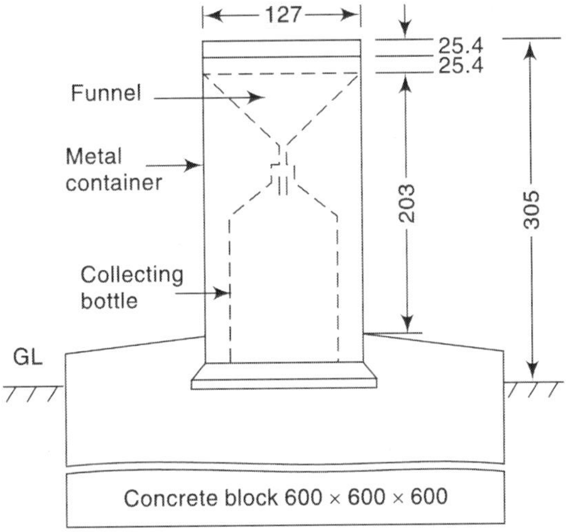

Fig. 2.5 ***Nonrecording Raingauge (Symons' Gauge)***

vessel is measured by a suitably graduated measuring glass, with an accuracy up to 0.1 mm.

Recently, the Indian Meteorological Department (*IMD*) has changed over to the use of fibreglass reinforced polyester raingauges, which is an improvement over the *Symons' gauge.* These come in different combinations of collector and bottle. The collector is in two sizes having areas of 200 and 100 cm^2 respectively. Indian standard (IS: 5225–1969) gives details of these new raingauges.

For uniformity, the rainfall is measured every day at 8.30 a.m. (*IST*) and is recorded as the rainfall of that day. The receiving bottle normally does not hold more than 10 cm of rain and as such, in the case of heavy rainfall, the measurements must be done more frequently and entered. However, the last reading must be taken at 8.30 a.m. and the sum of the previous readings in the past 24 hours entered as the total of that day. Proper care, maintenance and inspection of raingauges, especially during dry weather to keep the instrument free from dust and dirt, is very necessary. The details of installation of nonrecording raingauges and measurement of rain are specified in Indian Standard (IS: 4986–1968).

This raingauge can also be used to measure snowfall. When snow is expected, the funnel and receiving bottle are removed and the snow is allowed to collect in the outer metal container. The snow is then melted and the depth of resulting water measured. Antifreeze agents are sometimes used to facilitate melting of snow. In areas where considerable snowfall is expected, special snowgauges with shields (for minimising the wind effect) and storage pipes (to collect snow over longer durations) are used.

2. Recording Gauges

Recording gauges produce a continuous plot of rainfall against time and provide valuable data of intensity and duration of rainfall for hydrological analysis of storms. The following are some of the commonly used recording raingauges.

(a) Tipping-Bucket Type

This is a 30.5 cm size raingauge adopted for use by the *US* Weather Bureau. The catch from the funnel falls onto one of a pair of small buckets. These buckets are so balanced that when 0.25 mm of rainfall collects in one bucket, it tips and brings the other one in position. The water from the tipped bucket is collected in a storage can. The tipping actuates an electrically driven pen to trace a record on the clockwork-driven chart. The water collected in the storage can is measured at regular intervals to provide the total rainfall and also serve as a check. It may be noted that the record from the tipping bucket gives data on the intensity of rainfall. Further, the instrument is ideally suited for digitalising of the output signal.

(b) Weighing-Bucket Type

In this raingauge, the catch from the funnel empties into a bucket mounted on a weighing scale. The weight of the bucket and its contents are recorded on a clockwork-driven chart. The clockwork mechanism has the capacity to run for as long as one week. This instrument gives a plot of the accumulated rainfall against the elapsed time, i.e. the mass curve of rainfall. In some instruments of this type, the recording unit is so constructed that the pen reverses its direction at every preset value, say 7.5 cm (3 in.) so that a continuous plot of storm is obtained.

(c) Natural-Syphon Type

This type of recording raingauge is also known as *float-type gauge*. Here, the rainfall collected by a funnel-shaped collector is led into a float chamber causing a float to rise. As the float rises, a pen attached to the float through a lever system records the elevation of the float on a rotating drum driven by a clockwork mechanism. A syphon arrangement empties the float chamber when the float has reached a preset maximum level. This type of raingauge is adopted as the standard recording-type raingauge in India and its details are described in Indian Standard (IS: 5235–1969).

A typical chart from this type of raingauge is shown in Fig. 2.6. This chart shows a rainfall of 53.8 mm in 30 h. The vertical lines in the pen-trace correspond to the sudden emptying of the float chamber by syphon action which resets the pen to zero level. It is obvious that the natural syphon-type recording raingauge gives a plot of the mass curve of rainfall.

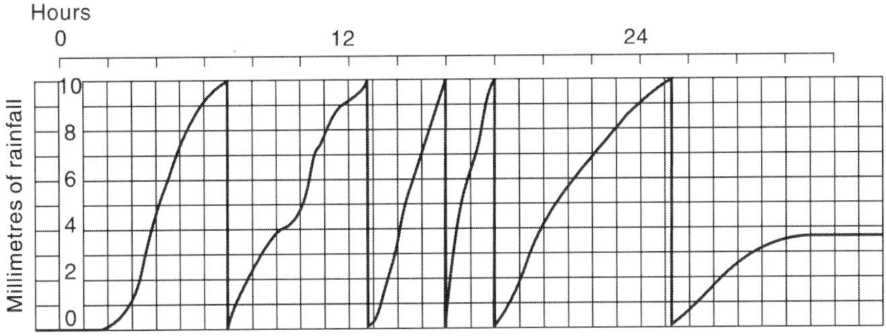

Fig. 2.6 ***Recording from a Natural Syphon-type Gauge (Schematic)***

3. Telemetering Raingauges

These raingauges are of the recording type and contain electronic units to transmit the data on rainfall to a base station both at regular intervals and on interrogation. The tipping-bucket type raingauge, being ideally suited, is usually adopted for this purpose. Any of the other types of recording raingauges can also be used equally effectively. Telemetering gauges are of utmost use in gathering rainfall data from mountainous and generally inaccessible places.

4. Radar Measurement of Rainfall

The meteorological radar is a powerful instrument for measuring the areal extent, location and movement of rain storms. Further, the amounts of rainfall over large areas can be determined through the radar with a good degree of accuracy.

The radar emits a regular succession of pulses of electromagnetic radiation in a narrow beam. When raindrops intercept a radar beam, it has been shown that

$$P_r = \frac{CZ}{r^2} \tag{2.1}$$

where P_r = average echopower, Z = radar-echo factor, r = distance to target volume and C = a constant. Generally, the factor Z is related to the intensity of rainfall as

$$Z = aI^b \tag{2.2}$$

where a and b are coefficients and I = intensity of rainfall in mm/h. The values a

and b for a given radar station have to be determined by calibration with the help of recording raingauges. A typical equation for Z is

$$Z = 200\ I^{1.60}$$

Meteorological radars operate with wavelengths ranging from 3 to 10 cm, the common values being 5 and 10 cm. For observing details of heavy flood-producing rains, a 10 cm radar is used while for light rain and snow, a 5 cm radar is used. The hydrological range of the radar is about 200 km. Thus, a radar can be considered to be a remote-sensing super gauge covering an areal extent of as much as 100,000 km^2. Radar measurement is continuous in time and space. Present-day developments in the field include (i) online processing of radar data on a computer, and (ii) Doppler-type radars for measuring the velocity and distribution of raindrops.

2.5.2 Snowfall

Snowfall as a form of precipitation differs from rainfall in that it may accumulate over a surface for some time before it melts and causes runoff. Further, evaporation from the surface of accumulated snow surface is a factor to be considered in analysis dealing with snow. Water equivalent of snowfall is included in the total precipitation amounts of a station to prepare seasonal and annual precipitation records.

1. Depth of Snowfall

Depth of snowfall is an important indicator for many engineering applications and in hydrology, it is useful for seasonal precipitation and long-term runoff forecasts. A graduated stick or staff is used to measure the depth of snow at a selected place. Average of several measurements in an area is taken as the depth of snow in a snowfall event. *Snow stakes* are permanent graduated posts used to measure total depth of accumulated snow at a place.

Snow boards are 40 cm side square boards used to collect snow samples. These boards are placed horizontally on a previous accumulation of snow and after a snowfall event, the snow samples are cut off from the board and depth of snow and water equivalent of snow are derived and recorded.

2. Water Equivalent of Snow

Water equivalent of snow is the depth of water that would result in melting of a unit of snow. This parameter is important in assessing the seasonal water resources of a catchment as well as in estimates of stream flow and floods due to melting of snow.

The amount of water present in a known depth of snow could be estimated if the information about the density of snow is available. The density of snow, however, varies quite considerably. Freshly fallen snow may have a density in the range of 0.07 to 0.15 with an average value of about 0.10. The accumulated snow, however, causes compaction and in regions of high accumulation, densities as high as 0.4 to 0.6 is not uncommon. Where specific data is not available, it is usual to assume the density of fresh snow as 0.10.

Water equivalent of snow is obtained in two ways:

(a) Snow Gauges Like rain gauges, *snow gauges* are receptacles to catch precipitation as it falls in a specified sampling area. Here, a large cylindrical receiver, 203 mm in diameter, is used to collect the snow as it falls. The height of the cylinder depends upon the snow storage needed at the spot as a consequence of accessibility, etc. and may range from 60 cm to several metres. The receiver is mounted on a tower to keep the rim of the gauge above the anticipated maximum depth of accumulated snow in the area. The top of the cylinder is usually a funnel like fulcrum of cone with side slopes not less than 1 H: 6 V, to minimise deposits of ice on the exterior of the gauge. Also, a windshield is provided at the top. Melting agents or heating systems are sometimes provided in the remote snow gauges to reduce the size of the containers. The snow collected in the cylinder is brought in to a warm room and the snow melted by adding a pre-measured quantity of hot water. Through weighing or by volume measurements, the water equivalent of snow is ascertained and recorded.

(b) Snow Tubes Water equivalent of accumulated snow is measured by means of *snow tubes* which are essentially a set of telescopic metal tubes. While a tube size of 40 mm diameter is in normal use, higher sizes up to 90 mm diameter are also in use. The main tube is provided with a cutter edge for easy penetration as well as to enable extracting of core sample. Additional lengths of tube can be attached to the main tube depending upon the depth of snow.

To extract a sample, the tube is driven into the snow deposit till it reaches the bottom of the deposit and then twisted and turned to cut a core. The core is extracted carefully and studied for its physical properties and then melted to obtain water equivalent of the snow core. Obviously, a large number of samples are needed to obtain representative values for a large area deposit. Usually, the sampling is done along an established route with specified locations called *snow course.*

2.6 *Raingauge Network*

2.6.1 Network Density

Since the catching area of a raingauge is very small compared to the areal extent of a storm, it is obvious that to get a representative picture of a storm over a catchment, the number of raingauges should be as large as possible, i.e. the catchment area per gauge should be small. On the other hand, economic considerations to a large extent and other considerations, such as topography, accessibility, etc. to some extent, restrict the number of gauges to be maintained. Hence, one aims at an optimum density of gauges from which reasonably accurate information about the storms can be obtained. Towards this, the World Meteorological Organisation (*WMO*) recommends the following densities.

- In flat regions of temperate, Mediterranean and tropical zones
  Ideal—1 station for 600–900 km^2
  Acceptable—1 station for 900–3000 km^2
- In mountainous regions of temperate, Mediterranean and topical zones
  Ideal—1 station for 100–250 km^2
  Acceptable—1 station for 25–1000 km^2
- In arid and polar zones: 1 station for 1500–10,000 km^2 depending on the feasibility.

Ten per cent of raingauge stations should be equipped with self-recording gauges to know the intensities of rainfall.

From practical considerations of Indian conditions, the Indian Standard (IS: 4987–1968) recommends the following densities as sufficient.

- In plains: 1 station per 520 km^2;
- In regions of average elevation of 1000 m: 1 station per 260–390 km^2; and
- In predominantly hilly areas with heavy rainfall: 1 station per 130 km^2.

2.6.2 Adequacy of Raingauge Stations

If there are already some raingauge stations in a catchment, the optimal number of stations that should exist to have an assigned percentage of error in the estimation of mean rainfall is obtained by statistical analysis as

$$N = \left(\frac{C_v}{\varepsilon}\right)^2 \tag{2.3}$$

where N = optimal number of stations, ε = allowable degree of error in the estimate of the mean rainfall and C_v = coefficient of variation of the rainfall values at the existing m stations (in per cent). If there are m stations in the catchment each recording rainfall values $P_1, P_2, \ldots, P_i, \ldots P_m$ in a known time, the coefficient of variation C_v is calculated as:

$$C_v = \frac{100 \times \sigma_{m-1}}{\bar{P}}$$

where $$\sigma_{m-1} = \sqrt{\left[\frac{\sum_1^m (P_i - \bar{P})^2}{m-1}\right]} = \text{standard deviation}$$

P_i = precipitation magnitude in the ith station

$$\bar{P} = \frac{1}{m}\left(\sum_1^m P_i\right) = \text{mean precipitation...}$$

Consider the existing m raingauges. They have a mean rainfall of $\bar{P}$ and a coefficient of variation of C_v. To know the percentage of error (ε_{ex}) of estimation of mean in the existing system of m raingauges, Eq. (2.3) can be rewritten (by changing N by m) as

$$m = \left(\frac{C_v}{\varepsilon_{ex}}\right)^2$$

$$\varepsilon_{ex} = \frac{C_v}{\sqrt{m}} \tag{2.3-a}$$

In the above Eq. (2.3-a), the term ε_{ex} represents the expected error (in percentage) in the estimation of the mean, $\bar{P}$. It is a measure of the accuracy of estimation of mean precipitation in the existing system and is called *standard error* in the estimation of the mean.

In calculating the number of raingauges N for a given level of error by using Eq. (2.3) it is usual to take ε_{ex} as 10%. It is seen that if the value of ε_{ex} is small, the number of raingauges required will be more. According to *WMO* recommendations, at least 10% of total raingauges should be of self-recording type.

Example 2.1 A catchment has six raingauge stations. In a year, the annual rainfall recorded by the gauges are as follows:

Station	A	B	C	D	E	F
Rainfall (cm)	82.6	102.9	180.3	110.3	98.8	136.7

(a) Determine the standard error in the estimation of mean rainfall in the existing set of raingauges.
(b) For a 10% error in the estimation of the mean rainfall, calculate the optimum number of raingauge stations in the catchment.

Solution

For the given data

Number of raingauges $m = 6$.

Mean annual rainfall $\bar{P} = 118.60$ cm

Standard deviation $= \sigma_{m-1} = 35.04$ cm

(calculated by using a calculator or a spreadsheet like MS Excel).

Coefficient of variation $C_v = \dfrac{100 \times \sigma_{m-1}}{\bar{P}} = \dfrac{100 \times 35.04}{118.60} = 29.54$

(a) Standard error in the estimation of the mean

$$= \varepsilon_{ex} = \frac{C_v}{\sqrt{m}} = \frac{29.54}{\sqrt{6}} = 12.06\%$$

(b) When the error is limited to 10%, $\varepsilon = 10$ and the optimum number of raingauges in the catchment is given by

$$N = \left(\frac{C_v}{\varepsilon}\right)^2 = \left(\frac{29.54}{10}\right)^2 = 8.73$$

Hence, optimum number of raingauges is 9 raingauges.

Thus, the number of additional raingauges required = (9 – 6) = 3.

2.7 *Preparation of Data*

Before using the rainfall data in application, it is necessary to first check the data for continuity and consistency. The continuity of record may be broken with missing data due to many reasons such as damage or fault in the raingauge during a period. If the conditions of installation of a raingauge undergo a change, the records of the period before and after the change would be inconsistent. Methods of pre-cleaning the records for such problems of continuity and consistency are indicated in the following sections.

2.7.1 Estimation of Missing Data

In the estimation of missing data from a raingauge station, performance of a group of neighbouring stations including the one with missing data are considered. A comparison of the recordings of these stations are made by using their *normal rainfall* as standard of comparison. The normal rainfall is the average value of rainfall at a particular date, month or year over a specified 30-year period. The 30-year normal is recomputed every decade. Thus, the term *normal annual precipitation* at a station *A* means the average annual precipitation at station A based on a specified 30-year of record. Insertion of missing data to a station record must be done sparingly. If too many data are estimated, the quality of the total data set may be diluted due to interpolation. sometimes, if too many gaps exist in a record, it may be worthwhile to neglect that station than to have a station record with too much of interpolated data. A *WMO* guideline states that not more than 5 to 10% of a record should have interpolated data. A commonly used procedure for estimating missing data of a station is given below.

Procedure of Missing Data Estimation

(a) Statement of the Missing Data Problem Given the annual precipitation values P_1, P_2, P_3,.....P_m at neighbouring M stations 1, 2, 3, ..., M respectively, it is required to find the missing annual precipitation P_x at a station X not included in the above M stations. Further, the normal annual precipitations N_1, N_2, ...N_i,...at each of the above (M+1) stations, including station X, are known.

(b) Procedure (i) If the normal annual precipitations at various stations are within about 10% of the normal annual precipitation at station X then a simple arithmetic average procedure is followed to estimate P_x. Thus,

$$P_x = \frac{1}{M}\left[P_1 + P_2 + ... + P_m\right] \tag{2.4}$$

(ii) If the normal annual precipitations vary considerably then P_x is estimated by weighing the precipitation at various stations by the ratios of normal annual precipitations. This method, known as the *normal ratio method,* gives P_x as

$$P_x = \frac{N_x}{M}\left[\frac{P_1}{N_1} + \frac{P_2}{N_2} + ... + \frac{P_m}{N_m}\right] \tag{2.5}$$

Example 2.2 The normal annual rainfall at stations A, B, C and D in a basin are 80.97, 67.59, 76.28 and 92.01 cm respectively. In the year 1985, the station D was inoperative and the stations A, B and C recorded annual precipitation of 91.11, 72.23 and 79.89 cm respectively. Estimate the rainfall at station D in that year.

Solution

Maximum difference in normal rainfall values from the normal value at station $D = 92.01 - 67.59 = 24.42$ cm

Percentage maximum difference with respect to normal at station

$$D = \frac{24.42}{92.01} \times 100 = 26.54\%.$$

As the normal rainfall values of the surrounding stations vary considerably more than 10%, the normal ratio method is adopted.

In the normal ratio method: $M = 3$. Using Eq. (2.5),

Estimated Rainfall at Station $D = P_D = \dfrac{N_D}{3}\left[\dfrac{P_A}{N_A} + \dfrac{P_B}{N_B} + \dfrac{P_C}{N_C}\right]$

$$P_D = \frac{92.01}{3}\left[\frac{91.11}{80.97} + \frac{72.23}{67.59} + \frac{79.89}{76.28}\right]$$

$$= 30.67 \times (1.128 + 1.069 + 1.047) = 99.49 \text{ cm}$$

2.7.2 Test for Consistency of Record

If the conditions relevant to the recording of a raingauge station have undergone a significant change during the period of record, inconsistency would arise in the rainfall data of that station. This inconsistency would be felt from the time the significant change took place. Some of the common causes for inconsistency of record are (i) shifting of a raingauge station to a new location, (ii) the neighbourhood of the station undergoing a marked change, (iii) change in the ecosystem due to calamities, such as forest fires, land slides, and (iv) occurrence of observational error from a certain date. The checking for inconsistency of a record is done by the *double-mass curve technique.* This technique is based on the principle that when each recorded data comes from the same parent population, they are consistent.

A group of 5 to 10 base stations in the neighbourhood of the problem station X is selected. The data of the annual (or monthly or seasonal mean) rainfall of the station X and also the average rainfall of the group of base stations covering a long period is arranged in the reverse chronological order (i.e. the latest record as the first entry and the oldest record as the last entry in the list). The accumulated precipitation of the station X (i.e. ΣP_x) and the accumulated values of the average of the group of base stations (i.e. ΣP_{av}) are calculated starting from the latest record. Values of ΣP_x are plotted against ΣP_{av} for various consecutive time periods (Fig. 2.7). A decided break in the slope of the resulting plot indicates a change in the precipitation regime of station X. The precipitation values at station X beyond the period of change of regime (point 63 in Fig. 2.7) is corrected by using the relation

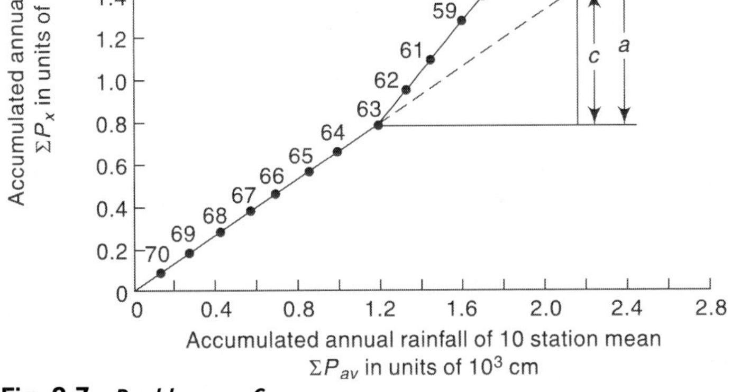

Fig. 2.7 ***Double-mass Curve***

$$P_{cx} = P_x \frac{M_c}{M_a} \tag{2.6}$$

where P_{cx} = corrected precipitation at any time period t_1 at station X

P_x = original recorded precipitation at time period t_1 at station X

M_c = corrected slope of the double-mass curve

M_a = original slope of the double-mass curve

In this way, the older records are brought to the new regime of the station. It is apparent that the more homogeneous the base station records are, the more accurate will be the corrected values at station X. A change in the slope is normally taken as significant only where it persists for more than five years. The double-mass curve is also helpful in checking systematic arithmetical errors in transferring rainfall data from one record to another.

Example 2.3 Annual rainfall data for station *M* as well as the average annual rainfall values for a group of ten neighbouring stations located in a meteorologically homogeneous region are given below.

Year	Annual Rainfall of Station *M* (mm)	Average Annual Rainfall of the group (mm)	Year	Annual Rainfall of Station *M* (mm)	Average Annual Rainfall of the group (mm)
1950	676	780	1965	1244	1400
1951	578	660	1966	999	1140
1952	95	110	1967	573	650
1953	462	520	1968	596	646
1954	472	540	1969	375	350
1955	699	800	1970	635	590
1956	479	540	1971	497	490
1957	431	490	1972	386	400
1958	493	560	1973	438	390
1959	503	575	1974	568	570
1960	415	480	1975	356	377
1961	531	600	1976	685	653
1962	504	580	1977	825	787
1963	828	950	1978	426	410
1964	679	770	1979	612	588

Test the consistency of the annual rainfall data of station *M* and correct the record if there is any discrepancy. Estimate the mean annual precipitation at station *M*.

Solution

The data is sorted in descending order of the year, starting from the latest year 1979. Cumulative values of station *M* rainfall (ΣP_m) and the ten station average rainfall values (ΣP_{av}) are calculated as shown in Table 2.1. The data is then plotted with ΣP_m on the Y-axis and ΣP_{av} on the X-axis to obtain a double-mass curve plot (Fig. 2.8). The value of the year corresponding to the plotted points is also noted on the plot. It is seen that the data plots as two straight lines with a break of grade at the year 1969. This represents a change in the regime of the station *M* after the year 1968. The slope of the best straight line for the

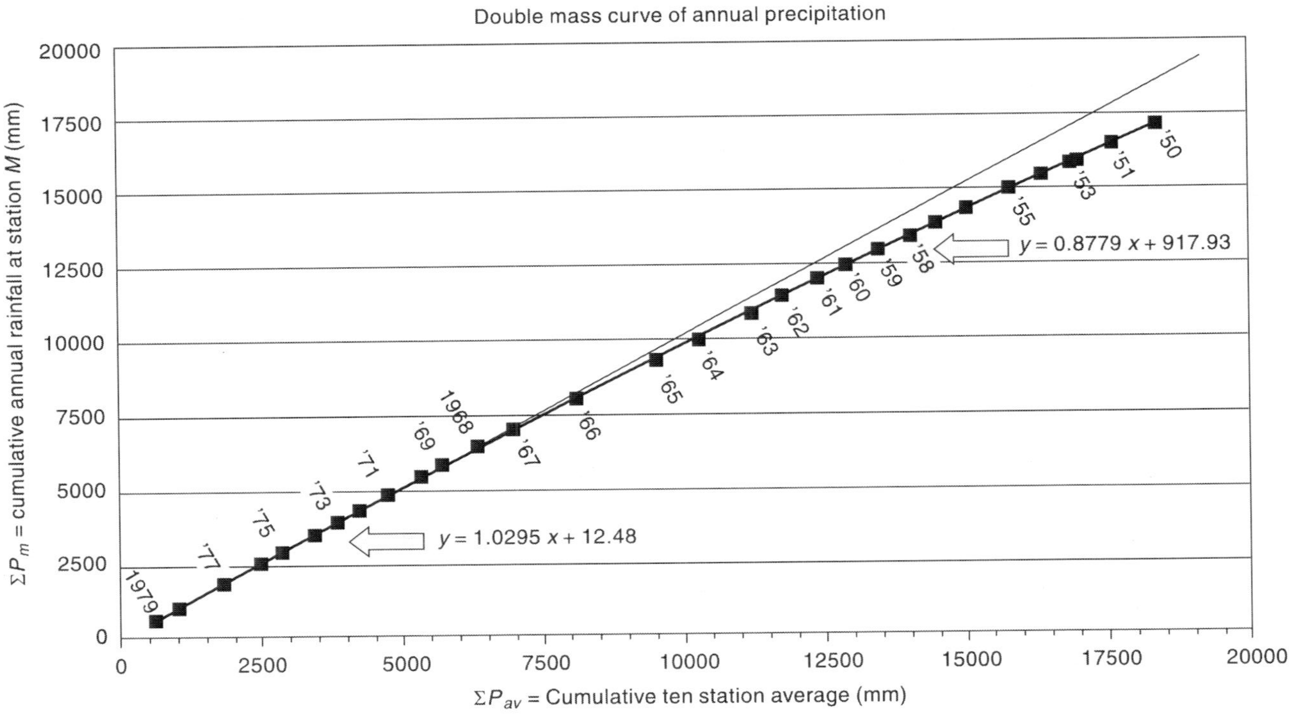

Fig. 2.8 ***Double-Mass Curve of Annual Rainfall at Station M***

period 1979–1969 is M_c = 1.0295 and the slope of the best straight line for the period 1968–1950 is M_a = 0.8779.

The correction ratio to bring the old records (1950–1968) to the current (post 1968) regime is = M_c/M_a = 1.0295/0.8779 = 1.173. Each of the pre 1969 annual rainfall value is multiplied by the correction ratio of 1.173 to get the adjusted value. The adjusted values at station *M* are shown in Col. 6 of Table 2.1. The finalised values of P_m (rounded off to nearest mm) for all the 30 years of record are shown in Col. 7.

The mean annual precipitation at station *M* (based on the corrected time series) = (19004/30) = 633.5 mm

Table 2.1 ***Calculation of Double-Mass Curve of Example 2.3***

1 Year	2 P_m (mm)	3 ΣP_m (mm)	4 P_{av} (mm)	5 ΣP_{av} (mm)	6 Adjusted values of P_m (mm)	7 Finalised values of P_m (mm)
1979	612	612	588	588		612
1978	426	1038	410	998		426
1977	825	1863	787	1785		825
1976	685	2548	653	2438		685
1975	356	2904	377	2815		356
1974	568	3472	570	3385		568
1973	438	3910	390	3775		438
1972	386	4296	400	4175		386
1971	497	4793	490	4665		497
1970	635	5428	590	5255		635
1969	375	5803	350	5605		375
1968	596	6399	646	6251	698.92	699
1967	573	6972	650	6901	671.95	672
1966	999	7971	1140	8041	1171.51	1172
1965	1244	9215	1400	9441	1458.82	1459
1964	679	9894	770	10211	796.25	796
1963	828	10722	950	11161	970.98	971
1962	504	11226	5801	11741	591.03	591
1961	531	11757	600	12341	622.70	623
1960	415	12172	480	12821	486.66	487
1959	503	12675	575	13396	589.86	590
1958	493	13168	560	13956	578.13	578
1957	431	13599	490	14446	505.43	505
1956	479	14078	540	14986	561.72	562
1955	699	14777	800	15786	819.71	820
1954	472	15249	540	16326	553.51	554
1953	462	15711	520	16846	541.78	542
1952	95	15806	110	16956	111.41	111
1951	578	16384	660	17616	677.81	678
1950	676	17060	780	18396	792.73	193

Total of P_m = 19004 mm
Mean of P_m = 633.5 mm

2.8 *Presentation Of Rainfall Data*

A few commonly used methods of presentation of rainfall data which have been found to be useful in interpretation and analysis of such data are given as follows:

2.8.1 Mass Curve of Rainfall

The mass curve of rainfall is a plot of the accumulated precipitation against time, plotted in chronological order. Records of float-type and weighing bucket-type gauges are of this form. A typical mass curve of rainfall at a station during a storm is shown in Fig. 2.9. Mass curves of rainfall are very useful in extracting the information on the duration and magnitude of a storm. Also, intensities at various time intervals in a storm can be obtained by the slope of the curve. For nonrecording raingauges, mass curves are prepared from a knowledge of the approximate beginning and end of a storm and by using the mass curves of adjacent recording gauge stations as a guide.

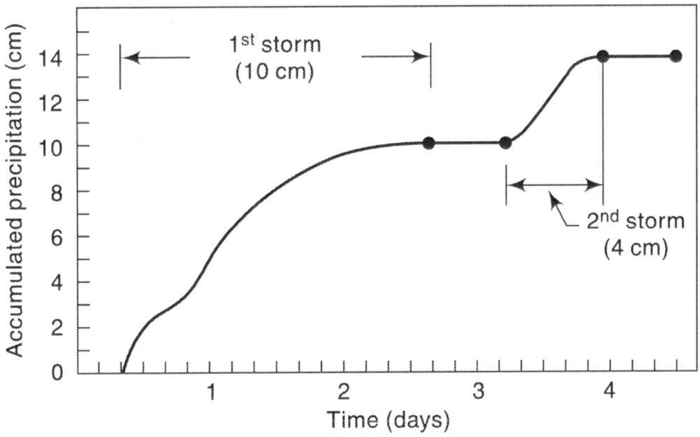

Fig. 2.9 ***Mass Curve of Rainfall***

2.8.2 Hyetograph

A hyetograph is a plot of the intensity of rainfall against the time interval. The hyetograph is derived from the mass curve and is usually represented as a bar chart (Fig. 2.10). It is a very convenient way of representing the characteristics of a storm and is particularly important in the development of design storms to predict extreme floods. The area under a hyetograph represents the total precipitation received in the period. The time interval used depends on the purpose, in urban-drainage problems small durations are used while in flood-flow computations in larger catchments the intervals are of about 6 h.

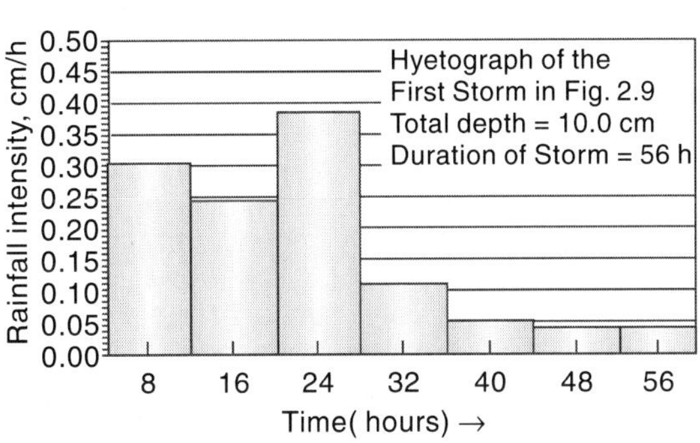

Fig. 2.10 ***Hyetograph of a Storm***

2.8.3 Point Rainfall

Point rainfall, also known as *station rainfall,* refers to the rainfall data of a station. Depending upon the need, data can be listed as daily, weekly, monthly, seasonal or annual values for various periods. Graphically, these data are represented as plots of magnitude *vs* chronological time in the form of a bar diagram. Such a plot, however, is not convenient for discerning a trend in the rainfall as there will be considerable variations in the rainfall values leading to rapid changes in the plot. The trend is often discerned by the method of *moving averages*, also known as *moving means*.

2.8.4 Moving Average

Moving average is a technique for smoothening out the high frequency fluctuations of a time series and to enable the trend, if any, to be noticed. The basic principle is that a window of time range m years is selected. Starting from the first set of m years of data, the average of the data for m years is calculated and placed in the middle year of the range m. The window is next moved sequentially one time unit (year) at a time and the mean of the m terms in the window is determined at each window location. The value of m can be 3 or more years; usually an odd value. Generally, the larger the size of the range m, the greater is the smoothening. There are many ways of averaging (and consequently the plotting position of the mean) and the method described above is called Central Simple Moving Average. Example 2.4 describes the application of the method of moving averages.

Example 2.4 Annual rainfall values recorded at station *M* for the period 1950 to 1979 is given in Example 2.3. Represent this data as a bar diagram with time in chronological order. (i) Identify those years in which the annual rainfall is (a) less than 20% of the mean, and (b) more than the mean. (ii) Plot the three-year moving mean of the annual rainfall time series.

Solution

(i) Figure 2.11 shows the bar chart with height of the column representing the annual

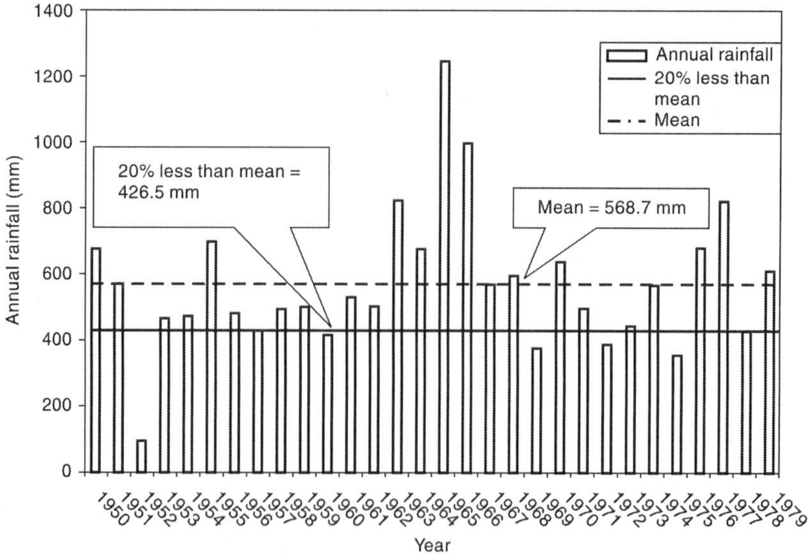

Fig. 2.11 ***Bar Chart of Annual Rainfall at Station M***

rainfall depth and the position of the column representing the year of occurrence. The time is arranged in chronological order.

The mean of the annual rainfall time series is 568.7 mm. As such, 20% less than the mean = 426.5 mm. Lines representing these values are shown in Fig. 2.11 as horizontal lines. It can be seen that in 6 years, viz. 1952, 1960, 1969, 1972, 1975 and 1978, the annual rainfall values are less than 426.5 mm. In thirteen years, viz. 1950, 1951, 1955, 1963, 1964, 1965, 1966, 1967, 1968, 1970, 1976, 1977 and 1978, the annual rainfall was more than the mean. (ii) Moving mean calculations are shown in Table 2.2. Three-year moving mean curve is

Table 2.2 ***Computation of Three-year Moving Mean***

1 Year	2 Annual Rainfall (mm) P_i	3 Three consecutive year total for moving mean $(P_{i-1} + P_i + P_{i+1})$	4 3-year moving mean (Col. 3/3)*
1950	676		
1951	578	676 + 578 + 95 = 1349	449.7
1952	95	578 + 95 + 462 = 1135	378.3
1953	462	95 + 462 + 472 = 1029	343.0
1954	472	462 + 472 + 699 = 1633	544.3
1955	699	472 + 699 + 479 = 1650	550.0
1956	479	699 + 479 + 431 = 1609	536.3
1957	431	479 + 431 + 493 = 1403	467.7
1958	493	431 + 493 + 503 = 1427	475.7
1959	503	493 + 503 + 415 = 1411	470.3
1960	415	503 + 415 + 531 = 1449	483.0
1961	531	415 + 531 + 504 = 1450	483.3
1962	504	531 + 504 + 828 = 1863	621.0
1963	828	504 + 828 + 679 = 2011	670.3
1964	679	828 + 679 + 1244 = 2751	917.0
1965	1244	679 + 1244 + 999 = 2922	974.0
1966	999	1244 + 999 + 573 = 2816	938.7
1967	573	999 + 573 + 596 = 2168	722.7
1968	596	573 + 596 + 375 = 1544	514.7
1969	375	596 + 375 + 635 = 1606	535.3
1970	635	375 + 635 + 497 = 1507	502.3
1971	497	635 + 497 + 386 = 1518	506.0
1972	386	497 + 386 + 438 = 1321	440.3
1973	438	386 + 438 + 568 = 1392	464.0
1974	568	438 + 568 + 356 = 1362	454.0
1975	356	568 + 356 + 685 = 1609	536.3
1976	685	356 + 685 + 825 = 1866	622.0
1977	825	685 + 825 + 426 = 1936	645.3
1978	426	825 + 426 + 162 = 1863	621.0
1979	612		

*The moving mean is recorded at the mid span of 3 years.

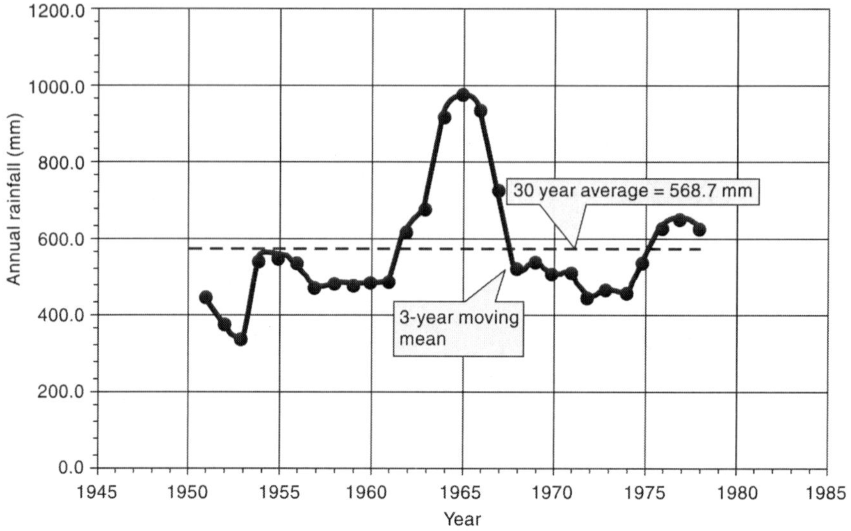

Fig. 2.12 ***Three-year Moving Mean***

shown plotted in Fig. 2.12 with the moving mean value as the ordinate and the time in chronological order as abscissa. Note that the curve starts from 1951 and ends in the year 1978. No apparent trend is indicated in this plot.

2.9 *Mean Precipitation Over An Area*

As indicated earlier, raingauges represent only point sampling of the areal distribution of a storm. In practice, however, hydrological analysis requires a knowledge of the rainfall over an area, such as over a catchment.

To convert the point rainfall values at various stations into an average value over a catchment, the following three methods are in use: (i) Arithmetical-mean method, (ii) Thiessen-polygon method, and (iii) Isohyetal method.

2.9.1 Arithmetical-Mean Method

When the rainfall measured at various stations in a catchment show little variation, the average precipitation over the catchment area is taken as the arithmetic mean of the station values. Thus, if $P_1, P_2, \ldots, P_i, \ldots P_n$ are the rainfall values in a given period in N stations within catchment then the value of the mean precipitation $\overline{P}$ over the catchment by the arithmetic-mean method is

$$\overline{P} = \frac{P_1 + P_2 + \ldots + P_i + \ldots + P_n}{N} = \frac{1}{N} \sum_{i=1}^{N} P_i \tag{2.7}$$

In practice, this method is used very rarely.

2.9.2 Thiessen-Mean Method

In this method, the rainfall recorded at each station is given a weightage on the basis of an area closest to the station. The procedure of determining the weighing area is as

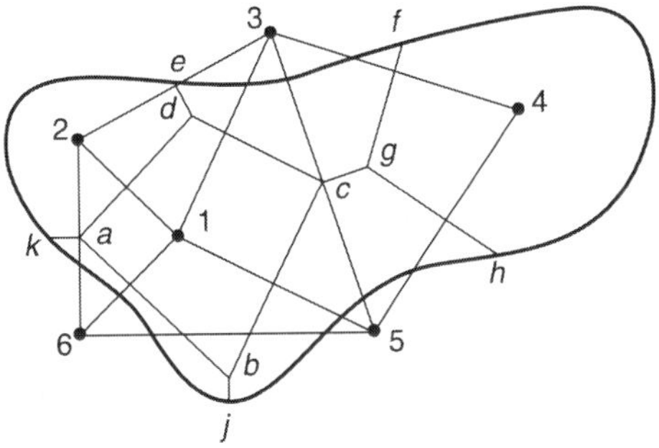

A = total catchment area

Station	Bounded by	Area	Weightage
1	abcd	A_1	A_1/A
2	kade	A_2	A_2/A
3	edcgf	A_3	A_3/A
4	fgh	A_4	A_4/A
5	hgcbj	A_5	A_5/A
6	jbak	A_6	A_6/A

Fig. 2.13 ***Thiessen Polygons***

follows: Consider the catchment area as in Fig. 2.13 containing six raingauge stations. There are three stations outside the catchment but in its neighbourhood. The catchment area is drawn to scale and the positions of the six stations marked on it. Stations 1 to 6 are joined to form a network of triangles. Perpendicular bisectors for each of the sides of the triangle are drawn. These bisectors form a polygon around each station. The boundary of the catchment, if it cuts the bisectors is taken as the outer limit of the polygon. Thus for station 1, the bounding polygon is *abcd*. For station 2, *kade* is taken as the bounding polygon. These bounding polygons are called *Thiessen polygons*. The areas of these six Thiessen polygons are determined either with a planimeter or by using an overlay grid. If P_1 P_2, … , P_6 are the rainfall magnitudes recorded by the stations 1, 2,…, 6 respectively, and A_1, A_2,…, A_6 are the respective areas of the Thiessen polygons then the average rainfall over the catchment $\overline{P}$ is given by

$$\overline{P} = \frac{P_1 A_1 + P_2 A_2 + \ldots + P_6 A_6}{(A_1 + A_2 + \ldots + A_6)}$$

Thus, in general, for M stations,

$$\overline{P} = \frac{\sum_{i=1}^{M} P_i A_i}{A} = \sum_{i=1}^{M} P_i \frac{A_i}{A} \qquad (2.8)$$

The ratio $\frac{A_i}{A}$ is called the *weightage factor* for each station.

The Thiessen-polygon method of calculating the average percipitation over an area is superior to the arithmetic-average method as some weightage is given to the various stations on a rational basis. Further, the raingauge stations outside the catchment are also used effectively. Once the weightage factors are determined, the calculation of $\overline{P}$ is relatively easy for a fixed network of stations.

2.9.3 Isohyetal Method

An *isohyet* is a line joining points of equal rainfall magnitude. In the isohyetal method, the catchment area is drawn to scale and the raingauge stations are marked. The recorded values for which areal average $\overline{P}$ is to be determined are then marked on the plot at appropriate stations. Neighbouring stations outside the catchment are

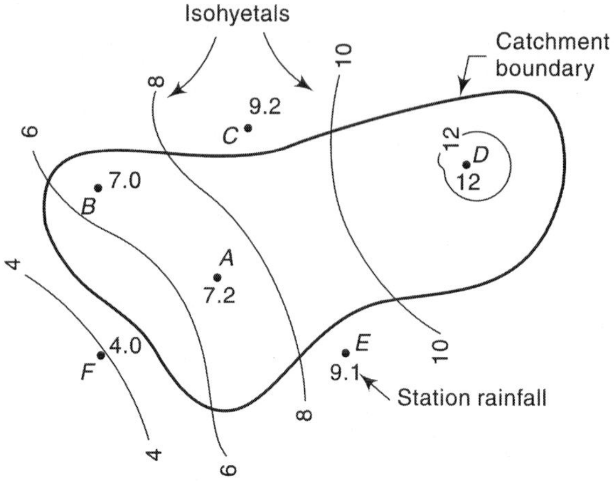

Fig. 2.14 ***Isohyetals of a Storm***

also considered. The isohyets of various values are then drawn by considering point rain-falls as guides and interpolating between them by the eye (Fig. 2.14). The procedure is similar to the drawing of elevation contours based on spot levels.

The area between two adjacent isohyets are then determined with a planimeter. If the isohyets go out of catchment, the catchment boundary is used as the bounding line. The average value of the rainfall indicated by two isohyets is assumed to be acting over the inter-isohyet area. Thus, $P_1, P_2, \ldots, P_n$ are the values of isohyets and if $a_1, a_2, \ldots, a_{n-1}$ are the inter-isohyet areas respectively, then the mean precipitation over the catchment of area A is given by

$$\bar{P} = \frac{a_1\left(\frac{P_1 + P_2}{2}\right) + a_2\left(\frac{P_2 + P_3}{2}\right) + \ldots + a_{n-1}\left(\frac{P_{n-1} + P_n}{2}\right)}{A} \tag{2.9}$$

The isohyet method is superior to the other two methods especially when the stations are large in number.

Example 2.5 For the catchment area shown in Fig. 2.15, the details of Thiessen polygons surrounding each raingauge and the recordings of the raingauges in the month of August 2011 are given below.

Raingauge Station	1	2	3	4	5	6
Thiessen polygon area (km^2)	720	380	440	1040	800	220
Recorded rainfall in mm during Aug. 2011	121	134	145	126	99	115

Determine the average depth of rainfall on the basin in August 2011 by (i) arithmetic mean method, and (ii) Thiessen mean method.

Solution

(i) *Arithmetic Mean:* Only raingauge stations 1, 2 and 4 are within the basin. Hence, the arithmetic mean is obtained by considering the readings of these three stations only.

Mean monthly rainfall on the basin in August 2011

$$= \frac{(121+134+126)}{3} = 127 \text{ mm}$$

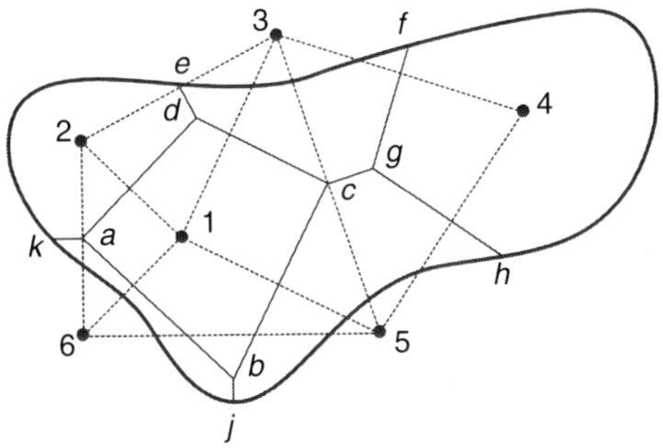

Fig. 2.15 ***Thiessen Polygons of Example 2.5***

(ii) *Thiessen Mean:* The calculations are performed in tabular form as shown below.

Table 2.3 ***Calculation of Thiessen Mean—Example 2.5***

Total area of basin = Total of Col.2 = 3600 km^2

1 Raingauge Station	2 Thiessen polygon area (km^2)	3 Thiessen weightage Factor (= Fraction of total area) = [Col.2/3600]	4 Station reading (=Monthly Rainfall in mm)	5 Weighted Station rainfall (mm) =[Col.3 × Col.4]
1	720	0.200	121	24.2
2	380	0.106	134	14.1
3	440	0.122	145	17.7
4	1040	0.289	126	36.4
5	800	0.222	99	22.0
6	220	0.061	115	7.0
Total catchment area=	3600	1.000		121.5

Thiessen mean of the station readings for the month of Aug. 2011
= Total of Col.5 = 121.5 mm

Example 2.6

In a catchment, whose shape can be approximated by a pentagon, four raingauge stations are situated inside the catchment. The coordinates of the corners of the catchment that define its boundaries and the coordinates of the four raingauge stations are given below. Also given are the annual rainfall recorded by the four stations in the year 2005. Determine the average annual rainfall over the catchment in that year by the Thiessen —mean method.

Distances are in Km.		Corner *a* is the origin of co-ordinates.				
Catchment Boundary	Corner	Corner *a*	Corner *b*	Corner *c*	Corner *d*	Corner *e*
	Co-ordinates	(0.0)	(120,0)	(120,80)	(60,140)	(0,80)
Rain gauge Station	Station	*P*	Q	*R*	*S*	
	Co-ordinates	(40,20)	(80,20)	(80,60)	(40,80)	
	Annual Rainfall (cm)	120	110	100	125	

Solution

The catchment area is drawn to scale and the stations are marked on it, Fig. 2.16. The stations *P*, *Q*, *R* and *S* are joined to form a set of triangles. Perpendicular bisectors of each side of the triangles are drawn. The polygons formed around each station by the bisectors and the boundaries of the catchment are identified and their areas determined. Knowing the Thiessen polygon area of each station, the weightage factor of each station is calculated and the average rainfall over the catchment area is determined by using the Thiessen mean equation, Eq. (2.8). The calculations are done in a tabular form as in Table 2.4.

Table 2.4 ***Calculation of Thiessen Mean—Example 2.6***

Rain gauge station	Boundary of Thiessen polygons as in Fig 2.15	Area (km^2)	Weightage Factor = (Fraction of total area) = [Col.3/13200]	Recorded station rainfall (cm)	Weighted station rainfall (cm)
1	2	3	4	5	6
P	301a	60 × 40 = 2400	0.1818	120	21.816
Q	102b	60 × 40 = 2400	0.1818	110	19.998
R	20dc	[(60 × 40) + (60 × 60 × 1/2)] = 4200	0.3182	100	31.82
S	30de	[(60 × 40) + (60 × 60 × 1/2)] = 4200	0.3182	125	39.775
Total area of Catchment =		13200	1.0000		113.41

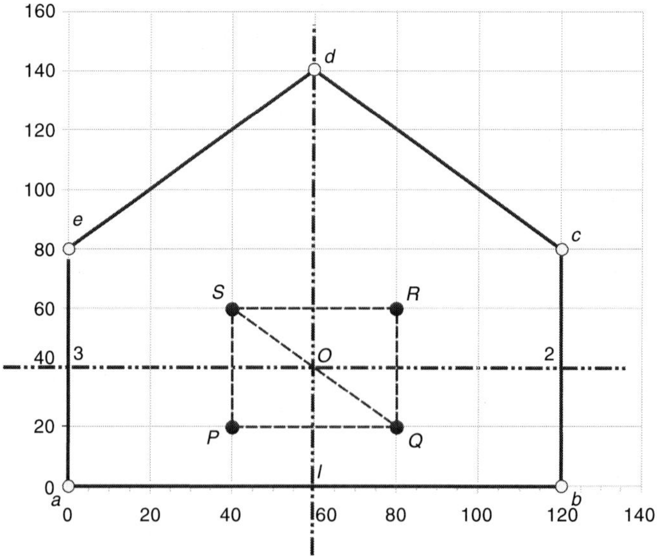

Fig. 2.16 ***Thiessen Polygons of Example 2.6***

Average precipitation over the catchment in 2005 = 113.41 cm

Example 2.7 The isohyets due to a storm in a catchment were drawn (Fig. 2.14) and the area of the catchment bounded by isohyets were tabulated as below.

Isohyets (cm)	Area (km^2)
Station—12.0	30
12.0—10.0	140
10.0—8.0	80
8.0—6.0	180
6.0—4.0	20

Estimate the mean precipitation due to the storm.

Solution

For the first area consisting of the station surrounded by a closed isohyet, a precipitation value of 12.0 cm is taken. For all other areas, the mean of two bounding isohyets are taken.

Table 2.5 *Calculation of Average Depth by Isohyetal Method–Example 2.7*

Isohytes	Average value of P (cm)	Area (km^2)	Fraction of total area (col. 3/450)	Weighted P (cm) (col. 2 × col. 4)
1	2	3	4	5
12.0	12.0	30	0.0667	0.800
12.0–10.0	11.0	140	0.3111	3.422
10.0–8.0	9.0	80	0.1778	1.600
8.0–6.0	7.0	180	0.4000	2.800
6.0–4.0	5.0	20	0.0444	0.222
Total		450	1.0000	8.844

Mean precipitation $\bar{P} = 8.84$ cm

2.10 *Depth-Area-Duration Relationships*

The areal distribution characteristics of a storm of given duration is reflected in its depth-area relationship. A few aspects of the interdependency of depth, area and duration of storms are discussed below.

2.10.1 Depth-Area Relation

For a rainfall of a given duration, the average depth decreases with the area in an exponential fashion given by

$$\bar{P} = P_0 \exp(-KA^n) \tag{2.10}$$

where $\bar{P}$ = average depth in cm over an area A km^2, P_0 = highest amount of rainfall in cm at the storm centre and K and n are constants for a given region. On the basis of

42 severemost storms in north India, Dhar and Bhattacharya[3] (1975) have obtained the following values for K and n for storms of different duration:

Duration	K	n
1 day	0.0008526	0.6614
2 days	0.0009877	0.6306
3 days	0.001745	0.5961

Since it is very unlikely that the storm centre coincides over a raingauge station, the exact determination of P_0 is not possible. Hence, in the analysis of large, area storms the highest station rainfall is taken as the average depth over an area of 25 km^2.

Equation (2.10) is useful in extrapolating an existing storm data over an area.

2.10.2 Maximum Depth-Area-Duration Curves

In many hydrological studies involving estimation of severe floods, it is necessary to have information on the maximum amount of rainfall of various durations occurring over various sizes of areas. The development of relationship, between maximum depth-area-duration for a region is known as *DAD* analysis and forms an important aspect of hydro-meteorological study. References 2 and 9 can be consulted for details on *DAD* analysis. A brief description of the analysis is given below.

First, the severemost rainstorms that have occurred in the region under question are considered. Isohyetal maps and mass curves of the storm are compiled. A depth-area curve of a given duration of the storm is prepared. Then from a study of the mass curve of rainfall, various durations and the maximum depth of rainfall in these durations are noted. The maximum depth-area curve for a given duration D is prepared by assuming the area distribution of rainfall for smaller duration to be similar to the total storm. The procedure is then repeated for different storms and the envelope curve of maximum depth-area for duration D is obtained. A similar procedure for various values of D results in a family of envelope curves of maximum depth *vs* area, with duration as the third parameter (Fig. 2.17). These curves are called *DAD curves.*

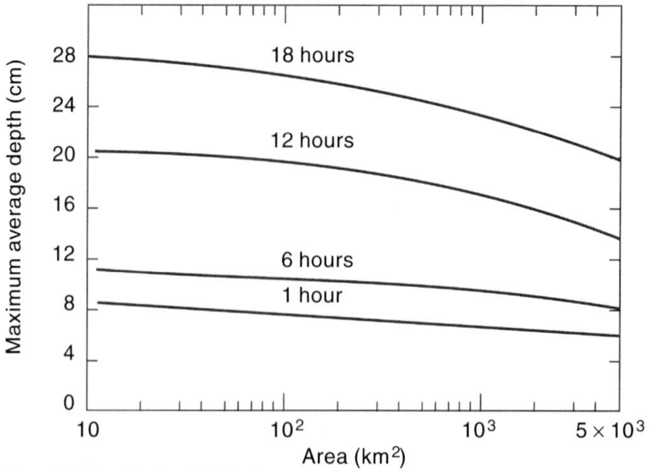

Fig. 2.17 ***Typical DAD Curves***

Figure 2.17 shows typical *DAD* curves for a catchment. In this, the average depth denotes the depth averaged over the area under consideration. It may be seen that the maximum depth for a given storm decreases with the area; for a given area, the maximum depth increases with the duration.

Preparation of *DAD* curves involves considerable computational effort and requires meteorological and topographical information of the region. Detailed data on severemost storms in the past are needed. *DAD* curves are essential to develop design storms for use in computing the design flood in the hydrological design of major structures such as dams.

Table 2.6 ***Maximum (Observed) Rain Depths (cm) over Plains of North India***[4, 5]

	Area in $km^2 \times 10^4$								
Duration	**.026**	**0.13**	**0.26**	**1.3**	**2.6**	**5.2**	**7.8**	**10.5**	**13.0**
1 day	81.0*	76.5*	71.1	47.2*	37.1 *	26.4	20.3†	18.0†	16.0†
2 days	102.9*	97.5*	93.2*	73.4*	58.7*	42.4*	35.6†	31.5†	27.9†
3 days	121.9†	110.7†	103.1†	79.2†	67.1†	54.6†	48.3†	42.7†	38.9†

Note: *—Storm of 17–18 September 1880 over north-west U.P.
†—Storm of 28–30 July 1927 over north Gujarat.

Maximum rain depths observed over the plains of north India are indicated in Table 2.6. These were due to two storms, which are perhaps the few severe most recorded rainstorms over the world.

Example 2.8 The data from an isohyetal map of a 24-hour storm is given below. Assuming that the storm centre had an area of 55 km^2 and the rainfall depth is 57 cm in the storm centre, obtain the depth-area curve of this storm. Using the depth-area curve, estimate the average depth of rainfall over an area of 3200 sq. km. It can be assumed that the storm centre is located at the centre of the area.

Isohyet (mm)	50	45	40	35	30	25	20	15	10
Enclosed area (km^2)	310	1300	2060	2700	2955	3600	4030	4800	6000

Solution

The rainfall is taken as constant at 57 cm in this storm centre area of 55 km^2.

The calculation of mean depth of rainfall for various isohyetal areas is performed in a tabular manner as shown in Table 2.7

- Each row represents the properties of one isohyetal of value given in Col.1.
- Column 2 represents the value of the area bounded by the isohyetal of value given in Col. 1.
- Column 3 represents the difference in areas of two adjacent isohyetals, the upper row and the current row.
- Column 4 is the average of isohyetal values of upper row and the current row.

Table 2.7 ***Computation of Data for Depth-Area curve: Example 2.8.***

Isohyetal (mm)	Area enclosed (km^2)	Net incremental area (km^2)	Average isohyetal value (mm)	Total incremental volume (km^2.mm) [Col. 3 × Col.4]	Total volume of rainfall (km^2.mm)	Average depth of rainfall (mm) [Col.6/ Col.2]
1	2	3	4	5	6	7
57	55	55	57	3135	3135	57
50	310	255	53.5	13643	16778	54
45	1300	990	47.5	47025	63803	49
40	2060	760	42.5	32300	96103	47
35	2700	640	37.5	24000	120103	44
30	2955	255	32.5	8288	128390	43
25	3600	645	27.5	17738	146128	41
20	4030	430	22.5	9675	155803	39
15	4800	770	17.5	13475	169278	35
10	6000	1200	12.5	15000	184278	31

- Column 5 is the incremental volume of rainfall between the isohyet of the current row and the isohyet of the row above it.
- Column 6 is the cumulative volume of rainfall from the storm center up to the current isohyet.
- Column 7 represents the average depth of rainfall corresponding to the area of the isohyet of the current row.

where $\overline{P}$ = Average depth in mm and A = area of bounding isohyet/catchment in km^2. By use of the correlation equation or from the use of Fig. 2.18, the average depth of rainfall over an area of 3200 km^2 is found to be 41.39 mm.

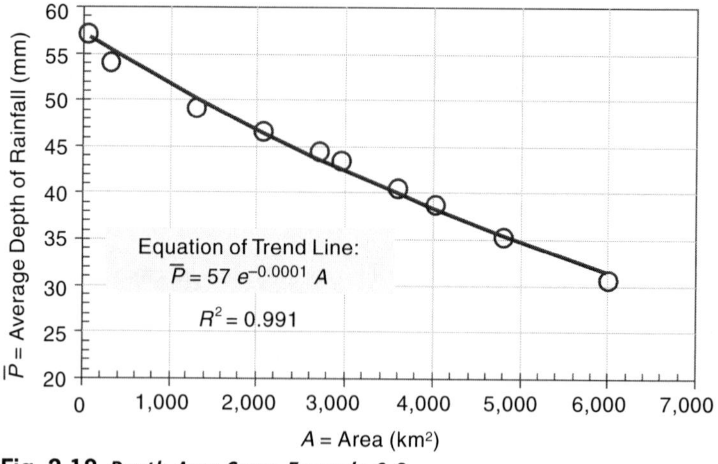

Fig. 2.18 ***Depth-Area Curve-Example 2.8***

2.11 *Frequency of Point Rainfall*

In many hydraulic-engineering applications such as those concerned with floods, the probability of occurrence of a particular extreme rainfall, e.g. a 24-h maximum rainfall, will be of importance. Such information is obtained by the frequency analysis of the point-rainfall data. The rainfall at a place is a random hydrologic process and a sequence of rainfall data at a place when arranged in chronological order constitutes a time series. One of the commonly used data series is the annual series composed of annual values such as annual rainfall. If the extreme values of a specified event occurring in each year is listed, it also constitutes an annual series. Thus, for example, one may list the maximum 24-h rainfall occurring in a year at a station to prepare an annual series of 24-h maximum rainfall values. The probability of occurrence of an event in this series is studied by frequency analysis of this annual data series. A brief description of the terminology and a simple method of predicting the frequency of an event is described in this section and for details, the reader is referred to standard works on probability and statistical methods. The analysis of annual series, even though described with rainfall as a reference is equally applicable to any other random hydrological process, e.g. stream flow.

First, it is necessary to correctly understand the terminology used in frequency analysis. The probability of occurrence of an event of a random variable (e.g. rainfall) whose magnitude is equal to or in excess of a specified magnitude X is denoted by P. The *recurrence interval* (also known as *return period*) is defined as

$$T = 1/P \tag{2.11}$$

This represents the average interval between the occurrence of a rainfall of magnitude equal to or greater than X. Thus, if it is stated that the return period of rainfall of 20 cm in 24 h is 10 years at a certain station A, it implies that on an average, rainfall magnitudes equal to or greater than 20 cm in 24 h occur once in 10 years, i.e. in a long period of say 100 years, 10 such events can be expected. However, it does not mean that every 10 years one such event is likely, i.e. periodicity is not implied. The probability of a rainfall of 20 cm in 24 h occurring in anyone year at station A is $1/T = 1/10 = 0.1$.

If the probability of an event occurring is P, the probability of the event *not occurring* in a given year is $q = (1-P)$. The binomial distribution can be used to find the probability of occurrence of the event r times in n successive years. Thus,

$$P_{r,n} = {}^nC_r P_r q^{n-r} = \frac{n!}{(n-r)!\,r!} P^r q^{n-r} \tag{2.12}$$

where $P_{r,n}$ = probability of a random hydrologic event (rainfall) of given magnitude and exceedence probability P occurring r times in n successive years. Thus, for example,

(a) The probability of an event of exceedence probability P occurring 2 times in n successive years is

$$P_{2,n} = \frac{n!}{(n-2)!\,2!} P^2 q^{n-2}$$

(b) The probability of the event not occurring at all in n successive years is

$$P_{0,n} = q^n = (1-P)^n$$

(c) The probability of the event occurring at least once in n successive years

$$P_1 = 1 - q^n = 1 - (1 - P)^n \tag{2.13}$$

Example 2.9 Analysis of data on maximum one -day rainfall depth at Chennai indicated that a depth of 300 mm had a return period of 50 years. Determine the probability of a one-day rainfall depth equal to or greater than 300 mm at Chennai occurring (a) once in 20 successive years, (b) two times in 15 successive years, and (c) at least once in 20 successive years.

Solution

Here, $P = \dfrac{1}{50} = 0.02, \qquad q = (1-P) = (1-0.02) = 0.98$

By using Eq. (2.12): $P_{r,n} = \dfrac{n!}{(n-r)!r!} P^r q^{n-r}$

(a) $n = 20, \mathrm{r} = 1: \; P_{1,20} = \dfrac{20!}{(20-1)!1!} P^1 q^{20-1} = \dfrac{20!}{19!1!} \times 0.02 \times (0.98)^{19}$

$$= 20 \times 0.02 \times 0.68123 = 0.272$$

(b) $n = 15, r = 2: \; P_{2,15} = \dfrac{15!}{(15-2)!2!} P^2 q^{15-2} = \dfrac{15!}{13!2!} \times (0.02)^2 \times (0.98)^{13}$

$$= \frac{15 \times 14}{2} \times 0.0004 \times 0.769 = 0.323$$

(c) By Eq. (2.13), $P_1 = 1 - (1–0.02)^{20} = 0.32$

2.11.1 Plotting Position

The purpose of the frequency analysis of an annual series is to obtain a relation between the magnitude of the event and its probability of exceedence. The probability analysis may be made either by empirical or by analytical methods.

A simple empirical technique is to arrange the given annual extreme series in descending order of magnitude and to assign an order number m. Thus, for the first entry $m = 1$, for the second entry $m = 2$, and so on, till the last event for which $m = N$ = number of years of record. The probability P of an event equalled to or exceeded is given by the *Weibull formula*

$$P = \left(\frac{m}{N+1}\right) \tag{2.14}$$

Equation (2.14) is an empirical formula and there are several other such empirical formulae available to calculate the probability P. Table 2.8 lists a few popular empirical formulae for calculating P. The exceedence probability of the event obtained by the use of the empirical formula, such as Equation (2.14) is called *plotting position.* Equation (2.14) is the most popular plotting position formula and hence only this formula is used in further sections of this book.

Having calculated P (and hence, T) for all events in the series, the variation of the rainfall magnitude is plotted against the corresponding T on a semi-log graph or on

Table 2.8 ***Plotting-Position Formulae***

m = order number, N = number of Years of record

Name of the Method	Equation for P
California	$P = \frac{m}{N}$
Hazen	$P = \frac{(m-0.5)}{N}$
Weibull	$P = \frac{m}{(N+1)}$
Chegodayev	$P = \frac{(m-0.3)}{(N+0.4)}$
Blom	$P = \frac{(m-0.44)}{(N+0.12)}$
Gringoten	$P = \frac{(m-0.375)}{(N+0.25)}$

a log-log graph. By suitable extrapolation of this plot, within appropriate limits, the rainfall magnitude of specific duration for any recurrence interval can be estimated.

This simple empirical procedure can give good results for small extrapolations and the errors increase with the amount of extrapolation. For accurate work, various analytical calculation procedures using frequency factors are available. Gumbel's extreme value distribution and Log Pearson Type III method are two commonly used analytical methods and are described in Chapter 7 of this book.

If P is the probability of exceedance of a variable having a magnitude M, a common practice is to designate the magnitude M as having (100 P) percent dependability. For example "75% dependable annual rainfall" at a station means the value of annual rainfall at the station that can be expected to be equalled to or exceeded 75% times (i.e., on an average 30 times out of 40 years). Thus, 75% dependable annual rainfall means the value of rainfall in the annual rainfall time series that has $P = 0.75$, i.e., $T = 1/P = 1.333$ years.

Example 2.10 The record of annual rainfall at station A covering a period of 22 years is given below. (a) Estimate the annual rainfall with return periods of 10 years and 50 years. (b) What would be the probability of an annual rainfall of magnitude equal to or exceeding 100 cm occurring at station A? (b) What is the 75% dependable annual rainfall at station A?

Year	Annual rainfall (cm)	Year	Annual rainfall (cm)
1960	130.0	1971	90.0
1961	84.0	1972	102.0
1962	76.0	1973	108.0

(Contd.)

(Contd.)

1963	89.0	1974	60.0
1964	112.0	1975	75.0
1965	96.0	1976	120.0
1966	80.0	1977	160.0
1967	125.0	1978	85.0
1968	143.0	1979	106.0
1969	89.0	1980	83.0
1970	78.0	1981	95.0

Solution

The data are arranged in descending order and the rank number assigned to the recorded events. The probability P of the event being equalled to or exceeded is calculated by using Weibull formula (Eq. 2.14). Calculations are shown in Table 2.9. It may be noted that when two or more events have the same magnitude (as for m = 13 and 14 in Table 2.9) the probability P is calculated for the largest m value of the set. The return period T is calculated as $T = 1/P$.

Table 2.9 ***Calculation of Return Periods–Example 2.10***

N = 22 years

m	Annual Rainfall (cm)	Probability $P = m/(N+1)$	Return Period $T = 1/P$ (years)	m	Annual Rainfall	Probability $P = m/(N+1)$	Return Period $T = 1/P$ (Years)
1	160.0	0.043	23.000	12	90.0	0.522	1.917
2	143.0	0.087	11.500	13	89.0	0.565	
3	130.0	0.130	7.667	14	89.0	0.609	1.643
4	125.0	0.174	5.750	15	85.0	0.652	1.533
5	120.0	0.217	4.600	16	84.0	0.696	1.438
6	112.0	0.261	3.833	17	83.0	0.739	1.353
7	108.0	0.304	3.286	18	80.0	0.783	1.278
8	106.0	0.348	2.875	19	78.0	0.826	1.211
9	102.0	0.391	2.556	20	76.0	0.870	1.150
10	96.0	0.435	2.300	21	75.0	0.913	1.095
11	95.0	0.478	2.091	22	60.0	0.957	1.045

A graph is plotted between the annual rainfall magnitude as the ordinate (on arithmetic scale) and the return period T as the abscissa (on logarithmic scale), (Fig. 2.19).

The equation of the best-fit line (trend line) is obtained as $P_r = 29.326 \ln(T) + 72.024$. In this, P_r = annual precipitation and T = return period. Use of a spreadsheet like MS Excel is very convenient for this type of problem: for computation, plotting of graph and for obtaining the trend line. The trendline equation is used for calculating the various required values:

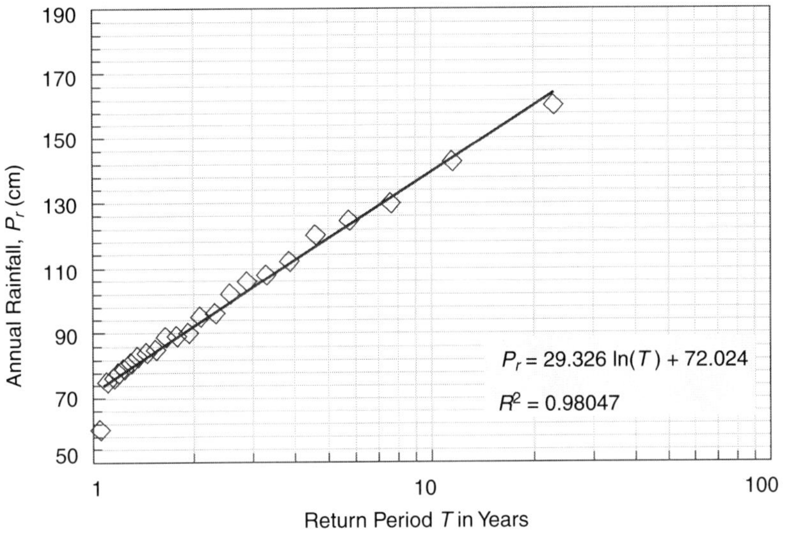

Fig. 2.19 ***Return Period of Annual Rainfall at Station—Example 2.10***

(a) (i) For $T = 10$ years, by using the equation of the derived trend line,
$P_r = 29.326 \ln(10) + 72.024 = 139.5$ cm

(ii) For $T = 50$ years, by using the equation of the derived trend line,
$P_r = 29.326 \ln(50) + 72.024 = 186.7$ cm.

(b) Return period of an annual rainfall magnitude (P_r) equal to or exceeding 100 cm, by using the equation of the derived trend line,
$100 = 29.326 \ln(T) + 72.024$
$T = 2.596$ years and the exceedance probability $P = 1/2.596 = 0.385$.

(c) 75% dependable annual rainfall at station A = Annual rainfall P_r with probability of exceedance $P = 0.75$, i.e., $T = 1/P = 1/0.75 = 1.333$ years.
By using the equation of the derived trend line,
$P_r = 29.326 \ln(1.333) + 72.024 = 80.4$ cm.

2.12 *Maximum Intensity/Depth-Duration-Frequency Relationship*

2.12.1 Maximum Intensity-Duration Relationship

In any storm, the actual intensity as reflected by the slope of the mass curve of rainfall varies over a wide range during the course of the rainfall. If the mass curve is considered divided into N segments of time interval Δt such that the total duration of the storm $D = N\,\Delta t$ then the intensity of the storm for various subdurations $t_j = (1.\,\Delta t), (2.\,\Delta t), (3.\,\Delta t), \ldots (j.\,\Delta t) \ldots$ and $(N.\,\Delta t)$ could be calculated. It will be found that for each duration (say t_j), the intensity will have a maximum value and this could be analysed to obtain a relationship for the variation of the maximum intensity with duration for the storm.

This process is basic to the development of maximum intensity-duration-frequency relationship for the station discussed later on.

Briefly, the procedure for analysis of a mass curve of rainfall for developing maximum intensity-duration relationship of the storm is as follows.

- Select a convenient time step Δt such that duration of the storm $D = N. \Delta t$.
- For each duration (say $t_j = j.\Delta t$), the mass curve of rainfall is considered to be divided into consecutive segments of duration t_j. For each segment, the incremental rainfall d_j in duration t_j is noted and intensity $I_j = d_j/t_j$ obtained.
- Maximum value of the intensity (I_{mj}) for the chosen t_j is noted.
- The procedure is repeated for all values of $j = 1$ to N to obtain a data set of I_{mj} as a function of duration t_j. Plot the maximum intensity I_m as function of duration t.
- It is common to express the variation of I_m with t as

$$I_m = \frac{c}{(t + a)^b}$$

where a, b and c are coefficients obtained through regression analysis. Example 2.11 describes the procedure in detail.

2.12.2 Maximum Depth-Duration Relationship

Instead of the maximum intensity I_m in a duration t, the product $(I_m. t) = d_m$ = maximum depth of precipitation in the duration t could be used to relate it to the duration. Such a relationship is known as the maximum depth-duration relationship of the storm. The procedure of developing this relationship is essentially same as that for maximum intensity-duration relationship described earlier.

Example 2.11 describes the procedure in detail

2.12.3 Maximum Intensity-Duration-Frequency Relationship

If the rainfall data from a self-recording raingauge is available for a long period, the frequency of occurrence of maximum intensity occurring over a specified duration can be determined. A knowledge of maximum intensity of rainfall of specified return period and of duration equal to the critical time of concentration is of considerable practical importance in evaluating peak flows related to hydraulic structures.

Briefly, the procedure to calculate the intensity-duration-frequency relationship for a given station is as follows.

- M numbers of significant and heavy storms in a particular year Y_1 are selected for analysis. Each of these storms are analysed for maximum intensity duration relationship as described in Sec. 2.12.1
- This gives the set of maximum intensity I_m as a function of duration for the year Y_1.
- The procedure is repeated for all the N years of record to obtain the maximum intensity $I_m (D_j)_k$ for all $j = 1$ to M and $k = 1$ to N.
- Each record of $I_m (D_j)_k$ for $k = 1$ to N constitutes a time series which can be analysed to obtain frequencies of occurrence of various $I_m (D_j)$ values. Thus there will be M time series generated.

- The results are plotted as maximum intensity *vs* return period with the *Duration* as the third parameter (Fig. 2.20). Alternatively, maximum intensity *vs* duration with frequency on return period as the third variable can also be adopted (Fig. 2.21).

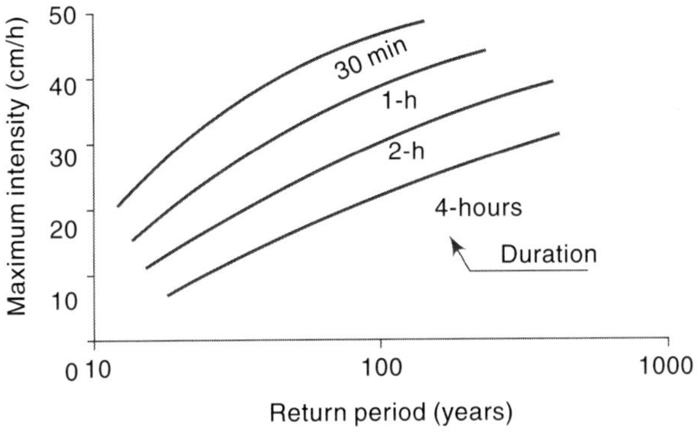

Fig. 2.20 ***Maximum Intensity-Return Period-Duration Curves***

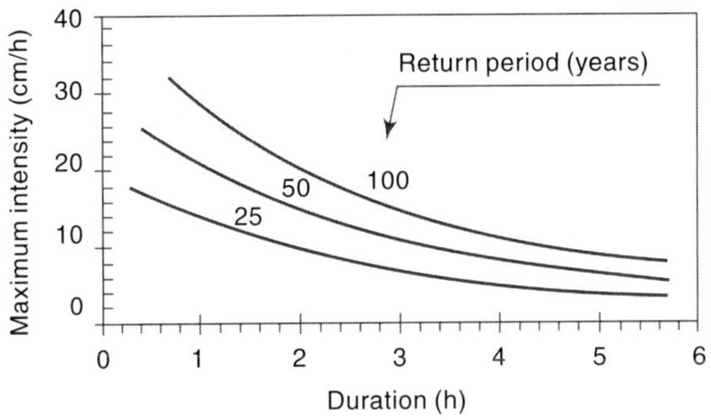

Fig. 2.21 ***Maximum Intensity-Duration-Frequency Curves***

Analytically, these relationships are commonly expressed in a condensed form by general form

$$i = \frac{KT^X}{(D+a)^n} \tag{2.15}$$

where i = maximum intensity (cm/h), T = return period (years), D = duration (hours) K, x, a and n are coefficients for the area represented by the station.

Sometimes, instead of maximum intensity, maximum depth is used as a parameter and the results are represented as a plot of maximum depth *vs* duration with return period as the third variable (Fig. 2.22).

[*Note:* While maximum intensity is expressed as a function of duration and return period, it is customary to refer this function as intensity-duration-frequency relationship. Similarly, in the depth-duration-frequency relationship deals with maximum depth in a given duration.]

Rambabu et al. (1979)[10] have analysed the self-recording raingauge rainfall records of 42 stations in the country and have obtained the values of coefficients K, x, a, and n

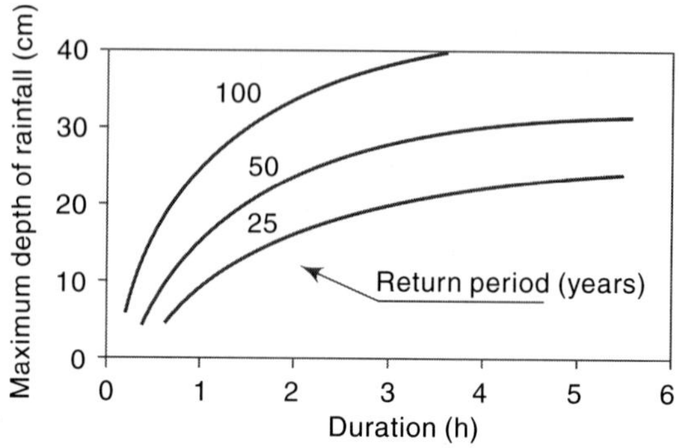

Fig. 2.22 ***Maximum Depth-Duration-Frequency Curves***

of Eq. 2.15. Some typical values of the coefficients for a few places in India are given in Table 2.10.

Table 2.10 ***Typical values of Coefficients K, x, a and n in Eq. (2.15)***

[Ref. 10]

Zone	Place	K	x	a	n
Northern Zone	Allahabad	4.911	0.1667	0.25	0.6293
	Amritsar	14.41	0.1304	1.40	1.2963
	Dehradun	6.00	0.22	0.50	0.8000
	Jodhpur	4.098	0.1677	0.50	1.0369
	Srinagar	1.503	0.2730	0.25	1.0636
	Average for the zone	5.914	0.1623	0.50	1.0127
Central Zone	Bhopal	6.9296	0.1892	0.50	0.8767
	Nagpur	11.45	0.1560	1.25	1.0324
	Raipur	4.683	0. 1389	0.15	0.9284
	Average for the zone	7.4645	0.1712	0.75	0.9599
Western Zone	Aurangabad	6.081	0.1459	0.50	1.0923
	Bhuj	3.823	0.1919	0.25	0.9902
	Veraval	7.787	0.2087	0.50	0.8908
	Average for the zone	3.974	0.1647	0.15	0.7327
Eastern Zone	Agarthala	8.097	0.1177	0.50	0.8191
	Kolkata (Dumdum)	5.940	0.1150	0.15	0.9241
	Gauhati	7.206	0.1157	0.75	0.9401
	Jarsuguda	8.596	0.1392	0.75	0.8740
	Average for the zone	6.933	0.1353	0.50	0.8801
Southern Zone	Bangalore	6.275	0.1262	0.50	1.1280
	Hyderabad	5.250	0.1354	0.50	1.0295
	Chennai	6.126	0.1664	0.50	0.8027
	Trivandrum	6.762	0.1536	0.50	0.8158
	Average for the zone	6.311	0.1523	0.50	0.9465

Extreme-point rainfall values of different durations and for different return periods have been evaluated by *IMD* and the *iso-pluvial* (lines connecting equal depths of rainfall) maps covering the entire country have been prepared. These are available for rainfall durations of 15 min, 30 min, 45 min, 1 h, 3 h, 6 h, 9 h, 15 h and 24 h for return periods of 2, 5, 10, 25, 50 and 100 years. A typical 50 year 24 h maximum rainfall map of the southern peninsula is given in Fig. 2.23. The 50 year l h maximum rainfall depths over India and the neighbourhood are shown in Fig. 2.24. Isopluvial maps of the maximum rainfall of various durations and of 50-year return periods covering the entire country are available in Ref. 1.

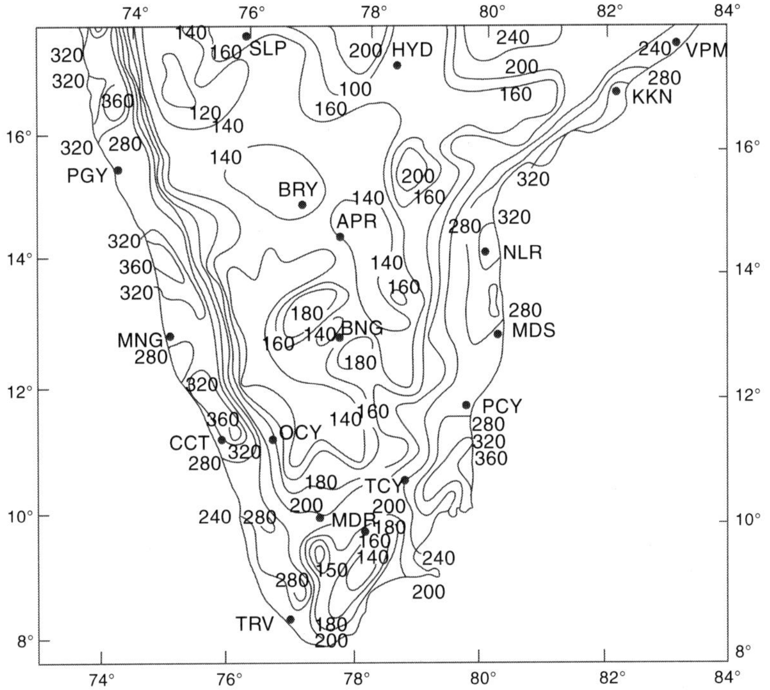

Fig. 2.23 ***Isopluvial Map of 50 yr 24 h Maximum Rainfall (mm)***

(Reproduced with permission from India Meteorological Department)

Based upon Survey of India map with the permission of the Surveyor General of India, © Government of India Copyright 1984

The territorial waters of India extend into the sea to a distance of 12 nautical miles measured from the appropriate baseline

Responsibility for the correctness of the internal details on the map rests with the publisher.

Example 2.11 The mass curve of rainfall in a storm of total duration 270 minutes is given below. (a) Draw the hyetograph of the storm at 30 minutes time step. (b) Plot the maximum intensity-duration curve for this storm. (c) Plot the maximum depth-duration curve for the storm.

Times since Start in Minutes	0	30	60	90	120	150	180	210	240	270
Cumulative Rainfall (mm)	0	6	18	21	36	43	49	52	53	54

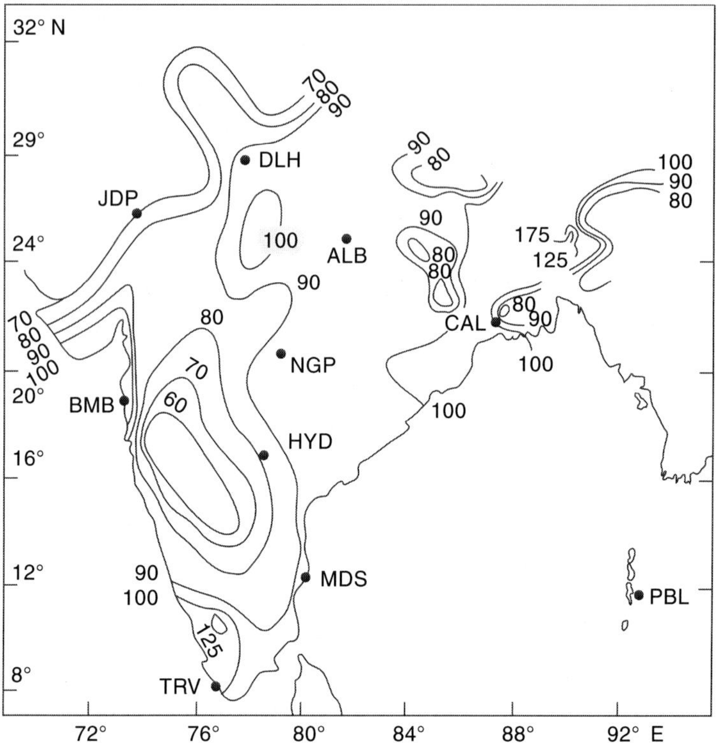

Fig. 2.24 ***Isopluvial Map of 50 yr–l h Maximum Rainfall (mm)***

(Reproduced from Natural Resources of Humid Tropical Asia—Natural Resources Research, XII. © UNESCO, 1974, with permission of UNESCO)

Based upon Survey of India map with the permission of the Surveyor General of India © Government of India Copyright 1984

The territorial waters of India extend into the sea to a distance of 12 nautical miles measured from the appropriate baseline

Responsibility for the correctness of the internal details on the map rests with the publisher.

Solution

(a) Hyetograph: The intensity of rainfall at various time durations is calculated as shown below:

Time since Start (min)	30	60	90	120	150	180	210	240	270
Cumulative Rainfall (mm)	6.0	18.0	21.0	36.0	43.0	49.0	52.0	53.0	54.0
Incremental depth of rainfall in the interval (mm)	6.0	12.0	3.0	15.0	7.0	6.0	3.0	1.0	1.0
Intensity (mm/h)	12.0	24.0	6.0	30.0	14.0	12.0	6.0	2.0	2.0

The hyetograph of the storm is shown in Fig. 2.25

(b) Various durations $\Delta t = 30, 60, 90 \ldots 240, 270$ minutes are chosen. For each duration Δt a series of running totals of rainfall depth is obtained by starting from various points of the mass curve. This can be done systematically as shown in Table 2.11(a and b). By inspection the maximum depth for each t_j is identified and corresponding maximum intensity is calculated. In Table 2.11(a) the maximum depth is marked by bold letters

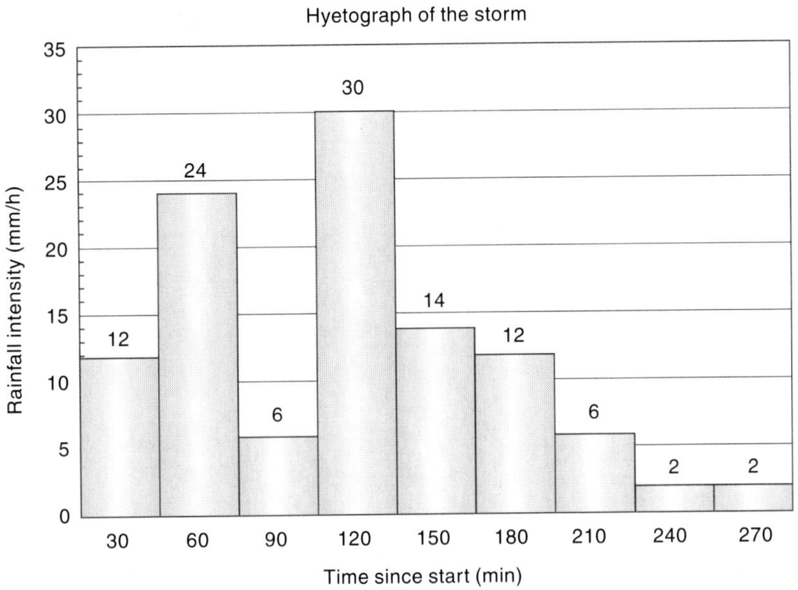

Fig. 2.25 ***Hyetograph of the Storm — Example 2.11***

and maximum intensity corresponding to a specified duration is shown in Row No. 3 of Table 2.11(b). The data obtained from the above analysis is plotted as maximum depth *vs* duration and maximum intensity *vs* duration as shown in Fig. 2.26.

Table 2.11(a) ***Maximum Intensity-Duration Relation—Example 2.11***

Incremental depth of rainfall (mm) in various durations										
		Durations(min)								
Time (min.)	*Cumulative Rainfall (mm)*	*30*	*60*	*90*	*120*	*150*	*180*	*210*	*240*	*270*
0	0									
30	6	6								
60	18	12	18							
90	21	3	15	21						
120	36	**15**	18	**30**	36					
150	43	7	**22**	25	**37**	43				
180	49	6	13	28	31	**43**	**49**			
210	52	3	9	16	31	34	46	**52**		
240	53	1	4	10	17	32	35	47	**53**	
270	54	1	2	5	11	18	33	36	48	**54**

Table 2.11(b) ***Maximum Intensity-Maximum Depth-Duration Relation–Example 2.11***

Maximum Intensity (mm/h)	30.0	22.0	20.0	18.5	17.2	16.3	14.9	13.3	12.0
Duration in min.	30	60	90	120	150	180	210	240	270
Maximum Depth (mm)	15.0	22.0	30.0	37.0	43.0	49.0	52.0	53.0	54.0

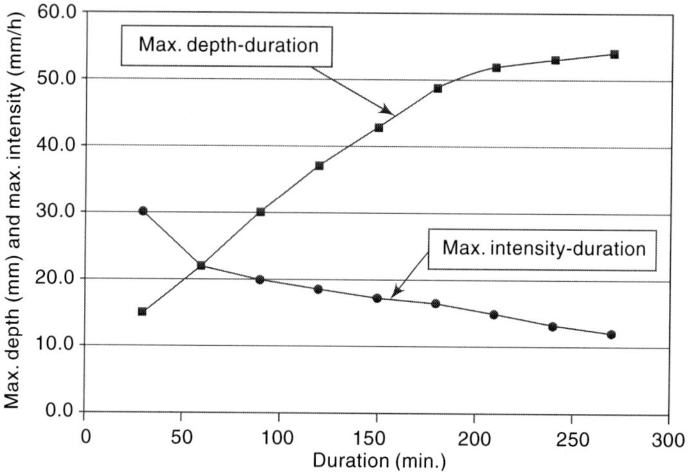

Fig. 2.26 ***Maximum Intensity-Duration and Maximum Depth-Duration Curves for the Storm of Example 2.11***

2.13 *Probable Maximum Precipitation (PMP)*

In the design of major hydraulic structures such as spillways in large dams, the hydrologist and hydraulic engineer would like to keep the failure probability as low as possible, i.e. virtually zero. This is because the failure of such a major structure will cause very heavy damages to life, property, economy and national morale. In the design and analysis of such structures, the maximum possible precipitation that can reasonably be expected at a given location is used. This stems from the recognition that there is a physical upper limit to the amount of precipitation that can fall over a specified area in a given time.

The Probable Maximum Precipitation (*PMP*) is defined as the greatest or extreme rainfall for a given duration that is physically possible over a station or basin. From the operational point of view, *PMP* can be defined as that rainfall over a basin which would produce a flood flow with virtually no risk of being exceeded. The development of *PMP* for a given region is an involved procedure and requires the knowledge of an experienced hydrometeorologist. Basically, two approaches are used (i) Meteorological methods, and (ii) the statistical study of rainfall data. Details of meteorological methods that use storm models are available in published literature.[8]

Statistic studies indicate that *PMP* can be estimated as

$$PMP = \overline{P} + K\sigma \tag{2.16}$$

where P = mean of annual maximum rainfall series, σ = standard deviation of the series and K = a frequency factor which depends upon the statistical distribution of the series, number of years of record and the return period. The value of K is usually in the neighbourhood of 15. Generalised charts for one-day *PMP* prepared on the basis of the statistical analysis of 60 to 70 years of rainfall data in the North-Indian plain area (Lat. 20° N to 32° N, Long. 68° E to 89° E) are available in Refs 4 and 5. It is found that *PMP* estimates for North-Indian plains vary from 37 to 100 cm for one-day rainfall. Maps depicting isolines of 1-day *PMP* over different parts of India are available in the *PMP* atlas published by the Indian Institute of Tropical Meteorology.[6]

2.14 World's Greatest Observed Rainfall

Based upon the rainfall records available all over the world, a list of world's greatest recorded rainfalls of various duration can be assembled. When this data is plotted on a log-log paper, an enveloping straight line drawn to the plotted points obeys the equation.

$$P_m = 42.16\, D^{0.475} \qquad (2.17)$$

where P_m = extreme rainfall depth in cm and D = duration in hours. The values obtained from this Eq. (2.17) are of use in *PMP* estimations.

2.15 Monsoon Studies in India

The economy of the country is very heavily dependent on the monsoon rains that bring in over 75% of rainfall over a major portion of the country. Monsoon is crucial for farm output and economic growth as about 60% of arable land is rain-fed, and farm sector accounts for about 15% of India's economy. While the farm sector contributes about only 15% to the country's Gross Domestic Product (*GDP*), it employs about 60 per cent of India's population.

A deficit in monsoon rainfall or abnormal spatial distribution not only affects the agricultural production but also the purchasing power of a large section of people of the country. This is particularly so for those dependent on the agricultural operations directly or indirectly. Further, the power production in the country, which has a hydro to thermal mix of 25: 75 gets severely impacted by any shortages in the reservoirs due to deficit monsoon. A shortage of power not only affects industrial production but also has a direct impact on agriculture as a substantial part of agricultural production, especially in drought years, is critically dependent on groundwater. These effects coupled with drought relief measures pulls down the *GDP* in the bad monsoon years.

In view of the great importance attached to the monsoon phenomenon in India, monsoon studies to understand and explain the phenomenon form an important component of the activities of *IMD* (the India Meteorological Department) and numerous research and educational institutions in the country. There is a great demand for prediction and different levels of forecasts relating to monsoon rainfall over the country. This comes from a vast variety of interested sectors that span practically the entire economic activity of the country. Reference 11 gives a good review of some aspects related Indian monsoon and an extensive reference related to monsoon studies.

Prediction or forecast of monsoon rainfall is an important component of the monsoon studies. Predicting *ISMR* (Indian Summer Monsoon rainfall) is one of the important mandates of *IMD*. To appreciate the present practice of monsoon prediction the following definitions of terminologies are needed.

1. ISMR (Indian Summer Monsoon Rainfall)

A very useful index of summer rainfall over the Indian region in any year is the All India Summer Monsoon Rainfall (*ISMR*). This is a weighted average of June-September rainfall of 306 well-distributed raingauge stations across the country. It is a convenient measure of monsoon rainfall in a year.

Figure 2.27 shows the variation of *ISMR* during the last century, 1900–2000. The ordinate is the anomaly defined as the difference between the *ISMR* of a year minus the long-term average. For the period under study, the mean is 85 cm. It is interesting to see that the variation about the mean is very small and the standard deviation is about 10% of the mean. The El Nino years are marked in dark. Note that there is a very good correlation between occurrence of El Nino and deficit monsoon rainfall.

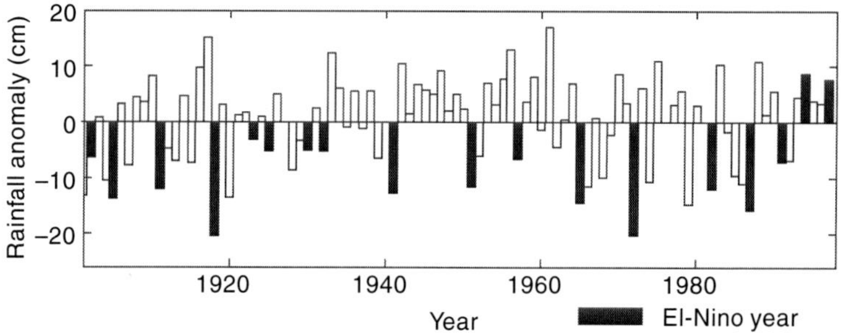

Fig. 2.27 ***Inter Annual variation of ISMR Anomaly (Ref.11)***

2. Southern Oscillation

It is a seesaw pattern of reversing surface air pressures between the eastern and western tropical pacific. Tahiti in Pacific ocean and Darwin in Australia are chosen as reference points. When the surface pressures are high in eastern tropical Pacific, it is low in western tropical Pacific, and vice versa. The changes in pressures are directly related to the surface currents and surface temperatures. The usual abbreviation for southern oscillation is *SO*.

3. El Nino

It is an abnormal warming of surface ocean waters in east tropical Pacific and is a part of *SO*. Because the ocean surface warming and pressure reversals are simultaneous (*coupled*) events, the phenomenon of El Nino and *SO* are together denoted as *ENSO*. El Nino meaning *the little Boy* in Spanish, refers to the Christ child from the normal occurrence of the phenomenon around Christmas time. When this warming occurs for only seven to nine months, it is classified as El Niño "*conditions*"; when it occurs for more than that period, it is classified as El Nino "*episodes*". El Nino events have a frequency of about 3 to 8 years and each event lasts anywhere from seven to twenty months.

It has been observed that there is a connection between the El Nino and *ISMR*.

This association that gives a strong inverse correlation relationship is known as *teleconnection.*

El Nino events affect weather pattern in large parts of the world and some of its major impacts are

(i) Weakening of trade winds with consequent drought conditions in *SE* Asia and parts of Africa
(ii) Decrease in monsoon rainfall (*ISMR*) in India as El Nino is monsoon unfriendly
(iii) Severe droughts in Argentina and north–east Brazil
(iv) Increase in the number of storms on the Pacific coast of *USA*
(v) Decrease in the number of hurricanes in Atlantic coast of *USA*

4. La Nina

This is the counterpart of El Nino, causing decrease in the surface ocean temperatures in eastern tropical Pacific. The effects are also, in general, opposite to that of El Nino. La Nina means *the little girl* in Spanish. The significant effects of La Nina are

(i) Less number of hurricanes in the northern hemisphere
(ii) Malaysia, Indonesia and Australia receive larger than normal rainfall.
(iii) Also, above average normal rains in Peru and Bolivia.

5. Prediction of Monsoon Rainfall

Prediction of monsoon rainfall in India is being practiced since 1886. The earlier models were based on relatively simple correlations of parameters then believed to be chief ones controlling the Indian monsoon phenomenon. After the discovery of teleconnection between El Nino and Indian monsoon, the empirical models improved considerably. Many new parameters were tried and incorporated. By 1988, *IMD* was using a 16 parameter operational model. The parameters included, among others, El Nino, *SO*, snow over Himalayas and Eurasia. This was a power regression parametric model and was in use till 2002. The performance of the model was of mixed nature. Unsatisfactory predictions in 1997 and 2002 led to search for better models and methodologies.

Currently, *IMD* uses a new statistical forecasting method based on ensemble technique. This model has 8 parameters and the forecast is given in two phases. In April, a forecast is given to the country as a whole and the update is issued in June as a mid-course correction. Continuous improvements in technologies related to data acquisition and large scale data analysis techniques of the present day makes one optimistic of more reliable monsoon forecasts in the near future.

2.16 *Rainfall Data In India*

Rainfall measurement in India began in the eighteenth century. The first recorded data were obtained at Calcutta (1784) and it was followed by observations at Madras (1792), Bombay (1823) and Simla (1840). The India Meteorological Department (*IMD*) was established in 1875 and the rainfall resolution of the Government of India in 1930 empowered *IMD* to have overall technical control of rainfall registration in the country. According to this resolution, which is still the basis, the recording, collection and publication of rainfall data is the responsibility of the state government whereas the technical control is under *IMD*. The state governments have the obligation to supply daily, monthly and annual rainfall data to *IMD* for compilation of its two important annual publications entitled *Daily Rainfall of India* and *Monthly Rainfall of India.*

India has a network of observatories and raingauges maintained by *IMD*. Currently (2012), *IMD* has 701 hydrometeorological observatories and 219 agrometeorological observatories. In addition, there are 8579 raingauge stations out of which 3540 stations report their data to *IMD*. A fair amount of these gauges are of self-recording type and *IMD* operates nearly 400 self-recording raingauges.

A set of 21 snow gauges, 10 ordinary raingauges and 6 seasonal snow poles form part of glaciological observatories of the country.

In addition to the above, a large number of raingauges are maintained by different governmental agencies such as Railways, State Departments of Agriculture, Forestry and Irrigation and also by private agencies like coffee and tea plantations. Data from these stations, though recorded regularly, are not published and as such are not easily available for hydrological studies.

REFERENCES

1. Central Water Commission, India, *Estimation of Design Flood Peak,* Flood Estimation Directorate, Report No. 1/73, New Delhi, 1973.
2. Chow, V.T. (Ed), *Handbook of Applied Hydrology,* McGraw-Hill, New York, N.Y., 1964.
3. Dhar, O.N. and B.K. Bhattacharya, "A study of depth area duration statistics of the severemost storms over different meteorological divisions of North India", *Proc. Nat. Symp on Hydrology,* Roorkee, India, 1975, pp. G-4–11.
4. Dhar, O.N. and A.K. Kulkarni, "Estimation of probable maximum, precipitation for some selected stations in and near Himalayas", *Proc. Nat. Symp. on Hydrology,* Roorkee, India, 1975, pp. G-12–16.
5. Dhar, O.N. and P. Rakecha. "A review of hydrometeorological studies of Indian rainfall", *Proc. 2nd World Congress on Water Resources,* New Delhi, Vol. III, 1975, pp. 449–462.
6. Indian Institute of Tropical Meteorology, *Probable Maximum Precipitation Atlas, IITM*, Pune, India, March 1989.
7. Rakecha, P.K. and Deshpande, N.R., Precipitation Network Design, *Jal Vigyan Sameeksha (Hydrology Review),* Vol. II, No. 2, Dec. 1987, pp. 56–75.
8. Weisner, C.J., *Hydrometeorolgy,* Chapman and Hall, London, 1970.
9. World Meteorological Organisation, *Manual for Depth-Area-Duration Analysis of Storm Precipitation, WMO* No. 237, TP 129, Geneva, Switzerland, 1969.
10. Ram Babu et al. *Rainfall Intensity-Duration-Return Period Equations and Nomographs of India,* Bull. No. 3, *CSWCRTI*, Dehradun, India, 1979.
11. Gadgil, S., *The Indian Monsoon and its variability, Annual review*, Earth planet Sci., 2003, 31; 429-67. http://caos.iisc.ernet.in/faculty/sulo/annrev.pdf.

REVISION QUESTIONS

2.1 Describe the different methods of recording of rainfall.

2.2 Discuss the current practice and status of rainfall recording in India.

2.3 Describe the salient characteristics of precipitation on India.

2.4 Explain the different methods of determining the average rainfall over a catchment due to a storm. Discuss the relative merits and demerits of the various methods.

2.5 Explain a procedure for checking a rainfall data for consistency.

2.6 Explain a procedure for supplementing the missing rainfall data.

2.7 Explain briefly the following relationships relating to the precipitation over a basin:
(a) Depth-Area Relationship
(b) Maximum Depth-Area-Duration Curves
(c) Intensity Duration Frequency Relationship.

2.8 What is meant by Probable Maximum Precipitation (*PMP*) over a basin? Explain how PMP is estimated.

2.9 Consider the statement: The 50 year–24 hour maximum rainfall at Bangalore is 160 mm. What do you understand by this statement?

2.10 Write short notes on (i) El Nino, (ii) *ENSO*, (iii) Indian Summer Monsoon Rainfall (*ISMR*), and (iv) La Nina.

2.11 Describe briefly the procedures adopted to measure the water equivalent of snowfall.

2.12 Write brief notes on (i) moving average, (ii) Thiessen polygon, (iii) isohyetals, and isopluvial maps.

PROBLEMS

2.1 A catchment area has seven raingauge stations. In a year, the annual rainfall recorded by the gauges are as follows:

Station	*P*	*Q*	*R*	*S*	*T*	*U*	*V*
Rainfall (cm)	130.0	142.1	118.2	108.5	165.2	102.1	146.9

(a) Determine the standards error in the estimation of mean rainfall in the existing set of raingauges. (b) For a 5% error in the estimation of the mean rainfall, calculate the minimum number of additional raingauge stations to be established in the catchment.

2.2 The normal annual precipitation of five raingauge stations *P, Q, R, S* and *T* are respectively 125, 102, 76, 113 and 137 cm. During a particular storm, the precipitation recorded by stations *P*, *Q*, *R*, and *S* are 13.2, 9.2, 6.8 and 10.2 cm respectively. The instrument at station *T* was inoperative during that storm. Estimate the rainfall at station *T* during that storm.

2.3 Test the consistency of the 22 years of data of the annual precipitation measured at station *A*. Rainfall data for station *A* as well as the average annual rainfall measured at a group of eight neighbouring stations located in a meteorologically homogeneous region are given as follows.

Year	Annual Rainfall of Station *A* (mm)	Average Annual Rainfall of 8 Station groups (mm)	Year	Annual Rainfall of Station *A* (mm)	Average Annual Rainfall of 8 Station groups (mm)
1946	177	143	1957	158	164
1947	144	132	1958	145	155
1948	178	146	1959	132	143

(Contd. Table)

(Contd. Table)

1949	162	147	1960	95	115
1950	194	161	1961	148	135
1951	168	155	1962	142	163
1952	196	152	1963	140	135
1953	144	117	1964	130	143
1954	160	128	1965	137	130
1955	196	193	1966	130	146
1956	141	156	1967	163	161

(a) In what year is a change in regime indicated?

(b) Adjust the recorded data at station *A* and determine the mean annual precipitation.

2.4 In a storm of 210 minutes duration, the incremental rainfall at various time intervals is given below.

Time since start of the storm (minutes)	30	60	90	120	150	180	210
Incremental rainfall in the time interval (cm)	1.75	2.25	6.00	4.50	2.50	1.50	0.75

(a) Obtain the ordinates of the hyetograph and represent the hyetograph as a bar chart with time in chronological order in the *x*-axis.

(b) Obtain the ordinates of the mass curve of rainfall for this storm and plot the same. What is the average intensity of storm over the duration of the storm?

2.5 Assuming the density of water as 998 kg/m^3, determine the internal diameter of a tubular snow sample such that 0.1 *N* of snow in the sample represents 10 mm of water equivalent.

2.6 Represent the annual rainfall data of station *A* given below as a bar chart with time in chronological order. If the annual rainfall less than 75% of long-term mean is taken to signify meteorological drought, identify the drought years and suitably display the same in the bar chart.

Year	1961	1962	1963	1964	1965	1966	1967	1968	1969	1970
Annual rain (mm)	760	750	427	380	480	620	550	640	624	500
Year	1971	1972	1973	1974	1975	1976	1977	1978	1979	1980
Annual rain (mm)	400	356	700	580	520	102	525	900	600	400

2.7 The watershed of a stream has five raingauge stations inside the basin. When Thiessen polygons were constructed, three more stations lying outside the watershed were found to have weightages. The details of Thiessen polygons surroundings each raingauge and the recordings of the raingauges in the month of July 2012 are given below:

Raingauge station	*A*	*B*	*C*	*D*	*E*	*F*	*G*	*H*
Thiessen Polygon area (km^2)	720	1380	1440	1040	900	2220	419	1456
Recorded rainfall in mm during July 2012	135	143	137	128	102	115	99	101

Stations B, D and F are outside the watershed. Determine the average depth of rainfall on the watershed in July 2012 by (i) arithmetic mean method, and (ii) Thiessen mean method.

2.8 For a drainage basin of 600 km^2, isohyetals drawn for a storm gave the following data:

Isohyetals (interval) (cm)	15–12	12–9	9–6	6–3	3–1
Inter-isohyetal area (km^2)	92	128	120	175	85

Estimate the average depth of precipitation over the catchment.

2.9 There are 10 raingauge stations available to calculate the rainfall characteristics of a catchment whose shape can be approximately described by straight lines joining the following coordinates (distances in kilometres):
(30, 0), (80, 10), (110, 30), (140, 90), (130, 115), (40, 110), (15, 60). Coordinates of the raingauge stations and the annual rainfall recorded in them in the year 2001 are given below.

Station	1	2	3	4	5
Co-ordinates	(0, 40)	(50, 0)	(140, 30)	(140, 80)	(90, 140)
Annual Rainfall (cm)	132	136	93	81	85
Station	**6**	**7**	**8**	**9**	**10**
Co-ordinates	(0, 80)	(40, 50)	(90, 30)	(90, 90)	(40, 80)
Annual Rainfall (cm)	124	156	128	102	128

Determine the average annual rainfall over the catchment by using isohyetals.

2.10 Figure 2.28 shows a catchment with seven raingauge stations inside it and three stations outside. The rainfall recorded by each of these stations are indicated in the figure. Draw the figure to an enlarged scale and calculate the mean precipitation by (a) Thiessen-mean method, (b) isohyetal method, and by (c) arithmetic-mean method.

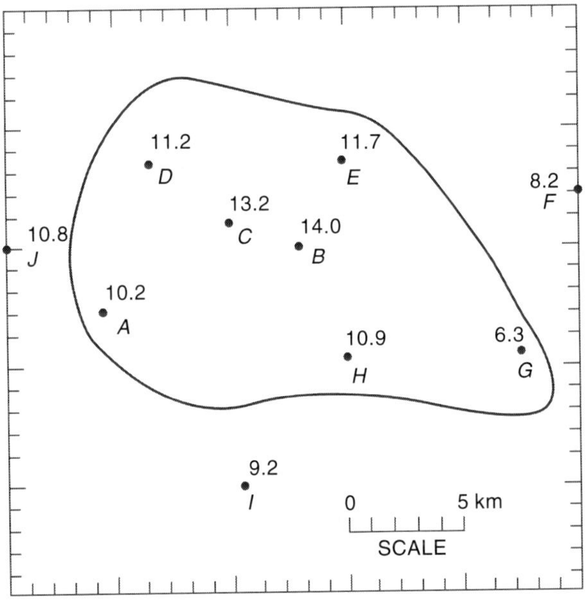

Fig. 2.28 ***Problem 2.10***

2.11 Annual rainfall at a point M is needed. At five points surrounding the point M the values of recorded rainfall together with the coordinates of these stations with respect to a set of axes at point M are given below. Estimate the annual rainfall at point M by using the *USNWS* method.

Station	Rainfall P (cm)	Coordinates of station (in units)	
		X	Y
A	102	2.0	1.0
B	120	2.0	2.0
C	126	3.0	1.0
D	108	1.5	1.0
E	131	4.5	1.5

[*Hint:* In the US National Weather Service (USNWS) method, the weightage to the stations are inversely proportional to the square of the distance of the station from the station M. If the co-ordinate of any station is (x, y) then $D^2 = x^2 + y^2$ and weightage $W = 1/D^2$. Then rainfall at $M = P_m = \frac{\Sigma PW}{\Sigma W}$].

2.12 The data from an isohyetal map of a 24-hour storm is given below. Assuming that the storm centre had an area of 25 km^2, obtain the Depth-Area curve of this store. Using the depth-area curve, estimate the average depth of rainfall over an area of 2400 sq. km. It can be assumed that the storm centre is located at the centre of the area.

Isohyet (mm)	21	20	19	18	17	16	15	14	13	12
Enclosed area (km^2)	543	1345	2030	2545	2955	3280	3535	3710	3880	3915

2.13 Following are the data of a storm as recorded in a self-recording rain gauge at a station:

Time from the beginning of storm (minutes)	10	20	30	40	50	60	70	80	90
Cumulative rainfall (mm)	19	41	48	68	91	124	152	160	166

(a) Plot the hyetograph of the storm.
(b) Plot the maximum intensity-duration curve of the storm.

2.14 Prepare the maximum depth-duration curve for the 90-minute storm given below:

Time (minutes)	0	10	20	30	40	50	60	70	80	90
Cumulative rainfall (mm)	0	8	15	25	30	46	55	60	64	67

2.15 The mass curve of rainfall in a storm of total duration 90 minutes is given below. (a) Draw the hyetograph of the storm at 10 minutes time step. (b) Plot the maximum intensity-duration curve for this storm. (c) Plot the maximum depth-duration curve for the storm.

Time (Minutes)	0	10	20	30	40	50	60	70	80	90
Cumulative rainfall (mm)	0	2.1	6.3	14.5	21.7	27.9	33.0	35.1	36.2	37.0

2.16 The record of annual rainfall at a place is available for 25 years. Plot the curve of recurrence interval *vs* annual rainfall magnitude and by suitable interpolation estimate the magnitude of rainfall at the station that would correspond to a recurrence interval of (a) 50 years, and (b) 100 years.

Year	Annual Rainfall (cm)	Year	Annual Rainfall (cm)
1950	113.0	1963	68.6
1951	94.5	1964	82.5
1952	76.0	1965	90.7
1953	87.5	1966	99.8
1954	92.7	1967	74.4
1955	71.3	1968	66.6
1956	77.3	1969	65.0
1957	85.1	1970	91.0
1958	122.8	1971	106.8
1959	69.4	1972	102.2
1960	81.0	1973	87.0
1961	94.5	1974	84.0
1962	86.3		

2.17 The annual rainfall values at a station *P* for a period of 20 years are as follows:

Year	Annual Rainfall	Year	Annual Rainfall (cm)
1975	120.0	1985	101.0
1976	84.0	1986	109.0
1977	68.0	1987	106.0
1978	92.0	1988	115.0
1979	102.0	1989	95.0
1980	92.0	1990	90.0
1981	95.0	1991	70.0
1982	88.0	1992	89.0
1983	76.0	1993	80.0
1984	84.0	1994	90.0

Determine

(a) The value of annual rainfall at *P* with a recurrence interval of 15 years.

(b) The probability of occurrence of an annual rainfall of magnitude 100 cm at station *P*.

(c) 75% dependable annual rainfall at the station.

[*Hint:* If an event (rainfall magnitude in the present case) occurs more than once, the rank m = number of times the event is equalled + number of times it is exceeded.]

2.18 Plot the three-year and the five-year moving means for the data of Problem 2.16. Comment on the effect of increase in the period of the moving mean. Is there any apparent trend in the data?

2.19 On the basis of isopluvial maps the 50-year–24 hour maximum rainfall at Bangalore is found to be 16.0 cm. Determine the probability of a 24 h rainfall of magnitude 16.0 cm occurring at Bangalore:
(a) Once in ten successive years
(b) Twice in ten successive years
(c) At least once in ten successive years

2.20 A one-day rainfall of 20.0 cm at a place *X* was found to have a period of 100 years. Calculate the probability that a one-day rainfall of magnitude equal to or larger than 20.0 cm:
(a) Will not occur at station *X* during the next 50 years.
(b) Will occur in the next year.

2.21 When long records are not available, records at two or more stations are combined to get one long record for the purposes of recurrence interval calculation. This method is known as *station-year method.*
The number of times a storm of intensity 6 cm/h was equalled or exceeded in three different rain gauge stations in a region were 4, 2 and 5 for periods of records of 36, 25 and 48 years. Find the recurrence interval of the 6 cm/h storm in that area by the *station–year method.*

2.22 Annual precipitation values at a place having 70 years of record can be tabulated as follows:

Range (cm)	Number of years
< 60.0	6
60.0–79.9	6
80.0–99.9	22
100.0–119.9	25
120.0–139.9	8
> 140.0	3

Calculate the probability of having
(a) an annual rainfall equal to or larger than 120 cm
(b) two successive years in which the annual rainfall is equal to or greener than 140 cm
(c) an annual rainfall less than 60 cm

OBJECTIVE QUESTIONS

2.1 A tropical cyclone is a
(a) low-pressure zone that occurs in the northern hemisphere only
(b) high-pressure zone with high winds
(c) zone of low pressure with clockwise winds in the northern hemisphere
(d) zone of low pressure with anticlockwise winds in the northern hemisphere

2.2 A tropical cyclone in the northern hemisphere is a zone of
(a) low pressure with clockwise wind
(b) low pressure with anticlockwise wind
(c) high pressure with clockwise wind
(d) high pressure with anticlockwise wind

2.3 Orographic precipitation occurs due to air masses being lifted to higher altitudes by
(a) the density difference of air masses
(b) a frontal action
(c) the presence of mountain barriers
(d) extratropical cyclones.

2.4 The average annual rainfall over the whole of India is estimated as
(a) 189 cm (b) 319 cm (c) 89 cm (d) 118 cm

2.5 Variability of annual rainfall in India is
(a) least in regions of scanty rainfall (b) largest in regions of high rainfall
(c) least in regions of high rainfall (d) largest in coastal areas.

2.6 The standard Symons' type raingauge has a collecting area of diameter
(a) 12.7 cm (b) 10 cm (c) 5.08 cm (d) 25.4 cm

2.7 The standard recording raingauge adopted in India is of
(a) weighing bucket type (b) natural siphon type
(c) tipping bucket type (d) telemetry type

2.8 The following recording raingauges does not produce the mass curve of precipitation as record:
(a) Symons' raingauge (b) tipping-bucket type gauge
(c) weighing-bucket type gauge (d) natural siphon gauge

2.9 When specific information about the density of snowfall is not available, the water equivalent of snowfall is taken as
(a) 50% (b) 30% (c) 10% (d) 90%

2.10 The normal annual rainfall at stations *A*, *B* and *C* situated in meteorologically homogeneous region are 175 cm, 180 cm and 150 cm respectively. In the year 2000, station *B* was inoperative and stations *A* and *C* recorded annual precipitations of 150 cm and 135 cm respectively. The annual rainfall at station *B* in that year could be estimated to be nearly
(a) 150 cm (b) 143 cm (c) 158 cm (d) 168 cm

2.11 The monthly rainfall at a place *A* during September 1982 was recorded as 55 mm above normal. Here, the term *normal* means
(a) the rainfall in the same month in the previous year
(b) the rainfall was normally expected based on previous month's data
(c) the average rainfall computed from past 12 months' record
(d) the average monthly rainfall for September computed from a specific 30 years of past record

2.12 The double-mass curve technique is adopted to
(a) check the consistency of raingauge records
(b) to find the average rainfall over a number of years
(c) to find the number of rainguages required
(d) to estimate the missing rainfall data

2.13 The mass curve of rainfall of a storm is a plot of
(a) rainfall depths for various equal durations plotted in decreasing order
(b) rainfall intensity *vs* time in chronological order
(c) accumulated rainfall intensity *vs* time
(d) accumulated precipitation *vs* time in chronological order

2.14 A plot between rainfall intensity *vs* time is called as
(a) hydrograph (b) mass curve (c) hyetograph (d) isohyet

2.15 A hyetograph is a plot of
(a) cumulative rainfall *vs* time (b) rainfall intensity *vs* time
(c) rainfall depth *vs* duration (d) discharge *vs* time

2.16 The Thiessen polygon is
(a) a polygon obtained by joining adjoining raingauge stations
(b) a representative area used for weighing the observed station precipitation
(c) an area used in the construction of depth-area curves
(d) the descriptive term for the shape of a hydrograph

2.17 An isohyet is a line joining points having
(a) equal evaporation value (b) equal barometric pressure
(c) equal height above the MSL (d) equal rainfall depth in a given duration.

2.18 By *DAD* analysis the maximum average depth over an area of 10^4 km^2 due to one day storm is found to be 47 cm. For the same area the maximum average depth for a three day storm can be expected to be
(a) < 47 cm (b) > 47 cm (c) = 47 cm
(d) inadequate information to conclude.

2.19 Depth-area-duration curves of precipitation are drawn as
(a) minimising envelopes through the appropriate data points
(b) maximising envelopes through the appropriate data point
(c) best-fit mean curves through the appropriate data points
(d) best-fit straight lines through the appropriate data points

2.20 Depth-area-duration curves of precipitation at a station would normally be
(a) curves, concave upwards, with duration increasing outward
(b) curves, concave downwards, with duration increasing outward
(c) curves, concave upwards, with duration decreasing outward
(d) curves, concave downwards, with duration decreasing outward

2.21 A study of the isopluvial maps revealed that at Kolkata a maximum rainfall depth of 200 mm in 12 h has a return period of 50 years. The probability of a 12 h rainfall equal to or greater than 200 mm occurring at Kolkata at least once in 30 years is
(a) 0.45 (b) 0.60 (c) 0.56 (d) 1.0

2.22 A 6-h rainfall of 6 cm at a place *A* was found to have a return period of 40 years. The probability that at *A* a 6-h rainfall of this or larger magnitude will occur at least once in 20 successive years is
(a) 0.397 (b) 0.603 (c) 0.309 (d) 0.025

2.23 The probability of a 10 cm rain in 1 hour occurring at a station *B* is found to be 1/60. What is the probability that a 1 hour rain of magnitude 10 cm or larger will occur in station *B* once in 30 successive years is
(a) 0.396 (b) 0.307 (c) 0.604 (d) 0.500

2.24 A one-day rainfall of 18 hours at station *C* was found to have a return period of 50 years. The probability that a one-day rainfall of this or larger magnitude will not occur at station *C* during next 50 years is
(a) 0.636 (b) 0.020 (c) 0.364 (d) 0.371

2.25 If the maximum depth of a 50 years–15 h-rainfall depth at Bhubaneshwar is 260 mm, the 50 year-3 h-maximum rainfall depth at the same place is
(a) < 260 mm (b) > 260 mm (c) = 260 mm
(d) inadequate date to conclude anything

2.26 The probable maximum depth of precipitation over a catchment is given by the relation *PMP* =
(a) $\bar{P} + KA^n$ (b) $\bar{P} + K\sigma$ (c) $P \exp(-KA^n)$ (d) mP

CHAPTER 3

ABSTRACTIONS FROM PRECIPITATION

3.1 *INTRODUCTION*

In engineering hydrology, runoff due to a storm event is often the major subject of study. All abstractions from precipitation, viz. those due to evaporation, transpiration, infiltration, surface detention and storage, are considered as losses in the production of runoff. Chief components of abstractions from precipitation, knowledge of which are necessary in the analysis of various hydrologic situations, are described in this chapter.

Evaporation from water bodies and soil masses together with transpiration from vegetation is termed *evapotranspiration*. Various aspects of evaporation from water bodies and evapotranspiration from a basin are discussed in detail in Secs 3.2 through 3.12. Interception and depression storages, which act as 'losses' in the production of runoff, are discussed in Secs 3.13 and 3.14. Infiltration process, which is a major abstraction from precipitation and an important process in groundwater recharge and in increasing soil moisture storage, is described in detail in Secs 3.15 through 3.20.

A: EVAPORATION

3.2 *Evaporation Process*

Evaporation is the process in which a liquid changes to the gaseous state at the free surface, below the boiling point through the transfer of heat energy. Consider a body of water in a pond. The molecules of water are in constant motion with a wide range of instantaneous velocities. An addition of heat causes this range and average speed to increase. When some molecules possess sufficient kinetic energy, they may cross over the water surface. Similarly, the atmosphere in the immediate neighborhood of the water surface contains water molecules within the water vapour in motion and some of them may penetrate the water surface. The net escape of water molecules from the liquid state to the gaseous state constitutes evaporation. Evaporation is a cooling process in that the latent heat of vaporization (at about 585 cal/g of evaporated water) must be provided by the water body. The rate of evaporation is dependent on (i) the vapour pressures at the water surface and air above, (ii) air and water temperatures, (iii) wind speed, (iv) atmospheric pressure, (v) quality of water, and (vi) size of the water body.

1. Vapour Pressure

The rate of evaporation is proportional to the difference between the saturation vapour pressure at the water temperature, e_w and the actual vapour pressure in the air, e_a. Thus,

$$E_L = C(e_w - e_a) \tag{3.1}$$

where E_L = rate of evaporation (mm/day) and C = a constant; e_w and e_a are in mm of mercury. Equation (3.1) is known as *Dalton's law of evaporation* after John Dalton (1802) who first recognised this law. Evaporation continues till $e_w = e_a$. If $e_w > e_a$ condensation takes place.

2. Temperature

Other factors remaining the same, the rate of evaporation increases with an increase in the water temperature. Regarding air temperature, although there is a general increase in the evaporation rate with increasing temperature, a high correlation between evaporation rate and air temperature does not exist. Thus, for the same mean monthly temperature, it is possible to have evaporation to different degrees in a lake in different months.

3. Wind

Wind aids in removing the evaporated water vapour from the zone of evaporation and consequently creates greater scope for evaporation. However, if the wind velocity is large enough to remove all the evaporated water vapour, any further increase in wind velocity does not influence the evaporation. Thus, the rate of evaporation increases with the wind speed up to a critical speed beyond which any further increase in the wind speed has no influence on the evaporation rate. This critical wind-speed value is a function of the size of the water surface. For large water bodies, high-speed turbulent winds are needed to cause maximum rate of evaporation.

3. Atmospheric Pressure

Other factors remaining same, a decrease in the barometric pressure, as in high altitudes, increases evaporation.

4. Soluble Salts

When a solute is dissolved in water, the vapour pressure of the solution is less than that of pure water and hence causes reduction in the rate of evaporation. The percent reduction in evaporation approximately corresponds to the percentage increase in the specific gravity. Thus, for example, under identical conditions, evaporation from sea water is about 2–3% less than that from fresh water.

5. Heat Storage in Water Bodies

Deep water bodies have more heat storage than shallow ones. A deep lake may store radiation energy received in summer and release it in winter causing less evaporation in summer and more evaporation in winter compared to a shallow lake exposed to a similar situation. However, the effect of heat storage is essentially to change the seasonal evaporation rates and the annual evaporation rate is seldom affected.

3.3 *Evaporimeters*

Estimation of evaporation is of utmost importance in many hydrologic problems associated with planning and operation of reservoirs and irrigation systems. In arid zones, this estimation is particularly important to conserve the scarce water resources. However, the exact measurement of evaporation from a large body of water is indeed one of the most difficult tasks.

The amount of water evaporated from a water surface is estimated by the following methods: (i) using evaporimeter data, (ii) empirical evaporation equations, and (iii) analytical methods.

3.3.1 Types of Evaporimeters

Evaporimeters are water-containing pans which are exposed to the atmosphere and the loss of water by evaporation is measured in them at regular intervals. Meteorological data, such as humidity, wind movement, air and water temperatures, and precipitation are also noted along with evaporation measurement.

Many types of evaporimeters are in use and a few commonly used pans are described here.

1. Class A Evaporation Pan

It is a standard pan of 1210 mm diameter and 255 mm depth used by the *US* Weather Bureau and is known as Class *A* Land Pan. The depth of water is maintained between 18 cm and 20 cm (Fig. 3.1). The pan is normally made of unpainted galvanised iron sheet. Monel metal is used where corrosion is a problem. The pan is placed on a wooden platform of 15 cm height above the ground to allow free circulation of air below the pan. Evaporation measurements are made by measuring the depth of water with a hook gauge in a stilling well.

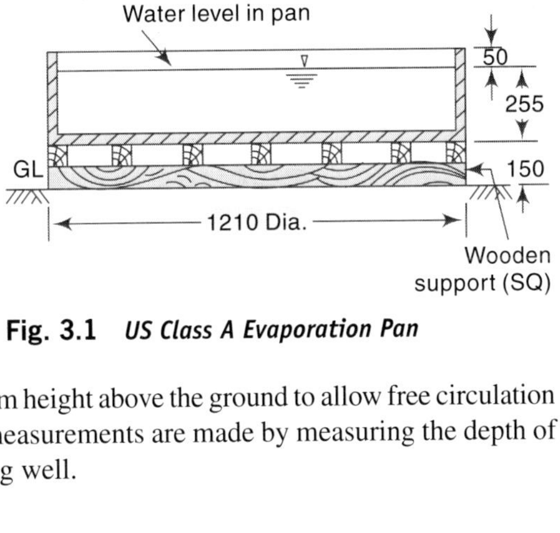

Fig. 3.1 ***US Class A Evaporation Pan***

2. ISI Standard Pan

This pan evaporimeter specified by IS: 5973–1970, also known as modified Class A Pan, consists of a pan 1220 mm in diameter with 255 mm of depth. The pan is made of copper sheet of 0.9 mm thickness, tinned inside and painted white outside (Fig. 3.2). A fixed point gauge indicates the level of water. A calibrated cylindrical measure is used to add or remove water maintaining the water level in the pan to a fixed mark. The top of the pan is covered fully with a hexagonal wire netting of galvanised iron to protect the water in the pan from birds. Further, the presence of a wire mesh makes the water temperature more uniform during day and night. The evaporation from this pan is found to be less by about 14% compared to that from an unscreened pan. The pan is placed over a square wooden platform of 1225 mm width and 100 mm height to enable circulation of air underneath the pan.

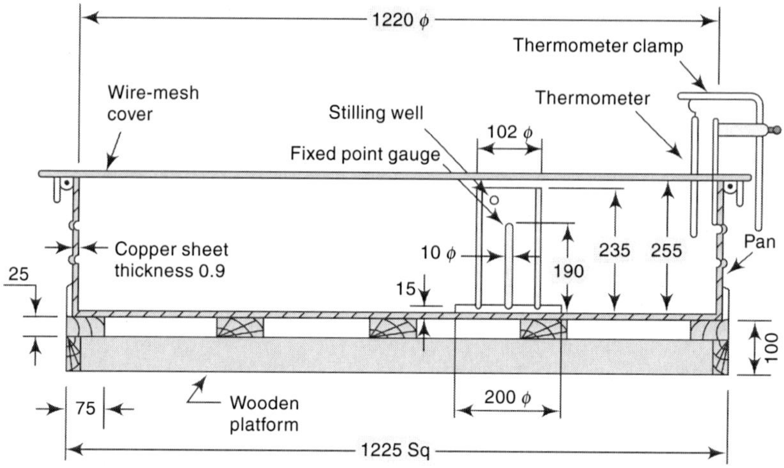

Fig. 3.2 ***ISI Standard Evaporation Pan***

3. Colorado Sunken Pan

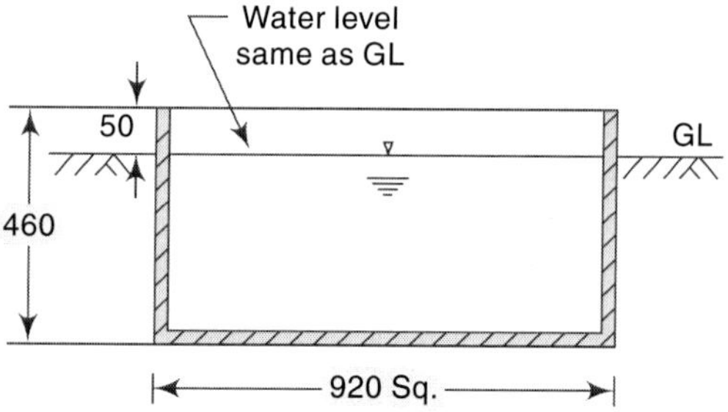

Fig. 3.3 ***Colorado Sunken Evaporation Pan***

This pan, 920 mm square and 460 mm deep, is made up of unpainted galvanised iron sheet and buried into the ground within 100 mm of the top (Fig. 3.3). The chief advantage of the sunken pan is that radiation and aerodynamic characteristics are similar to those of a lake. However, it has the following disadvantages: (i) difficult to detect leaks, (ii) extra care is needed to keep the surrounding area free from tall grass, dust, etc., and (iii) expensive to install.

4. US Geological Survey Floating Pan

With a view to simulate the characteristics of a large body of water, this square pan (900 mm side and 450 mm depth) supported by drum floats in the middle of a raft (4.25 m × 4.87 m) is set afloat in a lake. The water level in the pan is kept at the same level as the lake leaving a rim of 75 mm. Diagonal baffles provided in the pan reduce the surging in the pan due to wave action. Its high cost of installation and maintenance together with the difficulty involved in performing measurements are its main disadvantages.

3.3.2 Pan Coefficient C_p

Evaporation pans are not exact models of large reservoirs and have the following principal drawbacks:

1. They differ in the heat-storing capacity and heat transfer from the sides and bottom. The sunken pan and floating pan aim to reduce this deficiency. As a result of this factor, the evaporation from a pan depends to a certain extent on its size. While a pan of 3 m diameter is known to give a value which is about the

same as from a neighbouring large lake, a pan of 1.0 m diameter size indicates about 20% excess evaporation than that of the 3 m diameter pan.

2. The height of the rim in an evaporation pan affects the wind action over the surface. Also, it casts a shadow of variable magnitude over the water surface.
3. The heat-transfer characteristics of the pan material is different from that of the reservoir.

In view of the above, the evaporation observed from a pan has to be corrected to get the evaporation from a lake under similar climatic and exposure conditions. Thus, a coefficient is introduced as

$$\text{Lake evaporation} = C_p \times \text{Pan evaporation}$$

in which C_p = pan coefficient. The values of C_p in use for different pans are given in Table 3.1.

Table 3.1 *Values of Pan Coefficient C_p*

S.No.	Types of pan	Average value	Range
1.	Class A Land Pan	0.70	0.60–0.80
2.	ISI Pan (modified Class A)	0.80	0.65–1.10
3.	Colorado Sunken Pan	0.78	0.75–0.86
4.	*USGS* Floating Pan	0.80	0.70–0.82

3.3.3 Evaporation Stations

It is usual to instal evaporation pans in such locations where other meteorological data are also simultaneously collected. The *WMO* recommends the minimum network of evaporimeter stations as below:

1. *Arid zones*—One station for every 30,000 km^2
2. *Humid temperate climates*—One station for every 50,000 km^2
3. *Cold regions*—One station for every 100,000 km^2

Currently, about 220 pan-evaporimeter stations are being maintained by the India Meteorological Department.

A typical hydrometeorological station contains the following: ordinary raingauge; recording raingauge; Stevenson box with maximum and minimum thermometer and dry and wet bulb thermometers; wind anemometer, wind direction indicator, sunshine recorder, thermohydrograph and pan evaporimeter.

3.4 *Empirical Evaporation Equations*

A large number of empirical equations are available to estimate lake evaporation using commonly available meteorological data. Most formulae are based on the Dalton-type equation and can be expressed in the general form

$$E_L = K f(u)(e_w - e_a) \tag{3.2}$$

where E_L = lake evaporation in mm/day, e_w = saturated vapour pressure at the water-surface temperature in mm of mercury, e_a = actual vapour pressure of overlying air at a specified height in mm of mercury, $f(u)$ = wind-speed correction function and

K = a coefficient. The term e_a is measured at the same height at which wind speed is measured. Two commonly used empirical evaporation formulae are the following:

3.4.1 Meyer's Formula (1915)

$$E_L = K_M(e_w - e_a)\left(1 + \frac{u_9}{16}\right) \tag{3.3}$$

in which E_L, e_w, e_a are as defined in Eq. (3.2), u_9 = monthly mean wind velocity in km/h at about 9 m above ground and K_M = coefficient accounting for various other factors with a value of 0.36 for large deep waters and 0.50 for small, shallow waters.

3.4.2 Rohwer's Formula (1931)

Rohwer's formula considers a correction for the effect of pressure in addition to the wind-speed effect and is given by

$$E_L = 0.771(1.465 - 0.000732\, p_a)(0.44 + 0.0733\, u_0)\, (e_w - e_a) \tag{3.4}$$

in which E_L, e_w, and e_a are as defined earlier in Eq. (3.2),

p_a = mean barometric reading in mm of mercury

u_0 = mean wind velocity in km/h at ground level, which can be taken to be the velocity at 0.6 m height above ground.

These empirical formulae are simple to use and permit the use of standard meteorological data. However, in view of the various limitations of the formulae, they can at best be expected to give an approximate magnitude of the evaporation. References 2 and 3 list several other popular empirical formulae.

In using the empirical equations, the saturated vapour pressure at a given temperature (e_w) is found from a table of e_w *vs* temperature in °C, such as Table 3.3. Often, the wind-velocity data would be available at an elevation other than that needed in the particular equation. However, it is known that in the lower part of the atmosphere, up to a height of about 500 m above the ground level, the wind velocity can be assumed to follow the 1/7 power law as

$$u_h = Ch^{1/7} \tag{3.5}$$

where u_h = wind velocity at a height h above the ground and C = constant. This equation can be used to determine the velocity at any desired level ifu_h is known.

Example 3.1 (a) A reservoir with a surface area of 250 hectares had the following average values of climate parameters during a week: Water temperature = 20°C,
Relative humidity = 40%, Wind velocity at 1.0 m above ground surface = 16 km/h.
Estimate the average daily evaporation from the lake by using Meyer's formula.

(b) An *ISI* Standard evaporation pan at the site indicated a pan coefficient of 0.80 on the basis of calibration against controlled water-budgeting method. If this pan indicated an evaporation of 72 mm in the week under question, (i) estimate the accuracy if Meyer's method relative to the pan evaporation measurements. (ii) Also, estimate the volume of water evaporated from the lake in that week.

Solution

(a) From Table 3.3

e_w = 17.54 mm of Hg $\qquad e_a = 0.4 \times 17.54 = 7.02$ mm of Hg

u_9 = wind velocity at a height of 9.0 m above ground = $u_1 \times (9)^{1/7}$ = 21.9 km/h

By Meyer's Formula [Eq. (3.3)],

$$E_L = 0.36\,(17.54 - 7.02)\left(1 + \frac{21.9}{16}\right) = 8.97 \text{ mm/day}$$

(b) (i) Daily evaporation as per pan eaporimeter = $\left(\frac{72.00}{7}\right) \times 0.8 = 8.23$ mm

Error by Meyer's formula = (8.23 – 8.97) = –0.74 mm. Hence, Meyer's method overestimates the evaporation relative to the pan.

Percentage over estimation by Meyer's formula = (0.74/8.23) × 100 = 9%

(ii) Considering the pan measurements as the basis, volume of water evaporated from the lake in 7 days = 7 × (8.23/1000) × 250 × 10^4 = 144,025 m^3

[The lake area is assumed to be constant at 250 ha throughout the week.]

3.5 *Analytical Methods of Evaporation Estimation*

The analytical methods for the determination of lake evaporation can be broadly classified into three categories as

1. Water-budget method,
2. Energy-balance method, and
3. Mass-transfer method.

3.5.1 Water-Budget Method

The water-budget method is the simplest of the three analytical methods and is also the least reliable. It involves writing the hydrological continuity equation for the lake and determining the evaporation from a knowledge or estimation of other variables. Thus, considering the daily average values for a lake, the continuity equation is written as

$$P + V_{is} + V_{ig} = V_{os} + V_{og} + E_L + \Delta S + T_L \tag{3.6}$$

where P = daily precipitation

V_{is} = daily surface inflow into the lake

V_{ig} = daily groundwater inflow

V_{os} = daily surface outflow from the lake

V_{og} = daily seepage outflow

E_L = daily lake evaporation

ΔS = increase in lake storage in a day

T_L = daily transpiration loss

All quantities are in units of volume (m^3) or depth (mm) over a reference area. Equation (3.6) can be written as

$$E_L = P + (V_{is} - V_{os}) + (V_{ig} - V_{og}) - T_L - \Delta S \tag{3.7}$$

In this, the terms P, V_{is}, V_{os} and ΔS can be measured. However, it is not possible to measure V_{ig}. V_{og} and T_L and therefore these quantities can only be estimated. Transpiration losses can be considered to be insignificant in some reservoirs. If the unit of time is kept large, say weeks or months, better accuracy in the estimate of E_L is possible. In view of the various uncertainties in the estimated values and the possibilities of errors in measured variables, the water-budget method cannot be expected to give very accurate results. However, controlled studies such as at Lake Hefner in *USA* (1952) have given fairly accurate results by this method.

3.5.2 Energy-Budget Method

The energy-budget method is an application of the law of conservation of energy. The energy available for evaporation is determined by considering the incoming energy, outgoing energy and energy stored in the water body over a known time interval.

Considering the water body as in Fig. 3.4, the energy balance to the evaporating surface in a period of one day is give by

$$H_n = H_a + H_e + H_g + H_s + H_i \tag{3.8}$$

where H_n = net heat energy received by the water surface

$= H_c(1 - r) - H_b$

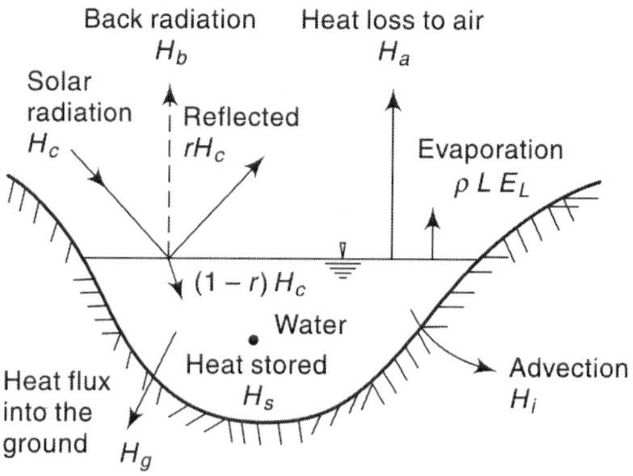

Fig. 3.4 ***Energy Balance in a Water Body***

in which $H_c(1 - r)$ = incoming solar radiation into a surface of reflection coefficient (albedo) r

H_b = back radiation (long wave) from water body

H_a = sensible heat transfer from water surface to air

H_e = heat energy used up in evaporation

$= \rho LE_L$, where ρ = density of water, L = latent heat of evaporation and E_L = evaporation in mm

H_g = heat flux into the ground

H_s = heat stored in water body

H_i = net heat conducted out of the system by water flow (advected energy)

All the energy terms are in calories per square mm per day. If the time periods are short, the terms H_s and H_i can be neglected as negligibly small. All the terms except H_a can either be measured or evaluated indirectly. The sensible heat term H_a which cannot be readily measured is estimated using *Bowen's ratio* β given by the expression

$$\beta = \frac{H_a}{\rho LE_L} = 6.1 \times 10^{-4} \times p_a \frac{T_w - T_a}{e_w - e_a} \tag{3.9}$$

where p_a = atmospheric pressure in mm of mercury, e_w = saturated vapour pressure in mm of mercury, e_a = actual vapour pressure of air in mm of mercury, T_w = temperature of water surface in °C and T_a = temperature of air in °C. From Eqs (3.8) and (3.9) E_L can be evauated as

$$E_L = \frac{H_n - H_g - H_s - H_i}{\rho L(1+\beta)} \tag{3.10}$$

Estimation of evaporation in a lake by the energy balance method has been found to give satisfactory results, with errors of the order of 5% when applied to periods less than a week. Further details of the energy-budget method are available in Refs 2, 3 and 5.

3.5.3 Mass-Transfer Method

This method is based on theories of turbulent mass transfer in boundary layer to calculate the mass water vapour transfer from the surface to the surrounding atmosphere. However, the details of the method are beyond the scope of this book and can be found in published literature[2, 5]. With the use of quantities measured by sophisticated (and expensive) instrumentation, this method can give satisfactory results.

Example 3.2 Following observation were made for conducting the water budget of a reservoir over a period of one month of 30 days.

Average surface area = 10 km^2	Rainfall = 10 cm
Mean surface inflow rate = 10 m^3/s	Mean surface outflow rate = 15 m^3/s
Fall in reservoir level = 1.50 m	Pan evaporation = 20 cm
Assume pan evaporation coefficient = 0.70	

Estimate the average seepage discharge during that month.

Solution

Area of reservoir = 10×10^6 m^2

Inflow		Outflow	
Item	**Volume (m^3)**	**Item**	**Volume (m^3)**
Surface inflow = 15 × 3600 × 24 × 30	25,920,000	Surface outflow = 15 × 3600 × 24 × 30	38,880,000
Rainfall = $10 \times 10^6 \times 10/100$	1,000,000	Evaporation = 0.7 × $10 \times 10^6 \times 20/100$	1,400,000
		Seepage volume = S_e	S_e
Total inflow volume	26,920,000	Total outflow	40,280,000 + S_e
Reduction in storage = $10 \times 10^6 \times 1.5$	15,000,000		

Total outflow volume – Total inflow volume = Reduction in storage

$(40{,}280{,}000 + S_e) - (26{,}920{,}000) = 15{,}000{,}000$

$S_e = 1{,}640{,}000 = 1.64$ Mm^3

3.6 Reservoir Evaporation and Methods for its Reduction

Any of the methods mentioned above may be used for the estimation of reservoir evaporation. Although analytical methods provide better results, they involve parameters that are difficult to assess or expensive to obtain. Empirical equations can at best give approximate values of the correct order of magnitude. Therefore, the pan measurements find general acceptance for practical application. Mean monthly and annual evaporation data collected by *IMD* are very valuable in field estimations. The water volume lost due to evaporation from a reservoir in a month is calculated as

$$V_E = A\, E_{pm}\, C_p \tag{3.11}$$

where V_E = volume of water lost in evaporation in a month (m^3)
A = average reservoir area during the month (m^2)
E_{pm} = pan evaporation loss in metres in a month (m)
= E_L in mm/day × No. of days in the month × 10^{-3}
C_p = relevant pan coefficient

Evaporation from a water surface is a continuous process. Typically, under Indian conditions, evaporation loss from a water body is about 160 cm in a year with enhanced values in arid regions. The quantity of stored water lost by evaporation in a year is indeed considerable as the surface area of many natural and human-made lakes in the country are very large. While a small sized tank (lake) may have a surface area of about 20 ha, large reservoirs such as Narmada Sagar have a surface area of about 90,000 ha. Table 3.2 (a) indicates surface areas and capacities of ten largest Indian reservoirs.

Table 3.2(a) ***Surface Areas and Capacities of Some Indian Reservoirs***

[*Source:* National Register of large dams, *CWC*, India]

Sl. No	Name	River	Year of Completion	Gross Storage Capacity (km^3)	Reservoir area (km^2)
1	Indira Sagar (Narmada Sagar)	Narmada	2006	12.20	913
2	Nagarjuna Sagar	Krishna	1960	11.56	285
3	Rihand	Rihand	1962	10.62	468
4	Bhakra	Satluj	1963	9.62	168
5	Sardar Sarovar	Narmada	UC	9.50	*370
6	SriSailam	Krishna	1984	8.72	616
7	Pong	Beas	1974	8.57	260
8	Ukai	Tapi	*1971	8.51	*520
9	Hirakud	Mahanadi	1957	8.14	743
10	Gandhisagar	Chambal	1960	7.32	660

* = Data from other sources, UC= under construction

Using evaporation data from 29 major and medium reservoirs in the country, the National Commission for Integrated Water Resources Development (1999)[8] has estimated the national annual water loss due to evaporation at various time horizons as below:

Table 3.2(b) ***Annual Water Loss due to Evaporation (Volume in km^3)***

Sl. No.	Particular	1997	2010	2025	2050
1.	Live Capacity–Major storage	173.7	211.4	249.2	381.5
2.	Live Capacity–Minor storage	34.7	42.3	49.8	76.3
3.	Evaporation for Major storage	26.1	31.7	37.4	57.2
	Reservoirs @ 15% of live capacity				
4.	Evaporation for Minor storage	8.7	10.6	12.5	19.1
	Reservoirs @ 25% of live capacity				
5.	**Total Evaporation loss/year**	**35**	**42**	**50**	**76**

Roughly, a quantity equivalent to entire live capacity of minor storages is lost annually by evaporation. As the construction of various reservoirs as a part of water resources developmental effort involve considerable inputs of money, which is a scarce resource, evaporation from such water bodies signifies an economic loss. In semi-arid zones where water is scarce, the importance of conservation of water through reduction of evaporation is obvious.

3.6.1 Methods to Reduce Evaporation Losses

Various methods available for reduction of evaporation losses can be considered in three categories:

1. Reduction of Surface Area

Since the volume of water lost by evaporation is directly proportional to the surface area of the water body, the reduction of surface area wherever feasible reduces evaporation losses. Measures like having deep reservoirs in place of wider ones and elimination of shallow areas can be considered under this category.

2. Mechanical Covers

Permanent roofs over the reservoir, temporary roofs and floating roofs such as rafts and lightweight floating particles can be adopted wherever feasible. Obviously, these measures are limited to very small water bodies such as ponds, etc.

3. Chemical Films

This method consists of applying a thin chemical film on the water surface to reduce evaporation. Currently this is the only feasible method available for reduction of evaporation of reservoirs up to moderate size.

Certain chemicals such as *cetyl alcohol* (hexadecanol) and *stearyl alcohol* (octadecanol) form monomolecular layers on a water surface. These layers act as evaporation inhibitors by preventing the water molecules to escape past them. The thin film formed has the following desirable features:

1. The film is strong and flexible and does not break easily due to wave action.
2. If punctured due to the impact of raindrops or by birds, insects, etc., the film closes back soon after.
3. It is pervious to oxygen and carbon dioxide; the water quality is therefore not affected by its presence.
4. It is colourless, odourless and nontoxic.

Cetyl alcohol is found to be the most suitable chemical for use as an evaporation inhibitor. It is a white, waxy, crystalline solid and is available as lumps, flakes or powder. It can be applied to the water surface in the form of powder, emulsion or solution in mineral turpentine. Roughly about 3.5 N/hectare/day of cetyl alcohol is needed for effective action. The chemical is periodically replenished to make up the losses due to oxidation, wind sweep of the layer to the shore and its removal by birds and insects. Evaporation reduction can be achieved to a maximum if a film pressure of 4×10^{-2} N/m is maintained.

Controlled experiments with evaporation pans have indicated an evaporation reduction of about 60% through use of cetyl alcohol. Under field conditions, the reported values of evaporation reduction range from 20 to 50%. It appears that a reduction of 20–30% can be achieved easily in small size lakes (≤1000 hectares) through the use of these monomolecular layers. The adverse effect of heavy wind appears to be the only major impediment affecting the efficiency of these chemical films.

B: EVAPOTRANSPIRATION

3.7 Transpiration

Transpiration is the process by which water leaves the body of a living plant and reaches the atmosphere as water vapour. The water is taken up by the plant-root system and escapes through the leaves. The important factors affecting transpiration are: atmospheric vapour pressure, temperature, wind, light intensity and characteristics of the plant, such as the root and leaf systems. For a given plant, factors that affect the free-water evaporation also affect transpiration. However, a major difference exists between transpiration and evaporation. Transpiration is essentially confined to daylight hours and the rate of transpiration depends upon the growth periods of the plant. Evaporation, on the other hand, continues all through the day and night although the rates are different.

3.8 Evapotranspiration

While transpiration takes place, the land area in which plants stand also lose moisture by the evaporation of water from soil and water bodies. In hydrology and irrigation practice, it is found that evaporation and transpiration processes can be considered advantageously under one head as *evapotranspiration*. The term *consumptive use* is also used to denote this loss by evapotranspiration. For a given set of atmospheric conditions, evapotranspiration obviously depends on the availability of water. If sufficient moisture is always available to completely meet the needs of vegetation fully covering the area, the resulting evapotranspiration is called *potential evapotranspiration* (*PET*). Potential evapotranspiration no longer critically depends on the soil and plant factors but depends essentially on the climatic factors. The real evapotranspiration occurring in a specific situation is called *actual evapotranspiration* (*AET*).

It is necessary to introduce at this stage two terms: *field capacity* and *permanent wilting point*. Field capacity is the maximum quantity of water that the soil can retain against the force of gravity. Any higher moisture input to a soil at field capacity simply drains away. Permanent wilting point is the moisture content of a soil at which the moisture is no longer available in sufficient quantity to sustain the plants. At this stage, even though the soil contains some moisture, it will be so held by the soil grains that the roots of the plants are not able to extract it in sufficient quantities to sustain the plants and consequently the plants wilt. The field capacity and permanent wilting point depend upon the soil characteristics. The difference between these two moisture contents is called *available water*, the moisture available for plant growth.

If the water supply to the plant is adequate, soil moisture will be at the field capacity and *AET* will be equal to *PET*. If the water supply is less than *PET*, the soil dries out and the ratio *AET/PET* would then be less than unity. The decrease of the ratio *AET/PET* with available moisture depends upon the type of soil and rate of drying. Generally, for clayey soils, *AET/PET* = 1.0 for nearly 50% drop in the available moisture. As can be expected, when the soil moisture reaches the permanent wilting point, the *AET* reduces to zero (Fig. 3.5). For a catchment in a given period of time, the hydrologic budget can be written as

$$P - R_s - G_o - E_{\text{act}} = \Delta S \tag{3.12}$$

where P = precipitation, R_s = surface runoff, G_o = subsurface outflow, E_{act} = actual evapotranspiration (*AET*) and ΔS = change in the moisture storage. This water budgeting can be used to calculate E_{act} by knowing or estimating other elements of Eq. (3.12). Generally, the sum of R_s and G_o can be taken as the stream flow at the basin outlet without much error. Method of estimating *AET* is given in Sec. 3.12.

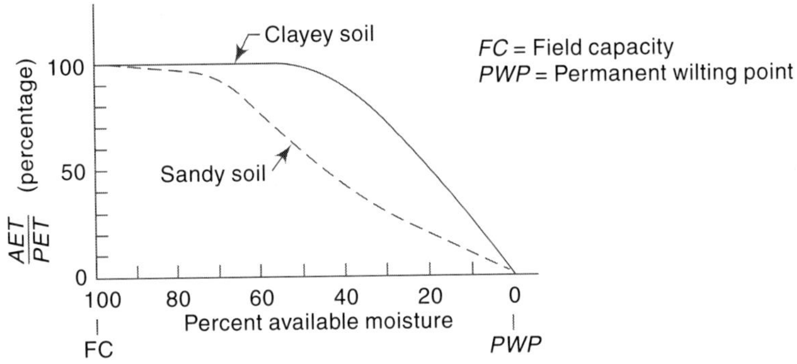

Fig. 3.5 ***Variation of AET***

Except in a few specialised studies, all applied studies in hydrology use *PET* (not *AET*) as a basic parameter in various estimations related to water utilisations connected with evapotranspiration process. It is generally agreed that *PET* is a good approximation for lake evaporation. As such, where pan evaporation data is not available, *PET* can be used to estimate lake evaporation.

3.9 *Measurement of Evapotranspiration*

The measurement of evapotranspiration for a given vegetation type can be carried out in two ways: either by using *lysimeters* or by the use of field plots.

3.9.1 Lysimeters

A lysimeter is a special watertight tank containing a block of soil and set in a field of growing plants. The plants grown in the lysimeter are the same as in the surrounding field. Evapotranspiration is estimated in terms of the amount of water required to maintain constant moisture conditions within the tank measured either volumetric ally or gravimetrically through an arrangement made in the lysimeter. Lysimeters should be designed to accurately reproduce the soil conditions, moisture content, type and size of the vegetation of the surrounding area. They should be so buried that the soil is at the same level inside and outside the container. Lysimeter studies are time-consuming and expensive.

3.9.2 Field Plots

In special plots all the elements of the water budget in a known interval of time are measured and the evapotranspiration determined as

Evapotranspiration = [Precipitation + Irrigation input – Runoff
– Increase in soil storage–groundwater loss]

Measurements are usually confined to precipitation, irrigation input, surface runoff and soil moisture. Groundwater loss due to deep percolation is difficult to measure and can be minimised by keeping the moisture condition of the plot at the field capacity. This method provides fairly reliable results.

3.10 *Evapotranspiration Equations*

The lack of reliable field data and the difficulties of obtaining reliable evapotranspiration data have given rise to a number of methods to predict *PET* by using climatological data. Large number of formulae are available: they range from purely empirical ones to those backed by theoretical concepts. Two useful equations are given below.

3.10.1 Penman's Equation

Penman's equation is based on sound theoretical reasoning and is obtained by a combination of the energy-balance and mass-transfer approach. Penman's equation, incorporating some of the modifications suggested by other investigators is

$$PET = \frac{AH_n + E_a \gamma}{A + \gamma} \tag{3.13}$$

where PET = daily potential evapotranspiration in mm per day

A = slope of the saturation vapour pressure *vs* temperature curve at the mean air temperature, in mm of mercury per °C (Table 3.3)

H_n = net radiation in mm of evaporable water per day

E_a = parameter including wind velocity and saturation deficit

γ= psychrometric constant = 0.49 mm of mercury/°C

The net radiation is the same as used in the energy budget [Eq. (3.8)] and is estimated by the following equation:

$$H_n = H_a (1 - r)\left(a + b\frac{n}{N}\right) - \sigma T_a^4 (0.56 - 0.092\sqrt{e_a})\left(0.10 + 0.90\frac{n}{N}\right) \quad (3.14)$$

where H_a = incident solar radiation outside the atmosphere on a horizontal surface, expressed in mm of evaporable water per day (it is a function of the latitude and period of the year as indicated in Table 3.4)

a = a constant depending upon the latitude ϕ and is given by $a = 0.29 \cos \phi$

b = a constant with an average value of 0.52

n = actual duration of bright sunshine in hours

N = maximum possible hours of bright sunshine (it is a function of latitude as indicated in Table 3.5)

r = reflection coefficient (albedo). Usual ranges of values of r are given below.

Surface	Range of r values
Close ground corps	0.15–0.25
Bare lands	0.05–0.45
Water surface	0.05
Snow	0.45–0.95

σ = Stefan-Boltzman constant = 2.01×10^{-9} mm/day

T_a = mean air temperature in degrees kelvin = 273 + °C

e_a = actual mean vapour pressure in the air in mm of mercury

The parameter E_a is estimated as

$$E_a = 0.35\left(1 + \frac{u_2}{160}\right)(e_w - e_a) \quad (3.15)$$

in which

u_2 = mean wind speed at 2 m above ground in km/day

e_w = saturation vapour pressure at mean air temperature in mm of mercury (Table 3.3)

e_a = actual vapour pressure, defined earlier

For the computation of *PET*, data on n, e_a, u_2, mean air temperature and nature of surface (i.e. value of r) are needed. These can be obtained from actual observations or through available meteorological data of the region. Equations (3.13), (3.14) and (3.15) together with tables 3.3, 3.4, and 3.5 enable the daily *PET* to be calculated. It may be noted that Penman's equation can be used to calculate evaporation from a water surface by using r = 0.05. Penman's equation is widely used in India, the *UK*, Australia and in some parts of *USA*. Further details about this equation are available elsewhere[2,5,7]. Tables 3.3, 3.4 and 3.5 are helpful in the use of Penman's equation.

Table 3.3 ***Saturation Vapour Pressure of Water***

Temperature (°C)	Saturation vapour pressure e_w (mm of Hg)	A (mm/°C)
0	4.58	0.30
5.0	6.54	0.45
7.5	7.78	0.54
10.0	9.21	0.60
12.5	10.87	0.71
15.0	12.79	0.80
17.5	15.00	0.95
20.0	17.54	1.05
22.5	20.44	1.24
25.0	23.76	1.40
27.5	27.54	1.61
30.0	31.82	1.85
32.5	36.68	2.07
35.0	42.81	2.35
37.5	48.36	2.62
40.0	55.32	2.95
45.0	71.20	3.66

$e_w = 4.584 \exp\left(\frac{17.27\,t}{237.3 + t}\right)$ mm of Hg, where t = temperature in °C.

Table 3.4 ***Mean Monthly Solar Radiation at Top of Atmosphere, H_a in mm of Evaporable Water/Day***

North latitude	Jan	Feb	Mar	Apr	May	Jun	Jul	Aug	Sep	Oct	Nov	Dec
0°	14.5	15.0	15.2	14.7	13.9	13.4	13.5	14.2	14.9	15.0	14.6	14.3
10°	12.8	13.9	14.8	15.2	15.0	14.8	14.8	15.0	14.9	14.1	13.1	12.4
20°	10.8	12.3	13.9	15.2	15.7	15.8	15.7	15.3	14.4	12.9	11.2	10.3
30°	8.5	10.5	12.7	14.8	16.0	16.5	16.2	15.3	13.5	11.3	9.1	7.9
40°	6.0	8.3	11.0	13.9	15.9	16.7	16.3	14.8	12.2	9.3	6.7	5.4
50°	3.6	5.9	9.1	12.7	15.4	16.7	16.1	13.9	10.5	7.1	4.3	3.0

Table 3.5 ***Mean Monthly Values of Possible Sunshine Hours, N***

North latitude	Jan	Feb	Mar	Apr	May	Jun	Jul	Aug	Sep	Oct	Nov	Dec
0°	12.1	12.1	12.1	12.1	12.1	12.1	12.1	12.1	12.1	12.1	12.1	12.1
10°	11.6	11.8	12.1	12.4	12.6	12.7	12.6	12.4	12.9	11.9	11.7	11.5
20°	11.1	11.5	12.0	12.6	13.1	13.3	13.2	12.8	12.3	11.7	11.2	10.9
30°	10.4	11.1	12.0	12.9	13.7	14.1	13.9	13.2	12.4	11.5	10.6	10.2
40°	9.6	10.7	11.9	13.2	14.4	15.0	14.7	13.8	12.5	11.2	10.0	9.4
50°	8.6	10.1	11.8	13.8	15.4	16.4	16.0	14.5	12.7	10.8	9.1	8.1

Example 3.3

Calculate the potential evapotranspiration from an area near New Delhi in the month of November by Penman's formula. The following data are available:

Latitude	:	28°4′N
Elevation	:	230 m (above sea level)
Mean monthly temperature	:	19° C
Mean relative humidity	:	75%
Mean observed sunshine hours	:	9 h
Wind velocity at 2 m height	:	85 km/day
Nature of surface cover	:	Close-ground green crop

Solution

From Table 3.3,

$$A = 1.00 \text{ mm/°C} \qquad e_w = 16.50 \text{ mm of Hg}$$

From Table 3.4,

$$H_a = 9.506 \text{ mm of water/day}$$

From Table 3.5,

$$N = 10.716 \text{ h} \qquad n/N = 9/10.716 = 0.84$$

From given data

$$e_a = 16.50 \times 0.75 = 12.38 \text{ mm of Hg}$$
$$a = 0.29 \cos 28° \, 4' = 0.2559$$
$$b = 0.52$$
$$\sigma = 2.01 \times 10^{-9} \text{ mm/day}$$
$$T_a = 273 + 19 = 292 \text{ K}$$
$$\sigma T_a^4 = 14.613$$

r = albedo fer close-ground green crop is taken as 0.25

From Eq. (3.14),

$$H_n = 9.506 \times (1 - 0.25) \times (0.2559 + (0.52 \times 0.84))$$
$$- 14.613 \times (0.56 - 0.092\sqrt{12.38}) \times (0.10 + (0.9 \times 0.84))$$
$$= 4.936 - 2.946 = 1.990 \text{ mm of water/day}$$

From Eq. (3.15),

$$E_a = 0.35 \times \left(1 + \frac{85}{160}\right) \times (16.50 - 12.38) = 2.208 \text{ mm/day}$$

From Eq. (3.13), noting the value of $\gamma = 0.49$.

$$PET = \frac{(1 \times 1.990) + (2.208 \times 0.49)}{(1.00 + 0.49)} = 2.06 \text{ mm/day}$$

Example 3.4 Using the data of Example 3.3, estimate the daily evaporation from a lake situated in that place.

Solution

For estimating the daily evaporation from a lake, Penman's equation is used with the albedo $r = 0.05$.

Hence,

$$H_n = 4.936 \times \frac{(1.0 - 0.05)}{(1.0 - 0.25)} - 2.946 = 6.252 - 2.946 = 3.306 \text{ mm of water/day}$$

$$E_a = 2.208 \text{ mm/day}$$

From Eq.(3.13),

$$PET = \text{Lake evaporation}$$

$$= \frac{(1.0 \times 3.306) + (2.208 \times 0.49)}{(1.0 - 0.49)} = 2.95 \text{ mm/day}$$

3.10.2 Reference Crop Evapotranspiration (ET_0)

In irrigation practice, the *PET* is extensively used in calculation of crop-water requirements. For purposes of standardisation, *FAO* recommends[3] a *reference crop evapotranspiration* or *reference evapotranspiration* denoted as ET_o. The reference surface is a hypothetical grass reference crop with an assumed crop height of 0.12 m, a defined fixed surface resistance of 70 s m^{-1} and an albedo of 0.23. The reference surface closely resembles an extensive surface of green, well-watered grass of uniform height, actively growing and completely shading the ground. The defined fixed surface resistance 70 s m^{-1} implies a moderately dry soil surface resulting from about a weekly irrigation frequency. The *FAO* recommends a method called *FAO Penman-Monteith method* to estimate ET_o by using radiation, air temperature, air humidity and wind speed data. Details of *FAO Penman-Monteith method* are available in Ref. 3.

The potential evapotranspiration of any other crop (*ET*) is calculated by multiplying the reference crop evapotranspiration by a coefficient *K*, the value of which changes with stage of the crop. Thus,

$$ET = K(ET_o) \tag{3.16}$$

The value of *K* varies from 0.5 to 1.3. Table 3.7 gives average values of *K* for some selected crops.

3.10.3 Empirical Formulae

A large number of empirical formulae are available for estimation of *PET* based on climatological data. These are not universally applicable to all climatic areas. They should be used with caution in areas different from those for which they were derived.

1. Blaney–Criddle Formula

This purely empirical formula is based on data from arid western United States. This formula assumes that the *PET* is related to hours of sunshine and temperature, which are taken as measures of solar radiation at an area. The potential evapotranspiration in a crop-growing season is given by

$$E_T = 2.54\, KF$$

and

$$F = \Sigma P_h \overline{T}_f / 100 \tag{3.17}$$

where E_T = *PET* in a crop season in cm

K = an empirical coefficient, depends on the type of the crop and stage of growth

F = sum of monthly consumptive use factors for the period

P_h = monthly percent of annual day-time hours, depends on the latitude of the place (Table 3.6)

and $\overline{T}_f$ = mean monthly temperature in °F

Values of K depend on the month and locality. Average value for the season for selected crops is given in Table 3.7. The Blaney–Criddle formula is largely used by irrigation engineers to calculate the water requirements of crops, which is taken as the difference between *PET* and effective precipitation. The Blaney–Morin equation is another empirical formula similar to Eq. (3.17) but with an additional correction for humidity.

Table 3.6 ***Monthly Daytime Hours Percentages, P_h, for use in Blaney–Criddle Formula (Eq. 3.17)***

North latitude (deg)	Jan	Feb	Mar	Apr	May	Jun	Jul	Aug	Sep	Oct	Nov	Dec
0	8.50	7.66	8.49	8.21	8.50	8.22	8.50	8.49	8.21	8.50	8.22	8.50
10	8.13	7.47	8.45	8.37	8.81	8.60	8.86	8.71	8.25	8.34	7.91	8.10
15	7.94	7.36	8.43	8.44	8.98	8.80	9.05	8.83	8.28	8.26	7.75	7.88
20	7.74	7.25	8.41	8.52	9.15	9.00	9.25	8.96	8.30	8.18	7.58	7.66
25	7.53	7.14	8.39	8.61	9.33	9.23	9.45	9.09	8.32	8.09	7.40	7.42
30	7.30	7.03	8.38	8.72	9.53	9.49	9.67	9.22	8.33	7.99	7.19	7.15
35	7.05	6.88	8.35	8.83	9.76	9.77	9.93	9.37	8.36	7.87	6.97	6.86
40	6.76	6.72	8.33	8.95	10.02	10.08	10.22	9.54	8.39	7.75	6.72	6.52

Table 3.7 ***Values of K for Selected Crops***

Crop	Average value of K	Range of monthly values
Rice	1.10	0.85–1.30
Wheat	0.65	0.50–0.75
Maize	0.65	0.50–0.80
Sugarcane	0.90	0.75–1.00
Cotton	0.65	0.50–0.90
Potatoes	0.70	0.65–0.75
Natural Vegetation:		
(a) Very dense	1.30	
(b) Dense	1.20	
(c) Medium	1.00	
(d) Light	0.80	

Example 3.5 Estimate the *PET* of an area for the season November to February in which wheat is grown. The area is in North India at a latitude of 30° N with mean monthly temperatures as below:

Month	Nov.	Dec.	Jan.	Feb.
Temp. (°C)	16.5	13.0	11.0	14.5

Use the Blaney—Criddle formula.

Solution

From Table 3.7, for wheat, $K = 0.65$. Values of P_h for 30° N is read from Table 3.6, the temperatures are converted to Fahrenheit and the calculations are performed in the following table.

Month	$\overline{T}_f$	P_h	$P_h \overline{T}_f/100$
Nov.	61.7	7.19	4.44
Dec.	55.4	7.15	3.96
Jan.	51.8	7.30	3.78
Feb.	58.1	7.03	4.08
		$\Sigma P_h \overline{T}_f/100 =$	16.26

By Eq. (3.17),

$$E_T = 2.54 \times 16.26 \times 0.65 = 26.85 \text{ cm}.$$

2. Thornthwaite Formula

This formula was developed from data of eastern USA and uses only the mean monthly temperature together with an adjustment for day-lengths. The PET is given by this formula as

$$E_T = 1.6\, L_a \left(\frac{10\overline{T}}{I_t}\right)^a \qquad (3.18)$$

where E_T = monthly *PET* in cm

L_a = adjustment for the number of hours of daylight and days in the month, related to the latitude of the place (Table 3.8)

Table 3.8 ***Adjustment Factor L_a for Use in Thornthwaite Formula (Eq. 3.18)***

North latitude (deg)	Jan	Feb	Mar	Apr	May	Jun	Jul	Aug	Sep	Oct	Nov	Dec
0	1.04	0.94	1.04	1.01	1.04	1.01	1.04	1.04	1.01	1.04	1.01	1.04
10	1.00	0.91	1.03	1.03	1.08	1.06	1.08	1.07	1.02	1.02	0.98	0.99
15	0.97	0.91	1.03	1.04	1.11	1.08	1.12	1.08	1.02	1.01	0.95	0.97
20	0.95	0.90	1.03	1.05	1.13	1.11	1.14	1.11	1.02	1.00	0.93	0.94
25	0.93	0.89	1.03	1.06	1.15	1.14	1.17	1.12	1.02	0.99	0.91	0.91
30	0.90	0.87	1.03	1.08	1.18	1.17	1.20	1.14	1.03	0.98	0.89	0.88
40	0.84	0.83	1.03	1.11	1.24	1.25	1.27	1.18	1.04	0.96	0.83	0.81

$\overline{T}$ = mean monthly air temperature °C

I_t = the total of 12 monthly values of heat index = $\sum_1^{12} i$,

where $i = (\overline{T}/5)^{1.514}$

a = an empirical constant

$= 6.75 \times 10^{-7} I_t^3 - 7.71 \times 10^{-5} I_t^2 + 1.792 \times 10^{-2} I_t + 0.49239$

3.11 *Potential Evapotranspiration Over India*

Using Penman's equation and the available climatological data, *PET* estimate[5] for the country has been made. The mean annual *PET* (in cm) over various parts of the country is shown in the form of *isopleths*—the lines on a map through places having equal depths of evapotranspiration [Fig. 3.6(a)]. It is seen that the annual *PET* ranges from 140 to 180 cm over most parts of the country. The annual *PET* is highest at Rajkot, Gujarat with a value of 214.5 cm. Extreme south-east of Tamil Nadu also show high average values greater than 180 cm. The highest *PET* for southern peninsula is at

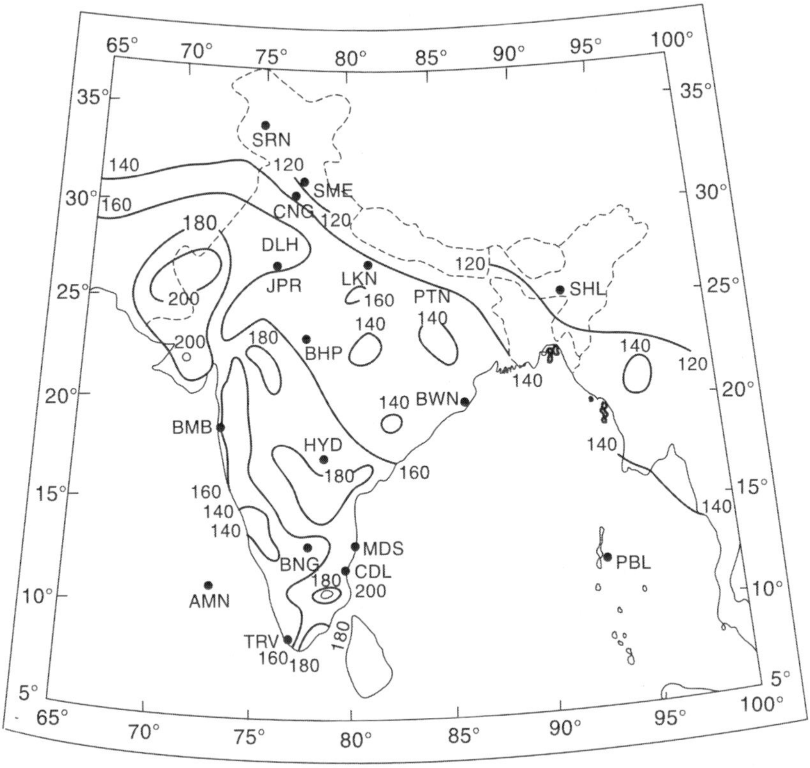

Fig. 3.6(a) ***Annual PET (cm) over India***

(Source : Scientific: Report No. 136, IMD, 1971, © Government of India Copyright)

Based upon survey of India map with the permission of the Surveyor General of India, © Government of India Copyright 1984

The territorial waters of India extend into the sea to a distance of 12 nautical miles measured from the appropriate baseline

Responsibility for the correctness of internal details on the map rests with the publisher.

Tiruchirapalli, Tamil Nadu, with a value of 209 cm. The variation of monthly *PET* at some stations located in different climatic zones in the country is shown in Fig. 3.6(b). Valuable *PET* data relevant to various parts of the country are available in Refs 4 and 7.

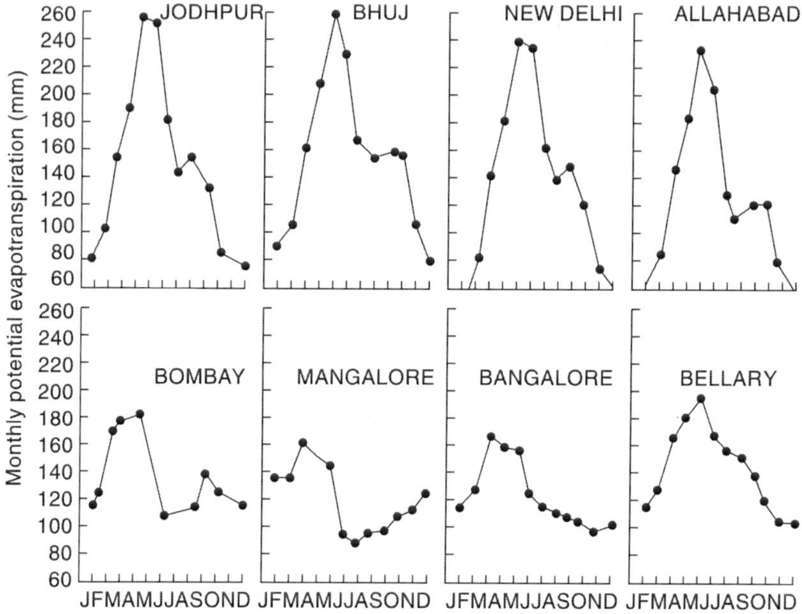

Fig. 3.6(b) ***Monthly Variation of PET (mm)***

(Source: Scientific Report No. 136, India Meteorological Department, 1971, © Government of India Copyright)

3.12 *Actual Evapotranspiration (AET)*

AET for hydrological and irrigation applications can be obtained through a process of water budgeting and accounting for soil-plant-atmosphere interactions. A simple procedure due to *Doorenbos* and *Pruit* is as follows:

1. Using available meteorological data, the reference crop evapotranspiration (ET_o) is calculated.
2. The crop coefficient K for the given crop (and stage of growth) is obtained from published tables such as Table 3.7. The potential crop evapotranspiration ET_c is calculated using Eq. 3.16 as $ET_c = K(ET_o)$.
3. The actual evapotranspiration (ET_a) at any time t at the farm having the given crop is calculated as below:
   - If $AASW \geq (1 - p)\ MASW$

$$ET_a = ET_o \qquad \text{(known as } \textit{potential condition}\text{)} \tag{3.19-a}$$

   - If $AASW < (1 - p)\ MASW$

$$ET_a = \left[\frac{AASW}{(1-p)\ MASW}\right] ET_c \tag{3.19-b}$$

where $MASW$ = total available soil water over the root depth

$AASW$ = actual available soil-water at time t over the root depth

p = soil-water depletion factor for a given crop and soil complex. (Values of p ranges from about 0.1 for sandy soils to about 0.5 for clayey soils)

[Note the equivalence of terms used earlier as $PET = ET_o$ and $AET = ET_a$]

Example 3.6 A recently irrigated field plot has on Day 1 the total available soil moisture at its maximum value of 10 cm. If the reference crop evapotranspiration is 5.0 mm/day, calculate the actual evapotranspiration on Day 1, Day 6 and Day 7. Assume soil-water depletion factor p = 0.20 and crop factor K = 0.8.

Solution

Here $ET_o = 5.0$ mm and $MASW = 100$ mm

$(1-p)\,MASW = (1-0.2) \times 100 = 80.0$ and $ET_c = 0.9 \times 5.0 = 4.5$ mm/day

Day 1: $AASW = 100 \text{ mm} > (1-p)\,MASW$

Hence potential condition exists and $ET_a = ET_c = 4.5$ mm/day

This rate will continue till a depletion of (100 – 80) = 20 mm takes place in the soil. This will take 20/4.5 = 4.44 days. Thus **Day 5** also will have $ET_a = ET_c = 4.5$ mm/day

Day 6: At the beginning of Day 6, $AASW = (100 - 4.5 \times 5) = 77.5$ mm

Since $AASW < (1-p)\,MASW$,

$$ET_a = \left[\frac{77.5}{80.0}\right] \times 4.5 = 4.36 \text{ mm}$$

Day 7: At the beginning of Day 7, $AASW = (77.5 - 4.36) = 73.14$ mm

Since $AASW < (1-p)\,MASW$

$$ET_a = \left[\frac{73.14}{80.0}\right] \times 4.5 = 4.11 \text{ mm.}$$

$AASW$ at the end of Day 7 = 73.14 – 4.11 = 69.03 mm.

C: *Initial Loss*

In the precipitation reaching the surface of a catchment, the major abstraction is from the infiltration process. However, two other processes, though small in magnitude, operate to reduce the water volume available for runoff and thus act as abstractions. These are (i) the *interception process*, and (ii) the *depression storage* and together they are called the *initial loss*. This abstraction represents the quantity of storage that must be satisfied before overland runoff begins. The following two sections deal with these two processes briefly.

3.13 *Interception*

When it rains over a catchment, not all the precipitation falls directly onto the ground. Before it reaches the ground, a part of it may be caught by the vegetation and subsequently evaporated. The volume of water so caught is called *interception.* The intercepted precipitation may follow one of the three possible routes:

1. It may be retained by the vegetation as surface storage and returned to the atmosphere by evaporation; a process termed *interception loss;*
2. It can drip off the plant leaves to join the ground surface or the surface flow; this is known as *throughfall*; and
3. The rainwater may run along the leaves and branches and down the stem to reach the ground surface. This part is called *stemflow.*

Interception loss is solely due to evaporation and does not include transpiration, throughfall or stemflow.

The amount of water intercepted in a given area is extremely difficult to measure. It depends on the species composition of vegetation, its density and also on the storm characteristics. It is estimated that of the total rainfall in an area during a plant-growing season the interception loss is about 10 to 20%. Interception is satisfied during the first part of a storm and if an area experiences a large number of small storms, the annual interception loss due to forests in such cases will be high, amounting to greater than 25% of the annual precipitation. Quantitatively, the variation of interception loss with the rainfall magnitude per storm for small storms is as shown in Fig. 3.7. It is seen that the interception loss is large for a small rainfall and levels off to a constant value for larger storms. For a given storm, the interception loss is estimated as

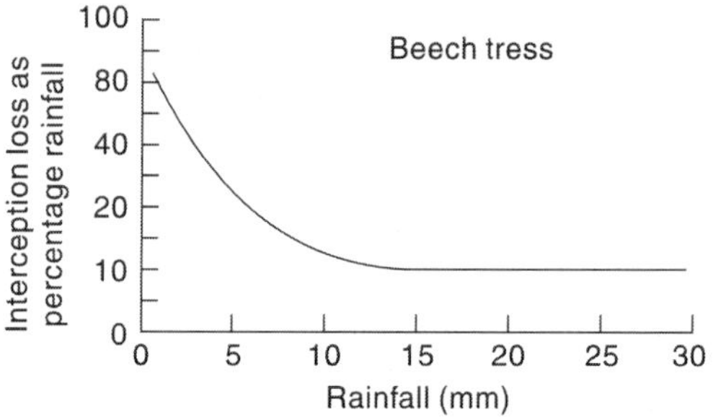

Fig. 3.7 ***Typical Interception Loss Curve***

$$I_i = S_i + K_i Et \tag{3.20}$$

where I_i = interception loss in mm, S_i = interception storage whose value varies from 0.25 to 1.25 mm depending on the nature of vegetation, K_i = ratio of vegetal surface area to its projected area, E = evaporation rate in mm/h during the precipitation and t = duration of rainfall in hours.

It is found that coniferous trees have more interception loss than deciduous ones. Also, dense grasses have nearly same interception losses as full-grown trees and can account for nearly 20% of the total rainfall in the season. Agricultural crops in their growing season also contribute high interception losses. In view of these the interception process has a very significant impact on the ecology of the area related to silvicultural aspects, in *in situ* water harvesting and in the water balance of a region. However, in hydrological studies dealing with floods, interception loss is rarely significant and is not separately considered. The common practice is to allow a lump-sum value as the initial loss to be deducted from the initial period of the storm.

3.14 *Depression Storage*

When the precipitation of a storm reaches the ground, it must first fill up all depressions before it can flow over the surface. The volume of water trapped in these depressions is called *depression storage.* This amount is eventually lost to runoff through processes of infiltration and evaporation and thus form a part of the initial loss. Depression storage depends on a vast number of factors, the chief of which are (i) the type of soil, (ii) the condition of the surface reflecting the amount and nature of depression, (iii) the slope of the catchment, and (iv) the antecedent precipitation, as a measure of the soil moisture. Obviously, general expressions for quantitative estimation of this loss are not available. Qualitatively, it has been found that antecedent precipitation has a very pronounced effect on decreasing the loss to runoff in a storm due to depression. Values of 0.50 cm in sand, 0.4 cm in loam and 0.25 cm in clay can be taken as representatives for depression-storage loss during intensive storms.

D: *Infiltration*

3.15 *Infiltration*

Infiltration is the flow of water into the ground through the soil surface. The distribution of soil moisture within the soil profile during the infiltration process is illustrated in Fig. 3.8. When water is applied at the surface of a soil, four moisture zones in the soil, as indicated in Fig. 3.8 can be identified.

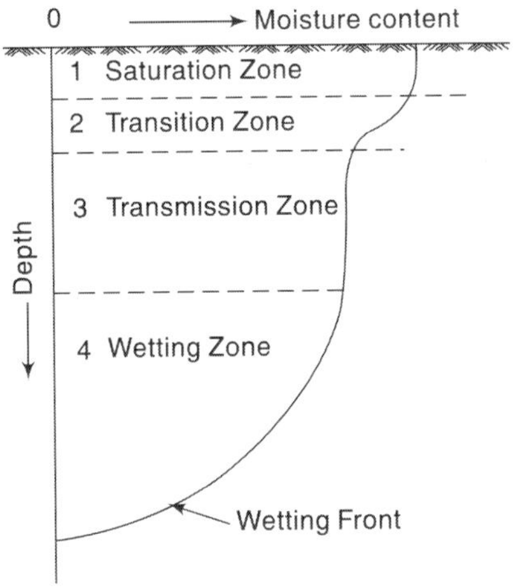

Fig. 3.8 ***Distribution of Soil Moisture in the Infiltration Process***

Zone 1 At the top, a thin layer of *saturated zone* is created.

Zone 2 Beneath zone 1, there is a *transition zone.*

Zone 3 The next lower zone is the *transmission zone* where the downward motion of the moisture takes place. The moisture content in this zone is above field capacity but below saturation. Further, it is characterised by unsaturated flow and fairly uniform moisture content.

Zone 4 The last zone is the *wetting zone*. The soil moisture in this zone will be at or near field capacity and the moisture content decreases with the depth. The boundary of the wetting zone is the wetting front where a sharp discontinuity exists between the newly wet soil and original moisture content of the soil. Depending upon the amount of infiltration and physical properties of the soil, the wetting front can extend from a few centimetres to metres.

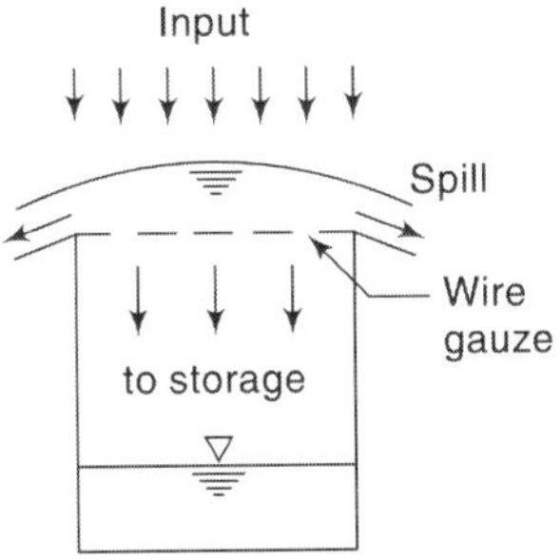

Fig. 3.9 ***An Analogy for Infiltration***

The infiltration process can be easily understood through a simple analogy. Consider a small container covered with wire gauze as in Fig. 3.9. If water is poured into the container, a part of it will go into the container and a part overflows. Further, the container can hold only a fixed quantity and when it is full no more flow into the container can take place. While this analogy is highly simplified, it underscores two important aspects; viz. (i) the maximum rate at which the ground can absorb water, the *infiltration capacity,* and (ii) the volume of water that the ground can hold, the *field capacity.* Since the infiltered water may contribute to the groundwater discharge in addition to increasing the soil moisture, the process can be schematically modelled as in Fig. 3.10(a) and (b) wherein two situations, viz. low-intensity rainfall and high-intensity rainfall are considered.

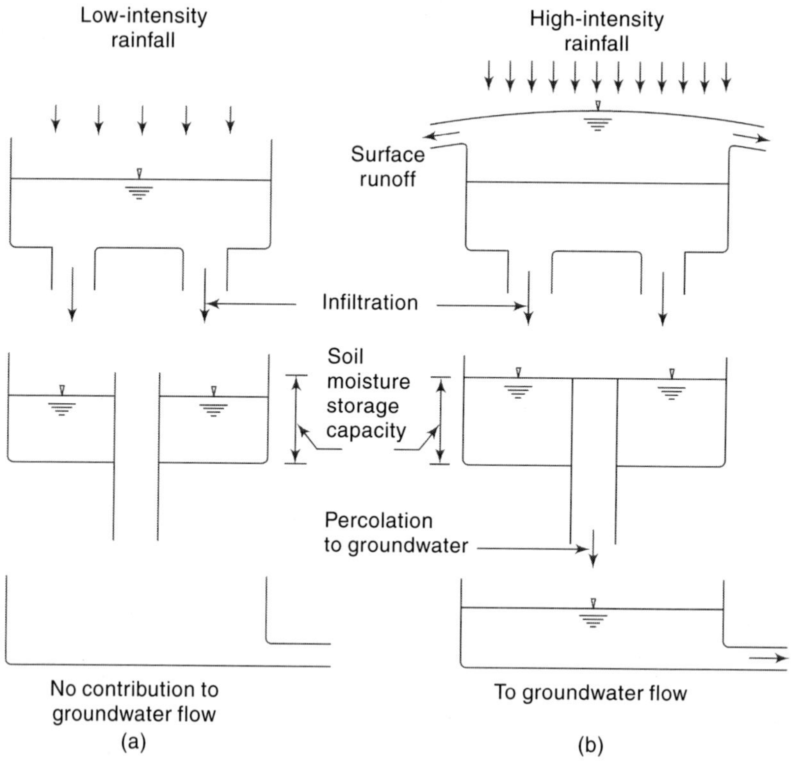

Fig. 3.10 ***An Infiltration Model***

3.16 *Infiltration Capacity*

3.16.1 Definition

The maximum rate at which a given soil at a given time can absorb water is defined as the *infiltration capacity.* It is designated as f_p and is expressed in units of cm/h. The actual rate of infiltration f can be expressed as

$$f = f_p \text{ when } i \geq f_p$$

and

$$f = i \text{ when } i < f_p \qquad (3.20)$$

where i = intensity of rainfall. The infiltration capacity of a soil is high at the beginning of a storm and has an exponential decay as the time elapses.

The typical variation of the infiltration capacity f_p of a soil with time is shown in Fig. 3.11. The infiltration capacity of an area is dependent on a large number of factors, chief of them are

- Characteristics of the soil (Texture, porosity and hydraulic conductivity)
- Condition of the soil surface
- Current moisture content
- Vegetative cover and
- Soil temperature

A few important factors affecting f_p are described below:

3.16.1 Characteristics of Soil

The type of soil, viz. sand, silt or clay, its texture, structure, permeability and underdrainage are the important characteristics under this category. A loose, permeable, sandy soil will have a larger infiltration capacity than a tight, clayey soil. A soil with good underdrainage, i.e. the facility to transmit the infiltered water downward to a groundwater storage would obviously have a higher infiltration capacity. When the soils occur in layers, the transmission capacity of the layers determines the overall infiltration rate. Also, a dry soil can absorb more water than one whose pores are already full (Fig. 3.11). The land use has a significant influence on f_p. For example, a forest soil rich in organic matter will have a much higher value of f_p under identical conditions than the same soil in an urban area where it is subjected to compaction.

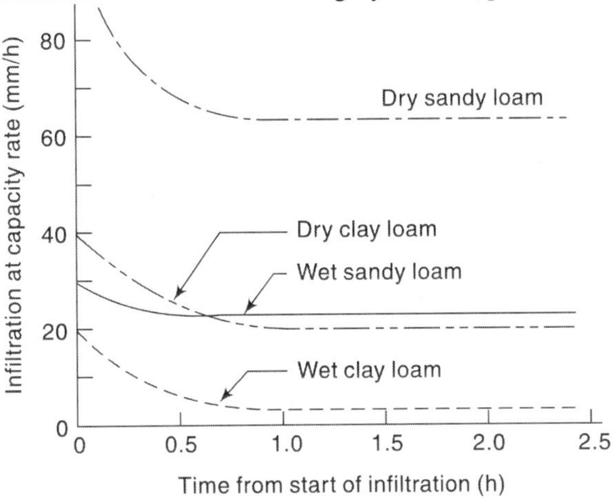

Fig. 3.11 ***Variation of Infiltration Capacity***

3.16.2 Surface of Entry

At the soil surface, the impact of raindrops causes the fines in the soil to be displaced and these in turn can clog the pore spaces in the upper layers of the soil. This is an important factor affecting the infiltration capacity. Thus, a surface covered with grass and other vegetation which can reduce this process has a pronounced influence on the value of f_p.

3.16.3 Fluid Characteristics

Water infiltrating into the soil will have many impurities, both in solution and in suspension. The turbidity of the water, especially the clay and colloid content, is an important factor and such suspended particles block the fine pores in the soil and reduce its infiltration capacity. The temperature of the water is a factor in the sense that it affects the viscosity of the water which in turn affects the infiltration rate. Contamination of the water by dissolved salts can affect the soil structure and in turn affect the infiltration rate.

3.17 *Measurement of Infiltration*

Infiltration characteristics of a soil at a given location can be estimated by

- Using flooding-type infiltrometers
- Measurement of subsidence of free water in a large basin or pond
- Rainfall simulator
- Hydrograph analysis

3.17.1 Flooding-Type Infiltrometer

Flooding-type infiltrometers are experimental devices used to obtain data relating to variation of infiltration capacity with time. Two types of flooding-type infiltrometers are in common use. They are (a) Tube-type (or Simple) infiltrometer, and (b) Double-ring infiltrometer.

1. Simple (Tube Type) Infiltrometer

This is a simple instrument consisting essentially of a metal cylinder, 30 cm diameter and 60 cm long, open at both ends. The cylinder is driven into the ground to a depth of 50 cm (Fig. 3.12(a)). Water is poured into the top part to a depth of 5 cm and a pointer is set to mark the water level. As infiltration proceeds, the volume is made up by adding water from a burette to keep the water level at the tip of the pointer. Knowing the volume of water added during different time intervals, the plot of the infiltration capacity *vs* time is obtained. The experiments are continued till a uniform rate of infiltration is obtained and this may take 2–3 hours. The surface of the soil is usually protected by a perforated disc to prevent formation of turbidity and its settling on the soil surface.

A major objection to the simple infiltrometer as above is that the infiltered water spreads at the outlet from the tube (as shown by dotted lines in Fig. 3.12(a)) and as such the tube area is not representative of the area in which infiltration is taking place.

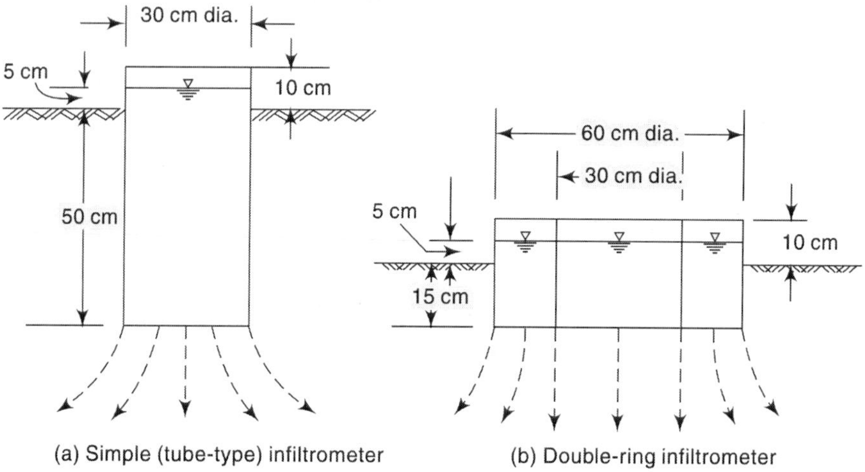

Fig. 3.12 ***Flooding-Type Infiltrometers***

2. Double-ring Infiltrometer

This most commonly used infiltrometer is designed to overcome the basic objection of the tube infiltrometer, viz. the tube area is not representative of the infiltrating area. In this, two sets of concentrating rings with diameters of 30 cm and 60 cm and of a minimum length of 25 cm, as shown in Fig. 3.12(b), are used. The two rings are inserted into the ground and water is applied into both the rings to maintain a constant depth of about 5.0 cm. The outer ring provides water jacket to the infiltering water from the inner ring and hence prevents the spreading out of the infiltering water of the inner ring. The water depths in the inner and outer rings are kept the same during the observation period. The measurement of the water volume is done on the inner ring only. The experiment is carried out till a constant infiltration rate is obtained. A perforated disc to prevent formation of turbidity and settling of fines on the soil surface is provided on the surface of the soil in the inner ring as well as in the annular space.

As the flooding-type infiltrometer measures the infiltration characteristics at a spot only, a large number of pre-planned experiments are necessary to obtain representative infiltration characteristics for an entire watershed. Some of the chief disadvantages of flooding-type infiltrometers are

1. the raindrop impact effect is not simulated;
2. the driving of the tube or rings disturbs the soil structure; and
3. the results of the infiltrometers depend to some extent on their size with the larger meters giving less rates than the smaller ones; this is due to the border effect.

3.17.2 Rainfall Simulator

In this, a small plot of land, of about 2 m × 4 m size, is provided with a series of nozzles on the longer side with arrangements to collect and measure the surface runoff rate. The specially designed nozzles produce raindrops falling from a height of 2 m and are capable of producing various intensities of rainfall. Experiments are conducted under controlled conditions with various combinations of intensities and durations and the surface runoff rates and volumes are measured in each case. Using the water-budget equation involving the volume of rainfall, infiltration and runoff, the infiltration rate and its variation with time are estimated. If the rainfall intensity is higher than the infiltration rate, infiltration capacity values are obtained.

Rainfall-simulator-type infiltrometers give lower values than flooding-type infiltrometers. This is due to effect of the rainfall impact and turbidity of the surface water present in the former.

3.17.3 Hydrograph Analysis

Reasonable estimation of the infiltration capacity of a small watershed can be obtained by analysing measured runoff hydrographs and corresponding rainfall records. If sufficiently good rainfall records and runoff hydrographs corresponding to isolated storms in a small watershed with fairly homogeneous soils are available, the water-budget equation can be applied to estimate the abstraction by infiltration. In this, the evapotranspiration losses are estimated by knowing the land cover/use of the watershed.

3.18 Modelling Infiltration Capacity

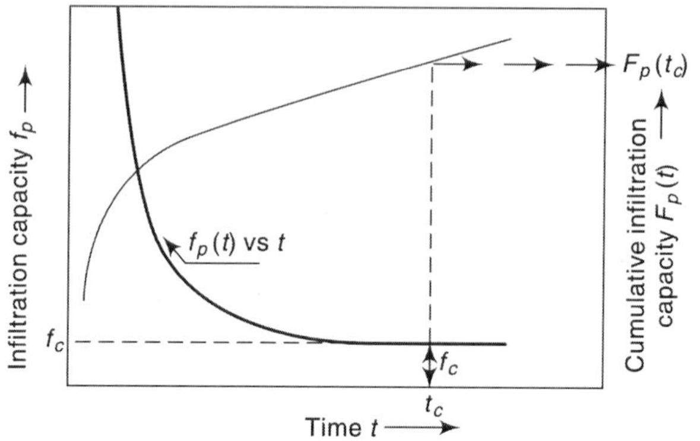

Fig. 3.13 ***Curves of Infiltration Capacity and Cumulative Infiltration Capacity***

Figure 3.13 shows a typical variation of infiltration capacity f_p with time. Cumulative infiltration capacity $F_p(t)$ is defined as the accumulation of infiltration volume over a ime period since the start of the process and is given by

$$F_p = \int_0^t f_p(t)\, dt \tag{3.22-a}$$

Thus, the curve $F_p(t)$ vs time in Fig. (3.13) is the mass curve of infiltratio. It may be noted that from Eq. (3.22-a), it follows that

$$f_p(t) = \frac{dF_p(t)}{dt} \tag{3.23-b}$$

Many equations have been proposed to express the curves $f_p(t)$ or $F_p(t)$ for use in hydrological analysis. In this section four such equations will be described.

[*Note:* Practically, all the infiltration equations relate infiltration capacity $f_p(t)$ or cumulative infiltration capacity $F_p(t)$ with time and other parameters. As such many authors find it convenient to drop the suffix p while denoting capacity. Thus, f_p is denoted as f and F_p as F.]

3.18.1 Horton's Equation (1933)

Horton expressed the decay of infiltration capacity with time as an exponential decay given by

$$f_p = f_c + (f_0 - f_c)\, e^{-K_h t} \qquad \text{for } 0 \geq t \leq t_c \tag{3.24}$$

where f_p = infiltration capacity at any time t from the start of the rainfall

f_0 = initial infiltration capacity at $t = 0$

f_c = final steady-state infiltration capacity occurring at $t = t_c$. Also, f_c is sometimes known as *constant rate* or *ultimate infiltration capacity.*

K_h = Horton's decay coefficient which depends upon soil characteristics and vegetation cover

The difficulty of determining the variation of the three parameters f_0, f_c and k_h with soil characteristics and antecedent moisture conditions preclude the general use of Eq. (3.23).

3.18.2 Philip's Equation (1957)

Philip's two-term model relates $F_p(t)$ as

$$F_p = st^{1/2} + Kt \tag{3.24}$$

where s = a function of soil suction potential called *sorptivity*
K = Darcy's hydraulic conductivity

By Eq. (3.22-b), infiltration capacity could be expressed as

$$f_p = \frac{1}{2} st^{-1/2} + K \tag{3.25}$$

3.18.3 Kostiakov Equation (1932)

The Kostiakov model expresses cumulative infiltration capacity as

$$F_p = at^b \tag{3.26}$$

where a and b are local parameters with $a > 0$ and $0 < b < 1$.
The infiltration capacity would now be expressed by Eq. (3.22-b) as

$$f_p = (ab)\, t^{(b-1)} \tag{3.27}$$

3.18.4 Green–Ampt Equation (1911)

Green and Ampt proposed a model for infiltration capacity based on Darcy's law as

$$f_p = K\left(1 + \frac{\eta S_c}{F_p}\right) \tag{3.28}$$

where η = porosity of the soil
S_c = capillary suction at the wetting front
K = Darcy's hydraulic conductivity

Equation (3.28) could be considered as

$$f_p = m + \frac{n}{F_p} \tag{3.29}$$

where m and n are Green–Ampt parameters of the infiltration model.

3.18.5 Estimation of Parameters of Infiltration Models

Data from infiltrometer experiments can be processed to generate data sets f_p and F_p values for various time t values. The following procedures are convenient to evaluate the parameters of the infiltration models.

1. Horton's Model

Value of f_c in a test is obtained from inspection of the delta. Equation (3.23) is rearranged to read as

$$(f_p - f_c) = (f_0 - f_c)\, e^{-K_h t} \tag{3.23-a}$$

Taking logarithms $\ln(f_p - f_c) = \ln(f_o - f_c) - K_h t$
Plot $\ln(f_p - f_c)$ against t and fit the best straight line through the plotted points. The intercept gives $\ln(f_0 - f_c)$ and the slope of the straight line is K_h.

2. Philip's Model

Use the expression for f_p as

$$f_p = \frac{1}{2} st^{-1/2} + K \tag{3.25}$$

Plot the observed values of f_p against ($t^{-0.5}$) on an arithmetic graph paper. The best fitting straight line through the plotted points gives K as the intercept and ($s/2$) as the slope of the line. While fitting Philip's model, it is necessary to note that K is positive and to achieve this, it may be necessary to neglect a few data points at the initial stages (viz. at small values of t) of the infiltration experiment. K will be of the order of magnitude of the asymptotic value of f_p.

3. Kostiakov Model

Kostiakov model relates F_p to t as

$$F_p = at^b \tag{3.26}$$

Taking logarithms of both sides of Eq. (3.26),

$$\ln(F_p) = \ln a + b \ln(t)$$

The data is plotted as $\ln(F_p)$ *vs* $\ln(t)$ on an arithmetic graph paper and the best fit straight through the plotted points gives $\ln a$ as intercept and the slope is b. Note that b is a positive quantity such that $0 < b < 1$.

4. Green–Ampt Model

Green–Ampt equation is considered in the form $f_p = m + \frac{n}{F_p}$. Values of f_p are plotted against ($1/F_p$) on a simple arithmetic graph paper and the best fit straight line is drawn through the plotted points. The intercept and the slope of the line are the coefficients m and n respectively. Sometimes values of f_p and corresponding F_p at very low values of t may have to be omitted to get best-fitting straight line with reasonably good correlation coefficient.

[*Note:* 1. Procedure for calculation of the best-fit straight line relating the dependent variable Y and independent variable X by the least-square error method is described in Section 4.9, Chapter 4.

2. Use of spreadsheets (for eg., MS Excel) greatly simplifies these procedures and the best values of parameters can be obtained by fitting regression equations. Further, various plots and the coefficient of correlation, etc. can be calculated with ease.]

Example 3.7 Infiltration capacity data obtained in a flooding-type infiltration test is given below:

Time since start (minutes)	5	10	15	25	45	60	75	90	110	130
Cumulative infiltration depth (cm)	1.75	3.00	3.95	5.50	7.25	8.30	9.30	10.20	11.28	12.36

(a) For this data, plot the curves of (i) infiltration capacity vs time, (ii) infiltration capacity vs cumulative infiltration, and (iii) cumulative infiltration vs time.

(b) Obtain the best values of the parameters in Horton's infiltration capacity equation to represent this data set.

Solution

Incremental infiltration values and corresponding infiltration intensities f_p at various data observation times are calculated as shown in the following table. Also other data required for various plots are calculated as shown in Table 3.9.

Table 3.9 ***Calculations for Example 3.7***

Time in Minutes	Cum. Depth (cm)	Incremental Depth in the interval (cm)	Infiltration rate, f_p(cm/h)	Ln $(f_p - f_c)$	Time in hours
0					
5	1.75	1.75	21.00	2.877	0.083
10	3.00	1.25	15.00	2.465	0.167
15	3.95	0.95	11.40	2.099	0.250
25	5.50	1.55	9.30	1.802	0.417
45.	7.25	1.75	5.25	0.698	0.750
60	8.30	1.05	4.20	–0.041	1.000
75	9.30	1.00	4.00	–0.274	1.250
90	10.20	0.90	3.60	–1.022	1.500
110	11.28	1.08	3.24		1.833
130	12.36	1.08	3.24		2.167

(a) Plots of f_p *vs* time and F_p *vs* time are shown in Fig. 3.14. Best-fitting curve for plotted points are also shown in the Fig. 3.14-a.

Plot of f_p vs F_p is shown in Fig. 3.14-b.

(b) By observation from Table 3.9, $f_c = 3.24$ cm/h

Ln$(f_p - f_c)$ is plotted against time t as shown in Fig. 3.14-c. The best-fit line through the plotted points is drawn and its equation is obtained as

$\ln(f_p - f_c) = 2.8868 - 2.6751\ t$

$-K_h$ = slope of the best fit line = –2.6751, thus $K_h = 2.6751\ \text{h}^{-1}$

$\ln(f - f_c)$ = intercept = 2.8868, thus $f_0 - f_c = 17.94$ and $f_0 = 21.18$ cm/h

Example 3.8 Values of infiltration capacities at various times obtained from an infiltration test are given below. Determine the parameters of (i) Green—Ampt equation, (ii) Philip's equation, and (iii) Kostiakov's equation that best fits this data.

Time since start (minutes)	5	10	15	20	25	30	60	90	120	150
Cumulative infiltration depth (cm)	1.0	1.8	2.5	3.1	3.6	4.0	6.1	8.1	9.9	11.6

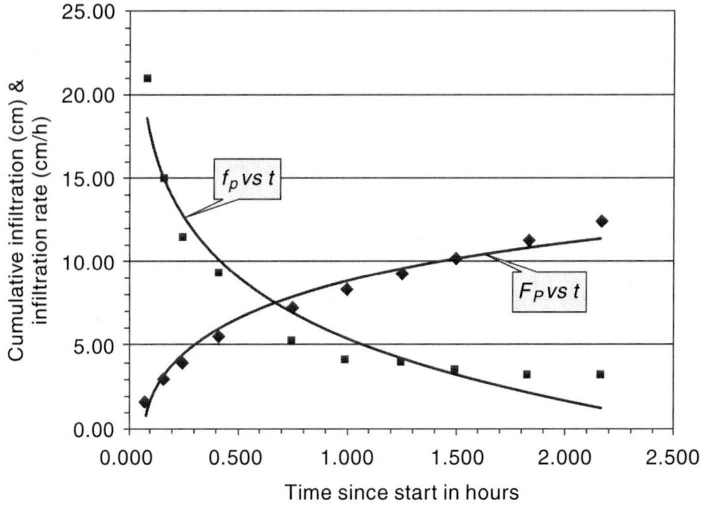

Fig. 3.14 (a) ***Plots of F_p vs Time and f_p vs Time***

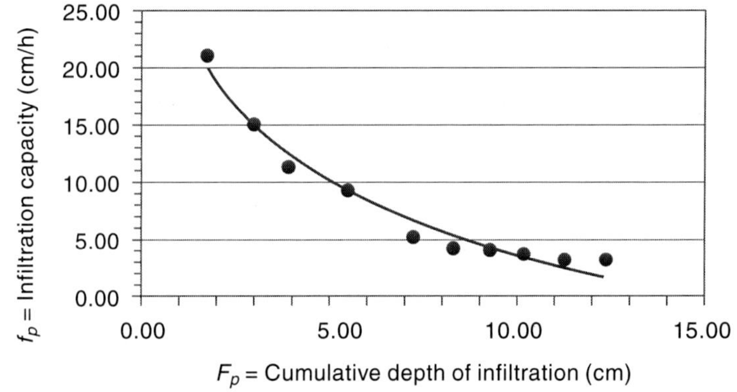

Fig. 3.14 (b) ***Plot of f_p vs F_p***

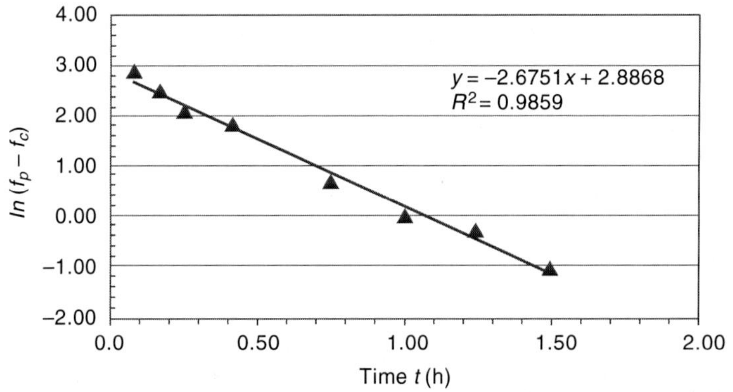

Fig. 3.14 (c) ***Horton's Equation. Plot of $ln(f_p - f_c)$ vs Time***

Solution

Incremental infiltration depth values and corresponding infiltration intensities f_p at various data observation times are calculated as shown in Table 3.10. Also, various parameters needed for plotting different infiltration models are calculated as shown in Table 3.10. The units used are f_p in cm/h, F_p in cm and t in hours.

Table 3.10 ***Calculations Relating to Example 3.8***

1 Time (min)	2 F_p (cm)	3 Incremental depth of infiltration (cm)	4 f_p (cm/h)	5 t in hours	6 $t^{-0.5}$	7 $1/F_p$	8 Ln F_p	9 Ln t
5	1.0	1.0	12.0	0.083	3.464	1.000	0.000	–2.485
10	1.8	0.8	9.6	0.167	2.449	0.556	0.588	–1.792
15	2.5	0.7	8.4	0.250	2.000	0.400	0.916	–1.386
20	3.1	0.6	7.2	0.333	1.732	0.323	1.131	–1.099
25	3.6	0.5	6.0	0.417	1.549	0.278	1.281	–0.875
30	4.0	0.4	4.8	0.500	1.414	0.250	1.386	–0.693
60	6.1	2.1	4.2	1.000	1.000	0.164	1.808	0.000
90	8.1	2.0	4.0	1.500	0.816	0.123	2.092	0.405
120	9.9	1.8	3.6	2.000	0.707	0.101	2.293	0.693
150	11.6	1.7	3.4	2.500	0.632	0.086	2.451	0.916

Green–Ampt Equation: $f_p = m + \frac{n}{F_p}$ (3.29)

Values of f_p (col. 4) are plotted against $1/F_p$ (col. 7) on an arithmetic graph paper (Fig. 3.15-a). The best-fit straight line through the plotted points is obtained as

$$f_p = 10.042\left(\frac{1}{F_p}\right) + 3.0256.$$

The coefficients of the Green–Ampt equations are $m = 3.0256$ and $n = 10.042$

Philip's Equation: The expression $f_p = \frac{1}{2}st^{-1/2} + K$ as in Eq. 3.25 is used. Values of f_p (Col. 4) are plotted against $t^{-0.5}$ (col. 6) on an arithmetic graph paper (Fig. 3.15-b). The best-fit straight line through the plotted points is obtained as

$$f_p = 3.2287\,t^{-0.5} + 1.23$$

The coefficients of Philip's equation are $s = 2 \times 3.2287 = 6.4574$ and $K = 1.23$

Kostiakov's Equation:

$$F_p = at^b \quad \text{Eq. (3.26),}$$

Taking logarithms of both sides of the equation (3.26), $\ln(F_p) = \ln a + b \ln(t)$.
The data set is plotted as ln (F_p) *vs* ln(t) on an arithmetic graph paper (Fig. 3.15-c) and the best-fit straight line through the plotted points is obtained as

$$\ln(F_p) = 1.8346 + 0.6966\ \text{in}(t).$$

The coefficients of Kostiakov equation are $b = 0.6966$ and $\ln a = 1.8346$ and hence $a = 6.2626$ Best fitting Kostiakov equation for the data is $F_p = 6.2626\,t^{0.6966}$

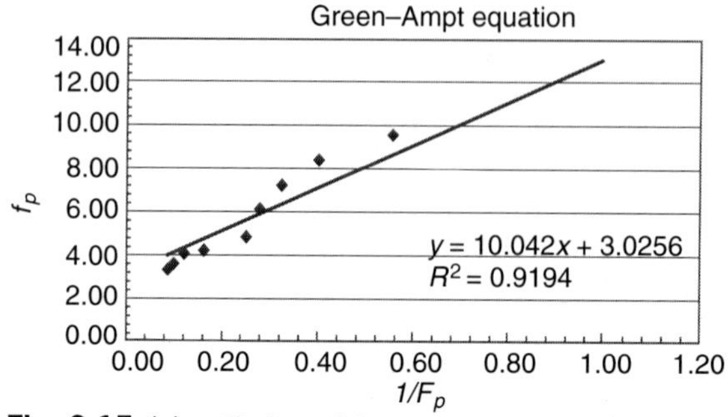

Fig. 3.15 (a) ***Fitting of Green–Ampt Equation***

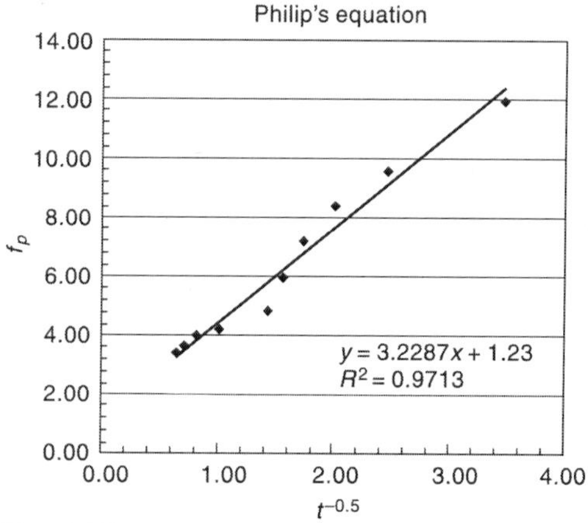

Fig. 3.15 (b) ***Fitting of Philip's Equation***

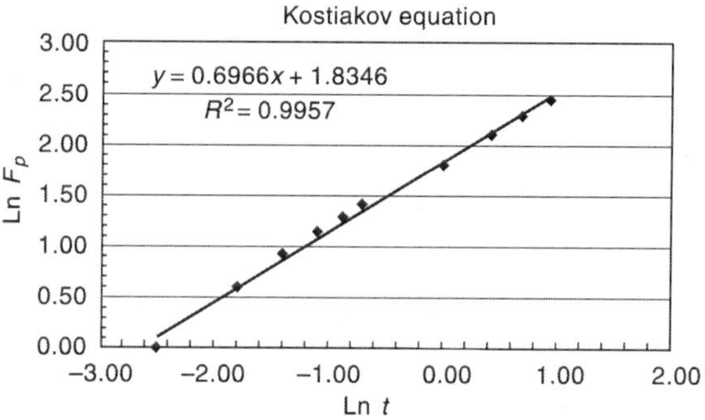

Fig. 3.15 (c) ***Fitting of Kostiakov Equation***

Example 3.9

The infiltration capacity in a basin is represented by Horton's equation as

$$f_p = 3.0 + e^{-2t}$$

where f_p is in cm/h and t is in hours. Assuming the infiltration to take place at capacity rates in a storm of 60 minutes duration, estimate the depth of infiltration in (i) the first 30 minutes and (ii) the second 30 minutes of the storm.

Solution

$$F_p = \int_0^t f_p\, dt \quad \text{and} \quad f_p = 3.0 + e^{-2t}$$

(i) In the first 0.5 hour,

$$F_{p1} = \int_{0}^{0.5} (3.0 + e^{-2t})\, dt = \left[3.0\,t - \frac{1}{2}e^{-2t}\right]_{0}^{0.5}$$

$$= [(3.0 \times 0.5) - (1/2)(e^{-(2\times 0.5)}] - [-(1/2)] = (1.5 - 0.184) + 0.5$$

$$= 1.816 \text{ cm}$$

(ii) In the second 0.5 hour,

$$F_{p2} = \int_{0.50}^{1.0} (3.0 + e^{-2t})\, dt = \left[3.0\,t - \frac{1}{2}e^{-2t}\right]_{0.5}^{1.0}$$

$$= [(3.0 \times 1.0) - (1/2)(e^{-2})] - [(3.0 \times 0.5) - (1/2)(e^{-(2\times 0.5)})]$$

$$= (3.0 - 0.0677) - (1.5 - 0.184) = 1.616 \text{ cm}$$

Example 3.10 The infiltration capacity of soil in a small watershed was found to be 6 cm/h before a rainfall event. It was found to be 1.2 cm/h at the end of 8 hours of storm. If the total infiltration during the 8 hours period of storm was 15 cm, estimate the value of the decay coefficient K_h in Horton's infiltration capacity equation.

Soution

Horton's equation is $f_p = f_c + (f_0 - f_c)\, e^{-K_h t}$

and

$$F_p = \int_0^t f_p(t)\, dt = f_c\, t + (f_0 - f_c)\int_0^t e^{-K_h t}\, dt$$

As $t \to \infty$, $\int_0^\infty e^{-K_h t}\, dt \to \frac{1}{K_h}$. Hence for large t values

$$F_p = f_c t + \frac{(f_0 - f_c)}{K_h}$$

Here, $F_p = 15.0$ cm, $f_0 = 6.0$ cm, $f_c = 1.2$ cm and $t = 8$ hours.

$$15.0 = (1.2 \times 8) + (6.0 - 1.2)/K_h$$

$$K_h = 4.8/5.4 = 0.888 \text{ h}^{-1}$$

3.19 *Classification of Infiltration Capacities*

For purposes of runoff-volume classification in small watersheds, one of the widely used methods is the *SCS–CN* method described in detail in Chapter 5. In this method, the soils are considered divided into four groups known as *hydrologic soil groups*. The steady-state infiltration capacity, being one of the main parameters in this soil classification, is divided into four infiltration classes as mentioned below.

Table 3.11 ***Classification of Infiltration Capacities***

Infiltration Class	Infiltration Capacity (mm/h)	Remarks
Very Low	< 2.5	Highly clayey soils
Low	2.5 to 25.0	Shallow soils, Clay soils, Soils low in organic matter
Medium	12.5 to 25.0	Sandy loam, Silt
High	>25.0	Deep sands, well drained aggregated soils

3.20 *Infiltration Indices*

In hydrological calculations involving floods, it is found convenient to use a constant value of infiltration rate for the duration of the storm. The defined average infiltration rate is called *infiltration index* and two types of indices are in common use.

3.20.1 φ–Index

The φ-index is the average rainfall above which the rainfall volume is equal to the runoff volume. The φ-index is derived from the rainfall hyetograph with the knowledge of the resulting runoff volume. The initial loss is also considered as infiltration. The φ value is found by treating it as a constant infiltration capacity. If the rainfall intensity is less than φ then the infiltration rate is equal to the rainfall intensity; however, if the rainfall intensity is larger than φ, the difference between the rainfall and infiltration in an interval of time represents the runoff volume as shown in Fig. 3.16. The amount of rainfall in excess of the index is called *rainfall excess*. In connection with runoff and flood studies, it is also known as *effective rainfall*, (details in Sec. 6.5, Chapter 6). The φ–index thus accounts for the total abstraction and enables magnitudes to be estimated for a given rainfall hyetograph. A detailed procedure for calculating φ–index for a given storm hyetograph and resulting runoff volume is as follows.

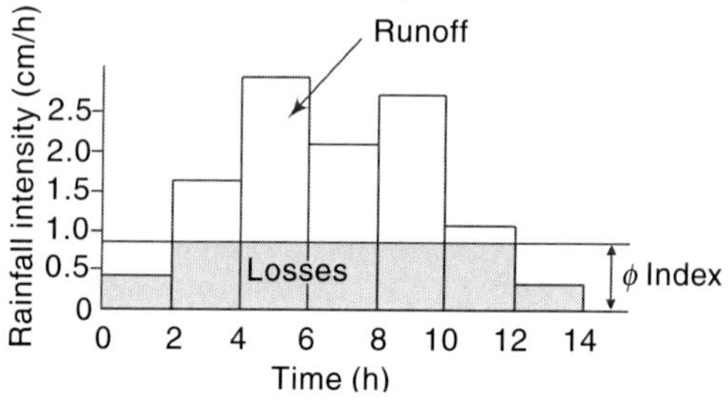

Fig. 3.16 *φ-Index*

Procedure for Calculation of φ-index

Consider a rainfall hyetograph of event duration D hours and having N pulses of time interval Δt such that

$$N \cdot \Delta t = D. \qquad \text{(In Fig. 3.16, } N = 7)$$

Let I_i be the intensity of rainfall in i^{th} pulse and R_d = total direct runoff.

$$\text{Total rainfall } P = \sum_1^N I_i \cdot \Delta t$$

If φ is the value of φ-index then $P - \varphi \cdot t_e = R_d$
where t_e = duration of rainfall excess.

If the rainfall hyetograph and total runoff depth R_d are given, the φ-index of the storm can be determined by trial-and-error procedure as given below.

1. Assume that out of given N pulses, M number of pulses have rainfall excess. (Note that $M \le N$). Select M number of pulses in decreasing order of rainfall intesity I_i.
2. Find the value of φ that satisfies the relation

$$R_d = \sum_1^M (I_i - \varphi)\Delta t$$

3. Using the value of φ of Step 2, find the number of pulses (M_c) which give rainfall excess. (Thus M_c = number of pulses with rainfall intensity $I_i \geq \varphi$).
4. If $M_c = M$ then φ of Step 2 is the correct value of φ-index. If not, repeat the procedure Step 1 onwards with new value of M. Result of Step 3 can be used as guidance to the next trial.

Example 3.11 illustrates this procedure in detail.

Example 3.11 A storm with 10 cm of precipitation produced a direct runoff of 5.8 cm. The duration of the rainfall was 16 hours and its time distribution is given below.

Estimate the φ-index of the storm.

Time from start (h)	0	2	4	6	8	10	12	14	16
Cumulative rainfall (cm)	0	0.4	1.3	2.8	5.1	6.9	8.5	9.5	10.0

Solution

Pulses of uniform time duration $\Delta t = 2\ h$ are considered. The pulses are numbered sequentially and intensity of rainfall in each pulse is calculated as shown below.

Table 3.12 ***Calculations for Example 3.11***

Pulse number	1	2	3	4	5	6	7	8
Time from start of rain (h)	2	4	6	8	10	12	14	16
Cumulative rainfall (cm)	0.4	1.3	2.8	5.1	6.9	8.5	9.5	10.0
Incremental rain (cm)	0.40	0.90	1.50	2.30	1.80	1.60	1.00	0.50
Intensity of rain (I_i) in cm/h.	0.20	0.45	0.75	1.15	0.90	0.80	0.50	0.25

Here, duration of rainfall $D = 16$ h, $\Delta t = 2$ h and $N = 8$.

Trial 1:

Assume $M = 8$, $\Delta t = 2$ h and hence $t_e = M \cdot \Delta t = 16$ hours.

Since $M = N$, all the pulses are included.

$$\text{Runoff } R_d = 5.8 \text{ cm} = \sum_1^8 (I_i - \varphi)\Delta t = \sum_1^8 I_i \times \Delta t - \varphi\,(8 \times 2)$$

$$5.8 = \{(0.20 \times 2) + (0.45 \times 2) + (0.75 \times 2) + (1.15 \times 2) + (0.90 \times 2) + (0.80 \times 2) + (0.50 \times 2) + (0.25 \times 2)\} - 16\,\varphi = 10.0 - 14\,\varphi$$

$$\varphi = 4.2/14 = 0.263 \text{ cm/h}$$

By inspection of row 5 of Table 3.12, M_c = number of pulses having $I_i \geq \varphi$, that is $I_i \geq 0.263$ cm/h is 6.

Thus, $M_c = 6 \neq M$. Hence assumed M is **not** correct. Try a new value of $M < 8$ in the next trial.

Trial 2:

Assume $M = 7$, $\Delta t = 2$ h and hence $t_e = M \cdot \Delta t = 14$ hours.

Select these 7 pules in decreasig order of I_i. Pulse 1 is omitted.

$$\text{Runoff } R_d = 5.8 \text{ cm} = \sum_1^7 (I_i - \varphi)\,\Delta t = \sum_1^7 (I_i \cdot \Delta t - \varphi\,(7 \times 2)$$

$$5.8 = \{(0.45 \times 2) + (0.75 \times 2) + (1.15 \times 2) + (0.90 \times 2) + (0.80 \times 2) + (0.50 \times 2) + (0.25 \times 2)\} - 14\,\varphi = 9.6 - 14\,\varphi$$

$$\varphi = 3.8/14 = 0.271 \text{ cm/h}$$

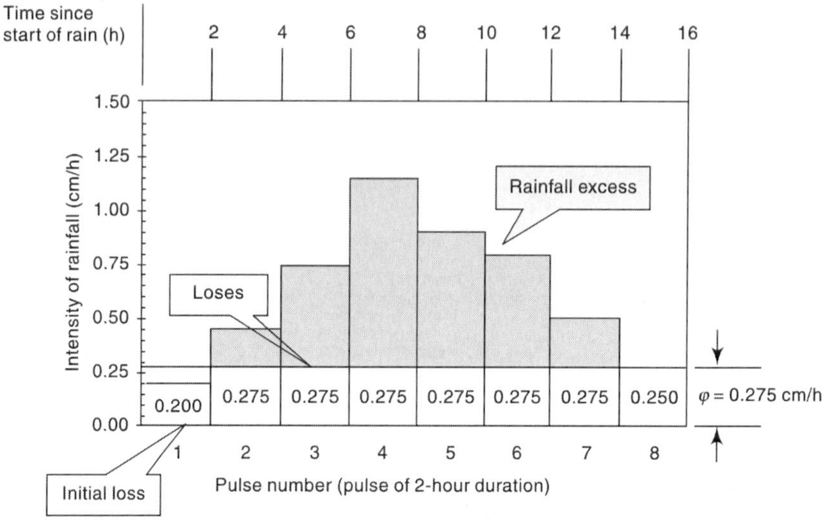

Fig. 3.17 ***Hyetograph and Rainfall Excess of the Storm–Example 3.11***

By inspection of row 5 of Table 3.12, M_c = number of pulses having $I_i \geq \varphi$, that is $I_i \geq 0.271$ cm/h is 6.

Thus, $M_c = 6 \neq M$. Hence, assumed M is **not** O.K. Try a new value of $M < 7$ in the next trial.

Trial 3:

Assume $M = 6 \cdot \Delta t = 2$ h and hence $t_e = M \cdot \Delta t = 12$ hours.

Select these 6 pulses in decreasing order of I_i. Pulse 1 and 8 are omitted.

Runoff $R_d = 5.8$ cm $= a$

$$5.8 = \{(0.45 \times 2) + (0.75 \times 2) + (1.15 \times 2) + (0.90 \times 2) + (0.80 \times 2) + (0.50 \times 2)\} - 12\,\varphi = 9.1 - 12\,\varphi$$

$$\varphi = 3.3/12 = 0.275 \text{ cm/h}$$

By inspection of row 5 of Table 3.12, M_c = number of pulses having $I_i \geq \varphi$, that is $I_i \geq 0.275$ cm/h is 6.

Thus, $M_c = 6 = M$. Hence, assumed M is *OK*.

The φ-index of the storm is 0.275 cm/h and the duration of rainfall excess = t_e = 12 hours. The hyetograph of the storm, losses, the rainfall excess and the duration of rainfall excess are shown in Fig. 3.17.

3.20.2 W-Index

In an attempt to refine the φ-index, the initial losses are separated from the total abstractions and an average value of infiltration rate, called W-index, is defined as

$$W = \frac{P - R - I_a}{t_e} \tag{3.29}$$

where P = total storm precipitation (cm)

R = total storm runoff (cm)

I_a = initial losses (cm)

t_e = duration of the rainfall excess, i.e. the total time in which the rainfall intensity is greater than W (in hours) and

W = defined average rate of infiltration (cm).

Since I_a rates are difficult to obtain, the accurate estimation of W-index is rather difficult.

The minimum value of the W-index obtained under very wet soil conditions, representing the constant minimum rate of infiltration of the catchment, is known as W_{min}. It is to be noted that both the φ-index and W-index vary from storm to storm.

Computation of W-index

To compute W-index from a given storm hyetograph with known values of I_a and runoff R, the following procedure is followed:

(i) Deduct the initial loss I_a from the storm hyetograph pulses starting from the first pulse.

(ii) Use the resulting hyetograph pulse diagram and follow the procedure indicated in Sec. 3.20 for ϕ-index.

Thus, the procedure is exactly same as in the determination of φ-index except for the fact that the storm hyetograph is appropriately modified by deducting I_a.

Example 3.12 The mass curve of rainfall of duration 180 minutes on a catchment is given below. The catchment had an initial loss of 0.5 cm. The ϕ - index of the catchment is known to be 0.4 cm/hour. Calculate the total surface runoff from the catchment due to this storm.

Time from start (min)	0	30	60	80	100	120	150	180
Cumulative rainfall (cm)	0	0.6	1.3	2.6	2.8	3.0	3.2	3.3

Solution

Runoff in each time step is equal to the incremental rainfall in that time step minus the infiltration loss calculated as = (ϕ. t). This number is always equal to or greater than zero, as the runoff is always a positive quantity. Note that the initial loss has already been accounted for in the definition (and hence in the calculation) of ϕ index. As such, the initial loss is redundant information in this problem. Calculations are shown in Table 3.13.

Table 3.13 ***Calculation of Runoff—Example 3.12.***

ϕ-index = 0.4 cm/hour. Time steps are as given in the data.

Time from start (Min)	Increment of time (Δt) hours	Cumulative rainfall (cm)	Incremental rainfall (cm)	Infiltration loss = ϕ.Δt (cm)	Runoff (cm)
0	0.000	0.00	0.000	0.000	0.000
30	0.500	0.60	0.600	0.200	0.400
60	0.500	1.30	0.700	0.200	0.500
80	0.333	2.00	0.700	0.133	0.567
100	0.333	2.80	0.800	0.133	0.667
120	0.333	3.00	0.200	0.133	0.067
150	0.500	3.20	0.200	0.200	0.000
180	0.500	3.30	0.100	0.200	0.000
				Total =	2.200

The total surface runoff due to the storm from the catchment is **2.20 cm**.

Example 3.13

Details related to an isolated 6-hour storm that occurred over a catchment are given. Estimate the runoff from the catchment due to the storm.

Sub-area	Areal Extent of sub area as % of catchment	φ-index	Rainfall (cm)		
			First 2-hours	Second 2-hours	Third 2-hours
A	0.25	0.3	0.80	1.60	1.30
B	0.45	0.5	0.90	1.40	1.10
C	0.30	0.4	0.86	1.30	0.90

Solution

The runoff from each sub-area is calculated separately and converted to depth of runoff over the whole catchment. These depths are in turn added to get the total depth of runoff over the catchment. The calculations are performed in a tabular form as indicated in Table 3.14.

Table 3.14 ***Calculation of Runoff – Example 3.13***

Sub-area	Areal Extant of Sub- area (% of catchment)	φ-Index of sub-area	Rainfall intensity (cm/h)	Du-ration (h)	Intensity of rainfall Excess (cm/h)	Depth of Rainfall excess (cm)	Runoff depth in sub-area (cm)	Runoff depth over total catchment area (cm)
A	0.25	0.30	0.40	2.00	0.10	0.20		
	0.25	0.30	0.80	2.00	0.50	1.00	1.90	0.475
	0.25	0.30	0.65	2.00	0.35	0.70		
B	0.45	0.50	0.45	2.00	0.00	0.00		
	0.45	0.50	0.70	2.00	0.20	0.40	0.50	0.225
	0.45	0.50	0.55	2.00	0.05	0.10		
C	0.30	0.40	0.43	2.00	0.03	0.06		
	0.30	0.40	0.65	2.00	0.25	0.50	0.66	0.198
	0.30	0.40	0.45	2.00	0.05	0.10		
							Total =	0.898

The total depth of runoff over the catchment due to the storm = **0.898 cm**

3.20.3 φ-index for Practical use

The φ-index for a catchment, during a storm, depends in general upon the soil type, vegetal cover, initial moisture condition, storm duration and intensity. To obtain complete information on the interrelationship between these factors, a detailed extensive study of the catchment is necessary. As such, for practical use in the estimation of flood magnitudes due to critical storms, a simplified relationship for

φ-index is adopted. As the maximum flood peaks are invariably produced due to long storms and usually in the wet season, the initial losses are assumed to be negligibly small. Further, only the soil type and rainfall are found to be critical in the estimate of the φ-index for maximum flood producing storms.

On the basis of rainfall and runoff correlations, *CWC*[1] has found the following relationships for the estimation of φ-index for flood producing storms and soil conditions prevalent in India

$$R = \alpha I^{1.2} \tag{3.31}$$

$$\varphi = \frac{I - R}{24} \tag{3.32}$$

where R = runoff in cm from a 24-h rainfall of intensity I cm/h and α = a coefficient which depends upon the soil type as indicated in Table 3.15. In estimating the maximum floods for design purposes, in the absence of any other data, a φ-index value of 0.10 cm/h can be assumed.

Table 3.15 ***Variation of Coefficient α in Eq. 3.30***

Sl. No.	Type of Soil	Coefficient α
1.	Sandy soils and sandy loam	0.17 to 0.25
2.	Coastal alluvium and silty loam	0.25 to 0.34
3.	Red soils, clayey loam, grey and brown alluvium	0.42
4.	Black–cotton and clayey soils	0.42 to 0.46
5.	Hilly soils	0.46 to 0.50

REFERENCES

1. Central Water Commission, India, *Estimation of Design Flood Peak*, Flood Estimation Directorate, Report No. 1/73, New Delhi, 1973.
2. Chow, V.T. (Ed.), *Handbook of Applied Hydrology,* McGraw-Hill, New York, NY., 1964.
3. Allen, R.G. et al., "Crop Evapotranspiration—Guidelines for Computing Crop–Water Requirements", *Irrigation and Drainage Paper 56, UN FAO*, Rome, Italy, 1988.
4. Gray, D.M., *Principles of Hydrology*, Water Inf. Center, Huntington, NY, 1970.
5. Rao, K.N. et al., "Potential Evapotranspiration (PE) over India", *Symp. on Water Resources,* I.I.Sc. Bangalore, India, 1971, 99 A 2-1-14. (Also *Sc. Rep. No. 136, IMD*, Feb. 1971).
6. Soil Conservation Div., *Handbook of Hydrology*, Min. of Agriculture, *GOI*, 1972.
7. Weisner, C.J., *Hydrometeorology*, Chapman and Hall, U.K., 1970.
8. Min. of Water Resources, GOI, Report *of The National Commission for Integrated Water Resources Development*, Vol.-1, New Delhi, Sept. 1999.
9. *CWC*, Evaporation Control in reservoirs, Basin Planning and Management Organization, *CWC*, Govt. of India, 2006. http://cwc.gov.in/main/downloads/Evaporation%20Control%20in%20reservoirs.pdf

REVISION QUESTIONS

3.1 Discuss briefly the various abstractions from precipitation.

3.2 Explain briefly the evaporation process.

3.3 Discuss the factors that affect the evaporation from a water body.

3.4 Describe a commonly used evaporimeter.

3.5 Explain the energy budget method of estimating evaporation from a lake.

3.6 Discuss the importance of evaporation control of reservoirs and possible methods of achieving the same.

3.7 Describe the factors affecting evapotranspiration process.

3.8 List the various data needed to use Penman's equation for estimating the potential evapotranspiration from a given area.

3.9 Describe briefly (a) reference crop evapotranspiration, and (b) actual evapotrans-piration.

3.10 Explain briefly the infiltration process and the resulting soil moisture zones in the soil.

3.11 Discuss the factors affecting the infiltration capacity of an area.

3.12 Describe the commonly used procedures for determining the infiltration characteristics of a plot of land. Explain clearly the relative advantages and disadvantages of the enumerated methods.

3.13 Describe various models adopted to represent the variation of infiltration capacity with time.

3.14 Explain a procedure for fitting Horton's infiltration equation for experimental data from a given plot.

3.15 Distinguish between
(a)Infiltration capacity and infiltration rate (b) Actual and potential evapotranspiration
(c)Field capacity and permanent wilting point (d) Depression storage and interception

PROBLEMS

3.1 Calculate the evaporation rate from an open water source, if the net radiation is 300 W/m^2 and the air temperature is 30° C. Assume value of zero for sensible heat, ground heat flux, heat stored in water body and advected energy. The density of water at 30° C = 996 kg/m^3.
[***Hint:*** Calculate latent heat of vapourisation *L* by the formula:
$L = 2501 - 2.37\,T$ (kJ/kg), where T = temperature in °C.]

3.2 A class A pan was set up adjacent to a lake. The depth of water in the pan at the beginning of a certain week was 195 mm. In that week there was a rainfall of 45 mm and 15 mm of water was removed from the pan to keep the water level within the specified depth range. If the depth of the water in the pan at the end of the week was 190 mm calculate the pan evaporation. Using a suitable pan coefficient estimate the lake evaporation in that week.

3.3 A reservoir has an average area of 50 km^2 over a year. The normal annual rainfall at the place is 120 cm and the class A pan evaporation is 240 cm. Assuming the land flooded by the reservoir has a runoff coefficient of 0.4, estimate the net annual increase or decrease in the streamflow as a result of the reservoir.

3.4 At a reservoir in the neighbourhood of Delhi, the following climatic data were observed. Estimate the mean monthly and annual evaporation from the reservoir using Meyer's formula.

Month	Temp. (°C)	Relative humidity (%)	Wind velocity at 2 m above GL (km/h)
Jan	12.5	85	4.0
Feb	15.8	82	5.0
Mar	20.7	71	5.0
Apr	27.0	48	5.0
May	31.0	41	7.8
Jun	33.5	52	10.0
Jul	30.6	78	8.0
Aug	29.0	86	5.5
Sep	28.2	82	5.0
Oct	28.8	75	4.0
Nov	18.9	77	3.6
Dec	13.7	73	4.0

3.5 For the lake in Prob. 3.4, estimate the evaporation in the month of June by (a) Penman formula, and (b) Thornthwaite equation by assuming that the lake evaporation is the same as *PET*, given latitude = 28° N and elevation = 230 m above MSL. Mean observed sunshine = 9 h/day.

3.6 A reservoir had an average surface area of 20 km^2 during June 1982. In that month the mean rate of inflow = 10 m^3/s, outflow = 15 m^3/s, monthly rainfall = 10 cm and change in storage = 16 million m^3. Assuming the seepage losses to be 1.8 cm, estimate the evaporation in that month.

3.7 For an area in South India (latitude = 12° N), the mean monthly temperatures are given.

Month	June	July	Aug	Sep	Oct
Temp (°C)	31.5	31.0	30.0	29.0	28.0

Calculate the seasonal consumptive use of water for the rice crop in the season June 16 to October 15, by using the Blaney–Criddle formula.

3.8 A catchment area near Mysore is at latitude 12° 18′ N and at an elevation of 770 m. The mean monthly temperatures are given below.

Month	Jan	Feb	Mar	Apr	May	Jun	Jul	Aug	Sep	Oct	Nov	Dec
Mean monthly temp. (°C)	22.5	24.5	27.0	28.0	27.0	25.0	23.5	24.0	24.0	24.5	23.0	22.5

Calculate the monthly and annual *PET* for this catchment using the Thornthwaite formula.

3.9 A wheat field has maximum available moisture of 12 cm. If the reference evapotranspiration is 6.0 mm/day, estimate the actual evapotranspiration on Day 2, Day 7 and Day 9 after irrigation. Assume soil-water depletion factor $p = 0.20$ and crop factor $K = 0.65$.

3.10 Results of an infiltrometer test on a soil are given below. Determine the Horton's infiltration capacity equation for this soil.

Time since start in (h)	0.25	0.50	0.75	1.00	1.25	1.50	1.75	2.0
Infiltration capacity in cm/h	5.6	3.20	2.10	1.50	1.20	1.10	1.0	1.0

3.11 Results of an infiltrometer test on a soil are given below. Determine the best values of the parameters of Horton's infiltration capacity equation for this soil.

Time since start in minutes	5	10	15	20	30	40	60	80	100
Cumulative infiltration in mm	21.5	37.7	52.2	65.8	78.4	89.5	101.8	112.6	123.3

3.12 Results of an infiltrometer test on a soil are as follows:

Time since start in minutes	5	10	15	20	30	40	60	120	150
Cumulative infiltration in mm	1.00	1.80	2.50	3.10	4.20	5.10	6.60	11.00	12.90

Determine the parameters of (i) Kostiakov's equation, (ii) Green–Ampt equation, and (iii) Philips equation

3.13 Determine the best values of the parameters of Horton's infiltration capacity equation for the following data pertaining to infiltration tests on a soil using double ring infiltrometer.

Time since start in minutes	5	10	15	25	40	60	75	90	110	130
Cumulative infiltration in mm	21.0	36.0	47.6	56.9	63.8	69.8	74.8	79.3	87.0	92.0

3.14 For the infiltration data set given below, establish (a) Kostiakov's equation, (b) Philips equation, and (c) Green–Ampt equation.

Time since start in minutes	10	20	30	50	80	120	160	200	280	360
Cumulative Infiltration in mm	9.8	18.0	25.0	38.0	55.0	76.0	94.0	110.0	137.0	163.0

3.15 The following table gives the values of a field study of infiltration using a flooding-type infiltrometer. (a) For this data, plot the curves of (i) infiltration capacity f_p (mm/h) *vs* time (h) on a log–log paper and obtain the equation of the best-fit line, and (ii) cumulative infiltration (mm) F_p *vs* time (h) on a semi-log paper and obtain the equation of the best fit line. (b) Establish Horton's infiltration capacity equation for this soil.

Time since start in minutes	2	10	30	60	90	120	240	360
Cumulative Infiltration in cm	7.0	20.0	33.5	37.8	39.5	41.0	43.0	45.0

3.16 The infiltration capacity of a catchment is represented by Horton's equation as

$$f_p = 0.5 + 1.2e^{-0.5t}$$

where f_p is in cm/h and t is in hours. Assuming the infiltration to take place at capacity rates in a storm of 4 hours duration, estimate the average rate of infiltration for the duration of the storm.

3.17 The infiltration process at capacity rates in a soil is described by Kostiakov's equation as $F_p = 3.0\,t^{0.7}$ where F_p is cumulative infiltration in cm and t is time in hours. Estimate the infiltration capacity at (i) 2.0 h, and (ii) 3.0 h from the start of infiltration.

3.18 The mass curve of an isolated storm in a 500 ha watershed is as follows:

Time from start (h)	0	2	4	6	8	10	12	14	16	18
Cumulative rainfall (cm)	0	0.8	2.6	2.8	4.1	7.3	10.8	11.8	12.4	12.6

If the direct runoff produced by the storm is measured at the outlet of the watershed as 0.340 Mm^3, estimate the φ-index of the storm and duration of rainfall excess.

3.19 The mass curve of an isolated storm over a watershed is given below.

Time from start (h)	0	0.5	1.0	1.5	2.0	2.5	3.0	3.5	4.0	4.5	5.0
Cummulative rainfall (cm)	0	0.25	0.50	1.10	1.60	2.60	3.50	5.70	6.50	7.30	7.70

If the storm produced a direct runoff of 3.5 cm at the outlet of the watershed, estimate the φ-index of the storm and duration of rainfall excess.

3.20 In a 140-min storm the following rates of rainfall were observed in successive 20-min intervals: 6.0, 6.0, 18.0, 13.0, 2.0, 2.0 and 12.0 mm/h. Assuming the φ-index value as 3.0 mm/h and an initial loss of 0.8 mm, determine the total rainfall, net runoff and *W*-index for the storm.

3.21 The mass curve of rainfall of 100 min duration is given below. If the catchment had an initial loss of 0.6 cm and a φ-index of 0.6 cm/h, calculate the total surface runoff from the catchment.

Time from start of rainfall (min)	0	20	40	60	80	100
Cummulative rainfall (cm)	0	0.5	1.2	2.6	3.3	3.5

3.22 An isolated 3-h storm occurred over a basin in the following fashion:

% of catchment area	φ–index (cm/h)	Rainfall (cm)		
		1st hour	2nd hour	3rd hour
20	1.00	0.8	2.3	1.5
30	0.75	0.7	2.1	1.0
50	0.50	1.0	2.5	0.8

Estimate the runoff from the catchment due to the storm.

OBJECTIVE QUESTIONS

3.1 If e_w and e_a are the saturated vapour pressures of the water surface and air respectively, the Dalton's law for evaporation E_L in unit time is given by $E_L =$
(a) $(e_w - e_a)$ (b) $K\, e_w\, e_a$ (c) $K\,(e_w - e_a)$ (d) $K\,(e_w + e_a)$

3.2 The average pan coefficient for the standard *US* Weather Bureau class A pan is
(a) 0.85 (b) 0.70 (c) 0.90 (d) 0.20

3.3 A canal is 80 km long and has an average surface width of 15 m. If the evaporation measured in a class A pan is 0.5 cm/day, the volume of water evaporated in a month of 30 days is (in m^3)
(a) 12600 (b) 18000 (c) 180000 (d) 126000

3.4 The *ISI* standard pan evaporimeter is the
(a) same as the *US* class A pan

(b) has an average pan coefficient value of 0.60
(c) has less evaporation than a *US* class A pan
(d) has more evaporation than a *US* class A pan

3.5 The chemical that is found to be most suitable as water evaporation inhibitor is
(a) ethyl alcohol (b) methyl alcohol
(c) cetyl alcohol (d) butyl alcohol

3.6 Wind speed is measured with
(a) a wind vane (b) a heliometer
(c) Stevenson box (d) anemometer

3.7 If the wind velocity at a height of 2 m above ground is 5.0 kmph, its value at a height of 9 m above ground can be expected to be in km/h about
(a) 9.0 (b) 6.2 (c) 2.3 (d) 10.6

3.8 Evapotranspiration is confined
(a) to daylight hours (b) night-time only
(c) land surfaces only (d) none of these

3.9 Lysimeter is used to measure
(a) infiltration (b) evaporation (c) evapotranspiration (d) vapour pressure.

3.10 The highest value of annual evapotranspiration in India is at Rajkot, Gujarat. Here the annual *PET* is about
(a) 150 cm (b) 150 mm (c) 210 cm (d) 310 cm

3.11 Interception losses
(a) include evaporation, through flow and stemflow
(b) consists of only evaporation loss
(c) includes evaporation and transpiration losses
(d) consists of only stemflow

3.12 The infiltration capacity of a soil was measured under fairly identical general conditions by a flooding type infiltrometer as f_f and by a rainfall simulator as f_r. One can expect
(a) $f_f = f_r$ (b) $f_f > f_r$ (c) $f_f < f_r$ (d) no fixed pattern

3.13 A watershed 600 ha in area experienced a rainfall of uniform intensity 2.0 cm/h for duration of 8 hours. If the resulting surface runoff is measured as 0.6 Mm^3, the average infiltration capacity during the storm is
(a) 1.5 cm/h (b) 0.75 cm/h (c) 1.0 cm/h (d) 2.0 cm/h

3.14 In a small catchment, the infiltration rate was observed to be 10 cm/h at the beginning of the rain and it decreased exponentially to an equilibrium value of 1.0 cm/h at the end of 9 hours of rain. If a total of 18 cm of water infiltered during 9 hours interval, the value of the decay constant K_h in Horton's infiltration equation in (h^{-1}) units is
(a) 0.1 (b) 0.5 (c) 1.0 (d) 2.0

3.15 In Horton's infiltration equation fitted to data from a soil, the initial infiltration capacity is 10 mm/h, final infiltration capacity is 5 mm/h and the exponential decay constant is 0.5 h^{-1}. Assuming the infiltration takes place at capacity rates, the total infiltration depth for a uniform storm of duration 8 hours is
(a) 40 mm (b) 60 mm (c) 80 mm (d) 50 mm

3.16 The rainfall on five successive days on a catchment was 2, 6, 9, 5 and 3 cm. If the φ-index for the storm can be assumed to be 3 cm/day, the total direct runoff from the catchment is
(a) 20 cm (b) 11 cm (c) 10 cm (d) 22 cm

3.17 A 6-hour storm had 6 cm of rainfall and the resulting runoff was 3.0 cm. If the φ-index remains at the same value the runoff due to 12 cm of rainfall in 9 h in the catchment is
(a) 9.0 cm (b) 4.5 cm (c) 6.0 cm (d) 7.5 cm

3.18 The fainfall in five successive days on a catchment was 3, 5, 9, 6, 1 cm respectively.

The ϕ-index for the storm can be assumed to be 3 cm/day. The total direct runoff from the catchment due to this storm was

(a) 8 cm (b) 11 cm (c) 20 cm (d) 23 cm

3.19 An isolated 4-hour storm occurred over a catchment as follows:

Time	1st Hour	2nd hour	3rd hour	4th hour
Rainfall (mm)	9	28	12	2

The ϕ-index for the catchment is 10 mm/h. Estimate runoff depth from the catchment due to the above storm.

(a) 10 mm (b) 16 mm (c) 20 mm (d) 23 mm

3.20 For a basin, in a given period Δt, there is no change in the groundwater and soil water status. If P = precipitation, R = total runoff, E = Evapotranspiration and ΔS = increase in the surface water storage in the basin, the hydrological water budget equation states

(a) $P = R - E \pm \Delta S$ (b) $R = P + E - \Delta S$

(c) $P = R + E + \Delta S$ (d) none of these

CHAPTER

4

STREAMFLOW MEASUREMENT

4.1 *INTRODUCTION*

Streamflow representing the runoff phase of the hydrologic cycle is the most important basic data for hydrologic studies. It was seen in the previous chapters that precipitation, evaporation and evapotranspiration are all difficult to measure exactly and the presently adopted methods have severe limitations. In contrast, the measurement of streamflow is amenable to fairly accurate assessment. Interestingly, streamflow is the only part of the hydrologic cycle that can be measured accurately.

A stream can be defined as a flow channel into which the surface runoff from a specified basin drains. Generally, there is considerable exchange of water between a stream and the underground water. Streamflow is measured in units of discharge (m^3/s) occurring at a specified time and constitutes historical data. The measurement of discharge in a stream forms an important branch of *Hydrometry*, the science and practice of water measurement. This chapter deals with only the salient streamflow measurement techniques to provide an appreciation of this important aspect of engineering hydrology. Excellent treatises[1, 2, 4, 5] and a bibliography[6] are available on the theory and practice of streamflow measurement and these are recommended for further details.

Streamflow measurement techniques can be broadly classified into two categories as (i) direct determination, and (ii) indirect determination. Under each category, there are a host of methods, the important ones are listed below:

1. Direct determination of stream discharge:
   - (a) Area-velocity methods,
   - (b) Dilution techniques,
   - (c) Electromagnetic method, and
   - (d) Ultrasonic method.
2. Indirect determination of stream-flow:
   - (a) Hydraulic structures, such as weirs, flumes and gated structures, and
   - (b) Slope-area method.

Barring a few exceptional cases, continuous measurement of stream discharge is very difficult. As a rule, direct measurement of discharge is a very time-consuming and costly procedure. Hence, a two-step procedure is followed. First, the discharge in a given stream is related to the elevation of the water surface (Stage) through a series of careful measurements. In the next step, the stage of the stream is observed routinely in a relatively inexpensive manner and the discharge is estimated by using the previously determined stage–discharge relationship. The observation of the stage is easy, inexpensive, and if desired, continuous readings can also be obtained. This method of discharge determination of streams is adopted universally.

4.2 *Measurement of Stage*

The stage of a river is defined as its water-surface elevation measured above a datum. This datum can be the Mean-Sea Level (*MSL*) or any arbitrary datum connected independently to the *MSL*.

4.2.1 Manual Gauges

1. Staff Gauge

The simplest of stage measurements are made by noting the elevation of the water surface in contact with a fixed graduated staff. The staff is made of a durable material with a low coefficient of expansion with respect to both temperature and moisture. It is fixed rigidly to a structure, such as an abutment, pier, wall, etc. The staff may be vertical or inclined with clearly and accurately graduated permanent markings. The markings are distinctive, easy to read from a distance and are similar to those on a surveying staff. Sometimes, it may not be possible to read the entire range of water-surface elevations of a stream by a single gauge and in such cases the gauge is built in sections at different locations. Such gauges are called *sectional gauges* (Fig. 4.1). When installing sectional gauges, care must be taken to provide an overlap between various gauges and to refer all the sections to the same common datum.

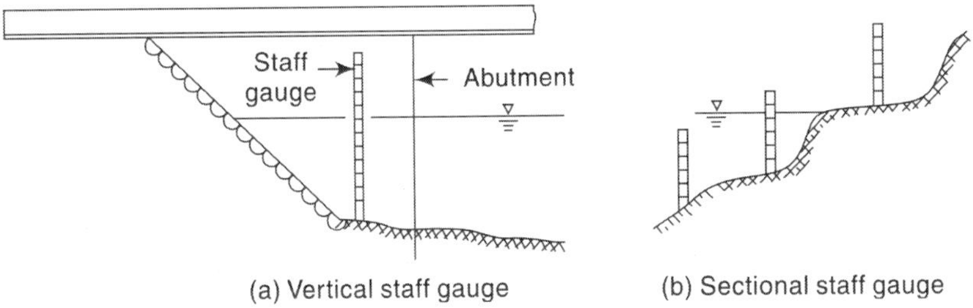

Fig. 4.1 ***Staff Gauge***

2. Wire Gauge

It is a gauge used to measure the water-surface elevation from above the surface such as from a bridge or similar structure. In this, a weight is lowered by a reel to touch the water surface. A mechanical counter measures the rotation of the wheel which is proportional to the length of the wire paid out. The operating range of this kind of gauge is about 25 m.

4.2.2 Automatic Stage Recorders

The staff gauge and wire gauge described earlier are manual gauges. While they are simple and inexpensive, they have to be read at frequent intervals to define the variation of stage with time accurately. Automatic stage recorders overcome this basic objection of manual staff gauges and find considerable use in stream-flow measurement practice. Two typical automatic stage recorders are described below.

1. Float-Gauge Recorder

The float-operated stage recorder is the most common type of automatic stage recorder in use. In this, a float operating in a stilling well is balanced by means of a counterweight over the pulley of a recorder. Displacement of the float due to the rising or lowering of the water-surface elevation causes an angular displacement of the pulley and hence of the input shaft of the recorder. Mechanical linkages convert this angular displacement to the linear displacement of a pen to record over a drum driven by clockwork. The pen traverse is continuous with automatic reversing when it reaches the full width of the chart. A clockwork mechanism runs the recorder for a day, week or fortnight and provides a continuous plot of stage *vs* time. A good instrument will have a large-size float and least friction. Improvements over this basic analogue model consists of models that give digital signals recorded on a storage device or transmit directly onto a central data-processing centre.

To protect the float from debris and to reduce the water surface wave effects on the recording, *stilling wells* are provided in all float-type stage recorder installations. Figure 4.2 shows a typical stilling well installation. Note the intake pipes that communicate with the river and flushing arrangement to flush these intake pipes off the sediment and debris occasionally. The water-stage recorder has to be located above the highest water level expected in the stream to prevent it from getting inundated during floods. Further, the instrument must be properly housed in a suitable enclosure to protect it from weather elements and vandalism. On account of these, the water-stage-recorder installations prove to be costly in most instances. A water-depth recorder is shown in Fig. 4.3.

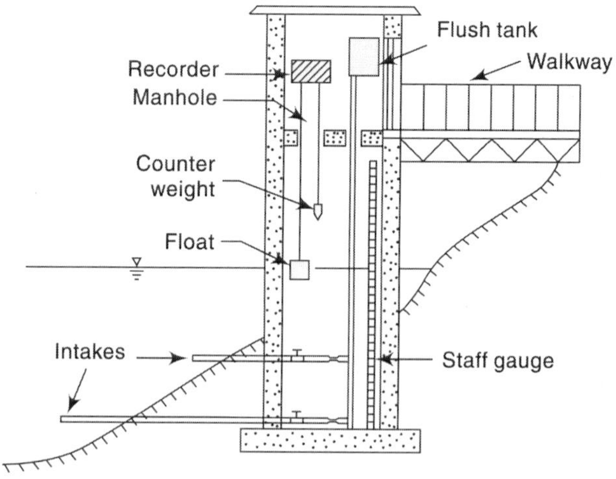

Fig. 4.2 ***Stilling well Installation***

Fig. 4.3 ***Water-depth recorder—Stevens Type F recorder***
(Courtesy: Leupold and Stevens, Inc. Beaverton, Oregon, USA)

2. Bubble Gauge

In this gauge, compressed air or gas is made to bleed out at a very small rate through an outlet placed at the bottom of the river [Figs. 4.4, 4.5 and 4.6]. A pressure gauge

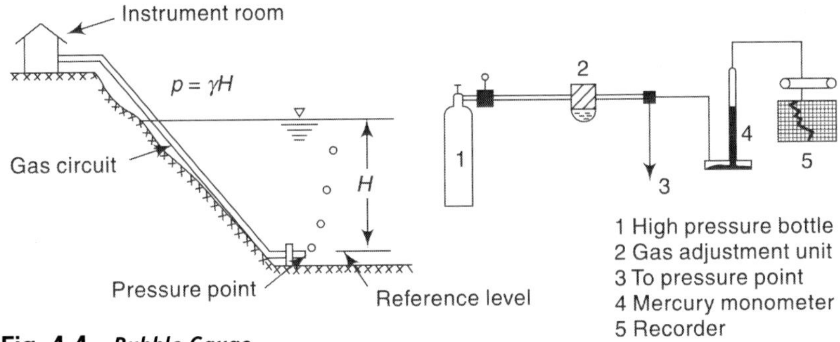

Fig. 4.4 ***Bubble Gauge***

Fig. 4.5 ***Bubble Gauge Installation—Telemnip***
(Courtesy: Neyrtec, Grenoble, France)

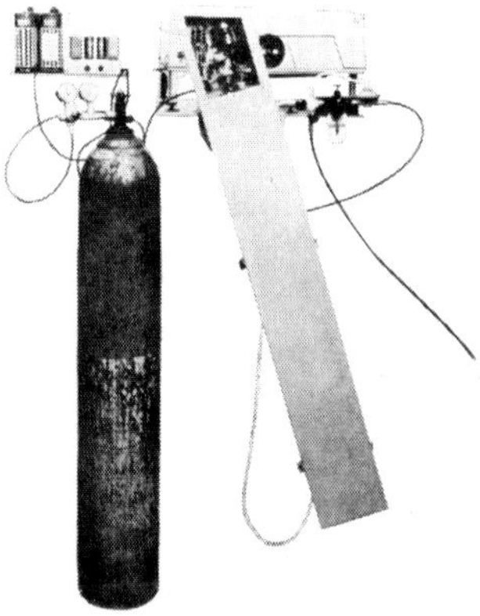

Fig. 4.6 ***Bubble Gauge—Stevens Manometer Servo***
(Courtesy: Leupold and Stevens, Inc. Beaverton, Oregon, USA)

measures the gas pressure which in turn is equal to the water column above the outlet. A small change in the water-surface elevation is felt as a change in pressure from the present value at the pressure gauge and this in turn is adjusted by a servo-mechanism to bring the gas to bleed at the original rate under the new head. The pressure gauge reads the new water depth which is transmitted to a recorder.

The bubble gauge has certain specific advantages over a float-operated water stage recorder and these can be listed as under:

1. There is no need for costly stilling wells;
2. A large change in the stage, as much as 30 m, can be measured;
3. The recorder assembly can be quite far away from the sensing point; and
4. Due to constant bleeding action there is less likelihood of the inlet getting blocked or choked.

4.2.3 Stage Data

The stage data is often presented in the form of a plot of stage against chronological time (Fig. 4.7) known as *stage hydrograph.* In addition to its use in the determination of stream discharge, stage data itself is of importance in design of hydraulic structures, flood warning and flood-protection works. Reliable long-term stage data corresponding to peak floods can be analysed statistically to estimate the design peak river stages for use in the design of hydraulic structures, such as bridges, weirs, etc. Historic flood stages are invaluable in the indirect estimation of corresponding flood discharges. In view of these multifarious uses, the river stage forms an important hydrologic parameter chosen for regular observation and recording.

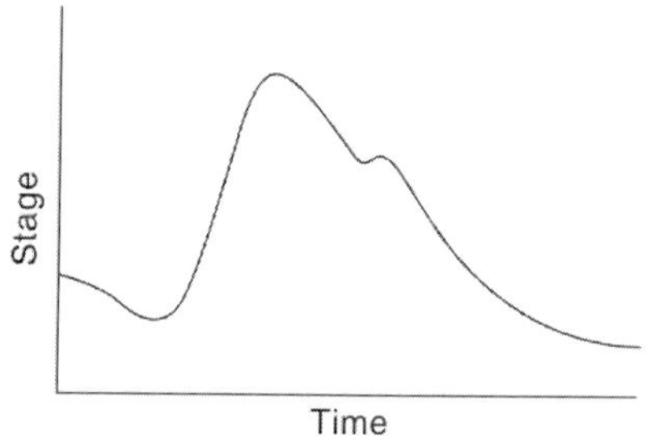

Fig. 4.7 ***Stage Hydrograph***

4.3 *Measurement of Velocity*

The measurement of velocity is an important aspect of many direct stream-flow measurement techniques. A mechanical device, called *current meter,* consisting essentially of a rotating element is probably the most commonly used instrument for accurate determination of the stream-velocity field. Approximate stream velocities can be determined by *floats.*

4.3.1 Current Meters

The most commonly used instrument in hydrometry to measure the velocity at a point in the flow cross-section is the current meter. It consists essentially of a rotating element which rotates due to the reaction of the stream current with an angular velocity proportional to the stream velocity. Historically, Robert Hooke (1663) invented a propeller-type current meter to measure the distance traversed by a ship. The present-day cup-type instrument and the electrical make-and-break mechanism were invented by Henry in 1868. There are two main types of current meters.

1. Vertical-axis meters, and
2. Horizontal-axis meters.

1. Vertical-Axis Meters

These instruments consist of a series of conical cups mounted around a vertical axis [Figs. 4.8 and 4.9]. The cups rotate in a horizontal plane and a cam attached to the vertical axial spindle records generated signals proportional to the revolutions of the cup assembly. The Price current meter and Gurley current meter are typical instruments under this category. The normal range of velocities is from 0.15 to 4.0 m/s. The accuracy of these instruments is about 1.50% at the threshold value and improves to about 0.30% at speeds in excess of 1.0 m/s. Vertical-axis instruments have the disadvantage that they cannot be used in situations where there are appreciable vertical components of velocities. For example, the instrument shows a positive velocity when it is lifted vertically in still water.

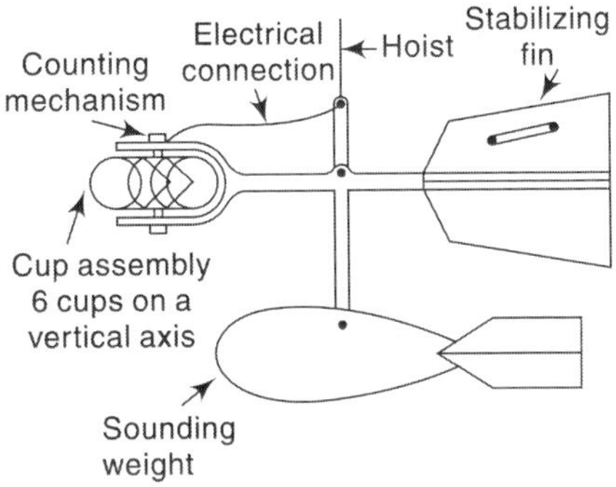

Fig. 4.8 ***Vertical-axis Current Meter***

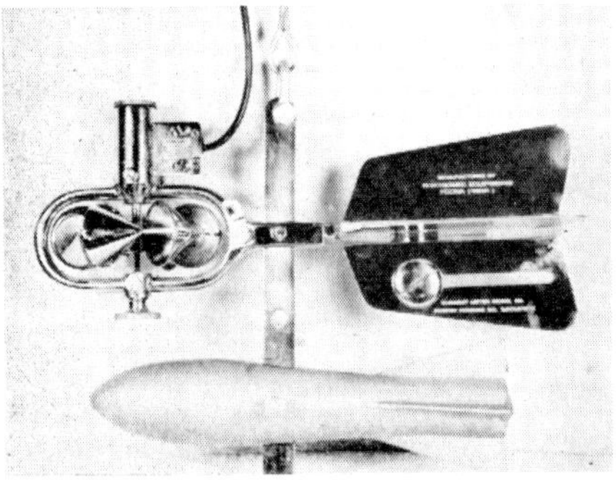

Fig. 4.9 ***Cup-type Current Meter with Sounding Weight—'Lynx' Type***
(Courtesy: Lawrence and Mayo (India) New Delhi)

2. Horizontal-Axis Meters

These meters consist of a propeller mounted at the end of horizontal shaft as shown in Fig. 4.10 and Fig. 4.11. These come in a wide variety of size with propeller diameters in the range 6 to 12 cm, and can register velocities in the range of 0.15 to 4.0 m/s. Ott, Neyrtec [Fig. 4.12] and Watt-type meters are typical instruments under this kind. These meters are fairly rugged and are not affected by oblique flows of as much as 15°. The accuracy of the instrument is about 1% at the threshold value and is about 0.25% at a velocity of 0.3 m/s and above.

Fig. 4.10 ***Propeller-type Current Meter—Neyrtec Type with Sounding Weight***

A current meter is so designed that its rotation speed varies linearly with the stream velocity v at the location of the instrument. A typical relationship is

$$v = aN_s + b \qquad (4.1)$$

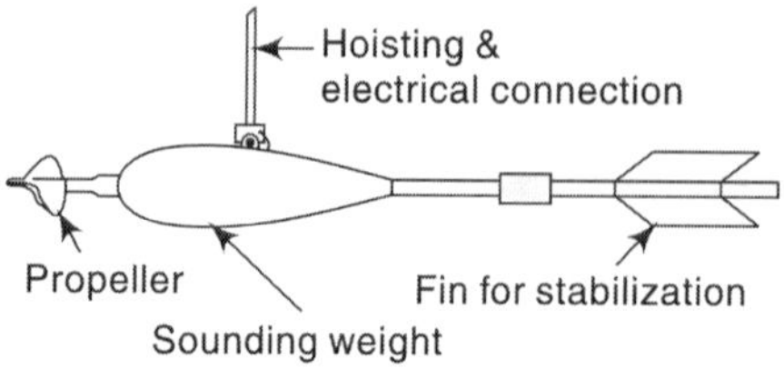

Fig. 4.11 ***Horizontal-axis Current Meter***

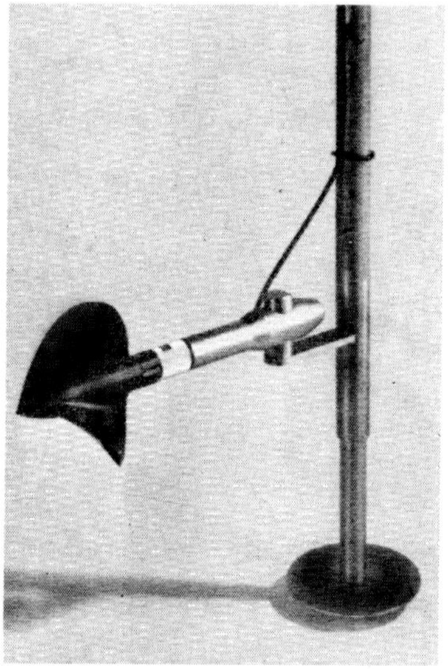

Fig. 4.12(a) ***Neyrtec Type Current Meter for use in Wading***

(Courtesy: Neyrtec, Grenoble, France)

Fig. 4.12(b) ***Neyrtec Type Meter in a Cableway***

where v = stream velocity at the instrument location in m/s, N_s = revolutions per second of the meter and a, b = constants of the meter. Typical values of a and b for a standard size 12.5 cm diameter Price meter (cup-type) is $a = 0.65$ and $b = 0.03$. Smaller meters of 5 cm diameter cup assembly called *pigmy meters* run faster and are useful in measuring small velocities. The values of the meter constants for them are of the order of $a = 0.30$ and $b = 0.003$. Further, each instrument has a threshold velocity below which Eq. (4.1) is not applicable. The instruments have a provision to count the number of revolutions in a known interval of time. This is usually accomplished by the making and breaking of an electric circuit, either mechanically or electromagnetically at each revolution of the shaft. In older model instruments, the breaking of the circuit would be counted through an audible sharp signal ("tick") heard on a headphone. The revolutions per second is calculated by counting the number of such signals in a known interval of time, usually about 100 s. Present-day models employ electromagnetic counters with digital signal processing capabilities.

3. Calibration

The relation between the stream velocity and revolutions per second of the meter as in Eq. (4.1) is called the *calibration equation*. The calibration equation is unique to each instrument and is determined by towing the instrument in a special tank. A *towing tank* is a long channel containing still water with arrangements for moving a carriage longitudinally over its surface at constant speed. The instrument to be calibrated is mounted on the carriage with the rotating element immersed to a specified depth in the water body in the tank. The carriage is then towed at a predetermined constant speed (v) and the corresponding average value of revolutions per second (N_s) of the instruments determined. This experiment is repeated over the complete range of velocities and a best-fit linear relation in the form of Eq. (4.1) obtained. The instruments are designed for rugged use and hence the calibration once done lasts for quite some time. However, from the point of view of accuracy, it is advisable to check the instrument calibration once in a while and whenever there is a suspicion that the instrument is damaged due to bad handling or accident. In India excellent towing-tank facilities for calibration of current meters exist at the Central Water and Power Research Station, Pune and the Indian Institute of Technology, Madras.

4. Field Use

The velocity distribution in a stream across a vertical section is logarithmic in nature. In a rough turbulent flow, the velocity distribution is given by

$$v = 5.75\, v_* \log_{10}\left(\frac{30\,y}{k_s}\right) \tag{4.2}$$

where v = velocity at a point y above the bed, v_* = shear velocity and k_s = equivalent sand-grain roughness. To accurately determine the average velocity in a vertical section, one has to measure the velocity at a large number of points on the vertical. As it is time-consuming, certain simplified procedures have been evolved.

- In shallow streams of depth up to about 3.0 m, the velocity measured at 0.6 times the depth of flow below the water surface is taken as the average velocity $\bar{v}$ in the vertical,

  $$\bar{v} = v_{0.6} \tag{4.3}$$

  This procedure is known as the single-point observation method.
- In moderately deep streams the velocity is observed at two points; (i) at 0.2 times the depth of flow below the free surface ($v_{0.2}$) and (ii) at 0.8 times the depth of flow below the free surface ($v_{0.8}$). The average velocity in the vertical $\bar{v}$ is taken as

  $$v = \frac{v_{0.2} + v_{0.8}}{2} \tag{4.4}$$

- In rivers having flood flows, only the surface velocity (v_s) is measured within a depth of about 0.5 m below the surface. The average velocity $\bar{v}$ is obtained by using a reduction factor K as

  $$v = Kv_s \tag{4.5}$$

  The value of K is obtained from observations at lower stages and lie in the range of 0.85 to 0.95.

In small streams of shallow depth, the current meter is held at the requisite depth below the surface in a vertical by an observer who stands in the water. The arrangement, called *wading* is quite fast but is obviously applicable only to small streams.

In rivers flowing in narrow gorges in well-defined channels, a cableway is stretched from bank to bank well above the flood level. A carriage moving over the cableway is used as the observation platform.

Bridges, while hydraulically not the best locations, are advantageous from the point of view of accessibility and transportation. Hence, railway and road bridges are frequently employed as gauging stations. The velocity measurement is performed on the downstream portion of the bridge to minimise the instrument damage due to drift and knock against the bridge piers.

For wide rivers, boats are the most satisfactory aids in current meter measurement. A cross-sectional line is marked by distinctive land markings and buoys. The position of the boat is determined by using two theodolites on the bank through an intersection method. Use of total station simplifies the work considerably.

5. Sounding Weights

Current meters are weighted down by lead weights called *sounding weights* to enable them to be positioned in a stable manner at the required location in flowing water. These weights are of streamlined shape with a fin in the rear (Fig. 4.8) and are connected to the current meter by a hangar bar and pin assembly. Sounding weights come in different sizes and the minimum weight is estimated as

$$W = 50\ \bar{v}\ d \tag{4.6}$$

where W = minimum weight in N, v = average stream velocity in the vertical in m/s and d = depth of flow at the vertical in metres.

4.3.2 Velocity Measurement by Floats

A floating object on the surface of a stream when timed can yield the surface velocity by the relation

$$v_s = \frac{S}{t} \tag{4.7}$$

where S = distance travelled in time t. This method of measuring velocities while primitive still finds applications in special circumstances, such as (i) a small stream in flood, (ii) small stream with a rapidly changing water surface, and (iii) preliminary or exploratory surveys. While any floating object can be used, normally specially made leakproof and easily identifiable floats are used (Fig. 4.13). A simple float moving on the stream surface is called a *surface float*. It is easy to use and the mean velocity is obtained by multiplying the observed surface velocity by a reduction coefficient as in Eq. (4.5). However, surface floats are affected by surface winds. To get the average velocity in the vertical directly, special floats in which part of the body is under water are used. *Rod float* (Fig. 4.13), in which a cylindrical rod is weighed so that it can float vertically, belongs to this category.

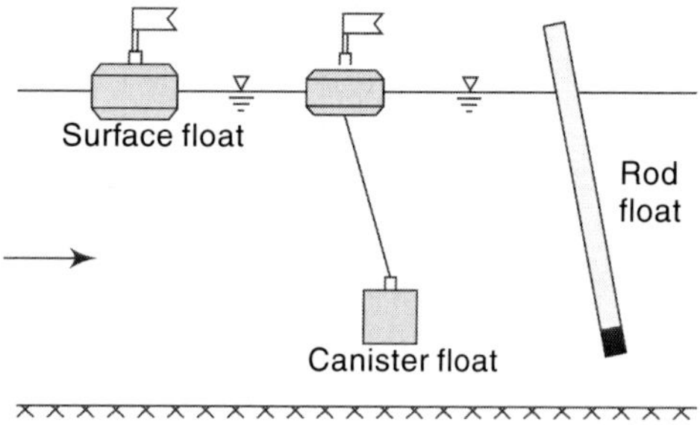

Fig. 4.13 ***Floats***

In using floats to observe the stream velocity, a large number of easily identifiable floats are released at fairly uniform spacings on the width of the stream at an upstream section. Two sections on a fairly straight reach are selected and the time to cross this reach by each float is noted and the surface velocity calculated.

4.4 *Area-Velocity Method*

This method of discharge measurement consists essentially of measuring the area of cross section of the river at a selected section called the *gauging site* and measuring the velocity of flow through the cross-sectional area. The gauging site must be selected with care to assure that the stage-discharge curve is reasonably constant over a long period of about a few years. Towards this, the following criteria are adopted.

- The stream should have a well-defined cross section which does not change in various seasons.
- It should be easily accessible all through the year.
- The site should be in a straight, stable reach.
- The gauging site should be free from backwater effects in the channel.

At the selected site, the section line is marked off by permanent survey markings and the cross section determined. Towards this, the depth at various locations are measured by sounding rods or sounding weights. When the stream depth is large or when quick and accurate depth measurements are needed, an electroacoustic instrument called *echo-depth recorder* is used. In this, a high frequency sound wave is sent down by a transducer kept immersed at the water surface and the echo reflected by the bed is also picked up by the same transducer. By comparing the time interval between the transmission of the signal and the receipt of its echo, the distance to the bed is obtained and is indicated or recorded in the instrument. Echo-depth recorders are particularly advantageous in high-velocity streams, deep streams and in streams with soft or mobile beds.

For purposes of discharge estimation, the cross section is considered to be divided into a large number of subsections by verticals (Fig. 4.14). The average velocity in these subsections are measured by current meters or floats. It is quite obvious that the accuracy of discharge estimation increases with the number of subsections used. However, the larger the number of segments, the larger is the effort, time and expenditure involved. The following are some of the guidelines to select the number of segments.

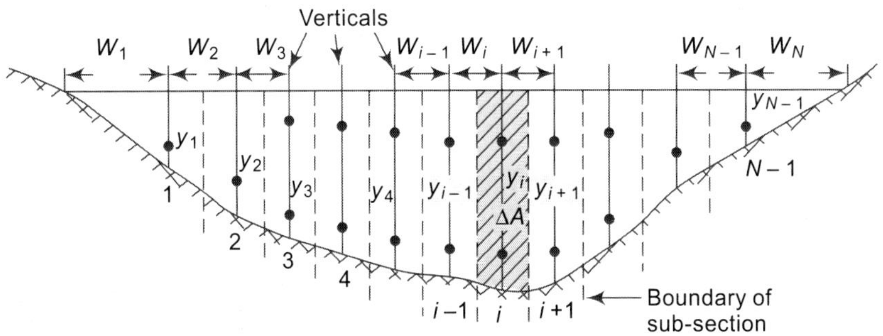

Fig. 4.14 ***Stream Section for Area-velocity Method***

- The segment width should not be greater than 1/15 to 1/20 of the width of the river.
- The discharge in each segment should be less than 10% of the total discharge.
- The difference of velocities in adjacent segments should not be more than 20%.

It should be noted that in natural rivers, the verticals for velocity measurement are not necessarily equally spaced. The area-velocity method as above using the current meter is often called the *standard current meter method.*

4.4.1 Calculation of Discharge

Figure 4.14 shows the cross section of a river in which $N-1$ verticals are drawn. The velocity averaged over the vertical at each section is known. Considering the total area to be divided into $N-1$ segments, the total discharge is calculated by the *method of mid-sections* as follows.

$$Q = \sum_{i=1}^{N-1} \Delta Q_i \tag{4.8}$$

where ΔQ_i = discharge in the ith segment

= (depth at the ith segment) × $\left(\frac{1}{2}\right.$ width to the left

$+\left.\frac{1}{2}\text{ width to right}\right)$ × (average velocity at the ith vertical)

$$\Delta Q_i = y_i \times \left(\frac{W_i}{2} + \frac{W_{i+1}}{2}\right) \times v_i \qquad \text{for } i = 2 \text{ to } (N-2) \tag{4.9}$$

For the first and last sections, the segments are taken to have triangular areas and area calculated as

$$\Delta A_1 = \overline{W}_1 \cdot y_1$$

where $\overline{W}_1 = \dfrac{\left(W_1 + \dfrac{W_2}{2}\right)^2}{2W_1}$ and $\Delta A_N = \overline{W}_{N-1} \cdot y_{N-1}$

where $\overline{W}_{N-1} = \dfrac{\left(W_N + \dfrac{W_{N-1}}{2}\right)^2}{2W_N}$

to get

$$\Delta Q_1 = \bar{v}_1 \cdot \Delta A_1 \text{ and } \Delta Q_{N-1} = \bar{v}_{N-1} \cdot \Delta A_{N-1} \tag{4.10}$$

Example 4.1 The data pertaining to a stream-gauging operation at a gauging site are given below.

The rating equation of the current meter is $v = 0.51\,N_s + 0.03$ m/s where N_s = revolutions per second. Calculate the discharge in the stream.

Distance from left water edge (m)	0	1.0	3.0	5.0	7.0	9.0	11.0	12.0
Depth (m)	0	1.1	2.0	2.5	2.0	1.7	1.0	0

(Contd.)

(Contd.)

Revolutions of a current meter kept at 0.6 depth	0	39	58	112	90	45	30	0
Duration of observation (s)	0	100	100	150	150	100	100	0

Solution:

The calculations are performed in a tabular form.

For the first and last sections,

Average width, $$\overline{W} = \frac{\left(1+\frac{2}{2}\right)^2}{2\times 1} = 2.0 \text{ m}$$

For the rest of the segments,

$$\overline{W} = \left(\frac{2}{2}+\frac{2}{2}\right) = 2.0 \text{ m}$$

Since the velocity is measured at 0.6 depth, the measured velocity is the average velocity at that vertical ($\bar{v}$).

The calculation of discharge by the mid-section method is shown in tabular form below:

Distance from left water edge (m)	Average width $\overline{W}$ (m)	Depth y (m)	N_s = Rev./second	Velocity $\bar{v}$ (m/s)	Segmental discharge ΔQ_i (m^3/s)
0	0	0			0.0000
1	2	1.10	0.390	0.2289	0.5036
3	2	2.00	0.580	0.3258	1.3032
5	2	2.50	0.747	0.4110	2.0549
7	2	2.00	0.600	0.3360	1.3440
9	2	1.70	0.450	0.2595	0.8823
11	2	1.00	0.300	0.1830	0.3660
12	0	0.00			0.0000
				Sum =	**6.45393**

Discharge in the stream = 6.454 m^3/s

4.4.2 Moving-Boat Method

Discharge measurement of large alluvial rivers, such as the Ganga, by the standard current meter method is very time-consuming even when the flow is low or moderate. When the river is in spate, it is almost impossible to use the standard current meter technique due to the difficulty of keeping the boat stationary on the fast-moving surface of the stream for observation purposes. It is in such circumstances that the moving-boat techniques prove very helpful.

In this method, a special propeller-type current meter which is free to move about a vertical axis is towed in a boat at a velocity v_b at right angles to the stream flow. If the flow velocity is v_f, the meter will align itself in the direction of the resultant velocity v_R making an angle θ with the direction of the boat (Fig. 4.15). Further, the meter will register the velocity v_R. If v_b is normal to v_f,

$$v_b = v_R \cos \theta \quad \text{and} \quad v_f = v_R \sin \theta$$

If the time of transit between two verticals is Δt then the width between the two verticals (Fig. 4.15) is

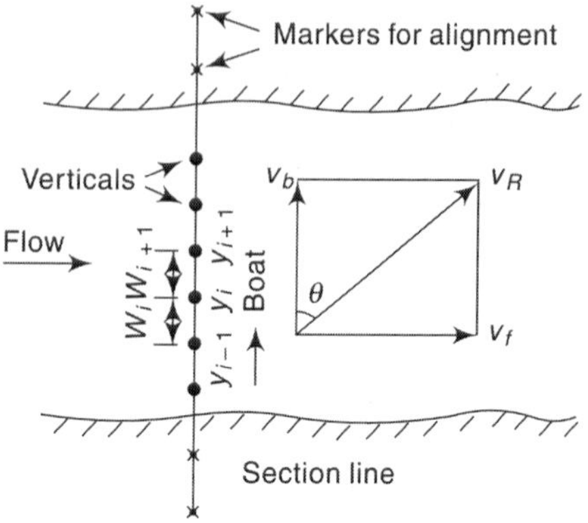

Fig. 4.15 ***Moving-boat Method***

$$W = v_b \Delta t$$

The flow in the sub-area between two verticals i and $i + 1$ where the depths are y_i and y_{i+1} respectively, by assuming the current meter to measure the average velocity in the vertical, is

$$\Delta Q_i = \left(\frac{y_i + y_{i+1}}{2}\right) W_{i+1} v_f$$

i.e.

$$\Delta Q_i = \left(\frac{y_i + y_{i+1}}{2}\right) v_R^2 \sin \theta \cdot \cos \theta \cdot \Delta t \qquad (4.11)$$

Thus, by measuring the depths y_i, velocity v_R and θ in a reach and the time taken to cross the reach Δt, the discharge in the sub-area can be determined. The summation of the partial discharges ΔQ_i over the whole width of the stream gives the stream discharge

$$Q = \Sigma \Delta Q_i \qquad (4.12)$$

In field application, a good stretch of the river with no shoals, islands, bars, etc. is selected. The cross-sectional line is defined by permanent landmarks so that the boat can be aligned along this line. A motorboat with different sizes of outboard motors for use in different river stages is selected. A special current meter of the propeller-type, in which the velocity and inclination of the meter to the boat direction θ in the horizontal plane can be measured, is selected. The current meter is usually immersed at a depth of 0.5 m from the water surface to record surface velocities. To mark the various vertical sections and know the depths at these points, an echo-depth recorder is used.

In a typical run, the boat is started from the water edge and aligned to go across the cross-sectional line. When the boat is in sufficient depth of water, the instruments are lowered. The echo-depth recorder and current meter are commissioned. A button on the signal processor when pressed marks a distinctive mark line on the depth vs time chart of the echo-depth recorder. Further, it gives simultaneously a sharp audio signal to enable the measuring party to take simultaneous readings of the velocity v_R and the inclination θ. A large number of such measurements are taken during the traverse of the boat to the other bank of the river. The operation is repeated in the return journey of the boat. It is important that the boat is kept aligned along the cross-sectional line and this requires considerable skill on the part of the pilot. Typically, a river of about 2 km stretch takes about 15 min for one crossing. A number of crossings are made to get the average value of the discharge.

The surface velocities are converted to average velocities across the vertical by applying a coefficient [Eq. (4.5)]. The depths y_i and time intervals Δt are read from the echo-depth recorder chart. The discharge is calculated by Eqs. (4.11) and (4.12). In practical use, additional coefficients may be needed to account for deviations from the ideal case and these depend upon the actual field conditions.

4.5 *Dilution Technique of Streamflow Measurement*

The *dilution method* of flow measurement, also known as the *chemical method*, depends upon the continuity principle applied to a tracer which is allowed to mix completely with the flow.

Consider a tracer which does not react with the fluid or boundary. Let C_0 be the small initial concentration of the tracer in the streamflow. At Section 1 a small quantity (volume $\mathcal{V}_1$)" of high concentration C_1 of this tracer is added as shown in Fig. 4.16. Let Section 2 be sufficiently far away on the downstream of Section 1 so that the tracer mixes thoroughly with the fluid due to the turbulent mixing process while passing through the reach. The concentration profile taken at Section 2 is schematically shown in Fig. 4.16. The concentration will have a base value of C_0, increases from time t_1 to a peak value and gradually reaches the base value of C_0 at time t_2. The stream flow is assumed to be steady. By continuity of the tracer material

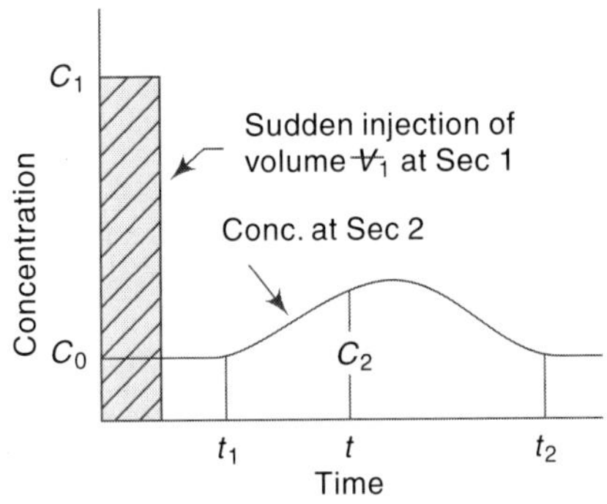

Fig. 4.16 ***Sudden-injection Method***

$$M_1 = \text{mass of tracer added at Section 1} = \mathcal{V}_1\, C_1$$

$$= \int_{t_1}^{t_2} Q(C_2 - C_0)\, dt + \frac{\mathcal{V}_1}{t_2 - t_1} \int_{t_1}^{t_2} (C_2 - C_0)\, dt$$

Neglecting the second term on the right-had side as insignificantly small,

$$Q = \frac{\mathcal{V}_1 C_1}{\int_{t_1}^{t_2} (C_2 - C_0)\, dt} \tag{4.13}$$

Thus the discharge Q in the stream can be estimated if for a known M_1 the variation of C_2 with time at Section 2 and C_0 are determined. This method is known as *sudden injection* or *gulp* or *integration method*.

Another way of using the dilution principle is to inject the tracer of concentration C_1 at a constant rate Q_t at Section 1. At Section 2, the concentration gradually rises from the background value of C_0 at time t_1 to a constant value C_2 as shown in Fig. 4.17. At the steady state, the continuity equation for the tracer is

$$Q_t C_1 + Q C_0 = (Q + Q_t) C_2$$

i.e.,
$$Q = \frac{Q_t (C_1 - C_2)}{(C_2 - C_0)} \qquad (4.14)$$

This technique in which Q is estimated by knowing C_1, C_2, C_0 and Q_t is known as *constant rate injection method* or *plateau gauging*.

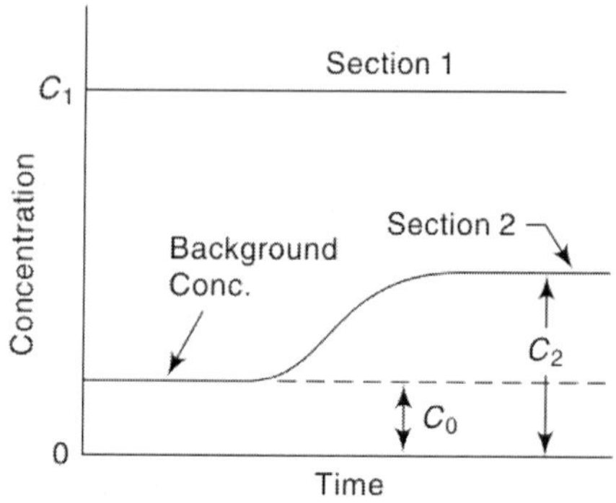

Fig. 4.17 ***Constant-Rate Injection Method***

It is necessary to emphasise here that the dilution method of gauging is based on the assumption of steady flow. If the flow is unsteady and the flow rate changes appreciably during gauging, there will be a change in the storage volume in the reach and the steady-state continuity equation used to develop Eqs. (4.13) and (4.14) is not valid. Systematic errors can be expected in such cases.

4.5.1 Tracers

The tracer used should have ideally the following properties:

1. It should not be absorbed by the sediment, channel boundary and vegetation. It should not chemically react with any of the above surfaces and also should not be lost by evaporation.
2. It should be non-toxic.
3. It should be capable of being detected in a distinctive manner in small concentrations.
4. It should not be very expensive.

The tracers used are of three main types:

1. Chemicals (common salt and sodium dichromate are typical)
2. Fluorescent dyes (Rhodamine-WT and Sulpho-Rhodamine B Extra are typical)
3. Radioactive materials (such as Bromine-82, Sodium-24 and Iodine-132).

Common salt can be detected with an error of ±1% up to a concentration of 10 ppm. Sodium dichromate can be detected up to 0.2 ppm concentrations. Fluorescent dyes have the advantage that they can be detected at levels of tens of nanograms per litre (~1 in 10^{11}) and hence require very small amounts of solution for injections. Radioactive tracers are detectable up to accuracies of tens of picocuries per litre (~1 in 10^{14}) and therefore permit large-scale dilutions. However, they involve the use of very sophisticated instruments and handling by trained personnel only. The availability of detection instrumentation, environmental effects of the tracer and overall cost of the operation are chief factors that decide the tracer to be used.

1. Length of Reach

The length of the reach between the dosing section and sampling section should be adequate to have complete mixing of the tracer with the flow. This length depends upon the geometric dimensions of the channel cross section, discharge and turbulence levels. An empirical formula suggested by Rimmar (1960) for estimation of mixing length for point injecion of a tracer in a straight reach is

$$L = \frac{0.13\, B^2 C\,(0.7C + 2\sqrt{g}\,)}{gd} \tag{4.15}$$

where L = mixing length (m), B = average width of the stream (m), d = average depth of the stream (m), C = Chezy coefficient of roughness and g = acceleration due to gravity. The value of L varies from about 1 km for a mountain stream carrying a discharge of about 1.0 m^3/s to about 100 km for river in a plain with a discharge of about 300 m^3/s. The mixing length becomes very large for large rivers and is one of the major constraints of the dilution method. Artificial mixing of the tracer at the dosing station may prove beneficial for small streams in reducing the mixing length of the reach.

2. Use

The dilution method has the major advantage that the discharge is estimated directly in an absolute way. It is a particularly attractive method for small turbulent streams, such as those in mountainous areas. Where suitable, it can be used as an occasional method for checking the calibration, stage-dischage curves, etc. obtained by other methods.

Example 4.2 A 25 g/l solution of a flourescent tracer was discharged into a stream at a constant rate of 10 cm^3/s. The background concentration of the dye in the stream water was found to be zero. At a downstream section sufficiently far away, the dye was found to reach an equilibrium concentration of 5 parts per billion. Estimate the stream discharge.

Solution

By Eq. (4.14), or the constant-rate injection method,

$$Q = \frac{Q_t\,(C_1 - C_2\,)}{C_2 - C_0}$$

$$Q_t = 10 \text{ cm}^3/\text{s} = 10 \times 10^{-6} \text{ m}^3/\text{s}$$

$$C_1 = 0.025,\ C_2 = 5 \times 10^{-9},\ C_0 = 0$$

$$Q = \frac{10 \times 10^{-6}}{5 \times 10^{-9}}\,(0.025 - 5 \times 10^{-9}) = 50 \text{ m}^3/\text{s}$$

Example 4.3 The dilution method with sudden injection procedure was used to measure the discharge of a rapid stream. A fluorescent dye was suddenly injected at station R at 08 hours. The weight of injected dye was 400 N. The data of concentration measurements at station S are given below. The base concentration of the dye in the stream was zero. Estimate the discharge in the stream.

Time (h)	8	9	10	11	12	13	14	15	16	17	18	19	20	21
Concentration at Station S in parts per 10^9 parts by weight	0	1	3	10.5	18	18.5	16	12	9	6	4	2	1	0

Solution

By Eq. 4.13, the discharge $Q = \dfrac{\forall_1 C_1}{\int_{t_1}^{t_2} (C_2 - C_0)\,dt}$

In the present case, $C_0 = 0$ and the denominator is the area of the time-concentration curve. The numerator is the weight of tracer introduced in a gulp and is equal to 400 N. Area of time-concentration curve at station S =

$[0 + 1 + 3 + 10.5 + 18.0 + 18.5 + 16 + 12 + 9 + 6 + 4 + 2 + 1 + 0] \times 60 \times 60 = 363,600$ units.

Taking the unit weight of water as 9790 N/m^3,

$363,600 \times 10^{-9} \times (9790) \times Q = 400$

$$Q = \frac{400}{3.5596} = 112 \text{ m}^3\text{/s}$$

4.6 *Electromagnetic Method*

The electromagnetic method is based on the Faraday's principle that an emf is induced in the conductor (water in the present case) when it cuts a normal magnetic field. Large coils buried at the bottom of the channel carry a current I to produce a controlled vertical magnetic field (Fig. 4.18). Electrodes provided at the sides of the channel section measure the small voltage produced due to flow of water in the channel. It has been found that the signal output E will be of the order of millivolt and is related to the discharge Q as

$$Q = K_1\left(\frac{Ed}{I} + K_2\right)^n \tag{4.16}$$

where d = depth of flow, I = current in the coil, and n, K_1 and K_2 are system constants.

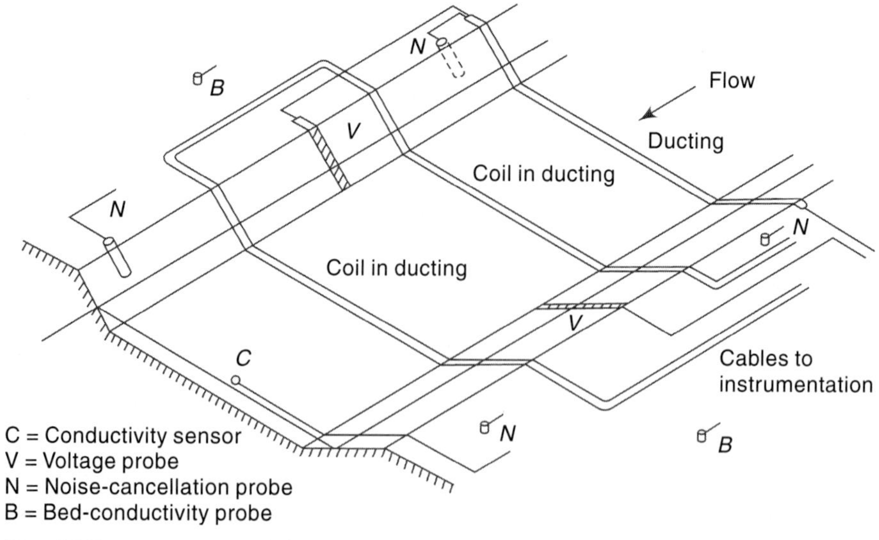

Fig. 4.18 ***Electromagnetic Method***

The method involves sophisticated and expensive instrumentation and has been successfully tried in a number of installations. The fact that this kind of set-up gives the total discharge when once it has been calibrated, makes it specially suited for field situations where the cross-sectional properties can change with time due to weed growth, sedimentation, etc. Another specific application is in tidal channels where the flow undergoes rapid changes both in magnitude as well as in direction. Present-day commercially available electromagnetic flowmeters can measure the discharge to an accuracy of ±3%, the maximum channel width that can be accommodated being 100 m. The minimum detectable velocity is 0.005 m/s.

4.7 Ultrasonic Method

This is essentially an area-velocity method with the average velocity being measured by using ultrasonic signals. The method was first reported by Swengel (1955), since then it has been perfected and complete systems are available commercially.

Consider a channel carrying a flow with two transducers *A* and *B* fixed at the same level *h* above the bed and on either side of the channel (Fig. 4.19). These transducers can receive as well as send ultrasonic signals. Let *A* send an ultrasonic signal to be received at *B* after an elapse time t_1 Similarly, let *B* send a signal to be received at *A* after an elapse time t_2. If C = velocity of sound in water,

$$t_1 = L/(C + v_p) \tag{4.17}$$

where L = length of path from *A* to *B* and v_p = component of the flow velocity in the sound path = $v \cos \theta$. Similarly, from Fig. 4.19 it is easy to see that

$$t_2 = \frac{L}{(C - v_p)} \tag{4.18}$$

Thus
$$\frac{1}{t_1} - \frac{1}{t_2} = \frac{2v_p}{L} = \frac{2v\cos\theta}{L}$$

or
$$v = \frac{L}{2\cos\theta}\left(\frac{1}{t_1} - \frac{1}{t_2}\right) \tag{4.19}$$

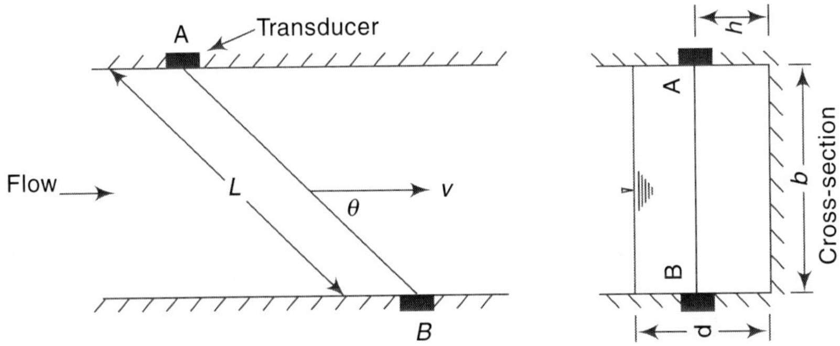

Fig. 4.19 ***Ultrasonic Method***

Thus, for a given L and θ, by knowing t_1 and t_2, the average velocity along the path AB, i.e., v can be determined. It may be noted that v is the average velocity at a height h above the bed and is not the average velocity V for the whole cross section. However, for a given channel, cross section v can be related to V and by calibration a relation between v/V and h can be obtained. For a given set-up, as the area of cross section is fixed, the discharge is obtained as a product of area and mean velocity V. Estimation of discharge by using one signal path as above is called *single-path gauging*. Alternatively, for a given depth of flow, multiple single paths can be used to obtain v for different h values. Mean velocity of flow through the cross section is obtained by averaging these v values. This technique is known as *multi-path gauging*.

Ultrasonic flowmeters using the above principal have frequencies of the order of 500 kHz. Sophisticated electronics are involved to transmit, detect and evaluate the mean velocity of flow along the path. In a given installation, a calibration (usually performed by the current-meter method) is needed to determine the system constants. Currently available commercial systems have accuracies of about 2% for the single-path method and 1% for the multipath method. The systems are currently available for rivers up to 500 m width.

The specific advantages of the ultrasonic system of river gauging are the following:

1. It is rapid and gives high accuracy.
2. It is suitable for automatic recording of data.
3. It can handle rapid changes in the magnitude and direction of flow, as in tidal rivers.
4. The cost of installation is independent of the size of rivers.

The accuracy of this method is limited by the factors that affect the signal velocity and averaging of flow velocity, such as (i) unstable cross section, (ii) fluctuating weed growth, (iii) high loads of suspended solids, (iv) air entrainment, and (v) salinity and temperature changes.

4.8 *Indirect Methods*

Under this category are included those methods which make use of the relationship between the flow discharge and the depths at specified locations. The field measurement is restricted to the measurements of these depths only.

Two broad classifications of these indirect methods are

1. Flow measuring structures, and
2. Slope area method.

4.8.1 Flow-Measuring Structures

Use of structures like notches, weirs, flumes and sluice gates for flow measurement in hydraulic laboratories is well known. These conventional structures are used in field conditions also but their use is limited by the ranges of head, debris or sediment load of the stream and the backwater effects produced by the installations. To overcome many of these limitations a wide variety of flow measuring structures with specific advantages are in use.

The basic principle governing the use of a weir, flume or similar flow-measuring structure is that these structures produce a unique *control section* in the flow. At these structures, the discharge Q is a function of the water-surface elevation measured at a specified upstream location,

$$Q = f(H) \tag{4.20}$$

where H = water-surface elevation measured from a specified datum. Thus, for example, for weirs, Eq. (4.20) takes the form

$$Q = KH^n \tag{4.21}$$

where H = head over the weir and K, n = system constants. Equation (4.20) is applicable so long as the downstream water level is below a certain limiting water level known as the *modular limit*. Such flows which are independent of the downstream water level are known as *free flows*. If the tail water conditions do affect the flow then the flow is known as *drowned* or *submerged flow*. Discharges under drowned condition are obtained by applying a reduction factor to the free flow discharges. For example, the submerged flow over a weir [Fig. 4.2(b)] is estimated by the *Villemonte formula*,

$$Q_s = Q_1\left[1-\left(\frac{H_2}{H_1}\right)^n\right]^{0.385} \tag{4.22}$$

where Q_s = submerged discharge, Q_1 = free flow discharge under head H_1, H_1 = upstream water-surface elevation measured above the weir crest, H_2 = downstream water-surface elevation measured above the weir crest, n = exponent of head in the free flow head discharge relationship [Eq. (4.21)]. For a rectangular weir, n = 1.5.

The various flow measuring structures can be broadly considered under three categories:

1. Thin-plate structures

Are usually made from a vertically set metal plate. The V-notch, rectangular full width and contracted notches are typical examples under this category.

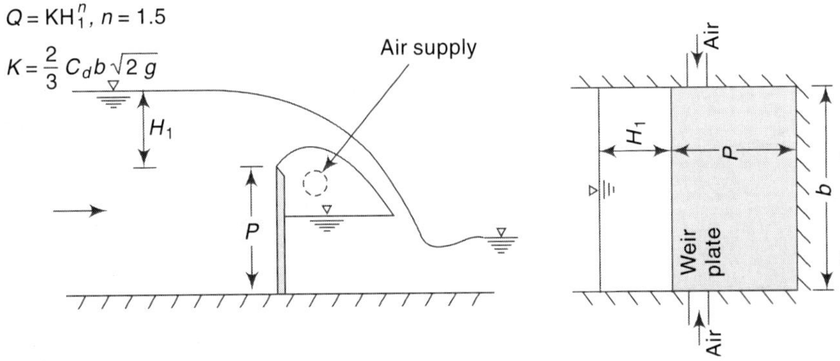

Fig. 4.20(a) ***Flow over a Weir: (a) Free Flow***

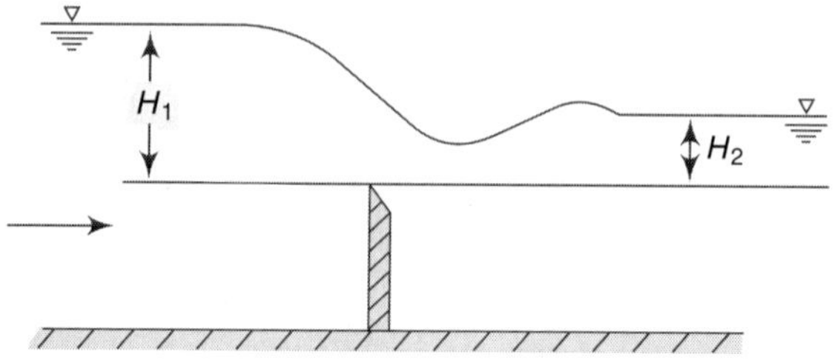

Fig. 4.20(b) ***Submerged Flow***

2. Long-base Weirs

Also known as broad-crested weirs are made of concrete or masonry and are used for large discharge values.

3. Flumes

Are made of concrete, masonry or metal sheets depending on their use and location. They depend primarily on the width constriction to produce a control section.

Details of the discharge characteristics of flow-measuring structures are available in Refs. 1, 2 and 7.

4.8.2 Slope-Area Method

The resistance equation for uniform flow in an open channel, e.g. Manning's formula can be used to relate the depths at either ends of a reach to the discharge. Figure 4.21 shows the longitudinal section of the flow in a river between two sections, 1 and 2. Knowing the water-surface elevations at the two sections, it is required to estimate the discharge. Applying the energy equation to Sections 1 and 2,

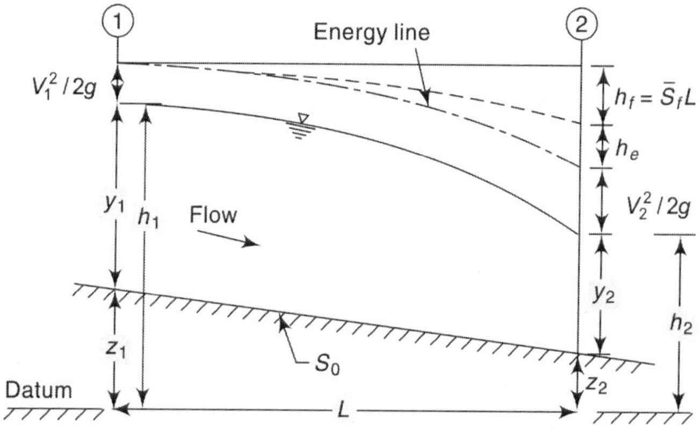

Fig. 4.21 ***Slope-area Method***

$$Z_1 + y_1 + \frac{V_1^2}{2g} = Z_2 + y_2 + \frac{V_2^2}{2g} + h_L$$

where h_L = head loss in the reach. The head loss h_L can be considered to be made up of two parts (i) frictional loss h_f, and (ii) eddy loss h_e. Denoting $z + y = h$ = water-surface elevation above the datum,

$$h_1 + \frac{V_1^2}{2g} = h_2 + \frac{V_2^2}{2g} + h_e + h_f$$

or $$h_f = (h_1 - h_2) + \left(\frac{V_1^2}{2g} - \frac{V_2^2}{2g}\right) - h_e \tag{4.23}$$

If L = length of the reach, by Manning's formula for uniform flow,

$$\frac{h_f}{L} = S_f = \text{energy slope} = \frac{Q^2}{K^2}$$

where K = conveyance of the channel = $\frac{1}{n} AR^{2/3}$

In nonuniform flow, an average conveyance is used to estimate the average energy slope and

$$\frac{h_f}{L} = S_f = \frac{Q^2}{K^2} \tag{4.24}$$

where $K = \sqrt{K_1 K_2}$; $K_1 = \frac{1}{n_1} A_1 R_1^{2/3}$ and $K_2 = \frac{1}{n_2} A_2 R_2^{2/3}$

n = Manning's roughness coefficient

The eddy loss h_e is estimated as

$$h_e = K_e \left|\frac{V_1^2}{2g} - \frac{V_2^2}{2g}\right| \tag{4.25}$$

where K_e = eddy-loss coefficient having values as below.

Cross-section characteristic of the reach	**Value of K_e**	
	Expansion	**Contraction**
Uniform	0	0
Gradual transition	0.3	0.1
Abrupt transition	0.8	0.6

Equations (4.23), (4.24) and (4.25) together with the continuity equation $Q = A_1 V_1 = A_2 V_2$ enable the discharge Q to be estimated for known values of h, channel cross-sectional properties and n.

The discharge is calculated by a trial-and-error procedure using the following sequence of calculations:

1. Assume $V_1 = V_2$. This leads to $V_1^2/2\text{ g} = V_2^2/2\text{ g}$ and by Eq. (4.23) $h_f = h_1 - h_2 = F$ = fall in the water surface between Sections 1 and 2.
2. Using Eq. (4.24), calculate discharge Q.
3. Compute $V_1 = Q/A_1$ and $V_2 = Q/A_2$. Calculate velocity heads and eddy-loss h_e.
4. Now calculate a refined value of h_f by Eq. (4.23) and go to step (2). Repeat the calculations till two successive calculations give values of discharge (or h_f) differing by a negligible margin.

This method of estimating the discharge is known as the *slope-area method*. It is a very versatile indirect method of discharge estimation and requires (i) the selection of a reach in which cross-sectional properties including bed elevations are known at its ends, (ii) the value of Manning's n, and (iii) water-surface elevations at the two end sections.

Example 4.4 During a flood flow, the depth of water in a 10 m wide rectangular channel was found to be 3.0 m and 2.9 m at two sections 200 m apart. The drop in the water-surface elevation was found to be 0.12 m. Assuming Manning's coefficient to be 0.025, estimate the flood discharge through the channel.

Solution

Using suffixes 1 and 2 to denote the upstream and downstream sections respectively, the cross-sectional properties are calculated as follows:

Section 1	Section 2
$y_1 = 3.0$ m	$y_2 = 2.90$ m
$A_1 = 30$ m^2	$A_2 = 29$ m^2
$P_1 = 16$ m	$P_2 = 15.8$ m
$R_1 = 1.875$ m	$R_2 = 1.835$ m
$K_1 = \dfrac{1}{0.025} \times 30 \times (1.875)^{2/3}$	$K_2 = \dfrac{1}{0.025} \times 29 \times (1.835)^{2/3}$
$= 1824.7$	$= 1738.9$

Average K for the reach $= \sqrt{K_1 K_2} = 1781.3$

To start with h_f = fall = 0.12 m is assumed.

Eddy loss $h_e = 0$

The calculations are shown in Table 4.1.

$$\bar{S}_f = h_f/L = h_f/200 \qquad Q = K\sqrt{\bar{S}_f} = 1781.3\sqrt{S_f}$$

$$\frac{V_1^2}{2g} = \left(\frac{Q}{30}\right)^2 /19.62, \quad \frac{V_2^2}{2g} = \left(\frac{Q}{29}\right)^2 /19.62$$

$$h_f = (h_1 - h_2) + \left(\frac{V_1^2}{2g} - \frac{V_2^2}{2g}\right)$$

$$h_f = \text{fall} + \left(\frac{V_1^2}{2g} - \frac{V_2^2}{2g}\right) = 0.12 + \left(\frac{V_1^2}{2g} - \frac{V_2^2}{2g}\right) \quad \text{(E-1)}$$

Table 4.1 ***Calculations for Example 4.4***

Trial	h_f (trial)	S_f (units of 10^{-4})	Q (m^3/s)	$V_1^2/2g$ (m)	$V_2^2/2g$ (m)	h_f by Eq. (E–1) (m)
1	0.1200	6.000	43.63	0.1078	0.1154	0.1124
2	0.1124	5.622	42.24	0.1010	0.1081	0.1129
3	0.1129	5.646	42.32	0.1014	0.1081	0.1129

(The last column is h_f by Eq. (E–1) and its value is adopted for the next trial.)

The discharge in the channel is 42.32 m^3/s.

Flood Discharge by Slope-Area Method

The slope-area method is of particular use in estimating the flood discharges in a river by past records of stages at different sections. Floods leave traces of peak elevations called high-water marks in their wake. Floating vegetative matter, such as grass, straw and seeds are left stranded at high water levels when the flood subsides and form excellent marks. Other high-water marks include silt lines on river banks, trace of erosion on the banks called *wash lines* and silt or stain lines on buildings. In connection with the estimation of very high floods, interviews with senior citizens living in the area, who can recollect from memory certain salient flood marks are valuable. Old records in archives often provide valuable information on flood marks and dates of occurrence of those floods. Various such information relating to a particular flood are cross-checked for consistency and only reliable data are retained. The slope-area method is then used to estimate the magnitude of the flood.

The selection of the reach is probably the most important aspect of the slope-area method. The following criteria can be listed towards this:

- The quality of high-water marks must be good.
- The reach should be straight and uniform as far as possible. Gradually contracting sections are preferred to an expanding reach.
- The recorded fall in the water-surface elevation should be larger than the velocity head. It is preferable if the fall is greater than 0.15 m.
- The longer the reach, the greater is the accuracy in the estimated discharge. A length greater than 75 times the mean depth provides an estimate of the reach length required.

Manning's roughness coefficient n for use in the computation of discharge is obtained from standard tables[7]. Sometimes a relation between n and the stage is prepared from measured discharges at a neighbouring gauging station and an appropriate value of n selected from it, with extrapolation if necessary.

4.9 *Stage-Discharge Relationship*

As indicated earlier, the measurement of discharge by the direct method involves a two-step procedure; the development of the stage-discharge relationship which forms the first step is of utmost importance. Once the stage-discharge $(G - Q)$ relationship is established, the subsequent procedure consists of measuring the stage (G) and reading the discharge (Q) from the $(G - Q)$ relationship. This second part is a routine operation. Thus, the aim of all current-meter and other direct-discharge measurements is to prepare a stage-discharge relationship for the given channel gauging section. The stage-discharge relationship is also known as the *rating curve.*

The measured value of discharges when plotted against the corresponding stages gives relationship that represents the integrated effect of a wide range of channel and flow parameters. The combined effect of these parameters is termed *control.* If the $(G - Q)$ relationship for a gauging section is constant and does not change with time, the control is said to be *permanent.* If it changes with time, it is called *shifting control.*

4.9.1 Permanent Control

A majority of streams and rivers, especially nonalluvial rivers exhibit permanent control. For such a case, the relationship between the stage and the discharge is a single-valued relation which is expressed as

$$Q = C_r (G - a)^{\beta} \qquad (4.26)$$

in which Q = stream discharge, G = gauge height (stage), a = a constant which represent the gauge reading corresponding to zero discharge, C_r and β are rating curve constants. This relationship can be expressed graphically by plotting the observed relative stage $(G - a)$ against the corresponding discharge values in an arithmetic or logarithmic plot [Fig. 4.22(a) and (b)]. Logarithmic plotting is advantageous as Eq. (4.26) plots as a straight line in logarithmic coordinates. In Fig. 4.22(b), the straight line is drawn to best represent the data plotted as Q *vs* $(G - a)$. Coefficients C_r and β need not be the same for the full range of stages.

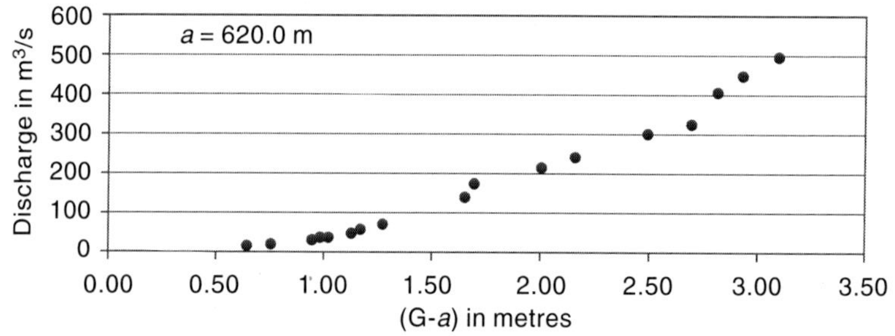

Fig. 4.22(a) ***Stage-Discharge Curve: Arithmetic Plot***

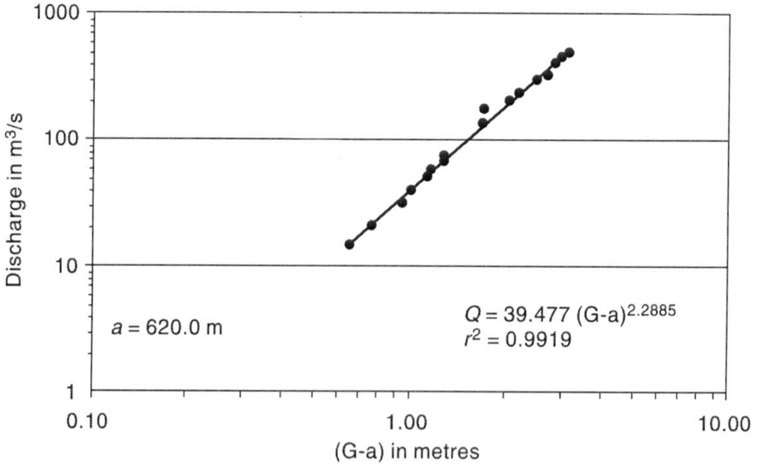

Fig. 4.22(b) ***Stage-Discharge Curve: Logarithmic Plot***

The best values of C_r and β in Eq. (4.26) for a given range of stage are obtained by the least-square-error method. Thus, by taking logarithms,

$$\log Q = \beta \log (G - a) + \log C_r \qquad (4.27)$$

or

$$Y = \beta X + b \qquad (4.27a)$$

in which the dependent variable $Y = \log Q$, independent variable $X = \log (G - a)$ and $b = \log C_r$. For the best-fit straight line of N observations of X and Y,by regressing $X = \log (G - a)$ on $Y = \log Q$

$$\beta = \frac{N(\Sigma XY) - (\Sigma X)(\Sigma Y)}{N(\Sigma X^2) - (\Sigma X)^2} \tag{4.28a}$$

and

$$b = \frac{\Sigma Y - \beta(\Sigma X)}{N} \tag{4.28b}$$

Pearson product moment correlation coefficient

$$r = \frac{N(\Sigma XY) - (\Sigma X)(\Sigma Y)}{\sqrt{[N(\Sigma X^2) - (\Sigma X)^2][N(\Sigma Y^2) - (\Sigma Y)^2]}} \tag{4.29}$$

Here, r reflects the extent of linear relationship between the two data sets.

For a perfect correlation, $r = 1.0$. If r is between 0.6 and 1.0, it is generally taken as a good correlation. It should be noted that in the present case, as the discharge Q increases with $(G - a)$, the variables Y and X are positively correlated and hence r is positive. Equation (4.26), viz.

$$Q = C_r(G - a)^{\beta}$$

is called the *rating equation* of the stream and can be used for estimating the discharge Q of the stream for a given gauge reading G within range of data used in its derivation.

4.9.2 Stage for Zero Discharge, *a*

In Eq. (4.26), the constant a representing the stage (gauge height) for zero discharge in the stream is a hypothetical parameter and cannot be measured in the field. As such, its determination poses some difficulties. The following alternative methods are available for its determination:

1. Plot Q *vs* G on an arithmetic graph paper and draw a best-fit curve. By extrapolating the curve by eye judgment, find a as the value of G corresponding to $Q = 0$. Using the value of a, plot $\log Q$ *vs* $\log (G - a)$ and verify whether the data plots as a straight line. If not, select another value in the neighbourhood of previously assumed value and by trial and error find an acceptable value of a which gives a straight line plot of $\log Q$ *vs* $\log (G - a)$.
2. A graphical method due to Running[8] is as follows. The Q *vs* G data are plotted to an arithmetic scale and a smooth curve through the plotted points are drawn. Three points A, B and C on the curve are selected such that their discharges are in geometric progression (Fig. 4.23), i.e.

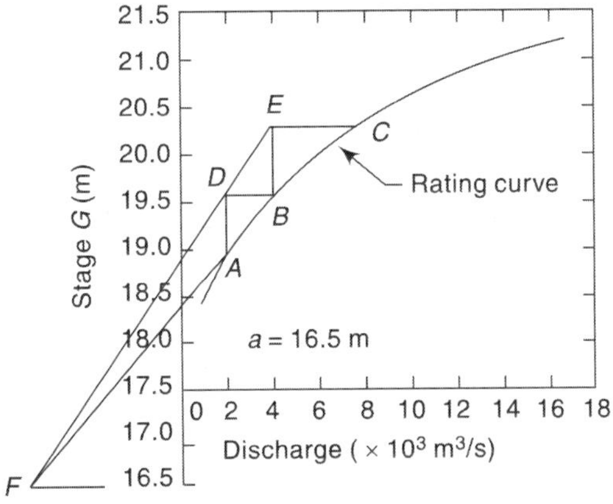

Fig. 4.23 ***Running's Method for Estimation of the Constant a***

$$\frac{Q_A}{Q_B} = \frac{Q_B}{Q_C}$$

At A and B, vertical lines are drawn and then horizontal lines are drawn at B and C to get D and E as intersection points with the verticals. Two straight lines ED and BA are drawn to intersect at F. The ordinate at F is the required value of a, the gauge height corresponding to zero discharge. This method assumes the lower part of the stage-discharge curve to be a parabola.

3. Plot Q *vs* G to an arithmetic scale and draw a smooth good-fitting curve by eye-judgement. Select three discharges Q_1, Q_2 and Q_3 such that $Q_1/Q_2 = Q_2/Q_3$ and note from the curve the corresponding values of gauge readings G_1, G_2 and G_3.From Eq. (4.26)

$$(G_1 - a)/(G_2 - a) = (G_2 - a)/(G_3 - a)$$

i.e.
$$a = \frac{G_1 G_3 - G_2^2}{(G + G_3) - 2G_2} \tag{4.30}$$

4. A number of optimisation procedures are available to estimate the best value of a. A trial-and-error search for a which gives the best value of the correlation coefficient is one of them.

Example 4.5 Following are the data of gauge and discharge collected at a particular section of the river by stream gauging operation. (a) Develop a gauge-discharge relationship for this stream at this section for use in estimating the discharge for a known gauge reading. What is the coefficient of correlation of the derived relationship? Use a value of a = 7.50 m for the gauge reading corresponding to zero discharge. (b) Estimate the discharge corresponding to a gauge reading of 10.5 m at this gauging section.

Gauge reading (m)	Discharge (m^3/s)	Gauge reading (m)	Discharge (m^3s)
7.65	15	8.48	170
7.70	30	8.98	400
7.77	57	9.30	600
7.80	39	9.50	800
7.90	60	10.50	1500
7.91	100	11.10	2000
8.08	150	11.70	2400

Solution:

(a) The gauge–discharge equation is $Q = C_r(G - a)^\beta$

Taking the logarithms $\log Q = \beta \log (G - a) + \log C_r$

or $Y = \beta X + b$

where $Y = \log Q$ and $X = \log (G - a)$.

Values of X, Y and XY are calculated for all the data as shown in Table 4.2.

Table 4.2 ***Calculations for Example 4.6***

$a = 7.5$ m $N = 14$

Stage (G) (metres)	($G - a$) (m)	Discharge (Q)(m^3/s)	log ($G - a$) $= X$	log Q $= Y$	XY	X^2	Y^2
7.65	0.15	15	–0.824	1.176	–0.969	0.679	1.383
7.70	0.20	30	–0.699	1.477	–1.032	0.489	2.182
7.77	0.27	57	–0.569	1.756	–0.998	0.323	3.083
7.80	0.30	39	–0.523	1.591	–0.832	0.273	2.531
7.90	0.40	60	–0.398	1. 778	–0.708	0.158	3.162
7.91	0.41	100	–0.387	2.000	–0.774	0.150	4.000
8.08	0.58	150	–0.237	2.176	–0.515	0.056	4.735
8.48	0.98	170	–0.009	2.230	–0.020	0.000	4.975
8.98	1.48	400	0.170	2.602	0.443	0.029	6.771
9.30	1.80	600	0.255	2.778	0.709	0.065	7.718
9.50	2.00	800	0.301	2.903	0.874	0.091	8.428
10.50	3.00	1500	0.477	3.176	1.515	0.228	10.088
11.10	3.60	2000	0.556	3.301	1.836	0.309	10.897
11.70	4.20	2400	0.623	3.380	2.107	0.388	11.426
		Sum	**–1.262**	**32.325**	**1.636**	**3.239**	**81.379**

From the above table:

$\Sigma X = -1.262$ $\Sigma Y = 32.325$ $\Sigma XY = 1.636$

$\Sigma X^2 = 3.239$ $\Sigma Y^2 = 81.39$

$(\Sigma X)^2 = 1.5926$ $(\Sigma Y)^2 = 1044.906$ $N = 14$

By using Eq. (4.28a),

$$\beta = \frac{N(\Sigma XY) - (\Sigma X)(\Sigma Y)}{N(\Sigma X^2) - (\Sigma X)^2} = \frac{(14 \times 1.636) - (-1.262)(32.325)}{(14 \times 3.239) - (-1.262)^2} = 1.4558$$

By Eq. (4.28b),

$$b = \frac{\Sigma Y - \beta(\Sigma X)}{N} = \frac{(32.325) - 1.4558 \times (-1.262)}{14} = 2.440$$

Hence, $C_r = 275.52$

The required gauge–discharge relationship is therefore

$$Q = 275.52\,(G - a)^{1.456}$$

By Eq. 4.29, coefficient of correlation

$$r = \frac{N(\Sigma XY) - (\Sigma X)(\Sigma Y)}{\sqrt{[N(\Sigma X^2) - (\Sigma X)^2][N(\Sigma Y^2) - (\Sigma Y)^2]}}$$

$$= \frac{(14 \times 1.636) - (-1.262)(32.325)}{\sqrt{[(14 \times 3.239) - (1.5926)][(14 \times 81.379) - (1044.906)]}} = 0.9913$$

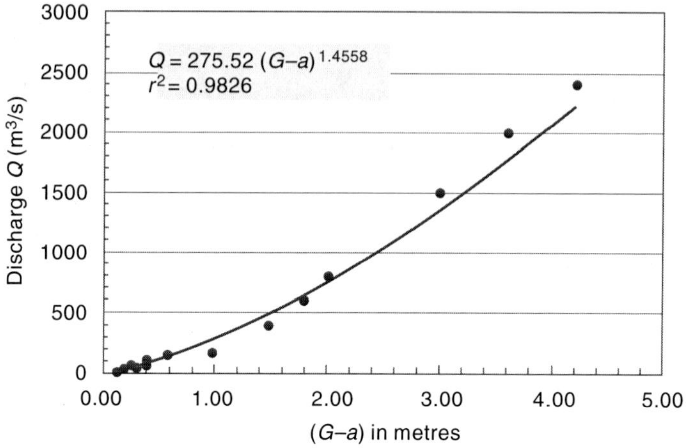

Fig. 4.24(a) ***Stage-discharge Relation (Arithmetic Plot)—Example 4.5***

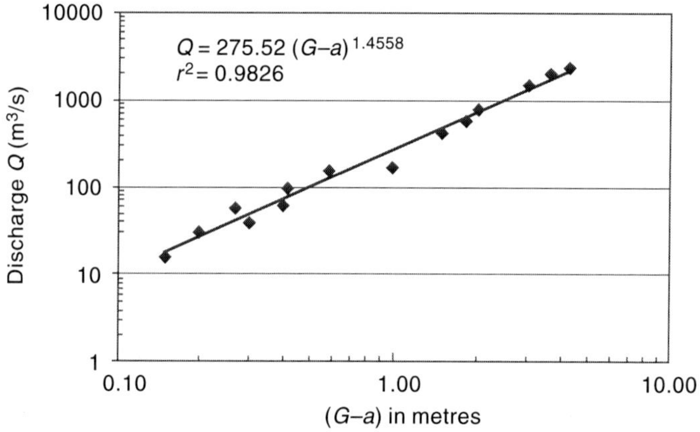

Fig. 4.24(b) ***Stage-discharge Relationship (Logarithmic Plot)—Example 4.5***

As the value of r is nearer to unity, the correlation is very good.
The variation of discharge (Q) with relative stage ($G - a$) is shown in Fig. 4.24(a)—arithmetic plot and in Fig. 4.24(b)—logarithmic plot.
(b) when $G = 10.05$: as $a = 7.5$ m

$$G = 275.52\ (10.05 - 7.50)^{1.456} = 1076\ \text{m}^3/\text{s}$$

Shifting Control

The control that exists at a gauging section giving rise to a unique stage-discharge relationship can change due to: (i) changing characteristics caused by weed growth, dredging or channel encroachment, (ii) aggradation or degradation phenomenon in an alluvial channel, (iii) variable backwater effects affecting the gauging section, and (iv) unsteady flow effects of a rapidly changing stage. There are no permanent corrective measure to tackle the shifting controls due to causes (i) and (ii) listed above. The only recourse in such cases is to have frequent current-meter gaugings and to update the rating curves. Shifting controls due to causes (iii) and (iv) are described below.

4.9.3 Backwater Effect

If the shifting control is due to variable backwater curves, the same stage will indicate different discharges depending upon the backwater effect. To remedy this situation another gauge, called the *secondary gauge* or *auxiliary gauge* is installed some distance downstream of the gauging section and readings of both gauges are taken. The difference between the main gauge and the secondary gauge gives the *fall* (F) of the water surface in the reach. Now, for a given main-stage reading, the discharge under variable backwater condition is a function of the fall F, i.e.

$$Q = f(G, F)$$

Schematically, this functional relationship is shown in Fig. 4.25. Instead of having a three-parameter plot, the observed data is normalised with respect to a constant fall value. Let F_0 be a normalising value of the fall taken to be constant at all stages and F the actual fall at a given stage when the actual discharge is Q. These two fall values are related as

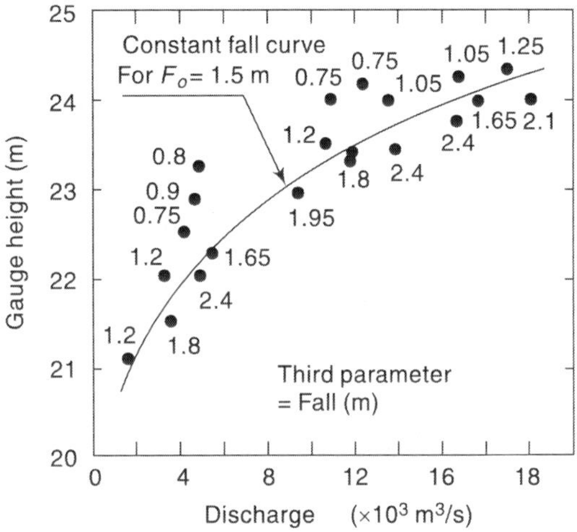

Fig. 4.25 ***Backwater Effect on a Rating Curve—Normalised Curve***

$$\frac{Q}{Q_0} = \left(\frac{F}{F_0}\right)^m \tag{4.31}$$

in which Q_0 = normalised discharge at the given stage when the fall is equal to F_0 and m = an exponent with a value close to 0.5. From the observed data, a convenient value of F_0 is selected. An approximate Q_0 *vs* G curve for a constant F_0 called *constant-fall curve* is drawn. For each observed data, Q/Q_0 and F/F_0 values are calculated and plotted as Q/Q_0 *vs* F/F_0 (Fig. 4.26). This is called the *adjustment curve*. Both the constant-fall curve and the

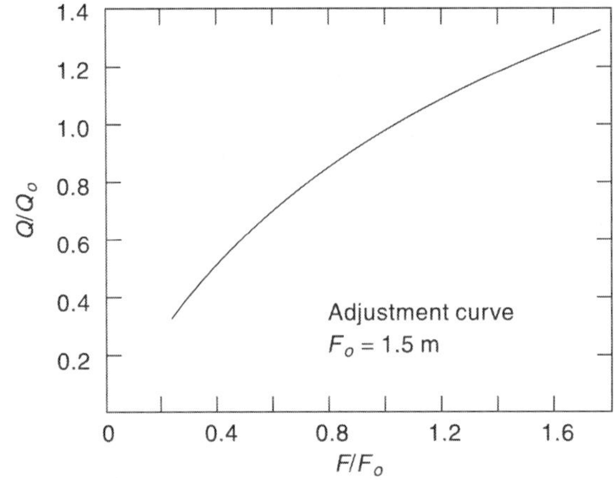

Fig. 4.26 ***Backwater Effect on a Rating Curve—Adjustment Curve***

adjustment curve are refined, by trial and error to get the best-fit curves. When finalised, these two curves provide the stage-discharge information for gauging purposes. For example, if the observed stage is G_1 and fall F_1, first by using the adjustment curve, the value of Q_1/Q_0 is read for a known value of F_1/F_0. Using the constant fall-rating curve, Q_0 is read for the given stage G_1 and the actual discharge calculated as $(Q_1/Q_0) \times Q_0$.

Unsteady-Flow Effect

When a flood wave passes a gauging station in the advancing portion of the wave, the approach velocities are larger than in the steady flow at corresponding stage. Thus, for the same stage, more-discharge than in a steady uniform flow occurs. In the retreating phase of the flood wave the converse situation occurs with reduced approach velocities giving lower discharges than in an equivalent steady flow case. Thus, the stage-discharge relationship for an unsteady flow will not be a single-valued relationship as in steady flow but it will be a looped curve as in Fig. 4.27. It may be noted that at the same stage, more discharge passes through the river during rising stages than in falling ones. Since the conditions for each flood may be different, different floods may give different loops.

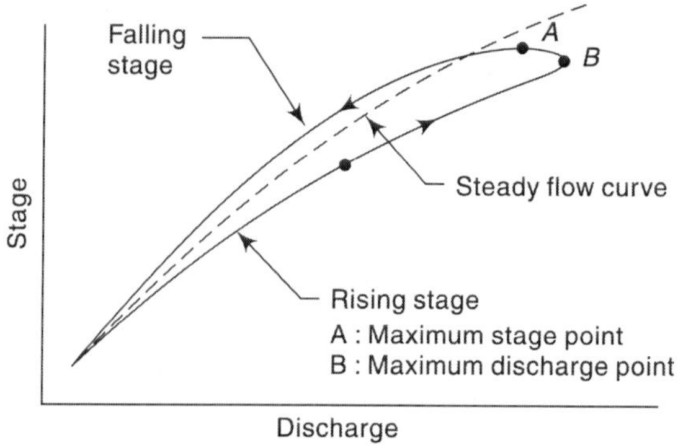

Fig. 4.27 ***Loop Rating Curve***

If Q_n is the normal discharge at a given stage under steady uniform flow ad Q_m is the measured (actual) unsteady flow the two are related as[7]

$$\frac{Q_M}{Q_n} = \sqrt{1 + \frac{1}{V_w S_0} \frac{dh}{dt}} \tag{4.32}$$

where S_0 = channel slope = water surface slope at uniform flow, dh/dt = rate of change of stage and V_w = velocity of the flood wave. For natural channels, V_w is usually assumed equal to 1.4 V, where V = average velocity for a given stage estimated by applying Manning's formula and the energy slope S_f. Also, the energy slope is used in place of S_0 in the denominator of Eq. (4.32). If enough field data about the flood magnitude and dh/dt are available, the term $(1/V_w S_0)$ can be calculated and plotted against the stage for use in Eq. (4.32). For estimating the actual discharge at an observed stage, Q_M/Q_n is calculated by using the observed data of dh/dt. Here Q_n is the discharge corresponding to the observed stage relationship for steady flow in the channel reach.

Example 4.6 An auxiliary gauge was used downstream of a main gauge in a river to provide corrections to the gauge-discharge relationship due to backwater effects. The following data were noted at a certain main gauge reading.

Main gauge (m above datum)	Auxiliary gauge (m above datum)	Discharge (m^3/s)
86.00	85.50	275
86.00	84.80	600

If the main gauge reading is still 86.00 m and the auxiliary gauge reads 85.30 m, estimate the discharge in the river.

Solution

Fall (F) = main gauge reading – auxiliary gauge reading.

When $F_1 = (86.00 - 85.50) = 0.50$ m $\quad Q_1 = 275\ m^3/s$

$F_2 = (86.00 - 84.80) = 1.20$ m $\quad Q_2 = 600\ m^3/s$

By Eq. (4.31), $(Q_1/Q_2) = (F_1/F_2)^m$ or $(275/600) = (0.50/1.20)^m$

Hence, $m = 0.891$

When the auxiliary gauge reads 85.30 m, at a main gauge reading of 86.00 m,

Fall $F = (86.00 - 85.30) = 0.70$ m and

$$Q = Q_2\,(F/F_2)^m = 600\,(0.70/1.20)^{0.891} = 371\ m^3/s$$

4.10 *Extrapolation of Rating Curve*

Most hydrological designs consider extreme flood flows. As an example, in the design of hydraulic structures, such as barrages, dams and bridges one needs maximum flood discharges as well as maximum flood levels. While the design flood discharge magnitude can be estimated from other considerations, the stage-discharge relationship at the project site will have to be used to predict the stage corresponding to design-flood discharges. Rarely will the available stage-discharge data include the design-flood range and hence the need for extrapolation of the rating curve.

Before attempting extrapolation, it is necessary to examine the site and collect relevant data on changes in the river cross-section due to flood plains, roughness and backwater effects. The reliability of the extrapolated value depends on the stability of the gauging section control. A stable control at all stages leads to reliable results. Extrapolation of the rating curve in an alluvial river subjected to aggradation and degradation is unreliable and the results should always be confirmed by alternate methods. There are many techniques of extending the rating curve and two well-known methods are described here.

4.10.1 Conveyance Method

The conveyance of a channel in nonuniform flow is defined by the relation

$$Q = K\sqrt{S_f} \tag{4.33}$$

where Q = discharge in the channel, S_f = slop of the energy line and K = conveyance. If Manning's formula is used,

$$K = \frac{1}{n} AR^{2/3} \tag{4.34}$$

where n = Manning's roughness, A = area of cross section and, R = hydraulic radius. Since A and R are functions of the stage, the values of K for various values of stage are calculated by using Eq. (4.34) and plotted against the stage. The range of the stage should include values beyond the level up to which extrapolation is desired. Then a smooth curve is fitted to the plotted points as shown in Fig. 4.28(a). Using the available discharge and stage data, values of S_f are calculated by using Eq. (4.33) as $S_f = Q^2/K^2$ and are plotted against the stage. A smooth curve is fitted through the plotted points as shown in Fig. 4.28(b). This curve is then extrapolated keeping in mind that S_f approaches a constant value at high stages.

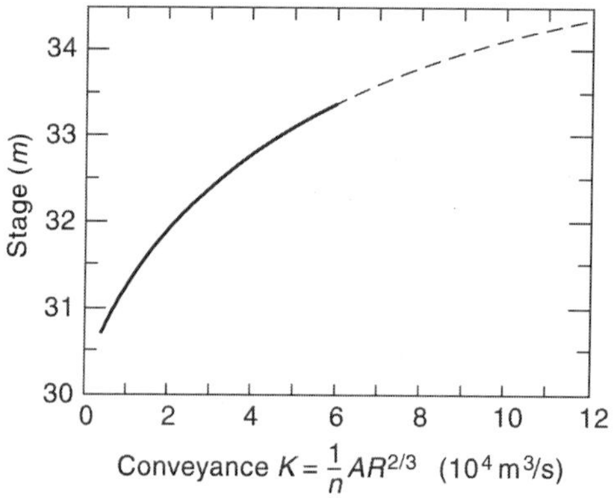

Fig. 4.28(a) ***Conveyance Method of Rating Curve Extension: K vs Stage***

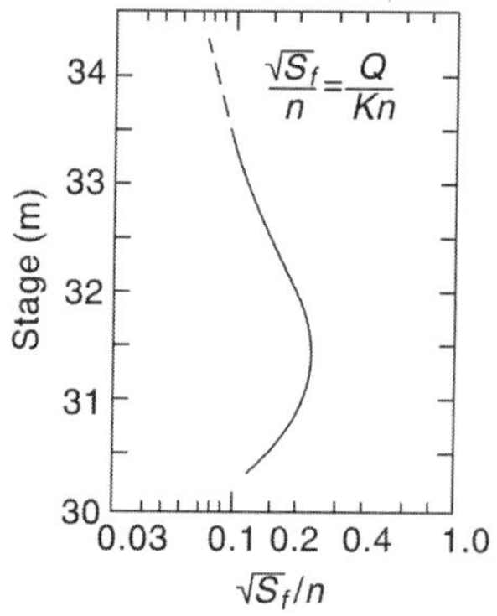

Fig. 4.28(b) ***Conveyance Method of Rating Curve Extension: S_f vs Stage***

Using the conveyance and slope curves, the discharge at any stage is calculated as $Q = K\sqrt{S_f}$ and a stage-discharge curve covering the desired range of extrapolation is constructed. With this extrapolated-rating curve, the stage corresponding to a design-flood discharge can be obtained.

4.10.2 Logarithmic-Plot Method

In this technique, the stage-discharge relationship given by Eq. (4.26) is made use of. The stage is plotted against the discharge on a log–log paper. A best-fit linear relationship is obtained for data points lying in the high-stage range and the line is extended to cover the range of extrapolation. Alternatively, coefficients of Eq. (4.26) are obtained by the least-square-error method by regressing X on Y in Eq. (4.27a). For this Eq. (4.27a) is written as

$$X = \alpha Y + C \tag{4.35}$$

where the dependent variable $X = \log (G - a)$ an $Y = \log Q$. The coefficients α and C are obtained as,

$$\alpha = \frac{N(\Sigma XY) - (\Sigma Y)(\Sigma X)}{N(\Sigma Y^2) - (\Sigma Y)^2} \tag{4.35a}$$

and
$$C = \frac{(\Sigma X) - \alpha(\Sigma Y)}{N} \tag{4.35b}$$

The relationship governing the stage and discharge is now
$$(G - a) = C_1 Q^{\alpha} \tag{4.36}$$

where C_1 = antilog C.

By the use of Eq. (4.36), the value of the stage corresponding to a design flood discharge is estimated.

Example 4.7 For the stage-discharge data of Example 4.4, fit a regression equation for use in estimation of stage for a known value of discharge. Use a value of 7.50 m as the gauge reading corresponding to zero discharge. Determine the stage for a discharge of 3500 m^3/s.

Solution

The regression equation is $X = \alpha Y + C$ (Eq. 4.35)

where $X = \log(G - a)$ and $Y = \log Q$. The value of α is given by Eq. (4.35a) as
$$\alpha = \frac{N(\Sigma XY) - (\Sigma Y)(\Sigma X)}{N(\Sigma Y^2) - (\Sigma Y)^2}$$

Values of X, Y and XY are the same as calculated for the data in Table 4.3. Thus,

$\Sigma X = -1.262$ $\quad\Sigma Y = 32.325$ $\quad\Sigma XY = 1.636$

$\Sigma X^2 = 3.239$ $\quad\Sigma Y^2 = 81.379$

$(\Sigma X)^2 = 1.5926$ $\quad(\Sigma Y)^2 = 1044.906$ $\quad N = 14$

Substituting these values in Eq (4.35)
$$\alpha = \frac{(14 \times 1.636) - (32.325)(-1.262)}{(14 \times 81.379) - (1044.906)} = 0.675$$

The coefficient C is given by Eq. (4.35b) as
$$C = \frac{(\Sigma X) - \alpha(\Sigma Y)}{N} = \frac{(-1.262) - 0.675(32.325)}{14} = -1.6486$$

C_1 = antilog $C = 0.02246$ leading to the gauge-discharge equation as

$$(G - a) = 0.02246\, Q^{0.675}$$

The variation of relative stage $(G - a)$ with discharge (Q) is shown in Fig. 4.29(a)—arithmetic plot and in Fig. 4.29(b)—logarithmic plot.

(b) When $Q = 3500$ m^3/s and given that $a = 7.50$ m
$$(G - 7.50) = 0.02246\,(3500)^{0.675} = 5.540 \text{ m}$$
$$G = 13.04 \text{ m}$$

4.11 *Hydrometry Stations*

As the measurement of discharge is of paramount importance in applied hydrologic studies, considerable expenditure and effort are being expended in every country to collect and store this valuable historic data. The *WMO* recommendations for the minimum number of hydrometry stations in various geographical regions are given in Table 4.3.

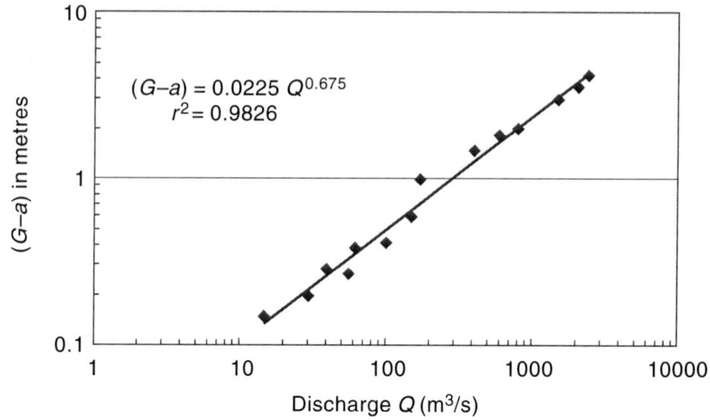

Fig. 4.29(a) ***Discharge-stage Relationship: Example 4.7 (Logarithmic Plot)***

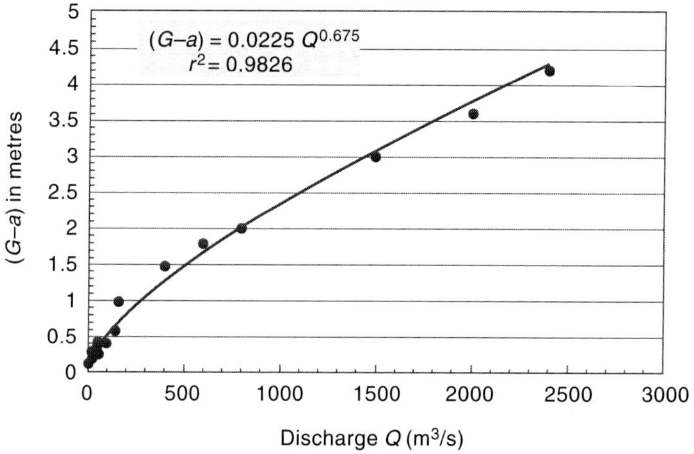

Fig. 4.29(b) ***Discharge-stage Relationship: Example 4.7 (Arithmetic Plot)***

Table 4.3 ***WMO Criteria for Hydrometry Station Density***

S. No.	Region	Minimum density (km^2/station)	Tolerable density under difficult conditions (km^2/station)
1.	Flat region of temperate, mediterranean and tropical zones	1,000 – 2,500	3,000 – 10,000
2.	Mountainous regions of temperate mediterranean and tropical zones	300 – 1,000	1,000 – 5,000
3.	Arid and polar zones	5,000 – 20,000	

Hydrometry stations must be sited in adequate number in the catchment area of all major streams so that the water potential of an area can be assessed as accurately as possible.

As a part of hydrological observation activities, *CWC* operates a vast network of 877 hydrological observation stations on various state and interstate rivers for collection of gauge, discharge, silt and water quality data which are stored after analysis in a central data bank. In addition to observation of river flow, *CWC* is also monitoring water quality, covering all the major river basins of India. The distribution of various kinds of *CWC* hydrological observation stations is as follows:

Type of Station	Number
Gauge observation only	236
Gauge—Discharge	281
Gauge—Discharge and Silt	41
Gauge—Discharge and water quality	80
Gauge—Discharge, water quality and Silt	239

In a few gauging stations on major rivers, moving-boat method facilities exist. Reports containing the gauge, discharge, sediment and water quality data are brought out by *CWC* every year as *Yearbooks*. In addition to the above, the state governments maintain nearly 800 gauging stations. Further, in most of the states institutional arrangements exist for collection, processing and analysis of hydrometric and hydrometeorological data and publication of the same.

REFERENCES

1. Ackers, P. et al., *Weirs and Flumes for Flow Measurement,* Wiley Interscience, John Wiley, Chichester, U.K., 1978.
2. Bos, M.G. (Ed.), *Discharge Measuring Structures*, Int. Inst. for Land Reclamation and Improvement, Wageningen, The Netherlands, Pb. No. 20, 1976.
3. Central Water Commission, *Water Resources of India*, *CWC* Pub. No. 30/88, *CWC*, New Delhi, India, 1988.
4. Chow, V.T. (Ed.), *Handbook of Applied Hydrology,* McGraw-Hill, New York, *USA*, 1964.
5. Hershey, R.W. (Ed.), *Hydrometry*, Wiley Interscience, John Wiley, Chichester, *U.K.*, 1978.
6. Kolupaila, S., *Bioliography of Hydrometry,* Univ. of Notre Dame, Notre Dame, *USA*, 1960.
7. Subramanya, K., *Flow in Open Channels,* 3 Ed. Tata McGraw-Hill, New Delhi, India, 2009.
8. Wisler, C.O., and E.F. Brater, *Hydrology,* John Wiley, New York, USA, 1959.
9. W.M.O. Manual on Stream gauging, *WMO*-No. 1044, Switzerland, 2010.

REVISION QUESTIONS

4.1 Explain the various commonly used methods of measurement of stage of a river. Indicate for each method its specific advantage and the conditions under which one would use it.

4.2 What factors should be considered in selecting a site for a stream gauging station?

4.3 Explain the salient features of a current meter. Describe briefly the procedure of using a current meter for measuring velocity in a stream.
4.4 List the qualities of a good tracer for use in dilution technique of flow measurement.
4.5 Explain briefly the dilution method of flow measurement.
4.6 Explain the streamflow measurement by area-velocity method.
4.7 Describe briefly the moving boat method of stream flow measurement.
4.8 Describe the slope-area method of measurement of flood discharge in a stream.
4.9 Explain the procedure for obtaining the stage-discharge relationship of a stream by using the stage-discharge data from a site with permanent control.
4.10 Describe briefly:
(a) Backwater effect on a rating curve.
(b) Unsteady flow effect on a rating curve
4.11 Describe a procedure for extrapolating a rating curve of a stream.
4.12 Discuss the advantages and disadvantages of the following relative to the flow measurement by using current meters:
(a) Electromagnetic method (b) Ultrasound method
4.13 Explain briefly the important aspects relating to the following instruments
(a) Float-gauge recorder (b) Bubble gauge
(c) Echo-depth recorder (d) Current meter
4.14 Explain briefly the procedures used for extrapolating of ratings curves.

PROBLEMS

4.1 The following data were collected during a stream-gauging operation in a river. Compute the discharge.

Distance from left water edge (m)	Depth (m)	Velocity (m/s)	
		at 0.2 *d*	at 0.8 *d*
0.0	0.0	0.0	0.0
1.5	1.3	0.6	0.4
3.0	2.5	0.9	0.6
4.5	1.7	0.7	0.5
6.0	1.0	0.6	0.4
7.5	0.4	0.4	0.3
9.0	0.0	0.0	0.0

4.2 The velocity distribution in a stream is usually approximated as $v/v_a = (y/a)^m$, where v and v_a are velocities at heights y and a above the bed respectively and m is a coefficient with a value between 1/5 to 1/8. (i) Obtain an expression for $v/\bar{v}$, where $\bar{v}$ is the mean velocity in terms of the depth of flow. (ii) If $m = 1/6$ show that (a) the measured velocity is equal to the mean velocity if the velocity is measured at 0.6 depth from the water surface, and (b) $\bar{v} = \frac{1}{2}(v_{0.2} + v_{0.82})$, where $v_{0.2}$ and $v_{0.82}$ are the velocities measured at 0.2 and 0.82 depths below the water surface respectively.

4.3 The following are the data obtained in a stream-gauging operation. A current meter with a calibration equation $V = (0.32N + 0.032)$ m/s, where N = revolutions per second was used to measure the velocity at 0.6 depth. Using the mid-section method, calculate the discharge in the stream.

Distance from right bank (m)	0	2	4	6	9	12	15	18	20	22	23	24
Depth (m)	0	0.50	1.10	1.95	2.25	1.85	1.75	1.65	1.50	1.25	0.75	0
Number of revolutions	0	80	83	131	139	121	114	109	92	85	70	0
Observation Time (s)	0	180	120	120	120	120	120	120	120	120	150	0

4.4 In the moving-boat method of discharge measurement, the magnitude (V_R) and direction (θ) of the velocity of the stream relative to the moving boat are measured. The depth of the stream is also simultaneously recorded. Estimate the discharge in a river that gave the following moving-boat data. Assume the mean velocity in a vertical to be 0.95 times the surface velocity measured by the instrument.

Section	V_R (m/s)	θ (degrees)	Depth (m)	Remark
0	—	—	—	Right bank.
1	1.75	55	1.8	
2	1.84	57	2.5	
3	2.00	60	3.5	θ is the angle made by V_R
4	2.28	64	3.8	with the boat direction
5	2.30	65	4.0	
6	2.20	63	3.8	The various sections are
7	2.00	60	3.0	spaced at a constant
8	1.84	57	2.5	distance of 75 m apart
9	1.70	54	2.0	
10	—	—	—	Left bank

4.5 The dilution method with gulp injection procedure was used to measure the discharge of a small but rapid stream. A sodium dichromate solution was suddenly injected at station A. In this, 60 litres of solution with concentration of 300 mg/cc was used. The data of concentration measurements at station B on the downstream indicated a time-concentration curve that could be approximated as a triangle with a base of 5 minutes and the maximum concentration of 12 parts per million by weight.. The base concentration of the dye in the stream was zero. Estimate the discharge in the stream.

4.6 A 500 g/*l* solution of sodium dichromate was used as chemical tracer. It was dosed at a constant rate of 4 *l*/s. At a downstream section, the equilibrium concentration was measured as 4 parts per million (ppm). Estimate the discharge in the stream.

4.7 A 200 g/*l* solution of common salt was discharged into a stream at a constant rate of 25 *l*/s. The background concentration of the salt in the stream water was found to be 10 ppm. At a downstream section where the solution was believed to have been completely mixed, the salt concentration was found to reach an equilibrium value of 45 ppm. Estimate the discharge in the stream.

4.8 It is proposed to adopt the dilution method of stream gauging for a river whose hydraulic properties at average flow are as follows: width = 45 m, depth = 2.0 m, discharge = 85 m^3/s, Chezy coefficient = 20 to 30. Determine the safe mixing length that has to be adopted for this stream.

4.9 During a high flow, water-surface elevations of a small stream were noted at two sections *A* and *B*, 10 km apart. These elevations and other salient hydraulic properties are given below.

Section	Water-surface elevation (m)	Area of cross-section (m^2)	Hydraulic radius (m)	Remarks
A	104.771	73.293	2.733	*A* is upsteam of *B*
B	104.500	93.375	3.089	$n = 0.020$

The eddy loss coefficients of 0.3 for gradual expansion and 0.1 for gradual contraction are appropriate. Estimate the discharge in the stream.

4.10 A small stream has a trapezoidal cross section with base width of 12 m and side slope 2 horizontal: 1 vertical in a reach of 8 km. During a flood the high water levels record at the ends of the reach are as follows.

Section	Elevation of bed (m)	Water surface elevation (m)	Remarks
Upstream	100.20	102.70	Manning's $n = 0.030$
Downstream	98.60	101.30	

Estimate the discharge in the stream.

4.11 The stage-discharge data of a river are given below. Establish the stage-discharge relationship to predict the discharge for a given stage. Assume the value of stage for zero discharge as 35.00 m. (2) What is the correlation coefficient of the relationship established above? (3) Estimate the discharge corresponding to stage values of 42.50 m and 48.50 m respectively.

Stage (m)	Discharge (m^3/s)	Stage (m)	Discharge (m^3/s)
35.91	89	39.07	469
36.90	230	41.00	798
37.92	360	43.53	2800
44.40	3800	48.02	5900
45.40	4560	49.05	6800
46.43	5305	49.55	6900
		49.68	6950

4.12 Downstream of a main gauging station, an auxiliary gauge was installed and the following readings were obtained.

Main gauge (m)	Auxiliary gauge (m)	Discharge (m^3/s)
121.00	120.50	300
121.00	119.50	580

What discharge is indicated when the main gauge reading is 121.00 m and the auxiliary gauge reads 120.10 m.

4.13 The following are the coordinates of a smooth curve drawn to best represent the stage-discharge data of a river.

Stage (m)	20.80	21.42	21.95	23.37	23.00	23.52	23.90
Discharge (m^3/s)	100	200	300	400	600	800	1000

Determine the stage corresponding to zero discharge.

4.14 The stage-discharge data of a river are given below. Establish a stage-discharge relationship to predict the stage for a known discharge. Assume the stage value for zero discharge as 20.50 m. Determine the stage of the river corresponding to a discharge of 2600 m^3/s.

Stage (m)	Discharge (m^3/s)	Stage (m)	Discharge (m^3/S)
21.95	100	24.05	780
22.45	220	24.55	1010
22.80	295	24.85	1220
23.00	400	25.40	1300
23.40	490	25.15	1420
23.75	500	25.55	1550
23.65	640	25.90	1760

(*Hint:* Use Eq. 4.35)

4.15 During a flood the water surface at a section in a river was found to increase at a rate of 11.2 cm/h. The slope of the river is 1/3600 and the normal discharge for the river stage read from a steady-flow rating curve was 160 m^3/s . If the velocity of the flood wave can be assumed as 2.0 m/s, determine the actual discharge.

OBJECTIVE QUESTIONS

4.1 The science and practice of water flow measurement is known as
(a) hypsometry (b) hydro-meteorology
(c) fluvimetry (d) hydrometry

4.2 The following is *not* a direct stream flow determination technique:
(a) Dilution method (b) Ultrasonic method
(c) Area-velocity method (d) Slope-area method

4.3 A stilling well is required when the stage measurement is made by employing a
(a) bubble gauge (b) float gauge recorder
(c) vertical staff gauge (d) inclined staff gauge

4.4 In a river carrying a discharge of 142 m^3/s, the stage at a station *A* was 3.6 m and the water surface slope was 1 in 6000. If during a flood the stage at *A* was 3.6 m and the water surface slope was 1/3000, the flood discharge (in m^3/s) was approximately
(a) 100 (b) 284 (c) 71 (d) 200

4.5 In a triangular channel, the top width and depth of flow were 2.0 m and 0.9 m respectively. Velocity measurements on the centre line at 18 cm and 72 cm below water surface indicated velocities of 0.60 m/s and 0.40 m/s respectively. The discharge in the channel (in m^3/s) is

(a) 0.90 (b) 1.80 (c) 0.45 (d) none of these.

4.6 In the moving-boat method of stream-flow measurement, the essential measurements are

(a) the velocity recorded by the current meter, the depths and the speed of the boat.
(b) the velocity and direction of the current meter, the depths and the time interval between depth readings
(c) the depth, time interval between readings, speed of the boat and velocity of the stream
(d) the velocity and direction of the current meter and the speed of the boat.

4.7 Which of the following instruments is *not* connected with stream flow measurement?

(a) Hygrometer (b) Echo-depth recorder
(c) Electromagnetic flow meter (d) Sounding weight

4.8 The surface velocity at any vertical section of a stream is

(a) not of any use in stream flow measurement
(b) smaller than the mean velocity in that vertical
(c) larger than the mean velocity in that vertical section
(d) equal to the velocity in that vertical at 0.6 times the depth

4.9 If a gauging section is having shifting control due to backwater effects then

(a) a loop rating curve results
(b) the section is useless for stream-gauging purposes
(c) the discharge is determined by area-velocity methods
(d) a secondary gauge downstream of the section is needed

4.10 The stage discharge relation in a river during the passage of a flood wave is measured. If Q_R = discharge at a stage when the water surface was rising and Q_F = discharge at the same stage when the water surface was falling then

(a) $Q_F = Q_R$ (b) $Q_R > Q_F$
(c) $Q_R < Q_F$ (d) Q_R/Q_F = constant at all stages

4.11 A large irrigation canal can be approximated as a wide rectangular channel and Manning's formula is applicable to describe the flow in it. If the gauge (G) is related to discharge (Q) as

$$Q = C_r(G - a)^\beta$$

where a = gauge height at zero discharge, the value of β is

(a) 1.67 (b) 1.50 (c) 2.50 (d) 0.67

4.12 The dilution method of stream gauging is ideally suited for measuring discharges in

(a) a large alluvial river
(b) flood flow in a mountain stream
(c) steady flow in a small turbulent stream
(d) a stretch of a river having heavy industrial pollution loads

4.13 A 400 g/*l* solution of common salt was discharged into a stream at a constant rate of 45 *l*/s. At a downstream section where the salt solution is known to have completely mixed with the stream flow the equilibrium concentration was read as 120 ppm. If a background concentration of 20 ppm is applicable, the discharge in the stream can be estimated to be, in m^3/s, as

(a) 150 (b) 180 (c) 117 (d) 889

4.14 In the gulp method of stream gauging by dilution technique, 60 litres of chemical X with concentration of 250 g/litre is introduced suddenly in to the stream at a section. At a downstream monitoring section the concentration profile of chemical X that crossed the section was found to be a triangle with a base of 10 hours and a peak of 0.01 ppm.

The discharge in the stream can be estimated to be about

(a) 83 m^3/s (b) 180 m^3/s (c) 15000 m^3/s (d) 833 m^3/s

4.15 The slope-area method is extensively used in

(a) development of rating curve
(b) estimation of flood discharge based on high-water marks
(c) cases where shifting control exists
(d) cases where backwater effect is present

4.16 For a given stream, the rating curve applicable to a section is available. To determine the discharge in this stream, the following-data are needed:

(a) Current meter readings at various verticals at the section
(b) Slope of the water surface at the section
(c) Stage at the section
(d) Surface velocity at various sections.

4.17 During a flood in a wide rectangular channel, it is found that at a section the depth of flow increases by 50% and at this depth the water-surface slope is half its original value in a given interval of time. This marks an approximate change in the discharge of

(a) –33% (b) +39% (c) +20% (d) no change

4.18 In a river, the discharge was 173 m^3/s, the water surface slope was 1 in 6000 and the stage at the station *X* was 10.00 m. If during a flood, the stage at station *X* was 10.00 and the water surface slope was 1/2000, the flood discharge was approximately

(a) 100 m^3/s (b) 519 m^3/s (c) 300 m^3/s (d) 371 m^3/s

4.19 During a flood, the water surface at a section was found to decrease at a rate of 10 cm/h. The slope of the river is 1/3600. Assuming the velocity of the flood wave as 2 m/s, the actual discharge in the stream can be estimated as

(a) 2.5% larger than the normal discharge
(b) 5% smaller than the normal discharge
(c) 2.5% smaller than the normal discharge
(d) Same as the normal discharge

where normal discharge is the discharge at a given stage under steady, uniform flow.

CHAPTER 5

RUNOFF

5.1 INTRODUCTION

5.1.1 Runoff Process

Runoff means the draining or flowing off of precipitation from a catchment area through a surface channel. It thus represents the output from the catchment in a given unit of time.

Consider a catchment area receiving precipitation. For a given precipitation, the evapotranspiration, initial loss, infiltration and detention storage requirements will have to be first satisfied before the commencement of runoff. When these are satisfied, the excess precipitation moves over the land surfaces to reach smaller channels. This portion of the runoff is called *overland flow* and involves building up of a storage over the surface and draining off of the same. Usually, the lengths and depths of overland flow are small and the flow is in the laminar regime. Flows from several small channels join bigger channels and flows from these in turn combine to form a larger stream, and so on, till the flow reaches the catchment outlet. The flow in this mode, where it travels all the time over the surface as overland flow and through the channels as open-channel flow and reaches the catchment outlet is called *surface runoff.*

A part of the precipitation that infilters moves laterally through upper crusts of the soil and returns to the surface at some location away from the point of entry into the soil. This component of runoff is known variously as *interflow, through flow, storm seepage, subsurface storm flow* or *quick return flow* (Fig. 5.1). The amount of interflow depends on the geological conditions of the catchment. A fairly pervious soil overlying a hard impermeable surface is conducive to large interflows. Depending upon the time delay between the infiltration and the outflow, the interflow is sometimes classified into *prompt interflow,* i.e. the interflow with the least time lag and *delayed interflow.*

Another route for the infiltered water is to undergo deep percolation and reach the groundwater storage in the soil. The groundwater follows a complicated and long path of travel and ultimately reaches the surface. The time lag, i.e. the difference in time between the entry into the soil and outflows from it is very large, being of the order of months and years. This part of runoff is called *groundwater runoff* or *groundwater flow.* Groundwater flow provides the dry-weather flow in perennial streams.

Based on the time delay between the precipitation and the runoff, the runoff is classified into two categories; as

1. Direct runoff, and
2. Base flow.

These are discussed below.

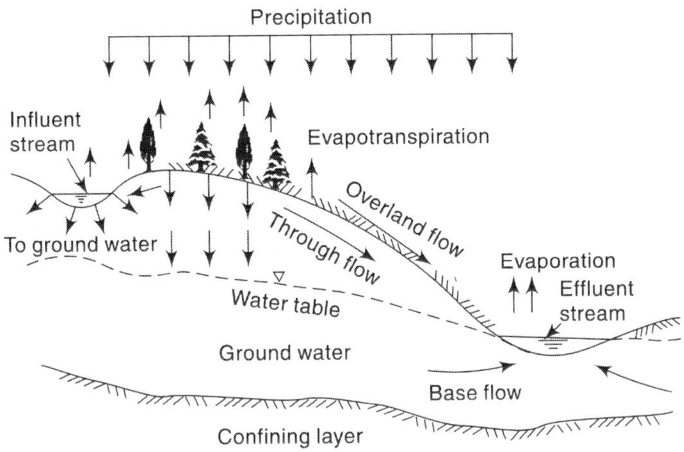

Fig. 5.1 ***Different Routes of Runoff***

5.1.2 Direct Runoff

It is that part of the runoff which enters the stream immediately after the rainfall. It includes surface runoff, prompt interflow and rainfall on the surface of the stream. In the case of snow-melt, the resulting flow entering the stream is also a direct runoff. Sometimes terms such as *direct storm runoff* and *storm runoff* are used to designate direct runoff. Direct runoff hydrographs are studied in detail in Chapter 6.

5.1.3 Base Flow

The delayed flow that reaches a stream essentially as groundwater flow is called *base flow.* Many times delayed interflow is also included under this category. In the annual hydrograph of a perennial stream (Fig. 5.2), the base flow is easily recognized as the slowly decreasing flow of the stream in rainless periods. Aspects relating to the identification of base flow in a hydrograph are discussed in Chapter 6.

5.1.4 Natural Flow

Runoff representing the response of a catchment to precipitation reflects the integrated effects of a wide range of catchment, climate and rainfall characteristics. True runoff is therefore stream flow in its natural condition, i.e. without human intervention. Such a stream flow unaffected by works of man, such as reservoirs and diversion structures on a stream, is called *natural flow* or *virgin flow.* When there exists storage or diversion works on a stream, the flow on the downstream channel is affected by the operational and hydraulic characteristics of these structures and hence does not represent the true runoff, unless corrected for the diversion of flow and return flow.

The natural flow (virgin flow) volume in time Δt at the terminal point of a catchment is expressed by water balance equation as

$$R_N = (R_o - V_r) + V_d + E + E_X + \Delta S \tag{5.1}$$

where R_N = natural flow volume in time Δt

R_o = observed flow volume in time Δt at the terminal site

V_r = volume of return flow from irrigation, domestic water supply and industrial use

V_d = volume diverted out of the stream for irrigation, domestic water supply and industrial use

E = net evaporation losses from reservoirs on the stream

E_X = net export of water from the basin

ΔS = change in the storage volumes of water storage bodies on the stream

In hydrological studies, one develops relations for natural flows. However, natural flows have to be derived based on observed flows and data on abstractions from the stream. In practice, however, the observed stream flow at a site includes return flow and is influenced by upstream abstractions. As such, natural flows have to be derived based on observed flows and data on abstractions from the stream. Always, it is the natural flow that is used in all hydrological correlations. Example 5.1 explains these aspects clearly.

Example 5.1 The following table gives values of measured discharges at a stream—gauging site in a year. Upstream of the gauging site a weir built across the stream diverts 3.0 Mm^3 and 0.50 Mm^3 of water per month for irrigation and for use in an industry respectively. The return flows from the irrigation is estimated as 0.8 Mm^3 and from the industry at 0.30 Mm^3 reaching the stream upstream of the gauging site. Estimate the natural flow. If the catchment area is 180 km^2 and the average annual rainfall is 185 cm, determine the runoff-rainfall ratio.

Month	1	2	3	4	5	6	7	8	9	10	11	12
Gauged flow (Mm^3)	2.0	1.5	0.8	0.6	2.1	8.0	18.0	22.0	14.0	9.0	7.0	3.0

Solution

In a month the natural flow volume R_N is obtained from Eq. (5.1) as

$$R_N = (R_o - V_r) + V_d + E + E_X + \Delta S$$

Here E, E_X and ΔS are assumed to be insignificant and of zero value.

V_r = volume of return flow from irrigation, domestic water supply and industrial use = 0.80 + 0.30 = 1.10 Mm^3

V_d = volume diverted out of the stream for irrigation, domestic water supply and industrial use = 3.0 + 0.5 = 3.5 Mm^3

The calculations are shown in the following table:

Month	1	2	3	4	5	6	7	8	9	10	11	12
$R_o(Mm^3)$	2.0	1.5	0.8	0.6	2.1	8.0	18.0	22.0	14.0	9.0	7.0	3.0
$V_d(Mm^3)$	3.5	3.5	3.5	3.5	3.5	3.5	3.5	3.5	3.5	3.5	3.5	3.5
$V_r(Mm^3)$	1.1	1.1	1.1	1.1	1.1	1.1	1.1	1.1	1.1	1.1	1.1	1.1
$R_N(Mm^3)$	4.4	3.9	3.2	3.0	4.5	10.4	20.4	24.4	16.4	11.4	9.4	5.4

$$\text{Total } R_N = 116.8 \text{ Mm}^3$$

Annual natural flow volume = Annual runoff volume = 116.8 Mm^3

Area of the catchment = 180 km^2 = 1.80×10^8

$$\text{Annual runoff depth} = \frac{1.168 \times 10^8}{1.80 \times 10^8} = 0.649 \text{ m} = 64.9 \text{ cm}$$

Annual rainfall = 185 cm (Runoff/Rainfall) = 64.9/185 = 0.35

5.2 *Hydrograph*

A plot of the discharge in a stream plotted against time chronologically is called a *hydrograph*. Depending upon the unit of time involved, we have

- Annual hydrographs showing the variation of daily or weekly or 10 daily mean flows over a year
- Monthly hydrographs showing the variation of daily mean flows over a month
- Seasonal hydrographs depicting the variation of the discharge in a particular season such as the monsoon season or dry season
- Flood hydrographs or hydrographs due to a storm representing stream flow due to a storm over a catchment

Each of these types have specific applications. Annual and seasonal hydrographs are of use in (i) calculating the surface water potential of stream, (ii) reservoir studies, and (iii) drought studies. Flood hydrographs are essential in analysing stream characteristics associated with floods. This chapter is concerned with the estimation and use of long-term runoffs. The study of storm hydrograph forms the subject matter of the next chapter.

5.2.1 Water Year

The annual hydrograph shows the seasonal variation of the discharge over a year and forms the basic data for further analysis through flow duration curve (*FDC*) and storage analysis. Figure 5.2 shows schematically a typical annual hydrograph of a

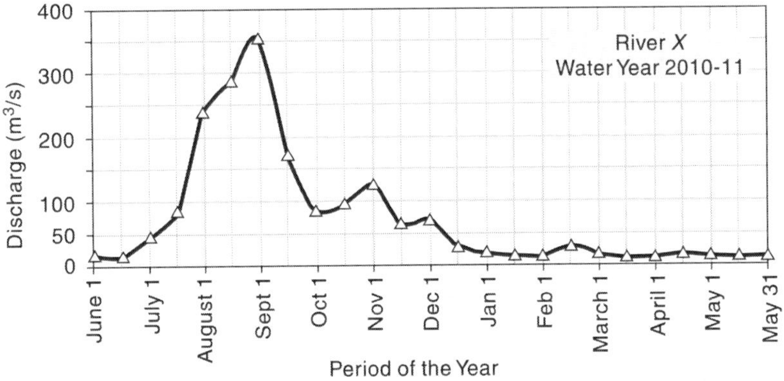

Fig. 5.2 ***Perennial stream***

perennial river. Observe the flow depicted in this figure starts from a relatively dry period and ends in a realtively dry period, at the end of a calendar year. Each country has fixed a definite 12-month period to begin and end in a relatvely dry season as the *water year*. In India, the water year starts from June 1st and ends on May 31st of the following calendar year. In *USA*, the water year is from October 1st to September 30th of the following year. All stream-flow records are maintained on water-year basis. In a water year, a complete cycle of climatic changes is expected and hence the annual data on the water-year basis is a representative statistic of the annual stream flow. Further, the water budget in a water year will have the least amount of carry over. In all stream-flow analysis, such as *FDC* and mass curve, the period of data should be integer multiple of water years and no fractional year data is permissible.

5.3 *Runoff Characteristics of Streams*

A study of the annual hydrographs of streams enables one to classify the streams in to three classes as (i) perennial, (ii) intermittent, and (iii) ephemeral.

A *perennial stream* is one that always carries some flow in it (Fig. 5.2). There is considerable amount of groundwater flow throughout the year. Even during the dry seasons, the water table will be above the bed of the stream

The *intermittent* stream has limited contribution from the groundwater. During the wet season, the water table is above the streambed and there is contribution of the base flow to the stream flow. However, during dry seasons the water table drops to a level lower than that of the streambed and the stream dries up. Excepting for an occasional storm that can produce a short-duration flow, the stream remains dry for the most part of the dry months (Fig. 5.3).

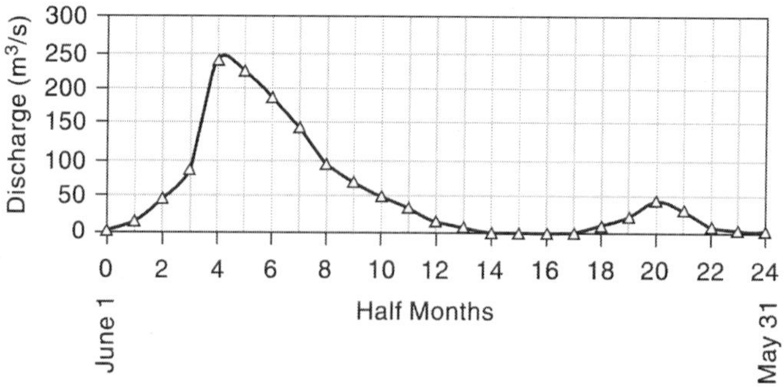

Fig. 5.3 ***Intermittent Stream***

An *ephemeral* stream is one that does not have any base-flow contribution. The annual hydrograph of such a stream shows a series of short-duration spikes marking flash flows in response to storms (Fig. 5.4). The stream becomes dry soon after the end of the storm flow. Typically, an ephemeral stream does not have a well-defined channel. Most of the rivers in arid zones are of the ephemeral kind.

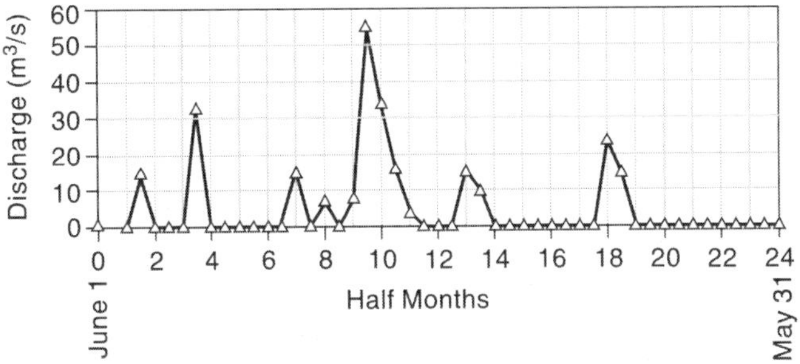

Fig. 5.4 ***Ephemeral Stream***

The flow characteristics of a stream depends upon

- The rainfall characteristics, such as magnitude, intensity, distribution according to time and space, and its variability
- Catchments characteristics such as soil, land use/cover, slope, geology, shape and slope
- Climate factors which influence evapotranspiration

The interrelationship of these factors is extremely complex. However, at the risk of oversimplification, the following points are significant

- The seasonal variation of rainfall is clearly reflected in the runoff. High stream discharges occur during monsoon months and low flow that is essentially due to the base flow is maintained during the rest of the year.
- The shape of the stream hydrograph and hence the peak flow is essentially controlled by the storm and the physical characteristics of the basin. Evapotranspiration plays a minor role in this.
- The annual runoff volume of a stream is mainly controlled by the amount of rainfall and evapotranspiration. The geology of the basin is significant to the extent of deep percolation losses. The land use/cover play an important role in creating infiltration and evaporation opportunities and retarding the runoff.

Some of the important characteristics of the catchment that affect the runoff due to a rainfall event are described briefly in the next section. The knowledge of these factors is useful in modeling the runoff process through watershed models and also in designing interventions in the watersheds for purposes of watershed management.

5.4 *Catchment Characteristics*

A catchment, also known as *drainage basin*, *drainage area* and *watershed*, is the basic hydrologic unit in the analysis of runoff phenomenon. Physical characteristics of the catchment such as area, shape, slope, and drainage channel pattern in the catchment are some of the major static characteristics that affect the volume of the surface runoff and shape of the runoff hydrograph from a catchment due to a storm. Study of the catchment characteristics forms an important component of the subject of geomorphology and is studied under the topic head of *quantitative geomorphology*. Some of the important geometrical parameters of the catchment as related to runoff process are briefly described below.

1. Area

Area is probably the most used parameter to represent the characteristics of a catchment. The area of the catchment is defined as the area of the closed curve forming the horizontal projection of the catchment boundary. The usual units are hectare (ha) for small catchments and square kilometer (km^2) for larger drainage areas.

2. Stream Order

It is a classification reflecting the pattern of branches that unite to form the trunk stream leaving the catchment. The smallest stream at the start of the network is designated as of '*order 1*'. Two channels of *order 1* when joined produce a stream of 'order 2'.Two channels of order 2 when joined produce a stream *of* order 3, and so on. Note that when a lower order stream (say order 2) meets a higher order stream (say order 3) the order of the resulting stream is still the higher order stream entering the confluence viz., order 3 in this case. The trunk stream discharging out of the drainage basin has the highest order. Fig. 5.5 shows schematically a drainage basin with a stream system of order 4.

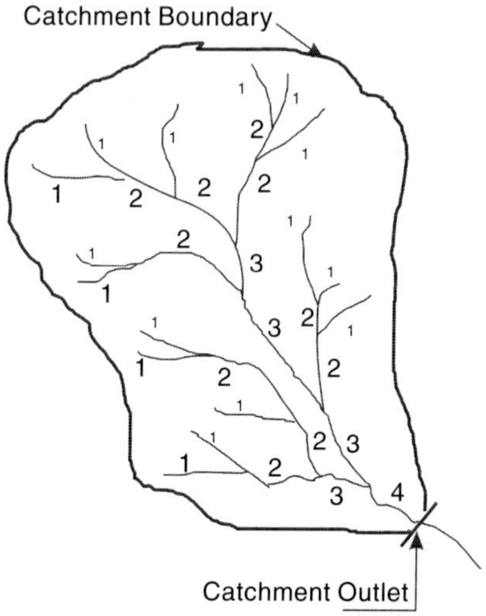

Fig. 5.5 ***Stream Channel Ordering in a Fourth-Order Stream***

Streams of order 1 to 3 are called *headwater streams* and constitute waterways in the upper reaches of the catchment. Nearly 80% of the world's waterways are of order 1 through 3. The size of a stream becomes larger as its stream order increases. Also, the slope of the stream decreases with its increase in the stream order number. Streams of size 4 to 6 are usually medium-size streams. Streams of order larger than 6 are called as rivers. The largest stream order known is 12; for example, the river Amazon has a stream order of 12.

Stream order is helpful in designating the nature of drainage pattern of a catchment and is of use in locating watershed treatment structures like check dams and Nala bunds.

3. Stream Density, S_d

The ratio of the number of streams (N_s) of all orders to the area of the basin (A) is known as *stream density* (S_d) of the basin. Thus, $S_d = \frac{N_s}{A}$. Stream density represents the number of streams per unit area and is indicative of the pattern of drainage channels in the catchment.

4. Drainage Density, D_d

Drainage density is the ratio of the total length of streams of all orders within a basin to its area. Thus, if L = length of a stream in a basin, then $D_d = \frac{\Sigma L}{A}$. The drainage density is a measure of how well the drainage basin is drained by the stream network.

The drainage density is called coarse, medium or fine depending on the value of D_d as follows: *Coarse* if $D_d \leq 5$ km/km^2; *Medium* if $D_d = 5 - 10$ km/km^2 and *Fine* if $D_d > 5$ km/km^2.

Drainage density depends upon the climate and physical characteristics and geology of the catchment. Generally, impermeable soil material in the catchment, thin land cover and steep slope cause more channels to move the runoff rapidly. Hence, a high drainage density may indicate one or more of the following:

- Existence of a mature, well developed channel system
- Surface runoff moves rapidly from the edge of the catchment
- Thin/deforested land cover
- Low infiltration soil or impervious geology.

Drainage density is a very useful numerical measure of the runoff potential and texture of the catchment and is measured fairly easily through use of *GIS*. This parameter finds considerable use in management of watershed erosion.

5. Relief

Maximum basin relief is the elevation difference (in meters) between the catchment outlet and the highest point on the basin perimeter.

6. Slope

Usually, a slope profile prepared along the main stream is used to characterize the slope of the catchment. For a given stretch of the stream, the ratio of horizontal distance between the two end parts of the stretch to the difference in elevation between these two points gives the slope of the stretch. As a rule, the catchment slope is the highest at the beginning of the stream and gradually decreases as one moves along the stream to the basin outlet. Slope is an important parameter in many watershed simulation models.

7. Length, *L*

The length of the catchment is defined as the length of the main stream measured from the basin outlet to the remotest point on the basin boundary. The main stream is identified by starting from the basin outlet and moving up the catchment. At any branching point the largest order branch is taken. If there is a branch of two streams of the same order, the one with the largest catchment area is taken as the main steam.

8. Shape

There are a number of ways of designating and quantifying the shape of a catchment. Some of the commonly used shape parameters are shown in Table 5.1 along with their definitions. In this table A = catchment area, L = catchment length and P = perimeter of the catchment.

9. Hypsometric Curve

It is a plot of horizontal cross-sectional drainage basin area to elevation. It is usual to plot the curve in a non-dimensional form as relative height h/H plotted against relative area a/A as shown in Fig. 5.6. In this h = height of a given contour, H = basin height (relief), a = basin area at contour h and A = total basin area.

Table 5.1 ***Different Catchment Shape Parameters***

Parameters	Definition	Formula
Form factor	$\frac{\text{Catchment area}}{(\text{Catchment length})^2}$	$\frac{A}{L^2}$
Shape factor	$\frac{(\text{Catchment length})^2}{\text{Catchment area}}$	$\frac{L^2}{A}$
Compactness coefficient	$\frac{\text{Perimeter of the catchment}}{\text{Perimeter of the circle whose area is that of the basin}}$	$\frac{0.2821P}{\sqrt{A}}$
Circulatory ratio	$\frac{\text{Catchment area}}{\text{Area if circle of catchment perimeter}}$	$\frac{12.57A}{P^2}$
Elongation ratio	$\frac{\text{Diameter of circle whose area is basin area}}{\text{Catchment length}}$	$\frac{1.128\sqrt{A}}{L}$

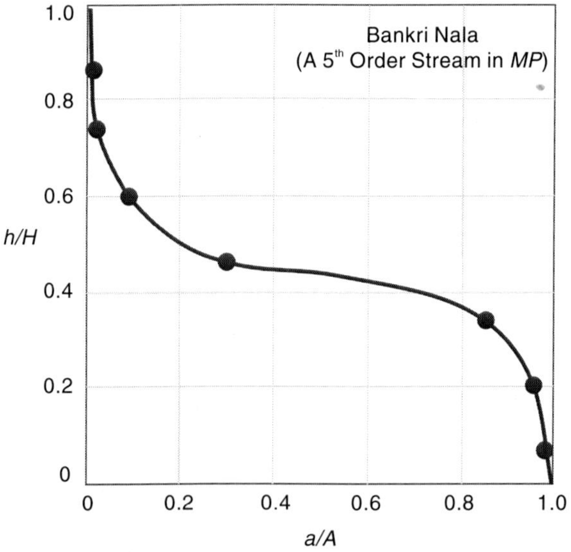

Fig. 5.6 ***Hypsometric curve of a 5th order Stream***
(Ref. INSA, Vol.40-A, No.1)

Hypsometric curve describes the proportion of basin area that is above a certain basin elevation. It represents an overall basin slope and embodies much of geomorphic information of the basin. These days, hypsometric curves are generated through use of *GIS* and its applications are many. Major applications include watershed treatment, planning and design of rainwater harvesting and erosion control programs. Further, the hypsometric curve finds use in some watershed models.

5.5 *Runoff Volume (Yield)*

5.5.1 Definition

The total quantity of surface water that can be expected in a given period from a stream at the outlet of its catchment is known as *yield* of the catchment in that period. Depending upon the period chosen we have annual yield and seasonal yield signifying yield of the catchment in an year and in a specified season respectively. Unless otherwise qualified, the term *yield* is usually used to represent annual yield. The term *yield* is used mostly by the irrigation engineering professionals in India.

The annual yield from a catchment is the end product of various processes such as precipitation, infiltration and evapotranspiration operating on the catchment. Due to the inherent nature of the various parameters involved in the processes, the yield is a random variable. A list of values of annual yield in a number of years constitutes an annual time series which can be analysed by methods indicated in Chapter 2 (Sec. 2.11) to assign probabilities of occurrences of various events. A common practice is to assign a dependability value (say 75% dependable yield) to the yield. Thus, 75% dependable annual yield is the value that can be expected to be equalled to or exceeded 75% times (i.e. on an average 15 times in a span of 20 years). Similarly, 50% dependable yield is the annual yield value that is likely to be equalled or exceeded 50% of times (i.e. on an average 10 times in 20 years).

It should be remembered that the yield of a stream is always related to the natural flow in the river. However, when water is diverted from a stream for use in activities such as irrigation, domestic water supply and industries, the non-consumptive part of the diverted water returns back to the hydrologic system of the basin. Such additional flow, known as *return flow*, is available for the suitable use and as such is added to the natural flow to estimate the yield. (Details pertaining to the return flow are available in Sec. 5.11). The annual yield of a basin at a site is thus taken as the annual natural water flow in the river at the site plus the return flow to the stream from different uses upstream of the site.

The yield of a catchment Y in a period Δt could be expressed by water balance equation (Eq. 5.1) as

$$Y = R_N + V_r = R_o + A_b + \Delta S \tag{5.1a}$$

where R_N = natural flow in time Δt

V_r = volume of return flow from irrigation, domestic water supply and industrial use

R_o = observed runoff volume at the terminal gauging station of the basin in time Δt.

A_b = abstraction in time, Δt for irrigation, water supply and industrial use and inclusive of evaporation losses in surface water bodies on the stream.

ΔS = change in the storage volumes of water storage bodies on the stream.

The calculation of natural runoff volume (and hence yield), is of fundamental importance in all surface water resources development studies. The most desirable basis for assessing the yield characteristics of a catchment is to analyse the actual flow records of the stream draining the catchment. However, in general, observed discharge data of sufficient length is unlikely to be available for many catchments. As such, other alternate methods such as the *empirical equations* and *watershed simulations* (described in Secs 5.5.3 to 5.5.4) are often adopted.

It should be noted that the observed stream flow at a site includes return flow. For small catchments and for catchments where water-resources developments are at a small scale, the return flow is likely to be a negligibly small part of the runoff. In the further parts of this chapter the term annual (or seasonal) runoff volume R and the term annual (or seasonal) yield are used synonymously with the implied assumption that the return flow is negligibly small. It is emphasized that when return flow is not negligible, it is the natural flow volume that is to be used in hydrological correlations with rainfall.

5.5.2 Rainfall–Runoff Correlation

The relationship between rainfall in a period and the corresponding runoff is quite complex and is influenced by a host of factors relating to the catchment and climate. Further, there is the problem of paucity of data which forces one to adopt simple correlations for adequate estimation of runoff. One of the most common methods is to correlate seasonal or annual measured runoff values *(R)* with corresponding rainfall *(P)* values. A commonly adopted method is to fit a linear regression line between R and P and to accept the result if the correlation coefficient is nearer unity. The equation of the straight-line regression between runoff R and rainfall P is

$$R = aP + b \tag{5.2}$$

and the values of the coefficent a and b are given by

$$a = \frac{N(\Sigma PR) - (\Sigma P)(\Sigma R)}{N(\Sigma P^2) - (\Sigma P)^2} \tag{5.3a}$$

and

$$b = \frac{\Sigma R - a(\Sigma P)}{N} \tag{5.3b}$$

in which N = number of observation sets R and P. The coefficient of correlation r can be calculated as

$$r = \frac{N(\Sigma PR) - (\Sigma P)(\Sigma R)}{\sqrt{[N(\Sigma P^2) - (\Sigma P)^2][N(\Sigma R^2) - (\Sigma R)^2]}} \tag{5.4}$$

The value of r lies between 0 and 1 as R can have only positive correlation with P. The value of $0.6 < r < 1.0$ indicates good correlation. Further, it should be noted that $R \geq 0$. The term r^2 is known as the *coefficient of determination.*

For large catchments, sometimes it is found advantageous to have exponential relationship as

$$R = \beta P^m \tag{5.5}$$

where β and m are constants, instead of the linear relationship given by Eq. (5.2). In that case Eq. (5.5) is reduced to linear form by logarithmic transformation as

$$\ln R = m \ln P + \ln \beta \tag{5.6}$$

and the coefficients m and $\ln \beta$ are determined by using methods indicated earlier.

Since rainfall records of longer periods than that of runoff data are normally available for a catchment, the regression equation [Eq. (5.2) or (5.5)] can be used to generate synthetic runoff data by using rainfall data. While this may be adequate for preliminary studies, for accurate results sophisticated methods are adopted for synthetic generation of runoff data. Many improvements of the above basic rainfall-runoff correlation by considering additional parameters such as soil moisture and antecedent rainfall have been attempted. Antecedent rainfall influences the initial soil moisture and hence the infiltration rate at the start of the storm. For calculation of the annual runoff from the annual rainfall a commonly used antecedent precipitation index P_a is given by

$$P_a = aP_i + bP_{i-1} + cP_{i-2} \tag{5.7}$$

where P_i, p_{i-1} and P_{i-2} are the annual precipitation in the ith, $(i-1)^{th}$ and $(i-2)^{th}$ year and i = current year, a, b and c are the coefficients with their sum equal to unity. The coefficients are found by trial and error to produce best results. There are many other types of antecedent precipitation indices in use to account for antecedent soil moisture condition. For example, in *SCS–CN* method (Sec. 5.6) the sum of past five-day rainfall is taken as the index of antecedent moisture condition.

Example 5.2 Annual rainfall and runoff values (in cm) of a catchment spanning a period of 21 years are given below. Analyse the data to (a) estimate the 75% and 50% dependable annual yield of the catchment, and (b) to develop a linear correlation equation to estimate annual runoff volume for a given annual rainfall value.

Year	Annual rainfall (cm)	Annual runoff (cm)	Year	Annual rainfall (cm)	Annual runoff (cm)
1975	118	54	1986	75	17
1976	98	45	1987	107	32
1977	112	51	1988	75	15
1978	97	41	1989	93	28
1979	84	21	1990	129	48
1980	91	32	1991	153	76
1981	138	66	1992	92	27
1982	89	25	1993	84	18
1983	104	42	1994	121	52
1984	80	11	1995	95	26
1985	97	32			

Solution

(*a*) The annual runoff values are arranged in descending order of magnitude and a rank (m) is assigned for each value starting from the highest value (Table 5.1). The exceedence probability p is calculate for each runoff value as $p = \dfrac{m}{N+1}$. In this m = rank number and N = number of data sets. (Note that in Table 5.1 three items have the same value of R = 32 cm and for this set p is calculated for the item having the highest value of m, i.e m = 12). For estimating 75% dependable yield, the value of p = 0.75 is read from Table 5.1 by linear interpolation between items having p = 0.773 and p = 0.727. By this method, the 75% dependable yield for the given annual yield time series is found to be R_{75} = 23.0 cm.

Similarly, the 50% dependable yield is obtained by linear interpolation between items having $p = 0.545$ and $p = 0.409$ as $R_{50} = 34.0$ cm.

(*b*) The correlation equation is written as $R = aP + b$

The coefficients of the best fit straight line for the data are obtained by the least square error method as mentioned in Table.

From Table 5.1,

$$\Sigma P = 2132 \qquad \Sigma R = 759 \qquad \Sigma PR = 83838$$

$$\Sigma P^2 = 224992 \qquad \Sigma R^2 = 33413$$

$$(\Sigma P)^2 = 4545424 \qquad (\Sigma R)^2 = 576081 \qquad N = 21$$

By using Eq. (5.3-a),

$$a = \frac{N(\Sigma PR) - (\Sigma P)(\Sigma R)}{N(\Sigma P^2) - (\Sigma R)^2} = \frac{(21 \times 83838) - (2132)(759)}{(21 \times 224992) - (2132)^2} = 0.7938$$

Table Calculations *for Example 5.2*

1	2	3	4	5	6	7	8	9
Year	*P* rainfall (cm)	*R* runoff (cm)	P^2	R^2	*PR*	rank, *m*	*R* (Sorted annual runoff) (cm)	Exceedance probability, *p*
1975	118	54	13924	2916	6372	1	76	0.045
1976	98	45	9604	2025	4410	2	66	0.091
1977	112	51	12544	2601	5712	3	54	0.136
1978	97	41	9409	1681	3977	4	52	0.182
1979	84	21	7056	441	1764	5	51	0.227
1980	91	32	8281	1024	2912	6	48	0.273
1981	138	66	19044	4356	9108	7	45	0.318
1982	89	25	7921	625	2225	8	42	0.364
1983	104	42	10816	1764	4368	9	41	0.409
1984	80	11	6400	121	880	10	32	
1985	97	32	9409	1024	3104	11	32	
1986	75	17	5625	289	1275	12	32	0.545
1987	107	32	11449	1024	3424	13	28	0.591
1988	75	15	5625	225	1125	14	27	0.636
1989	93	28	8649	784	2604	15	26	0.682
1990	129	48	16641	2304	6192	16	25	0.727
1991	153	76	23409	5776	11628	17	21	0.773
1992	92	27	8464	729	2484	18	18	0.818

(*Contd.*)

Table (*Contd.*)

1993	84	18	7056	324	1512	19	17	0.864
1994	121	52	14641	2704	6292	20	15	0.909
1995	95	26	9025	676	2470	21	11	0.955
SUM	**2132**	**759**	**224992**	**33413**	**83838**			

By Eq. (5.3-b)

$$b = \frac{\Sigma R - a(\Sigma P)}{N} = \frac{(759) - 0.7938 \times (2138)}{21} = -44.44$$

Hence, the required annual rainfall–runoff relationship of the catchment is given by

$R = 0.7938\ P - 44.44$ with both P and R being in cm and $R \geq 0$.

By Eq. (5.4) coefficient of correlation,

$$r = \frac{N(\Sigma PR) - (\Sigma P)(\Sigma R)}{\sqrt{[N(\Sigma P^2) - (\Sigma P)^2][N(\Sigma R^2) - (\Sigma R)^2]}}$$

$$= \frac{(21 \times 83838\ - (2132)(759)}{\sqrt{[(21 \times 224992) - (4545424)][(21 \times 33413) - (576081)}} = 0.949$$

As the value of r is nearer to unity the correlation is very good. Figure 5.5 represents the data points and the best fit straight line.

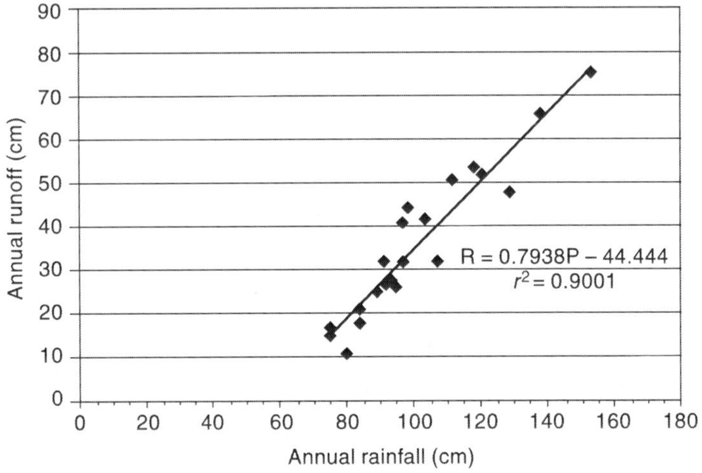

Fig. 5.7 ***Annual Rainfall–Runoff Correlation—Example 5.2***

5.5.3 Empirical Equations And Tables

The importance of estimating the water availability from the available hydrologic data for purposes of planning water-resource projects was recognised by engineers even in the nineteenth century. With a keen sense of observation in the region of their activity, many engineers of the past have developed empirical runoff estimation formulae. However, these formulae are applicable only to the region in which they were derived. These formulae are essentially rainfall–runoff relations with additional third or fourth parameters to account for climatic or catchment characteristics. Some of the important formulae used in various parts of India are given below.

1. Binnie's Percentages

Sir Alexander Binnie measured the runoff from a small catchment near Nagpur (Area of 16 km^2) during 1869 and 1872 and developed curves of cumulative runoff against cumulative rainfall. The two curves were found to be similar. From these, he established the percentages of runoff from rainfall. These percentages have been used in Madhya Pradesh and Vidarbha region of Maharashtra for the estimation of yield.

2. Barlow's Tables

Barlow, the first Chief Engineer of the Hydro-Electric Survey of India (1915) on the basis of his study in small catchments (area ~130 km^2) in Uttar Pradesh expressed runoff R as

$$R = K_b\, P \tag{5.8}$$

where K_b = runoff coefficient which depends upon the type of catchment and nature of monsoon rainfall. Values of K_b are given in Table 5.2.

Table 5.2 ***Barlow's Runoff Coefficient K_b in Percentage (Developed for use in UP)***

Class	Description of catchment	Values of K_b (percentage)		
		Season 1	Season 2	Season 3
A	Flat, cultivated and absorbent soils	7	10	15
B	Flat, partly cultivated, stiff soils	12	15	18
C	Average catchment	16	20	32
D	Hills and plains with little cultivation	28	35	60
E	Very hilly, steep and hardly any cultivation	36	45	81

Season 1: Light rain, no heavy downpour
Season 2: Average or varying rainfall, no continuous downpour
Season 3: Continuous downpour

3. Strange's Tables

Strange (1892) studied the available rainfall and runoff in the border areas of present-day Maharashtra and Karnataka and has obtained yield ratios as functions of indicators representing catchment characteristics. Catchments are classified as *good, average* and *bad* according to the relative magnitudes of yield they give. For example, catchments with good forest/vegetal cover and having soils of high permeability would be classified as *bad,* while catchments having soils of low permeability and having little or no vegetal cover is termed *good.* Two methods using tables for estimating the runoff volume in a season are given.

(a) Runoff Volume from Total Monsoon Season Rainfall A table giving the runoff volumes for the monsoon period (i.e. yield during monsoon season) for different total monsoon rainfall values and for the three classes of catchments (viz. *good, average and bad)* are given in Table 5.3-a. The correlation equations of best fitting lines relating percentage yield ratio (Y_r) to precipitation (P) could be expressed as

Table 5.3(a) ***Strange's Table of Total Monsoon Rainfall and estimated Runoff***

Total Monsoon rainfall (inches)	Total Monsoon rainfall (mm)	Percentage of Runoff to rainfall			Total Monsoon rainfall (inches)	Total Monsoon rainfall (mm)	Percentage of Runoff to rainfall		
		Good catchment	Average catchment	Bad catchment			Good catchment	Average catchment	Bad catchment
1.0	25.4	0.1	0.1	0.1	31.0	787.4	27.4	20.5	13.7
2.0	50.8	0.2	0.2	0.1	32.0	812.8	28.5	21.3	14.2
3.0	76.2	0.4	0.3	0.2	33.0	838.2	29.6	22.2	14.8
4.0	101.6	0.7	0.5	0.3	34.0	863.6	30.8	23.1	15.4
5.0	127.0	1.0	0.7	0.5	35.0	889.0	31.9	23.9	15.9
6.0	152.4	1.5	1.1	0.7	36.0	914.4	33.0	24.7	16.5
7.0	177.8	2.1	1.5	1.0	37.0	939.8	34.1	25.5	17.0
8.0	203.2	2.8	2.1	1.4	38.0	965.2	35.3	26.4	17.6
9.0	228.6	3.5	2.6	1.7	39.0	990.6	36.4	27.3	18.2
10.0	254.0	4.3	3.2	2.1	40.0	1016.0	37.5	28.1	18.7
11.0	279.4	5.2	3.9	2.6	41.0	1 041.4	38.6	28.9	19.3
12.0	304.8	6.2	4.6	3.1	42.0	1066.8	39.8	29.8	19.9
13.0	330.2	7.2	5.4	3.6	43.0	1092.2	40.9	30.6	20.4
14.0	355.6	8.3	6.2	4.1	44.0	1117.6	42.0	31.5	21.0
15.0	381.0	9.4	7.0	4.7	45.0	1143.0	43.1	32.3	21.5
16.0	406.4	10.5	7.8	5.2	46.0	1168.4	44.3	33.2	22.1
17.0	431.8	11.6	8.7	5.8	47.0	1193.8	45.4	34.0	22.7
18.0	457.2	12.8	9.6	6.4	48.0	1219.2	46.5	34.8	23.2
19.0	482.6	13.9	10.4	6.9	49.0	1244.6	47.6	35.7	23.8
20.0	508.0	15.0	11.3	7.5	50.0	1270.0	48.8	36.6	24.4
21.0	533.4	16.1	12.0	8.0	51.0	1295.4	49.9	37.4	24.9
22.0	558.8	17.3	12.9	8.6	52.0	1320.8	51.0	38.2	25.5
23.0	584.2	18.4	13.8	9.2	53.0	1346.2	52.1	39.0	26.0
24.0	609.6	19.5	14.6	9.7	54.0	1371.6	53.3	39.9	26.6
25.0	635.0	20.6	15.4	10.3	55.0	1397.0	54.4	40.8	27.2
26.0	660.4	21.8	16.3	10.9	56.0	1422.4	55.5	41.6	27.7
27.0	685.8	22.9	17.1	11.4	57.0	1447.8	56.6	42.4	28.3
28.0	711.2	24.0	18.0	12.0	58.0	1473.2	57.8	43.3	28.9
29.0	736.6	25.1	18.8	12.5	59.0	1498.6	58.9	44.4	29.41
30.0	762.0	26.3	19.7	13.1	60.0	1524.0	60.0	45.0	30.0

For *Good* catchment:

For $P < 250$ mm, $Y_r = 7 \times 10^{-5} P^2 - 0.0003\,P$ having $r^2 = 0.9994$ (5.9a)

For $250 < P < 760$ $Y_r = 0.0438\,P - 7.1671$ having $r^2 = 0.9997$ (5.9b)

For $760 < P < 1500$ $Y_r = 0.0443\,P - 7.479$ having $r^2 = 1.0$ (5.9c)

For *Average* catchment:

For $P < 250$ mm, $Y_r = 6 \times 10^{-5} P^2 - 0.0022\,P + 0.1183$ having $r^2 = 0.9989$ (5.10a)

For $250 < P < 760$ $Y_r = 0.0328\,P - 5.3933$ having $r^2 = 0.9997$ (5.10b)

For $760 < P < 1500$ $Y_r = 0.0333\,P - 5.7101$ having $r^2 = 0.9999$ (5.10c)

For *Bad* catchment:

For $P < 250$ mm, $Y_r = 4 \times 10^{-5} P^2 - 0.0011\,P + 0.0567$ having $r^2 = 0.9985$ (5.11a)

For $250 < P < 760$ $Y_r = 0.0219\,P - 3.5918$ having $r^2 = 0.9997$ (5.11b)

For $760 < P < 1500$ $Y_r = 0.0221\,P - 3.771$ having $r^2 = 1.0$ (5.11c)

where Y_r = Percentage yield ratio = ratio of seasonal runoff to seasonal rainfall in percentage and P = monsoon season rainfall in mm.

Since there is no appreciable runoff due to the rains in the dry (non-monsoon) period, the monsoon season runoff volume is recommended to be taken as annual yield of the catchment. This table could be used to estimate the monthly yields also in the monsoon season. However, it is to be used with the understanding that the table indicates relationship between cumulative monthly rainfall starting at the beginning of the season and cumulative runoff, i.e. a *double mass curve* relationship.

Example 5.3 illustrates this procedure.

(b) Estimating the Runoff Volume from Daily Rainfall In this method, Strange in a most intuitive way recognises the role of antecedent moisture in modifying the runoff volume due to a rainfall event in a given catchment. Daily rainfall events are considered and three states of antecedent moisture conditions prior to the rainfall event as *dry, damp* and *wet* are recognised. The classification of these three states is as follows:

Wetting Process

(a) **Transition from Dry to Damp**

- (i) 6 mm rainfall in the last 1 day
- (ii) 12 mm in the last 3 days
- (iii) 25 mm in the last 7 days
- (iv) 38 mm in the last 10 days

(b) **Transition from Damp to Wet**

- (i) 8 mm rainfall in the last 1 day
- (ii) 12 mm in the last 2 days
- (iii) 25 mm in the last 3 days
- (iv) 38 mm in the last 5 days

(c) **Direct Transition from Dry to Wet**

Whenever 64 mm rain falls on the *previous* day or on the *same* day.

Drying Process

(d) **Transition from Wet to Damp**

- (i) 4 mm rainfall in the last 1 day
- (ii) 6 mm in the last 2 days
- (iii) 12 mm in the last 4 days
- (iv) 20 mm in the last 5 days

(e) **Transition from Damp to Dry**

(i) 3 mm rainfall in the last 1 day
(ii) 6 mm in the last 3 days
(iii) 12 mm in the last 7 days
(iv) 15 mm in the last 10 days

The percentage daily rainfall that will result in runoff for *average* (yield producing) catchment is given in Table 5.3(b). For *good* (yield producing) and *bad* (yield producing) catchments *add* or *deduct* 25% of the yield corresponding to the *average* catchment.

Table 5.3(b) ***Strange's Table of Runoff Volume from Daily Rainfall for an Average Catchment***

Daily rainfall (mm)	Percentage of runoff volume to daily rainfall when original state of the ground was		
	Dry	Damp	Wet
6	—	—	8
13	—	6	12
19	—	8	16
25	3	11	18
32	5	14	22
38	6	16	25
45	8	19	30
51	10	22	34
64	15	29	43
76	20	37	55
102	30	50	70

Best fitting linear equations for the above table would read as below with K_s = runoff volume percentage and P = daily rainfall (mm):

For Dry *AMC*: $K_s = 0.5065\ P - 2.3716$ for $P > 20$ mm (5.12a)
with coefficient of determination $r^2 = 0.9947$

For Damp *AMC*: $K_s = 0.3259\ P - 5.1079$ for $P > 7$ mm (5.12b)
with coefficient of determination $r^2 = 0.9261$

For Wet *AMC*: $K_s = 0.6601\ P + 2.0643$ (5.12c)
with coefficient of determination $r^2 = 0.9926$

(c) Use of Strange's Tables Strange's monsoon rainfall-runoff table (Table 5.3-a) and Table (5.3-b) for estimating daily runoff corresponding to a daily rainfall event are in use in parts of Karnataka, Andhra Pradesh and Tamil Nadu. A calculation procedure using Table (5.3-a) to calculate monthly runoff volumes in a monsoon season using cumulative monthly rainfalls is shown in Example 5.3.

Example 5.3 Monthly rainfall values of the 50% dependable year at a site selected for construction of an irrigation tank is given below. Estimate the monthly and annual runoff volume of this catchment of area 1500 *ha*.

[Assume the catchment classification as Good catchment].

Month	June	July	Aug	Sept	Oct
Monthly rainfall (mm)	90	160	145	22	240

Solution

Calculations are shown in the Table 5.4 given below.

Table 5.4 ***Calculation of Monthly Yields by Strange's Method—Example 5.3***

No.	Month	June	July	August	September	October
1.	Monthly Rainfall (mm)	90	160	145	22	240
2.	Cumulative monthly rainfall (mm)	90	250	395	417	657
3.	Runoff/rainfall as % (From Strange's Table 5.3-a)	0.56	4.17	10.01	11.08	21.69
4.	Cumulative Runoff (mm)	0.50	10.43	39.54	46.20	142.50
5.	Monthly Runoff (mm)	0.50	9.92	29.11	6.66	96.30

Row No. 3 is obtained by using Strange's Tables 5.3(a). Note that cumulative monthly rainfall is used to get the cumulative runoff-ratio percentage at any month.

$$\text{Total monsoon runoff} = 142.50 \text{ mm} = (142.5/1000) \times (1500 \times 10^4)/10^6 \text{ Mm}^3$$

$$= 2.1375 \text{ Mm}^3$$

Annual runoff is taken as equal to monsoon runoff.

4. Inglis and DeSouza Formula

As a result of careful stream gauging in 53 sites in Western India, Inglis and DeSouza (1929) evolved two regional formulae between annual runoff R in cm and annual rainfall P in cm as follows:

1. For Ghat regions of western India

$$R = 0.85\,P - 30.5 \qquad (5.13)$$

2. For Deccan plateau

$$R = \frac{1}{254} P\,(P - 17.8) \qquad (5.14)$$

5. Khosla's Formula

Khosla (1960) analysed the rainfall, runoff and temperature data for various catchments in India and *USA* to arrive at an empirical relationship between runoff and rainfall. The time period is taken as a month. His relationship for monthly runoff is

$$R_m = P_m - L_m \qquad (5.15)$$

and $L_m = 0.48\ T_m$ for $T_m > 4.5°C$

where R_m = monthly runoff in cm and $R_m \geq 0$

P_m = monthly rainfall in cm

L_m = monthly losses in cm

T_m = mean monthly temperature of the catchment in °C

For $T_m \leq 4.5°C$, the loss L_m may provisionally be assumed as

T °C	4.5	–1	–6.5
L_m (cm)	2.17	1.78	1.52

Annual runoff = ΣR_m

Khosla's formulae is indirectly based on the water-balance concept and the mean monthly catchment temperature is used to reflect the losses due to evapotranspiration. The formulae has been tested on a number of catchments in India and is found to give fairly good results for the annual yield for use in preliminary studies.

Example 5.4 For a catchment in *UP*, India, the mean monthly temperatures are given. Estimate the annual runoff and annual runoff coefficient by Khosla's method.

Month	Jan	Feb	Mar	Apr	May	Jun	Jul	Aug	Sep	Oct	Nov	Dec
Temp°C	12	16	21	27	31	34	31	29	28	29	19	14
Rainfall (P_m)(cm)	4	4	2	0	2	12	32	29	16	2	1	2

Solution

In Khosla's formula applicable to the present case, $R_m = P_m - L_m$ with L_m = (0.48 × T °C) having a maximum value equal to corresponding P_m. The calculations are shown below:

Month	Jan	Feb	Mar	Apr	May	Jun	Jul	Aug	Sep	Oct	Nov	Dec
Rainfall (P_m) (cm)	4	4	2	0	2	12	32	29	16	2	1	2
Temp °C	12	16	21	27	31	34	31	29	28	29	19	14
L_m (cm)	4	4	2	0	2	12	14.9	13.9	13.4	2	1	2
Runoff (R_m)(cm)	0	0	0	0	0	0	17.1	15.1	2.6	0	0	0

Total annual runoff = 34.8 cm

Annual runoff coefficient = (Annual runoff/Annual rainfall) = (34.8/116.0) = 0.30

5.5.4 Watershed Simulation

The hydrologic water-budget equation for the determination of runoff for a given period is written as

$$R = R_s + G_0 = P - E_{et} - \Delta S \tag{5.16}$$

in which R_s = surface runoff, P = precipitation, E_{et} = actual evapotranspiration, G_0 = net groundwater outflow and ΔS = change in the soil moisture storage. The sum of R_s and G_0 is considered to be given by the total runoff R, i.e. streamflow.

Starting from an initial set of values, one can use Eq. (5.16) to calculate R by knowing values of P and functional dependence of E_{et}, ΔS and infiltration rates with catchment and climatic conditions. For accurate results, the functional dependence of various parameters governing the runoff in the catchment and values of P at short time intervals are needed. Calculations can then be done sequentially to obtain the runoff at any time. However, the calculation effort involved is enormous if attempted manually. With the availability of digital computers, the use of water budgeting as above to determine the runoff has become feasible. This technique of predicting the runoff, which is the catchment response to a given rainfall input is called *deterministic watershed simulation.* In this, the mathematical relationships describing the interdependence of various parameters in the system are first prepared and this is called the *model.* The model is then calibrated, i.e. the numerical values of various coefficients determined by simulating the known rainfall-runoff records. The accuracy of the model is further checked by reproducing the results of another string of rainfall data for which runoff values are known. This phase is known as *validation* or *verification* of the model. After this, the model is ready for use.

Crawford and Linsley (1959) pioneered this technique by proposing a watershed simulation model known as the Stanford Watershed Model (*SWM*). This underwent successive refinements and the Stanford Watershed Model-IV (*SWM-IV*) suitable for use on a wide variety of conditions was proposed in 1966. The flow chart of *SWM-IV* is shown in Fig. 5.8. The main inputs are hourly precipitation and daily evapotranspiration in addition to physical description of the catchment. The model considers the soil in three zones with distinct properties to simulate evapotranspiration, infiltration, overland flow, channel flow, interflow and baseflow phases of the runoff phenomenon. For calibration about 5 years of data are needed. In the calibration phase, the initial guess value of parameters are adjusted on a trial-and-error basis until the simulated response matches the recorded values. Using an additional length of rainfall-runoff of about 5 years duration, the model is verified for its ability to give proper response. A detailed description of the application of *SWM* to an Indian catchment is given in Ref. 11.

Based on the logic of *SWM-IV* many models and improved versions such as HSP (1966), *SSARR* (1968) and *KWM* (1970) were developed during late sixties and seventies. These models which simulate stream flow for long periods of time are called *continuous simulated models.* They permit generation of simulated long records for yield, drought and flood flow studies. In the early 1980s, there were at least 75 hydrologic simulation models that were available and deemed suitable for small watersheds. In the past two decades considerable effort has been directed towards the development of process-based, spatially explicit, and physically based models such as *MIKE SHE* (Refsgaard and Storm, 1955), and *GSSHA*—Gridded Surface/Subsurface Hydrologic Analysis (Downer et al., 2006). These are new generation of models that utilize *GIS* technology.

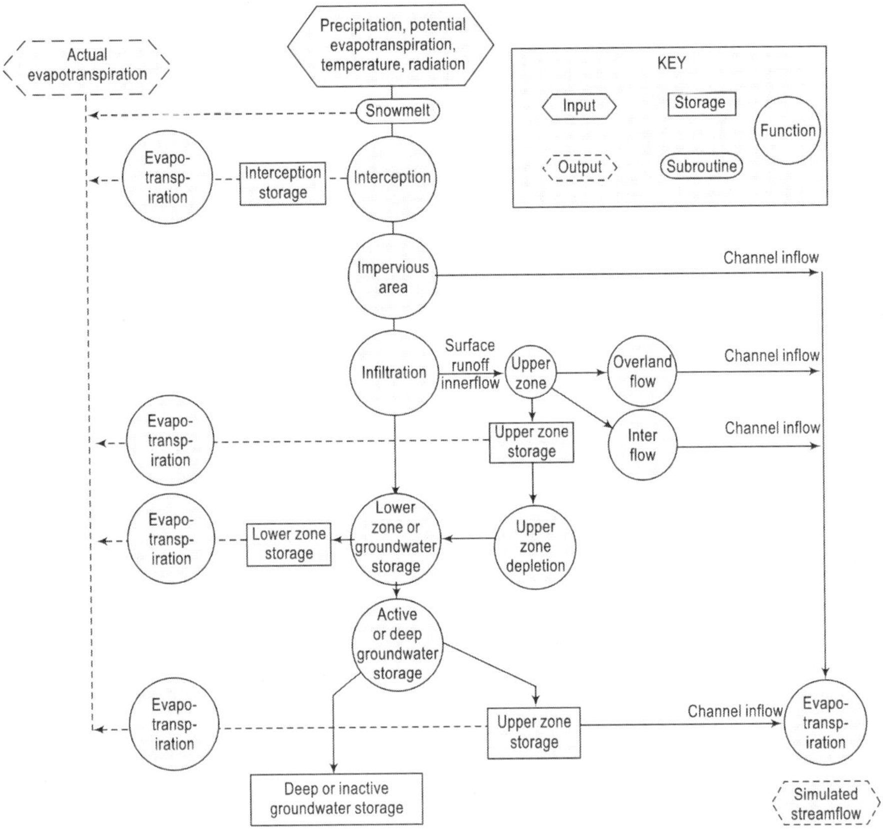

Fig. 5.8 ***Flow chart of SWM-IV***

5.6 *SCS-CN Method of Estimating Runoff Volume*

SCS-CN method, developed by Soil Conservation Services (*SCS*) of *USA* in 1969, is a simple, predictable, and stable conceptual method for estimation of direct runoff depth based on storm rainfall depth. It relies on only one parameter, *CN*. Currently, it is a well-established method, having been widely accepted for use in *USA* and many other countries. The details of the method are described in this section.

5.61 Basic Theory

The *SCS-CN* method is based on the water balance equation of the rainfall in a known interval of time Δt, which can be expressed as

$$P = I_a + F + Q \tag{5.17}$$

where P = total precipitation, I_a = initial abstraction, F = Cumulative infiltration excluding I_a and Q = direct surface runoff (all in units of volume occurring in time Δt). Two other concepts as below are also used with Eq. (5.17).

(i) The first concept is that the ratio of actual amount of direct runoff (Q) to maximum potential runoff ($= P - I_a$) is equal to the ratio of actual infiltration

(F) to the potential maximum retention (or infiltration), S. This proportionality concet cn be schematically shown as in Fig. 5.9.

Thus, $$\frac{Q}{P - I_a} = \frac{F}{S} \qquad (5.18)$$

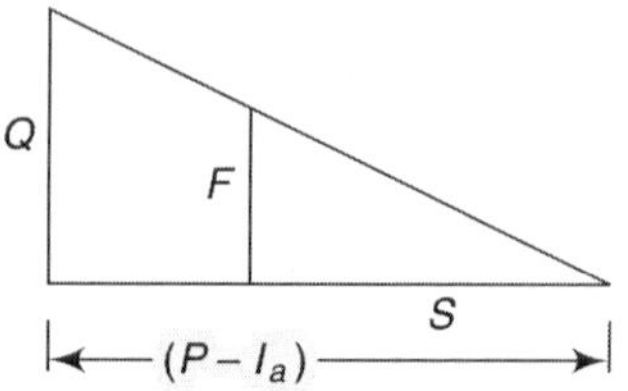

Fig. 5.9 ***Proportionality concept***

(ii) The second concept is that the amount of initial abstraction (I_a) is some fraction of the potential maximum retention (S).

Thus, $$I_a = \lambda S \qquad (5.19)$$

Combining Eqs. (5.18) and (5.19), and using (5.17),

$$Q = \frac{(P - I_a)^2}{P - I_a + S} = \frac{(P - \lambda S)^2}{P + (1 - \lambda)S} \quad = \text{ for } P > \lambda S \qquad (5.20a)$$

Further $$Q = 0 \text{ for } P \le \lambda S \qquad (5.20b)$$

For operation purposes, a time interval Δt = 1 day is adopted. Thus P = daily rainfall and Q = daily runoff from the catchment.

5.6.2 Curve Number (CN)

The parameter S representing the potential maximum retention depends upon the soil–vegetation–land use complex of the catchment and also upon the antecedent soil moisture condition in the catchment just prior to the commencement of the rainfall event. For convenience in practical application the Soil Conservation Services (*SCS*) of *USA* has expressed S (in mm) in terms of a dimensionless parameter CN (the Curve number) as

$$S = \frac{25400}{CN} - 254 = 254\left(\frac{100}{CN} - 1\right) \qquad (5.21)$$

The constant 254 is used to express S in mm.

The curve number CN is now related to S as

$$CN = \frac{25400}{S + 254} \qquad (5.22)$$

and has a range of $100 \ge CN \ge 0$. A CN value of 100 represents a condition of zero potential retention (i.e. impervious catchment) and CN = 0 represents an infinitely abstracting catchment with $S = \infty$. This curve number CN depends upon

- Soil type
- Antecedent moisture condition
- Land use/cover

Soils

In the determination of CN, the hydrological soil classification is adopted. Here, soils are classified into four classes A, B, C and D based upon the infiltration and other characteristics. The important soil characteristics that influence hydrological classification of soils are effective depth of soil, average clay content, infiltration characteristics and permeability. Following is a brief description of four hydrologic soil groups:

- **Group-*A*: (Low Runoff Potential)** Soils having high infiltration rates even when thoroughly wetted and consisting chiefly of deep, well to excessively drained sands or gravels. These soils have high rate of water transmission. [Example: Deep sand, Deep loess and Aggregated silt]
- **Group-*B*: (Moderately Low runoff Potential)** Soils having moderate infiltration rates when thoroughly wetted and consisting chiefly of moderately deep to deep, moderately well to well-drained soils with moderately fine to moderately coarse textures. These soils have moderate rate of water transmission. [Example: shallow loess, sandy loam, red loamy soil, red sandy loam and red sandy soil]
- **Group-*C*: (Moderately High Runoff Potential)** Soils having low infiltration rates when thoroughly wetted and consisting chiefly of moderately deep to deep, moderately well to well drained soils with moderately fine to moderately coarse textures. These soils have moderate rate of water transmission. [Example: clayey loam, shallow sandy loam, soils usually high in clay, mixed red and black soils]
- **Group-*D*: (High Runoff Potential)** Soils having very low infiltration rates when thoroughly wetted and consisting chiefly of clay soils with a high swelling potential, soils with a permanent high-water table, soils with a clay pan, or clay layer at or near the surface, and shallow soils over nearly impervious material. [Example: Heavy plastic clays, certain saline soils and deep black soils].

5.6.3 Antecedent Moisture Condition (*AMC*)

Antecedent Moisture Condition (*AMC*) refers to the moisture content present in the soil at the beginning of the rainfall-runoff event under consideration. It is well known that initial abstraction and infiltration are governed by *AMC*. For purposes of practical application three levels of *AMC* are recognized by *SCS* as follows:

AMC-I: Soils are dry but not to wilting point. Satisfactory cultivation has taken place.

AMC-II: Average conditions

AMC-III: Sufficient rainfall has occurred within the immediate past 5 days. Saturated soil conditions prevail.

The limits of these three *AMC* classes, based on total rainfall magnitude in the previous 5 days, are given in Table 5.5. It is to be noted that the limits also depend upon the seasons: two seasons, viz. growing season and dormant season are considered.

Table 5.5 ***Antecedent Moisture Conditions (AMC) for Determining the Value of CN***

AMC Type	Total Rain in Previous 5 days	
	Dormant Season	Growing Season
I	Less than 13 mm	Less than 36 mm
II	13 to 28 mm	36 to 53 mm
III	More than 28 mm	More than 53 mm

5.6.4 Land Use

The variation of CN under AMC-II, called CN_{II}, for various land-use conditions commonly found in practice are shown in Table 5.6(a, b and c).

Table 5.6(a) ***Runoff Curve Numbers [CN_{II}] for Hydrologic Soil Cover Complexes [Under AMC-II Conditions]***

Land Use	Cover		Hydrologic soil group			
	Treatment or practice	Hydrologic condition	*A*	*B*	*C*	*D*
Cultivated	Straight row		76	86	90	93
Cultivated	Contoured	Poor	70	79	84	88
		Good	65	75	82	86
Cultivated	Contoured &	Poor	66	74	80	82
	Terraced	Good	62	71	77	81
Cultivated	Bunded	Poor	67	75	81	83
		Good	59	69	76	79
Cultivated	Paddy		95	95	95	95
Orchards	With understory cover		39	53	67	71
	Without understory cover		41	55	69	73
Forest	Dense		26	40	58	61
	Open		28	44	60	64
	Scrub		33	47	64	67
Pasture	Poor		68	79	86	89
	Fair		49	69	79	84
	Good		39	61	74	80
Wasteland			71	80	85	88
Roads (dirt)			73	83	88	90
Hard surface areas			77	86	91	93

[Source: Ref.7]

Note: Sugarcane has a separate supplementary table of CN_{II} values [Table 5.6(b)].

The conversion of CN_{II} to other two AMC conditions can be made through the use of following correlation equations.[10]

$$\text{For } AMC\text{-I: } CN_I = \frac{CN_{II}}{2.281 - 0.01281\, CN_{II}} \tag{5.23}$$

Table 5.6(b) ***CN_{II} Values for Sugarcane***

[Source: Ref.7]

Cover and treatment	Hydrologic soil group			
	A	*B*	*C*	*D*
Limited cover, Straight Row	67	78	85	89
Partial cover, Straight row	49	69	79	84
Complete cover, Straight row	39	61	74	80
Limited cover, Contoured	65	75	82	86
Partial cover, Contoured	25	59	45	83
Complete cover, Contoured	6	35	70	79

Table 5.6(c) ***CN_{II} Values for Suburban and Urban Land Uses (Ref. 3)***

Cover and treatment	Hydrologic soil group			
	A	*B*	*C*	*D*
Open spaces, lawns, parks etc				
(i) In good condition, grass cover in more than 75% area	39	61	74	80
(ii) In fair condition, grass cover on 50 to 75% area	49	69	79	84
Commercial and business areas (85% impervious)	89	92	94	95
Industrial Districts (72% impervious)	81	88	91	93
Residential, average 65% impervious	77	85	90	92
Paved parking lots, paved roads with curbs, roofs, driveways, etc	98	98	98	98
Streets and rods				
Gravel	76	85	89	91
Dirt	72	82	87	89

For AMC-III: $$CN_{\text{III}} = \frac{CN_{II}}{0.427 + 0.00573\, CN_{\text{II}}} \tag{5.24}$$

The equations (5.23) and (5.24) are applicable in the CN_{II}, range of 55 to 95 which covers most of the practical range. Values of CN_{I} and CN_{III} covering the full range of CN_{II} are available in Refs 3 and 7. Procedures for evaluation of *CN* from data on small watersheds are available in Ref. 7.

Value of λ

On the basis of extensive measurements in small size catchments *SCS* (1985) adopted $\lambda = 0.2$ as a standard value. With this Eq. (5.20-a) becomes

$$Q = \frac{(P - 0.2\,S)^2}{P + 0.8\,S} \quad \text{for } P > 0.2S \tag{5.25}$$

where Q = daily runoff, P = daily rainfall and S = retention parameter, all in units of mm. Equation 5.25, which is well established, is called as the *Standard SCS-CN equation.*

5.6.5 *SCS-CN* Equation for Indian Conditions

Values of λ varying in the range $0.1 \leq \lambda \leq 0.4$ have been documented in a number of studies from various geographical locations, which include *USA* and many other countries. For use in Indian conditions $\lambda = 0.1$ and 0.3 subject to certain constraints of soil type and *AMC* type has been recommended (Ref. 7) as below:

$$Q = \frac{(P - 0.1\,S)^2}{P + 0.9\,S} \text{ for } P > 0.1\,S,$$

valid *for Black soils under AMC of Type II and III* (5.26)

$$Q = \frac{(P - 0.3\,S)^2}{P + 0.7\,S} \text{ for } P > 0.3S \text{ valid } \textit{for Black soils under}$$

AMC of Type 1 and for all other soils having AMC of types I, II and III (5.27)

Equations (5.26 & 5.27) along with Table 5.6 (a and b) are recommended (Ref. 7) for use in Indian conditions in place of the Standard *SCN-CN* equation.

5.6.6 Procedure for Estimating Runoff Volume from a Catchment

1. Land use/cover information of the catchment under study is derived based on interpretation of multi-season satellite images. It is highly advantageous if the *GIS* database of the catchment is prepared and land use/cover data is linked to it.
2. The soil information of the catchment is obtained by using soil maps prepared by National Bureau of Soil Survey and Land use planning (*NBSS & LUP*) (1966). Soil data relevant to the catchment is identified and appropriate hydrological soil classification is made and the spatial form of this data is stored in *GIS* database.
3. Available rainfall data of various raingauge stations in and around the catchment is collected, screened for consistency and accuracy and linked to the *GIS* database. For reasonable estimate of catchment yield it is desirable to have a rainfall record of at least 25 years duration.
4. Thiessen polygons are established for each identified raingauge station.
5. For each Thiessen cell, appropriate area weighted CN_{II} value is established by adequate consideration of spatial variation of land use and/cover and soil types. Further, for each cell, corresponding CN_I and CN_{III} values are determined by using Eqs. (5.23) and (5.24).
6. Using the relevant *SCS-CN* equations sequentially with the rainfall data, the corresponding daily runoff series is derived for each cell. From this, the needed weekly/monthly/annual runoff time series is derived. Further, by combining the results of various cells constituting the catchment, the corresponding catchment runoff time series is obtained.
7. Appropriate summing of the above time series, yields seasonal/annual runoff volume series and from this the desired dependable catchment yield can be estimated.

5.6.7 Current Status of *SCS-CN* Method

The *SCS-CN* method has received considerable applications and research study since its introduction in 1969. Recently, Ponce and Hawkins[10] (1996) have critically examined the method, clarified its capabilities, limitations and uses. There is a growing body of literature on this method and a good bibliography on this subject is available in Ref. 10. The chief advantages of *SCS-CN* method can be summed up as follows:

- It is a simple, predictable, and stable conceptual method for estimation of direct runoff depth based on storm rainfall depth, supported by empirical data.
- It relies on only one parameter, *CN*. Even though *CN* can have a theoretical range of 0–100, in practice it is more likely to be in the range 40–98.
- It features readily grasped and reasonably well-documented environmental inputs.
- It is a well-established method, having been widely accepted for use in *USA* and many other countries. The modifications suggested by the Ministry of Agriculture, Govt. of India[7], (1972), make its use effective for Indian conditions.

Example 5.5 In a 350 ha watershed the *CN* value was assessed as 70 for *AMC*-III. (a) Estimate the value of direct runoff volume for the following 4 days of rainfall. The *AMC* on July 1st was of category III. Use standard *SCS-CN* equations.

Date	July 1	July 2	July 3	July 4
Rainfall (mm)	50	20	30	18

(b) What would be the runoff volume if the CN_{III} value were 80?

Solution

(a) Given $CN_{III} = 70$ $\quad S = (25400/70) - 254 = 108.6$

$$Q = \frac{(P - 0.2\,S)^2}{P + 0.8\,S} \text{ for } P > 0.2S$$

$$= \frac{[P - (0.2 \times 108.86)]^2}{P + (0.8 \times 108.86)} = \frac{[P - 21.78]^2}{P + 87.09} \text{ for } P > 21.78 \text{ mm}$$

Date	*P* (mm)	*Q* (mm)
July 1	50	5.81
July 2	20	0
July 3	30	0.58
July 4	18	0
Total	**118**	**6.39**

Total runoff volume over the catchment $V_r = 350 \times 10^4 \times 6.39/(1000)$

$= \mathbf{22{,}365}\ \text{m}^3$

(b) Given $CN_{III} = 80$ $S = (25400/80) - 254 = 63.5$

$$Q = \frac{(P - 0.2\,S)^2}{P + 0.8\,S} \text{ for } P > 0.2\,S$$

$$= \frac{[P - (0.2 \times 63.5)]^2}{P + (0.8 \times 63.5)} = \frac{[P - 12.7]^2}{P + 50.8} \text{ for } P > 12.7 \text{ mm}$$

Date	*P* (mm)	*Q* (mm)
July 1	50	13.80
July 2	20	0.75
July 3	30	3.70
July 4	18	0.41
Total	**118**	**18.66**

Total runoff volume over the catchment $V_r = 350 \times 10^4 \times 18.66/(1000)$
$= \mathbf{65{,}310}$ m^3

Example 5.6 A small watershed near Nagpur is 250 ha in size and has group C soil. The land cover can be classified as 30% open forest and 70% poor quality pasture. Assuming AMC at average condition and the soil to be black soil, estimate the direct runoff volume due to a rainfall of 75 mm in one day. Use *SCS-CN* equation applicable to Indian conditions.

Solution

AMC = II. Hence, *CN* = *CN*(II). Soil = Black soil. Referring to Table (5.6-a) for C-group soil,

Land use	%	*CN*	Product
Open forest	30	60	1800
Pasture (poor)	70	86	6020
Total	**100**		**7820**

Average $CN = 7820/100 = 78.2$ $S = (25400/78.2) - 254 = 70.81$

The relevant runoff equation for Black soil and *AMC*-II is

$$Q = \frac{(P - 0.1\,S)^2}{P + 0.9\,S} = \frac{[75 - (0.1 \times 70.81)]^2}{75 + (0.9 \times 70.81)} = 33.25 \text{ mm}$$

Total runoff volume over the catchment $V_r = 250 \times 10^4 \times 33.25/(1000) = \mathbf{83{,}125}$ m^3

Example 5.7 The land use and soil characteristics of a 5000 ha watershed are as follows:

Soil: Not a black soil. Hydrologic soil classification: 60% is Group *B* and 40% is Group *C*

Land Use:

Hard surface areas = 10%
Waste Land = 05%
Orchard (without understory cover) = 30%
Cultivated (Terraced), poor condition = 55%

Antecedent rain: The total rainfall in past five days was 30 mm. The season is dormant season.
(a) Compute the runoff volume from a 125 mm rainfall in a day on the watershed
(b) What would have been the runoff if the rainfall in the previous 5 days was 10 mm?
(c) If the entire area is urbanised with 60% residential area (65% average impervious area), 10% of paved streets and 30% commercial and business area (85% impervious), estimate the runoff volume under AMC-II condition for one day rainfall of 125 mm.
Use *SCS-CN* equations applicable to Indian conditions.

Solution

(a) *Calculation of weighted CN*

From Table 5.5, *AMC* = Type III. Using Table (5.6-a), weighted CN_{II} is calculated as below:

Land use	Total (%)	Soil Group *B* (60%)			Soil Group C (40%)		
		%	*CN*	Product	%	*CN*	Product
Hard surface	10	6	86	516	4	91	364
Waste land	5	3	80	240	2	85	170
Orchard	30	18	55	990	12	69	828
Cultivated land	55	33	71	2343	22	77	1694
Total				**4089**			**3056**

$$\text{Weighted } CN = \frac{(4089 + 3056)}{100} = 71.45$$

By Eq. (5.24), $$CN_{III} = \frac{71.45}{0.427 + (0.00573 \times 71.45)} = 85.42$$

Since the soil is not a black soil, Eq. (5.27) is used to compute the surface runoff.

$$Q = \frac{(P - 0.3S)^2}{P + 0.7S} \text{ for } P > 0.3S \text{ and}$$

$$S = \frac{25400}{CN} - 254 = (25400/85.42) - 254 = 43.35$$

$$Q = \frac{[125 - (0.3 \times 43.35)]^2}{125 + (0.7 \times 43.35)} = 80.74 \text{ mm}$$

Total runoff volume over the catchment $V_r = 5000 \times 10^4 \times 80.74/(1000)$

$= \mathbf{4,037,000\ m^3 = 4.037\ Mm^3}$

(b) Here, *AMC* = Type I

Hence, $$CN_I = \frac{71.45}{2.281 - (0.01281 \times 71.45)} = 52.32$$

$$S = (25400/52.32) - 254 = 231.47$$

$$Q = \frac{[125 - (0.3 \times 231.47)]^2}{125 + (0.7 \times 231.47)} = 10.75 \text{ mm}$$

Total runoff volume over the catchment $V_r = 5000 \times 10^4 \times 10.75/(1000)$

$= \mathbf{537500\ m^3 = 0.5375\ Mm^3}$

(c) From Table 5.5, AMC = Type III. Using Table 5.6-c, weighted CN_{II} is calculated as below:

Land use (%)	Total %	Soil Group B (60%)			Soil Group C (40%)		
		%	CN	Product	%	CN	Product
Residential area (65% imp)	60	36	85	3060	24	90	2160
Commercial area (85% imp)	30	18	92	1656	12	94	1128
Paved roas	10	6	98	588	4	98	392
Total				**5304**			**3680**

$$\text{Weighted } CN_{II} = \frac{(5304 + 3680)}{100} = 89.8$$

By Eq. (5.24) $$CN_{III} = \frac{89.8}{0.427 + (0.00573 \times 89.8)} = 95.37$$

$$S = \frac{25400}{CN} - 254 = (25400/95.37) - 254 = 12.33$$

Since the soil is not a black soil, Eq. (5.27) is used to compute the surface runoff volume.

$$Q = \frac{(P - 0.3\,S)^2}{P + 0.7\,S} \text{ for } P > 0.3S \text{ and}$$

$$Q = \frac{[125 - (0.3 \times 12.33)]^2}{125 + (0.7 \times 12.33)} = 110.11 \text{ mm}$$

Total runoff volume over the catchment $V_r = 5000 \times 10^4 \times 110.11/(1000)$

$= \mathbf{5{,}505{,}500\ m^3 = 5.5055\ Mm^3}$

5.6.8 *CN* and *C* of Rational Formula

SCS-CN method estimates runoff volume while the rational formula (Chapter 7, Sec. 7.2) estimates runoff rate based on the runoff coefficient C. CN and C of are not easily related even though they depend on the same set of parameters. For an infinite sponge, C is 0 and CN is 0. Similarly, for an impervious surface, C is 1.0 and CN is 100. While the end points in the mapping are easily identifiable, the relationship between CN and C are nonlinear. In a general sense, high Cs are likely to be found where CN values are also high.

5.7 *Flow–Duration Curve*

It is well known that the streamflow varies over a water year. One of the popular methods of studying this streamflow variability is through flow-duration curves. A flow-duration curve of a stream is a plot of discharge against the per cent of time the flow was equalled or exceeded. This curve is also known as *discharge-frequency curve.*

The streamflow data is arranged in a descending order of discharges, using class intervals if the number of individual values is very large. The data used can be daily, weekly, ten daily or monthly values. If N number of data points are used in this listing, the plotting position of any discharge (or class value) Q is

$$P_p = \frac{m}{N+1} \times 100\% \qquad (5.28)$$

where m is the order number of the discharge (or class value), P_p = percentage probability of the flow magnitude being equalled or exceeded. The plot of the discharge Q against P_p is the flow-duration curve (Fig. 5.10). Arithmetic scale paper, or semi-log or log-log paper is used depending upon the range of data and use of the plot. The flow duration curve represents the cumulative frequency distribution and can be considered to represent the streamflow variation of an average year. The ordinate Q_p at any percentage probability P_p represents the flow magnitude in an average year that can be expected to be equalled or exceeded P_p per cent of time and is termed as P_p % dependable flow. In a perennial river Q_{100} = 100% dependable flow is a finite value. On the other hand, in an intermittent or ephemeral river the streamflow is zero for a finite part of the year and as such Q_{100} is equal to zero.

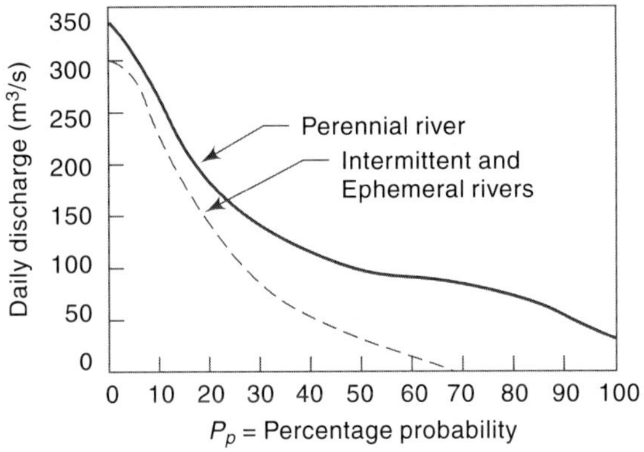

Fig. 5.10 ***Flow Duration Curve***

The following characteristics of the flow duration curve are of interest.

- The slope of a flow-duration curve depends upon the interval of data selected. For example, a daily stream flow data gives a steeper curve than a curve based on monthly data for the same stream. This is due to the smoothening of small peaks in the monthly data.
- The presence of a reservoir in a stream considerably modifies the virgin-flow duration curve depending on the nature of flow regulation. Figure 5.9 shows the typical reservoir regulation effect.
- The virgin-flow duration curve when plotted on a log probability paper plots as a straight line at least over the central region. From this property, various coefficients expressing the variability of the flow in a stream can be developed for the description and comparison of different streams.
- The chronological sequence of occurrence of the flow is masked in the flow-duration curve. A discharge of say 1000 m^3/s in a stream will have the same percentage P_p whether it has occurred in January or June. This aspect, a serious handicap, must be kept in mind while interpreting a flow-duration curve.

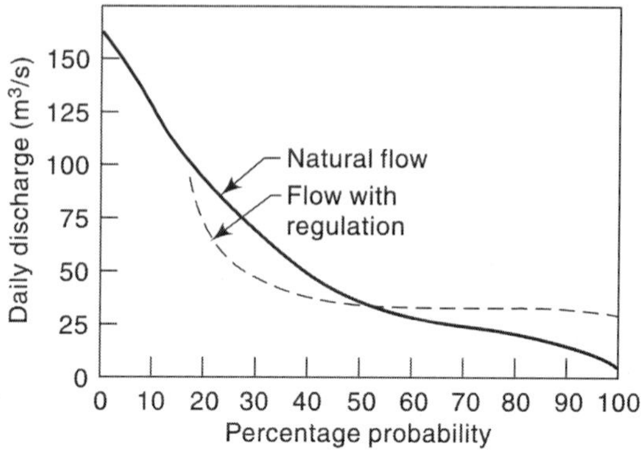

Fig. 5.11 ***Reservoir Regulation Effect***

- The flow-duration curve plotted on a log-log paper (Fig. 5.12) is useful in comparing the flow characteristics of different streams. A steep slope of the curve indicates a stream with a highly variable discharge. On the other hand, a flat slope indicates a slow response of the catchment to the rainfall and also indicates small variability. At the lower end of the curve, a flat portion indicates considerable base flow. A flat curve on the upper portion is typical of river basins having large flood plains and also of rivers having large snowfall during a wet season.

Flow-duration curves find considerable use in water resources planning and development activities. Some of the important uses are

1. In evaluating various dependable flows in the planning of water resources engineering projects
2. Evaluating the characteristics of the hydropower potential of a river
3. Designing of drainage systems
4. In flood-control studies
5. Computing the sediment load and dissolved solids load of a stream
6. Comparing the adjacent catchments with a view to extend the streamflow data

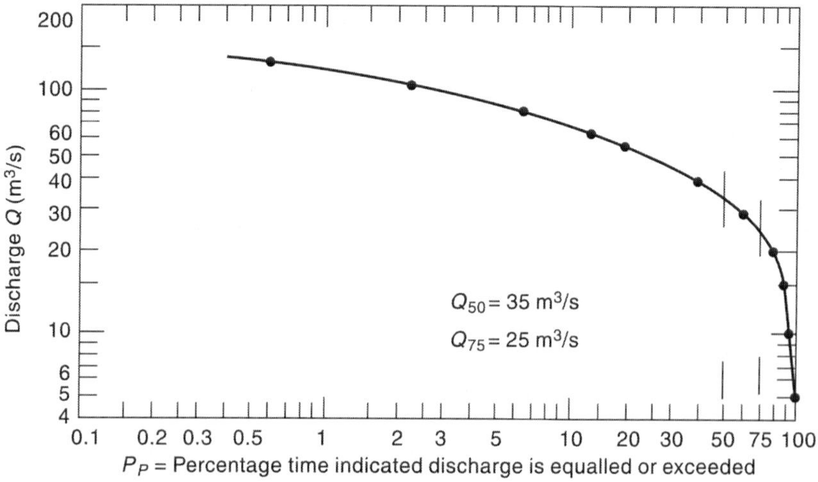

Fig. 5.12 ***Flow-Duration Curve—Example 5.8***

Example 5.8 The daily flows of a river for three consecutive years are shown in Table 5.7. For convenience the discharges are shown in class intervals and the number of days the flow belonged to the class is shown. Calculate the 50 and 75% dependable flows for the river.

Solution

The data are arranged in descending order of class value. In Table 5.7, column 5 shows the total number of days in each class. Column 6 shows the cumulative total of column 5, i.e. the number of days the flow is equal to or greater than the class interval. This gives the value of m. The percentage probability P_p the probability of flow in the class interval being equalled or exceeded is given by Eq. (5.28),

$$P_p = \frac{m}{N+1} \times 100\%$$

Table 5.7 *Calculation of Flow Duration Curve from Daily Flow Data—Example 5.8*

Daily mean discharge (m^3/s)	No. of days flow in each class interval			Total of columns 2, 3, 4 1961–64	Cumulative Total m	$P_p = \left(\frac{m}{N+1}\right) \times 100\%$
	1961–62	1962–63	1963–64			
1	2	3	4	5	6	7
140–120.1	0	1	5	6	6	0.55
120–100.1	2	7	10	19	25	2.28
100–80.1	12	18	15	45	70	6.38
80–60.1	15	32	15	62	132	12.03
60–50.1	30	29	45	104	236	21.51
50–40.1	70	60	64	194	430	39.19
40–30.1	84	75	76	235	665	60.62
30–25.1	61	50	61	172	837	76.30
25–20.1	43	45	38	126	963	87.78
20–15.1	28	30	25	83	1046	95.35
15–10.1	15	18	12	45	1091	99.45
10–5.1	5	—	—	5	1096	99.91
Total	**365**	**365**	**366**	**N = 1096**		

In the present case, N = 1096. The smallest value of the discharge in each class interval is plotted against P_p on a log-log paper (Fig. 5.10). From this figure Q_{50} = 50% dependable flow = 35 m^3/s and Q_{75} = 75% dependable flow = 26 m^3/s.

5.8 *Flow-Mass Curve*

The flow-mass curve is a plot of the cumulative discharge volume against time plotted in chronological order. The ordinate of the mass curve, V at any time t is thus

$$V = \int_{t_0}^{t} Q\,dt \tag{5.29}$$

where t_0 is the time at the beginning of the curve and Q is the discharge rate. Since the hydrograph is a plot of Q *vs* t, it is easy to see that the flow–mass curve is an integral curve (summation curve) of the hydrograph. The flow–mass curve is also known as *Rippl's mass curve* after Rippl (1882) who suggested its use first. Figure 5.9 shows a typical flow–mass curve. Note that the abscissa is chronological time in months in this figure. It can also be in days, weeks or months depending on the data being analysed. The ordinate is in units of volume in million m^3. Other units employed for ordinate include m^3/s day (cumec day), ha.m and cm over a catchment area.

The slope of the mass curve at any point represents $\frac{dV}{dt} = Q$ = rate of flow at that instant. If two points M and N are connected by a straight line, the slope of the line represents the average rate of flow that can be maintained between the times t_m and t_n if a reservoir of adequate storage is available. Thus, the slope of the line AB joining the starting point and the last points of a mass curve represents the average discharge over the whole period of plotted record.

5.8.1 Calculation of Storage Volume

Consider a reservoir on the stream whose mass curve is plotted in Fig. 5.13. If it is assumed that the reservoir is full at the beginning of a dry period, i.e. when the inflow rate is less than the withdrawal (demand) rate, the maximum amount of water drawn from the storage is the cumulative difference between supply and demand volumes from the beginning of the dry season. Thus, the storage required S is

$$S = \text{maximum of } (\Sigma V_D - \Sigma V_s)$$

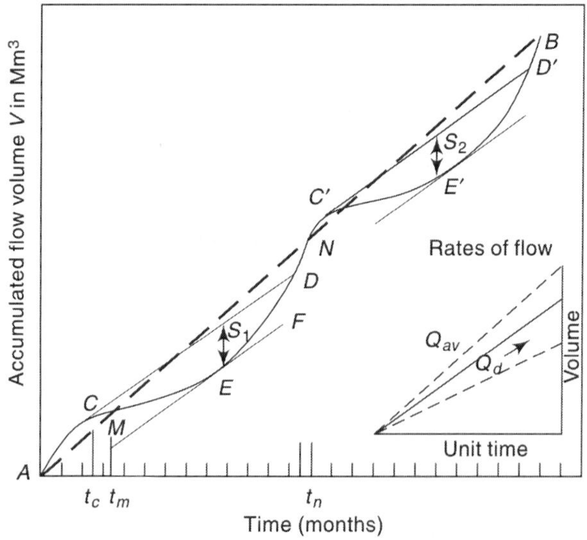

Fig. 5.13 ***Fow–Mass Curve***

where V_D = demand volume, V_S = supply volume. The storage, S which is the maximum cumulative deficiency in any dry season is obtained as the maximum difference in the ordinate between mass curves of supply and demand. The minimum storage volume required by a rescrvoir is the largest of such S values over different dry periods.

Consider the line CD of slope Q_d drawn tangential to the mass curve at a high point on a ridge. This represents a constant rate of withdrawal Q_d from a reservoir and is called *demand line*. If the reservoir is full at C (at time t_c) then from point C to E the demand is larger than the supply rate as the slope of the flow–mass curve is smaller than the demand line CD. Thus, the reservoir will be depleting and the lowest capacity is reached at E. The difference in the ordinates between the demand line CD and a line EF drawn parallel to it and tangential to the mass curve at E (S_1 in Fig. 5.11) represents the volume of water needed as storage to meet the demand from the time the reservoir was full. If the flow data for a large time period is available, the demand lines are drawn tangentially at various other ridges (e.g. $C'D'$ in Fig. 5.11) and the largest of the storages obtained is selected as the minimum storage required by a reservoir. Example 5.9 explains this use of the mass curve. Example 5.10 indicates, storage calculations by arithmetic calculations by adopting the mass-curve principle.

Example 5.9 The following table gives the mean monthly flows in a river during 1981. Calculate the minimum storage required to maintain a demand rate of 40 m^3/s.

Month	Jan	Feb	Mar	Apr	May	June	July	Aug	Sept	Oct	Nov	Dec
Mean Flow (m^3/s)	60	45	35	25	15	22	50	80	105	90	80	70

Solution

From the given data, the monthly flow volume and accumulated volumes and calculated as in Table 5.8. The actual number of days in the month are used in calculating of the monthly flow volume. Volumes are calculated in units of cumec. day (= 8.64×10^4 m^3).

Table 5.8 *Calculation of Mass Curve—Example 5.9*

Month	Mean flow (m^3/s)	Monthly flow volume (cumec-day)	Accumulated volume (cumec-day)
Jan	60	1860	1860
Feb	45	1260	3120
Mar	35	1085	4205
April	25	750	4955
May	15	465	5420
June	22	660	6080
July	50	1550	7630

(Contd.)

Table 5.8 (*Contd.*)

Aug	80	2480	10,110
Sep	105	3150	13,260
Oct	90	2790	16,050
Nov	80	2400	18,450
Dec	70	2170	20,620

A mass curve of accumulated flow volume against time is plotted (Fig. 5.14). In this figure, all the months are assumed to be of average duration of 30.4 days. A demand line with slope of 40 m^3/s is drawn tangential to the *hump* at the beginning of the curve; line *AB* in Fig. 5.14. A line parallel to this line is drawn tangential to the mass curve at the *valley* portion; line $A'B'$. The vertical distance S_1 between these parallel lines is the minimum storage required to maintain the demand. The value of S_1 is found to be 2100 cumec. Days = 181.4 million m^3.

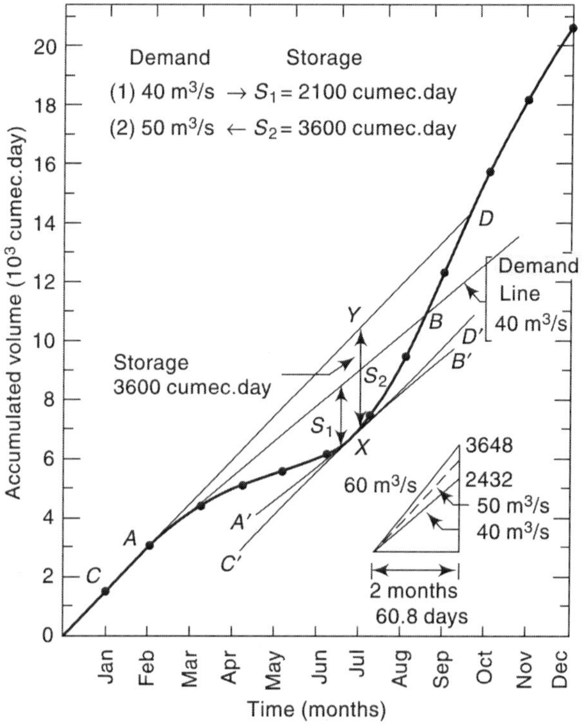

Fig. 5.14 ***Flow-Mass Curve—Example 5.9***

Example 5.10 Work out Example 5.9 through arithmetic calculation without the use of mass curve. What is the maximum constant demand that can be sustained by this river?

Table 5.9 *Calculation of Storage—Example 5.10*

1	2	3	4	5	6	7	8
Month	Mean inflow rate (m^3/s)	Inflow volume (cumec. day)	Demand rate (m^3/s)	Demand volume (cumec. day)	Departure [col. 3-(col. 5)	Cum. excess demand volume (cumec. day)	Cum. excess inflow volume (cumec. day)
Jan	60	1860	40	1240	620		620
Feb	45	1260	40	1120	140		760
Mar	35	1085	40	1240	–155	–155	
Apr	25	750	40	1200	–450	–605	
May	15	465	40	1240	–775	–1380	
Jun	22	660	40	1200	–540	–1920	
July	50	1550	40	1240	310		310
Aug	80	2480	40	1240	1240		1550
Sept	105	3150	40	1200	1950		3500
Oct	90	2790	40	1240	1550		5050
Nov	80	2400	40	1200	1200		6250
Dec	70	2170	40	1240	930		7180
	Monthly mean =	**1718.3**					

Solution

The inflow and demand volumes of each month are calculated as in Table 5.9. Column 6 indicating the departure of the inflow volume from the demand. The negative values indicate the excess of demand over the inflow and these have to be met by the storage. Column 7 indicates the cumulative excess demand (i.e., the cumulative excess negative departures). This column indicates the depletion of storage, the first entry of negative value indicates the beginning of *dry period* and the last value the end of the dry period. Column 8 indicates the filling up of storage and spillover (if any). Each dry period and each filling up stage is to be calculated separately as indicated in Table 5.9.

The maximum value in Col. 7 represents the minimum storage necessary to meet the demand pattern. In the present case, there is only one dry period and the storage requirement is 1920 cumec.day. Note that the difference between this value and the value of 2100 cumec.day obtained by using the mass curve is due to the curvilinear variation of inflow volumes obtained by drawing a smooth mass curve. The arithmetic calculation assumes a linear variation of the mass curve ordinates between two adjacent time units.

[**Note:** It is usual to take data pertaining to a number of *N* full years. When the analysis of the given data series of length *N* causes the first entry in Col. 7 to be a negative value and the last entry is also a negative value, then the calculation of the maximum deficit may pose some confusion. In such cases, repeating the data sequence by one more data cycle of *N* years in continuation with the last entry would overcome this confusion. (See Sec. 5.9, Bullet 2.) There are many other combinations of factors that may cause confusion in interpretation of the results and as such the use of *Sequent Peak Algorithm* described in Sec. 5.9 is recommended as the foolproof method that can be used with confidence in all situations.]

Column 8 indicates the cumulative excess inflow volume from each demand withdrawal from the storage. This indicates the filling up of the reservoir and volume in excess of the provided storage (in the present case, 1920 cumec.day) represent spillover. The calculation of this column is necessary to know whether the reservoir fills up after a depletion by meeting a critical demand and if so, when? In the present case the cumulative excess inflow volume will reach +1920 cumec.day in the beginning of September. The reservoir will be full after that time and will be spilling till end of February.

Average of the inflow volume per month = Annual inflow volume/12 = average monthly demand that can be sustained by this river = 1718.33 cumec.day.

5.8.2 Calculation of Maintainable Demand

The converse problem of determining the maximum demand rate that can be maintained by a given storage volume can also be solved by using a mass curve. In this case, tangents are drawn from the "ridges" of the mass curves across the next "valley" at various slopes. The demand line that requires just the given storage (u_1 v_1 in Fig. 5.15) is the proper demand that can be sustained by the reservoir in that dry period. Similar demand lines are drawn at other "valleys" in the mass curve (e.g. u_2 v_2 and the demand rates determined. The smallest of the various demand rates thus found denotes the maximum firm demand that can be sustained by the given storage. It may be noted that this problem involves a trial-and-error procedure for its solution. Example 5.10 indicates this use of the mass curve.

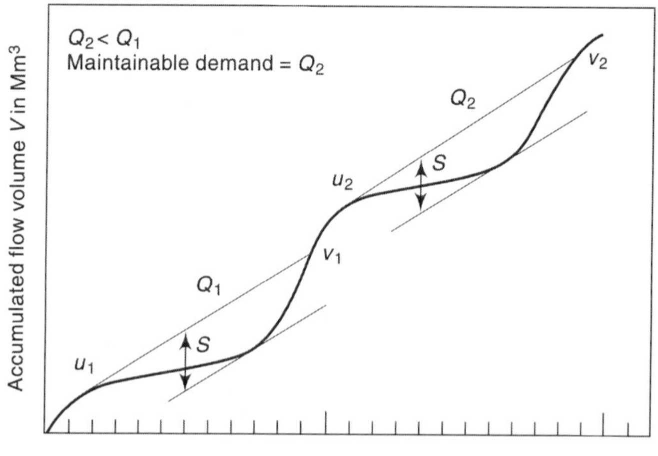

Fig. 5.15 ***Determination of Maintainable Demand***

The following salient points in the use of the mass curve are worth noting:

- The vertical distance between two successive tangents to a mass curve at the ridges (points v_1 and u_2 in Fig. 5.15) represent the water "wasted" over the spillway.
- A demand line must intersect the mass curve if the reservoir is to refill. Nonintersection of the demand line and mass curve indicates insufficient inflow.

Example 5.11 Using the mass curve of Example 5.9, obtain the maximum uniform demand rate that can be maintained by a storage of 3600 m^3/s days.

Solution

An ordinate *XY* of magnitude 3600 Cumec-days is drawn in Fig. 5.14 at an approximate lowest position in the dip of the mass curve and a line passing through *Y* and tangential to the "hump" of the mass curve at *C* is drawn (line *CYD* in Fig. 5.14). A line parallel to *CD* at the lowest position of the mass curve is now drawn and the vertical interval between the two measured. If the original guess location of *Y* is correct, this vertical distance will be 3600 m^3/s day. If not, a new location for *Y* will have to be chosen and the above procedure repeated.

The slope of the line *CD* corresponding to the final location of *XY* is the required demand rate. In this example this rate is found to be 50 m^3/s.

5.8.3 Variable Demand

In the examples given above, a constant demand rate was assumed to simplify the problem. In practice, it is more likely that the demand rate varies with time to meet various end uses, such as irrigation, power and water-supply needs. In such cases, a mass curve of demand, also known as *variable use line* is prepared and superposed on the flow–mass curve with proper matching of time. For example, the demand for the month of February must be against the inflow for the same month. If the reservoir is full at first point of intersection of the two curves, the maximum intercept between the two curves represents the needed storage of the reservoir (Fig. 5.16). Such a plot is sometimes known as *regulation diagram* of a reservoir.

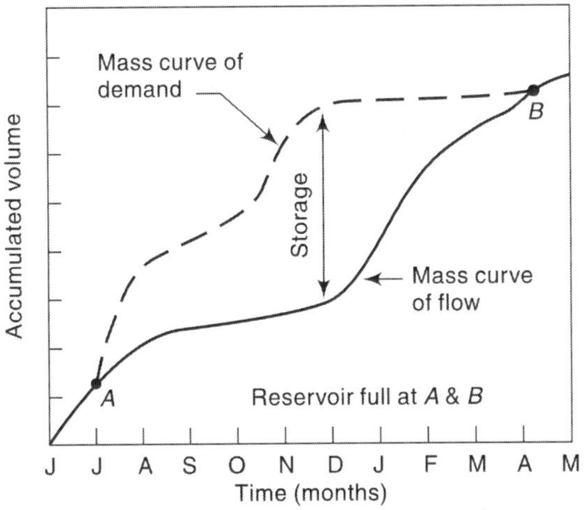

Fig. 5.16 ***Variable Use Line***

In the analysis of problems related to the reservoirs, it is necessary to account for evaporation, leakage and other losses from the reservoir. The losses in the volume of water in a known interval of time can either be included in the demand rates or deducted from inflow rates. Similarly, the contribution of the rainfall directly on to the reservoir surface is to be properly accounted for; either as contribution to the inflow or as deduction from the demand volumes. Example 5.12 gives an arithmetic calculation procedure for calculating storage under variable demand.

Example 5.12 For a proposed reservoir, the following data given below was calculated. The prior water rights required the release of natural flow or 5 m^3/s whichever is lower. Assuming an average reservoir area of 20 km^2, estimate the storage required to meet the demands listed. (Assume the runoff coefficient of the area submerged by the reservoir = 0.5)

Month	Mean Flow (m^3/s)	Demand (Mm^3)	Monthly Evaporation (cm)	Monthly Rainfall (cm)
Jun	40	36	16	15
July	80	30	12	24
Aug	70	30	12	20
Sept	60	30	12	19
Oct	20	35	12	9
Nov	20	66	11	5
Dec	10	66	17	2
Jan	10	66	12	2
Feb	10	66	13	2
Mar	5	66	17	1
Apr	5	66	18	1
May	4	66	20	1

Solution

Use actual number of days in a month for calculating the monthly flow and for prior right release. Let N = number of days in a month

(i) Evaporation volume = $\left(\dfrac{E}{100}\right) \times 20 \times 10^6 = 0.2E$ Mm^3

where E = Evaporation in cm.

(ii) Inflow volume = (Mean monthly flow rate in Mm^3) $\times N \times 8.64 \times 10^4$ m^3

(iii) Prior rights = ($5 \times N \times 8.64 \times 10^4$ m^3) or actual inflow volume whichever is lesser.

= $0.432\, N$ Mm^3 or actual inflow volume whichever is lesser.

(iv) Contribution of rainfall to withdrawal:

- If an area A_r of the catchment is covered by the water spread of the reservoir, all the rainfall on it will be reflected as additional volume.
- If the same area A_r is not covered by the water spread, and if C = runoff coefficient then due to a rainfall of P a volume equal to ($C \times A_r$) will go as runoff in to the stream and this will be reflected as inflow volume to the reservoir.

- Hence, the net effect of rainfall of P cm on a reservoir area of A_r is a net addition of flow in to the reservoir of magnitude = $(1 - C) A_r P$.
  In the present case, $C = 0.5$, $A = 20$ km^2, and the net volume addition to the reservoir = $\frac{P}{100} \times (1 - 0.5) \times 20 = 0.1P$ Mm^3. Note that this volume addition is taken as negative withdrawal in the calculations.

The calculations are performed in a tabular manner as shown in Table 5.10. The calculations can be performed very easily through use of a spreadsheet such as *MS* Excel.

Table 5.10 ***Calculation of Storage–Example 5.12***

Month	No. of Days in the Month	Inflow Volume (Mm^3)	Withdrawals				Total withdrawal (4 + 5 + 6 + 7) (Mm^3)	Departure = (3–8) (Mm^3)	Cumulative Excess Demand (Mm^3)	Cumulative Excess Inflow Volume (Mm^3)
			Demand (Mm^3)	Prio Rights (Mm^3)	Evapora-tion (Mm^3)	Contribution of Ranfall (Mm^3)				
1	2	3	4	5	6	7	8	9	10	11
Jun	30	103.68	36	12.96	3.20	– 1.50	50.66	53.02		53.02
July	31	214.27	30	13.39	2.40	– 2.40	43.39	170.88		223.90
Aug	31	187.49	30	13.39	2.40	– 2.00	43.79	143.70		367.60
Sept	30	155.52	30	12.96	2.40	– 1.90	43.46	112.06		479.66
Oct	31	53.57	35	13.39	2.40	– 0.90	49.89	3.68		483.33
Nov	30	51.84	66	12.96	2.20	– 0.50	80.66	– 28.82	– 28.82	
Dec	31	26.78	66	13.39	3.40	– 0.20	82.59	– 55.81	– 84.63	
Jan	31	26.78	66	13.39	2.40	– 0.20	81.59	– 54.81	– 139.44	
Feb	28	24.19	66	12.10	2.60	– 0.20	80.50	– 56.30	– 195.74	
Mar	31	13.39	66	13.39	3.40	– 0.10	82.69	– 69.30	– 265.04	
Apr	30	12.96	66	12.96	3.60	– 0.10	82.46	– 69.50	– 334.54	
May	31	10.71	66	10.71	4.00	– 0.10	80.61	– 69.90	– 404.44	

From Table 5.10, it is seen that the minimum storage required to meet the stipulated demand is 404.44 Mm^3 as indicated in the last entry of Col. 10. The maximum value of Col. 11 is 473.33 Mm^3 indicating that there is adequate inflow to fill the reservoir and the chosen storage is feasible.

The mass-curve method assumes a definite sequence of events and this is its major drawback. In reality, the runoff is subject to considerable time variations and definite sequential occurrences represent only an idealized situation. While the mass-curve analysis is adequate for small projects or for preliminary studies of large storage projects it is not suitable for final design of large storage projects. The latter require sophisticated methods such as *time-series analysis* of data in their final design.

5.9 Sequent Peak Algorithm

The mass curve method of estimating the minimum storage capacity to meet a specified demand pattern, described in the previous section has a number of different forms of use in its practical application. However, the following basic assumptions are made in all the adaptations of the mass-curve method of storage analysis.

- If N years of data are available, the inflows and demands are assumed to repeat in cyclic progression of N year cycles. It is to be noted that in historical data this leads to an implicit assumption that future flows will not contain a more severe drought than what has already been included in the data.
- The reservoir is assumed to be full at the beginning of a dry period. Thus, while using the mass curve method, the beginning of the dry period should be noted and the minimum storage required to pass each drought period calculated. Sometimes, for example in Problem 5.7, it may be necessary to repeat the given data series of N years sequentially for a minimum of one cycle, i.e. for additional N years, to arrive at the desired minimum storage requirement.

The mass curve method is widely used for the analysis of reservoir capacity-demand problems. However, there are many variations of the basic method to facilitate graphical plotting, handling of large data, etc. A variation of the arithmetical calculation described in Examples 5.10 and 5.12 called the *sequent peak algorithm* is particularly suited for the analysis of large data with the help of a computer. This procedure, first given by Thomas (1963), is described in this section.

Let the data be available for N consecutive periods not necessarily of uniform length. These periods can be year, month, day or hours depending upon the problem. In the ith period let x_i = inflow volume and D_i = demand volume. The surplus or deficit of storage in that period is the *net-flow volume* given by

$$\text{Net-flow volume} = \text{Inflow volume} - \text{Outflow volume}$$
$$= x_i - D_i$$

In the sequent peak algorithm a mass curve of cumulative net-flow volume against chronological time is used. This curve, known as *residual mass curve* (shown typically in Fig. 5.17), will have peaks (local maximums) and troughs (local minimums). For any peak P, the next following peak of magnitude greater than P, is called a *sequent peak*. Using two cycles of N periods, where N is the number of periods of the data series, the required storage volume is calculated by the following procedure:

1. Calculate the cumulative net-flow volumes, viz.

$$\sum^{t}(x_i - D_i) \qquad \text{for } t = 1, 2, 3 \ldots, 2N$$

2. Locate the first peak P_1 and the sequent peak P_2 which is the next peak of greater magnitude than P_1 (Fig. 5.17).
3. Find the *lowest trough* T_1 between P_1 and P_2 and calculate $(P_1 - T_1)$.
4. Starting with P_2 find the next sequent peak P_3 and the lowest through T_2 and calculate $(P_2 - T_2)$.

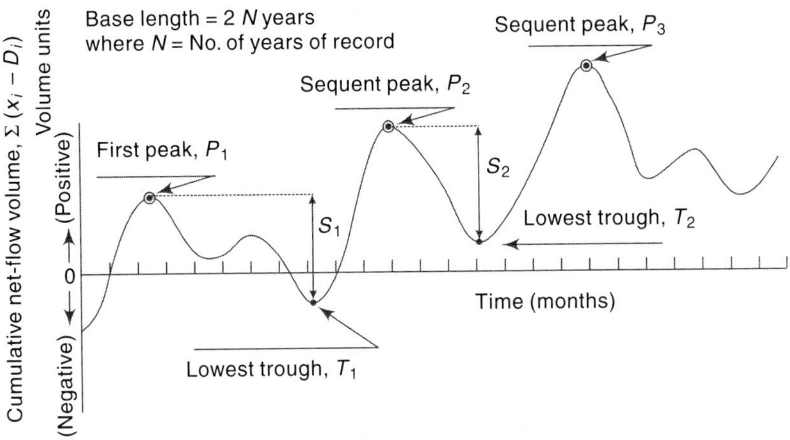

Fig. 5.17 ***Residual Mass Curve—Definition Sketch for Sequent Peak Algorithm***

5. Repeat the procedure for all the sequent peaks available in the $2N$ periods, i.e. determine the sequent peak P_j, the corresponding T_j and the jth storage $(P_j - T_J)$ for all j values.
6. The required reservoir storage capacity is

$$S = \text{maximum of } (P_j - T_j) \text{ values}$$

Example 5.13 Following is the stream-flow record of 2 consecutive critical years of a stream that is being considered for hydropower development. Estimate the minimum size of a reservoir on this stream to provide a constant downstream withdrawal of 7.0 m³/s. Use sequent peak algorithm.

Month (1st Year)	Monthly flow volume (M.m³)	Month (2nd Year)	Monthly flow volume (M.m³)
June	36.20	June	10.20
July	57.40	July	30.80
August	65.50	August	43.10
September	28.60	September	53.10
October	32.80	October	39.00
November	36.90	November	28.90
December	24.60	December	16.50
January	10.20	January	16.40
February	2.10	February	12.30
March	2.10	March	10.30
April	2.10	April	8.30
May	4.10	May	4.90

Solution:

The given data is for two years. And as such the sequent peak algorithm calculations are performed for 2 × 2 = 4 years. Calculations are shown in Tables: Table 5.11(a) and 5.11(b). In Table 5.11(a):

- Col.4 is the given inflow data
- Demand volume (Col. 5) = (constant withdrawal rate of 7.0 m^3/s)

$$\times (60 \times 60 \times 24) \times (\text{number of days in the month}) \times (1/10^6)\ Mm^3$$
$$= 0.6048 \times \text{Col.2}$$

- Col.6 = Col.4 – Col.5
- Col. 7 = Cumulative values of Col.6

Scanning the cumulative net inflow values obtained (Col.7) from the start, the first peak P_1 is identified as having a value of 152.57 Mm^3 . This occurs at the end of December of the first year, (Sl. No. 7). The sequent peak P_2 is the peak next to P_1 and of magnitude higher than 152.57 Mm^3. This P_2 is identified as having a magnitude of 176.27 Mm^3 and occurs at the end of November of second year, (Sl. No.18). Similarly the third and fourth sequent peaks are identified at Sl. no. 31 and 42 respectively and their values tabulated in Table 5.11(b).

Next, the lowest troughs between two sequent peaks are identified. There are three lowest troughs between four sequent peaks and their values are marked in Table 5.11(a) and also listed in Table 5.11(b).

Table 5.11(a) ***Computation sheet of Example 5.12***

1	2	3	4	5	6	7	8
Sl. No	Month	No. of days in the month	Inflow volume ($M.m^3$)	Demand volume ($M.m^3$)	Net Inflow volume ($M.m^3$)	Cumulative Net-Inflow volume ($M.m^3$)	Remarks
1	June	30	36.20	18.14	18.06	18.06	1st Year
2	July	31	57.40	18.75	38.65	56.71	
3	Aug	31	65.50	18.75	46.75	103.46	
4	Sept	30	28.60	18.14	10.46	113.91	
5	Oct	31	32.80	18.75	14.05	127.97	
6	Nov	30	36.90	18.14	18.76	146.72	
7	Dec	31	24.60	18.75	5.85	**152.57**	Peak-1, (P_1)
8	Jan	31	10.20	18.75	-8.55	144.02	
9	Feb	28	2.10	16.93	-14.83	129.19	
10	March	31	2.10	18.75	-16.65	112.54	
11	April	30	2.10	18.14	-16.04	96.50	
12	May	31	4.10	18.75	-14.65	81.85	
13	June	30	10.20	18.14	-7.94	**73.90**	2nd year Trough -1, (T_1)

(*Contd.*)

Table 5.11(a) (*Contd.*)

14	July	31	30.80	18.75	12.05	85.96	
15	Aug	31	43.10	18.75	24.35	110.31	
16	Sept	30	53.10	18.14	34.96	145.26	
17	Oct	31	39.00	18.75	20.25	165.51	
18	Nov	30	28.90	18.14	10.76	**176.27**	Peak-2, (P_2)
19	Dec	31	16.50	18.75	-2.25	174.02	
20	Jan	31	16.40	18.75	-2.35	171.67	
21	Feb	28	12.30	16.93	-4.63	167.04	
22	March	31	10.30	18.75	-8.45	158.59	
23	April	30	8.30	18.14	-9.84	148.74	
24	May	31	4.90	18.75	-13.85	**134.90**	Trough-2, (T_2)
25	June	30	36.20	18.14	18.06	152.95	3rd Year
26	July	31	57.40	18.75	38.65	191.60	
27	Aug	31	65.50	18.75	46.75	238.35	
28	Sept	30	28.60	18.14	10.46	248.81	
29	Oct	31	32.80	18.75	14.05	262.86	
30	Nov	30	36.90	18.14	18.76	281.62	
31	Dec	31	24.60	18.75	5.85	**287.47**	Peak-3, (P_3)
32	Jan	31	10.20	18.75	-8.55	278.92	
33	Feb	28	2.10	16.93	-14.83	264.09	
34	March	31	2.10	18.75	-16.65	247.44	
35	April	30	2.10	18.14	-16.04	231.39	
36	May	31	4.10	18.75	-14.65	216.74	
37	June	30	10.20	18.14	-7.94	**208.80**	4th Year Trough-3, (T_3)
38	July	31	30.80	18.75	12.05	220.85	
39	Aug	31	43.10	18.75	24.35	245.20	
40	Sept	30	53.10	18.14	34.96	280.16	
41	Oct	31	39.00	18.75	20.25	300.41	
42	Nov	30	28.90	18.14	10.76	**311.17**	Peak-4, (P_4)
43	Dec	31	16.50	18.75	-2.25	308.92	
44	Jan	31	16.40	18.75	-2.35	306.57	
45	Feb	28	12.30	16.93	-4.63	301.93	
46	March	31	10.30	18.75	-8.45	293.48	
47	April	30	8.30	18.14	-9.84	283.64	
48	May	31	4.90	18.75	-13.85	269.79	

Table 5.11(b) ***Results of computation, Example 5.13***

No	Peak (Mm^3)	Trough (Mm^3)	Storage $S_j = (P_j - T_j)$ (Mm^3)
1	152.573	73.904	78.669
2	176.270	134.896	41.374
3	287.469	208.800	78.669
4	311.166		

Required minimum storage = Maximum of $S_j = (P_j - T_j)$ values
= 78.669 Mm^3 = Say 78.67 Mm^3

5.10 *Droughts*

In the previous sections of this chapter, the variability of the stream flow was considered in the flow duration curve and flow-mass curve. However, the extremes of the stream flow as reflected in floods and droughts need special study. They are natural disasters causing large-scale human suffering and huge economic loss and considerable effort is devoted by the world over to control or mitigate the ill effects of these two hydrological extremes. The various aspects of floods are discussed in Chapters 7 and 8. The topic of drought, which is not only complex but also region specific is discussed, rather briefly, in this section. The classification of droughts and the general nature of drought studies are indicated with special reference to the Indian conditions. For further details, the reader is referred to References 1, 2, 4 and 6.

5.10.1 Definition and Classification

Drought is a climatic anomaly characterized by deficit supply of moisture. This may result from subnormal rainfall over large regions causing below normal natural availability of water over long periods of time. Drought phenomenon is a hydrological extreme like flood and is a natural disaster. However, unlike floods, the droughts are of the creeping kind; they develop in a region over a length of time and sometimes may extend to continental scale. The consequences of droughts on the agricultural production, hydropower generation and the regional economy in general is well known. Further, during droughts the quality of available water will be highly degraded resulting in serious environmental and health problems.

Many classifications of droughts are available in literature. The following classification into three categories proposed by the National Commission on Agriculture (1976) is widely adopted in the country:

- **Meteorological Drought** It is a situation where there is more than 25% decrease in precipitation from normal over an area.
- **Hydrological Drought** Meteorological drought, if prolonged, results in hydrological drought with marked depletion of surface water and groundwater. The consequences are the drying up of tanks, reservoirs, streams and rivers, cessation of springs and fall in the groundwater level.

- **Agricultural drought** This occurs when the soil moisture and rainfall are inadequate during the growing season to support healthy crop growth to maturity. There will be extreme crop stress and wilt conditions.

5.10.2 Meteorological Drought

The India Meteorological Department (*IMD*) has adopted the following criteria for sub-classification of meteorological droughts. A meteorological sub-division is considered to be affected by drought if it receives a total seasonal rainfall less than that of 75% of the normal value. Also, the drought is classified as *moderate* if the seasonal deficiency is between 26% and 50%. The drought is said to be *severe* if the deficiency is above 50% of the normal value. Further, a year is considered to be a *drought year* in case the area affected by moderate or severe drought either individually or collectively is more than 20% of the total area of the country.

If the drought occurs in an area with a probability $0.2 \le P \le 0.4$, the area is classified as *drought prone area*, if the probability of occurrence of drought at a place is greater than 0.4, such an area is called as *chronically drought-prone area.* Further, in India, the meteorological drought is in general related to the onset, breaks and withdrawal times of monsoon in the region. As such, the prediction of the occurrence of drought in a region in the country is closely related to the forecast of deficient monsoon season and its distribution. Accurate forecast of drought, unfortunately, is still not possible.

5.10.3 Hydrological Drought

From a hydrologist's point of view, drought means below average values of stream flow, contents in tanks and reservoirs, groundwater and soil moisture. Such a hydrological drought has four components:

(a) Magnitude (= amount of deficiency)
(b) Duration
(c) Severely (= cumulative amount of deficiency)
(d) Frequency of occurrence

The beginning of a drought is rather difficult to determine as drought is a creeping phenomenon. However, the end of the drought when adequate rainfall saturates the soil mass and restores the stream flow and reservoir contents to normal values is relatively easy to determine.

In the studies on hydrological drought, different techniques have to be adopted for study of (i) surface water deficit, and (ii) groundwater deficit. The surface water aspect of drought studies is essentially related to the stream and the following techniques are commonly adopted:

(a) Low-flow duration curve
(b) Low-flow frequency analysis
(c) Stream flow modelling.

Such studies are particularly important in connection with the design and operation of reservoirs, diversion of stream flow for irrigation, power and drinking water needs; and in all activities related to water quality.

5.10.4 Agricultural Drought and Drought Indices

Deficiency of rainfall has been the principal criteria for defining agricultural drought. However, depending on whether the study is at regional level, crop level or plant level there have been a variety of definitions. Considering the various phases of growth of a crop and its corresponding water requirements, the time scale for water deficiency in agricultural drought will have to be much smaller than in hydrological drought studies. Further, these will be not only regional specific but also crop and soil specific.

An *aridity index* (*AI*) is defined as

$$AI = \frac{PET - AET}{PET} \times 100 \tag{5.30}$$

where *PET* = *Potential evapotranspiration* and *AET* = *Actual evapotranspiration*. In this *AI* calculation, *AET* is calculated according to *Thornthwite's water balance technique*, taking in to account *PET*, actual rainfall and field capacity of the soil. It is common to calculate *AI* on weekly or bi-weekly basis. *AI* is used as an indicator of possible moisture stress experienced by crops. The departure of *AI* from its corresponding normal value, known as *AI anomaly*, represents moisture shortage. Based on *AI* anomaly, the intensity of agricultural drought is classified as follows (Table 5.12)

Table 5.12 ***Intensity of Agricultural Drought***

AI anomaly	Severity class
Zero or negative	Non-arid
1 –25	Mild arid
26 –50	Moderate arid
> 50	Severe arid

In addition to AI index, there are other indices such as *Palmer index* (PI) and *Moisture availability index* (*MAI*) which are used to characterise agricultural drought. *IMD* produces aridity index (*AI*) anomaly maps of India on a bi-weekly basis based on data from 210 stations representing different agro-climatic zones in the country. These are useful in planning and management of agricultural operations especially in the drought-prone areas. Remote sensing techniques using imageries offer excellent possibilities for monitoring agricultural drought over large areas.

5.10.5 Drought Management

The causes of drought are essentially due to temporal and spatial aberrations in the rainfall, improper management of available water and lack of soil and water conservation. Drought management involves development of both short-term and long-term strategies. *Short-term strategies* include early warning, monitoring and assessment of droughts The *long-term strategies* aim at providing drought mitigating measures through proper soil and water conservation, irrigation scheduling and cropping patterns. Figure 5.18 shows some impacts and possible modifications of various drought components. The following is a list of possible measures for making drought prone areas less vulnerable to drought associated problems:

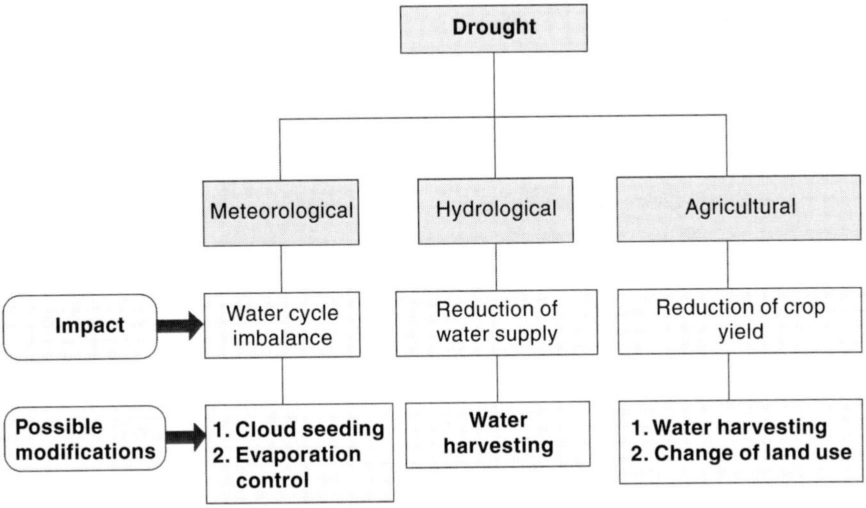

Fig. 5.18 ***Impact and Possible Modification of Drought Components***

- Creation of water storages through appropriate water resources development
- Inter-basin transfer of surface waters from surplus water areas to drought-prone areas
- Development and management of ground-water potential
- Development of appropriate water-harvesting practices
- *In situ* soil moisture conservation measures
- Economic use of water in irrigation through practices such as drip irrigation, sprinkler irrigation, etc.
- Reduction of evaporation from soil and water surfaces
- Development of afforestation, agro-forestry and agro-horticulture practices
- Development of fuelwood and fodder
- Sand dune stabilisation

Drought-proofing of a region calls for integrated approach, taking into account the multi-dimensional interlinkages between various natural resources, environment and local socio-economic factors.

Salient features of water harvesting, which forms an important component in modification of drought components is described in the next sub-section.

5.10.6 Water Harvesting

Water harvesting is a general term to include all systems that concentrate, collect and store runoff from small catchments for later use in smaller user areas. *FAO* defines water harvesting as, "*Water harvesting* is defined as the process of collecting and concentrating runoff water from a runoff area into a run-on area, where the collected water is either directly applied to the cropping area and stored in the soil profile for immediate use by the crop, i.e. runoff farming, or stored in an on-farm water reservoir for future productive uses, i.e. domestic use, livestock watering, aquaculture and irrigation." The collected water can also be used for groundwater recharge and storage in the aquifer, i.e. recharge enhancement. As a general rule, the catchment area from which the water is drawn is larger than the command area, where it is collected and used. The ratio of catchment, to command is inversely related to the amount and intensity of rainfall, the impermeability of soil, and the slope of the land on which it falls.

Water harvesting is essentially a traditional system used since many centuries, now being made over to meet present-day needs. Depending upon the nature of collecting surface and type of storages water harvesting is classified into several categories as mentioned in Fig. 5.19.

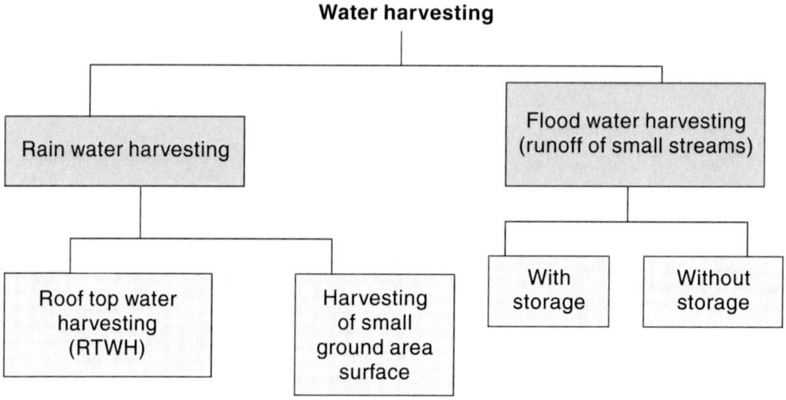

Fig. 5.19 ***Classification of Water Harvesting Techniques***

1. Rooftop Water Harvesting

The productive utilization of rainwater falling on rooftops of structures is known as *Rooftop Water Harvesting* (*RTWH*). In urban areas the rooftops are usually impervious and occupy considerable land area. Also, generally the municipal water supply is likely to be inadequate, inefficient or unreliable. In such situations, collection of runoff from roof tops of individual structures and storing them for later use has been found to be very attractive and economical proposition in many cases. Inadequacy of water availability and cost of supply has made many industries and large institutions in urban areas situated in arid and semi-arid regions to adopt *RTWH* systems in a big way. Factors like water quality, methods for efficient and economical collection and storage are some factors that have to be worked out in designing an efficient system to meet specific needs. The cost of adequate size storage is, generally, a constraint in economical *RTWH* design. In many cases, water collected from roof top is used for recharging the groundwater. Characteristics of the rainfall at the place, such as intensity, duration, nature of the rainfall season, average number of rainy days, determine the design of the *RTWH* design.

2. Micro Catchment System (Within the Field) of Rainwater Harvesting

In this system, the catchment is a small area which is not put for any productive purpose. The catchment length is usually between 1 and 30 metres and the overland flow from this during a storm is harvested by collecting and delivering it to a small cultivated plot. The ratio of catchment to the cultivated area is usually 1:1 to 3:1 and the runoff is stored in soil profile. Normally, there will be no provision for overflow.

Rainwater harvesting in micro catchments is sometimes referred to as *Within-Field Catchment System.*

Typical examples of such rainwater harvesting in micro catchments are

- Negarim micro catchments (for trees)
- Contour bunds (for trees)
- Contour ridges (for crops)
- Semi-circular bunds (for range and fodder)

Negarim micro catchment technique was originally developed in Israel; the word *Negarim* is derived from Hebrew word *Neger* meaning runoff. This technique consists of dividing the catchments into a large number of micro catchments in a diamond pattern along the slope. Each micro catchment is of square shape with a small earthen bund at its boundary and an infiltration pit is provided at the lowest corner as shown in Fig. 5.20. The pit is the cultivated area and usually a tree is grown in the pit. This arrangement of micro catchments of sizes 10 m^2 to 100 m^2, has been found to be very beneficial in arid and semiarid areas where rainfall can be as low as 150 mm.

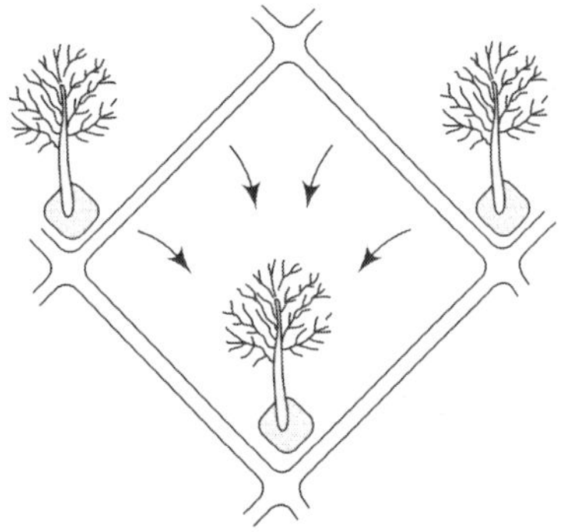

Fig. 5.20 ***Micro Catchment System: Negarim Micro Catchment for Trees***

3. Macro Catchment System (Within the Field) of Rainwater Harvesting

This system is designed for slightly larger catchment areas wherein overland flow and rill flow is collected behind a bund and allowed to be stored in the soil profile through infiltration. The catchment is usually 30 to 200 m long and the ratio of catchment to cultivated area is in the range 2:1 to 10:1. Typical arrangement consists of one row or two staggered rows of trapezoidal bunds with wing walls. Contour bunds made of piled up stones is also used in this system. It is usual to provide overflow arrangements for disposing of the excess runoff water. Infiltration area behind the bunds is used to grow crops.

4. Floodwater Farming (Floodwater Harvesting)

This system is used for larger catchments and the flow in the drainage is harvested. The catchment areas are several kilometres long and the ratio of catchment to command is larger than 10:1. Two sub-systems mentioned below are in common use:

1. Water harvesting using storage structures
2. Water harvesting through spreading of water over command

(a) Storage Structures Systems Small storage structures are built across the drainage to store a part of the runoff. While the stored surface water would serve as a source of utilisable water to the community for some time, the infiltration from this water body would provide valuable recharge to the groundwater. The commonly used structures are *check dams* and *nalabunds*. These structures have the additional advantage of arresting erosion products from the catchment. Further, these structures prevent the deepening and widening of gullies.

The check dams usually have a masonry overflow spillway and the flanks can be of either masonry construction or of earthen embankment. They are constructed on lower order streams (up to 3) with median slopes. Generally, check dams are proposed where water table fluctuations are high and the stream is influent.

Nalabunds are structures constructed across nalas (streams) for impounding runoff flow to cause a small storage. Increased water percolation and improving of soil moisture regime are its main objective. Nalabunds are of small dimension and are constructed by locally available material, usually an earthen embankment. In a nalabund the spillway is normally a stone lined or rock cut steep channel taking off from one of the ends of the bund at appropriate level. Structures similar to a nalabund but of larger dimension are known as *percolation tanks.* Nalabunds and percolation tanks are constructed in flat reach of a stream with slopes less than 2%.

The irrigation tanks of south India are also sometimes termed water-harvesting structures. Tanks on local streams form a significant source of irrigation in states of Andhra Pradesh, Karnataka, Maharashtra and Tamil Nadu. These are small storage structures formed by earthen bunds to store runoff, of a small stream. The embankment, surplus weir and a sluice outlet form the essential component of a tank. The tank system in a region, which can be a group of independent tanks or a set of tanks in cascade, form an important source of surface water for domestic use, drinking water for life stock, agriculture for growing food and fodder and recharge of subsurface aquifers.

(b) Spreading of Water In this method, a diversion across the drainage would cause the runoff to flow on to the adjacent land. Appropriate bunds either of rock or of earth would cause spreading the water over the command. The spread water infiltrates into the soil and is retained as soil moisture and this is used for growing crops. Provision for overflow spillway at the diversion structure, to pass excess water onto the downstream side of the diversion structure, is an important component of the diversion structure.

General The specific aspects related to the design of water-harvesting structures depends upon the rainfall in the region, soil characteristics and terrain slope. It is usual to take up water harvesting activity at a place as a part of integrated watershed management programme. Norms for estimating recharge from water harvesting structures are given in Sec. 9.14 of Chapter 9.

The water-harvesting methods described above are particularly useful in dry land agriculture and form important draught-management tool. Community participation in construction and management of water harvesting structure system is essential for economical and sustainable use of the system. Rehabilitation of old irrigation tanks through de-silting to bring it back to its original capacity is now recognized as a feasible and desirable activity in drought proofing of a region.

5.10.7 Droughts in India

1. Causes

Even though India receives a normal annual rainfall of 118 cm, the spatial and temporal variations lead to anomalies that lead to droughts and floods. Consequently, droughts have been an ever-present feature, at some part or other, of the country. While the variability of rainfall is a general reason, there are some additional factors also that are responsible for drought situation in the country. The major causes can be listed as below:

- The principal cause of drought may be attributed to erratic behavior of summer monsoon. The summer monsoon has a strangle hold on the Indian agriculture and Indian economy and, consequently, on the livelihood of a vast majority of people. An overwhelming majority of the cropped area in India falls within medium and low rainfall areas, (see Table 5.13)

Table 5.13 ***Cropped Area Under Various Ranges of Rainfall Distribution***

Annual Rainfall ranges	Classification	Percentage
Less than 750 mm	Low rainfall	33
750 mm to 1125 mm	Medium rainfall	35
1125 mm to 2000 mm	High rainfall	24
> 2000 mm	Very high rainfall	8

  Most parts of peninsular, central and north–west India are regions most prone to drought and they receive less than 1000 mm of rainfall. If the summer monsoon plays truant, these areas are the most vulnerable.
- Where irrigation by groundwater is practiced, due to absence of any legislative and administrative control, over-exploitation is taking place rapidly all over the country. When the surface water is scarce, as in a drought year, the groundwater becomes scarcer, the wells dry up and the water will not be available even for drinking purposes.
- Per-capita availability of water is decreasing steadily.
- Improper use of irrigation water in low rainfall areas chiefly; (i) attention on high water-consuming crops such as sugar cane and rice, (ii) no attention on recharge and conservation of groundwater, and (iii) little attention on sustainable drinking water supply. Traditional water-harvesting techniques like tanks, ponds etc., are being abandoned. Many tanks in peninsular India are encroached upon for alternate uses.

2. Status

In the past five decades, widespread and severe droughts have occurred in the years 1965-66, 1971-73, 1979-80, 1982-83, 1986-87, 1994-96, 1999-2000, 2001-2002 and 2008-2009. These droughts affected the agricultural production and the economy significantly and caused immense hardship and misery to a very large population. Since 1875 till 2012, India has faced 31 drought years; the 1918 being regarded as the worst year in which about 79% of the country was affected by drought. Analysis of drought records since 1801 reveals that nearly equal number of droughts occurred in 19th century and in 20th century and that there is a lower number of occurrences in the second quarter of both centuries.

Most of the drought-prone areas of the country lie in the states of Rajasthan, Karnataka, Andhra Pradesh, Maharashtra, Gujarat and Orissa. Roughly, 50% of the drought-prone area of the country lies in Deccan plateau. Further, while Rajasthan has a return period of two years for severe droughts, it is about three years in the Deccan plateau region. It is very difficult to estimate the economic losses of a drought, as it is a creeping phenomenon with wide spatial coverage. However, a wide spread drought in the country would cover agricultural areas of the order of 100 Lakh ha and the consequential and collateral losses with its diverse impacts on the economy could bring down the *GDP* significantly. It has been observed that most of the major droughts in India have been followed by recession.

5.11 *Surface-Water Resources of India*

5.11.1 Surface-Water Resources

Natural (virgin) flow in a river basin is reckoned as surface resource of a basin. In view of prior water resources development activities, such as construction of storage reservoirs in basin, assessment of natural flow is a very complex activity. In most of the basins of the country, water resources have already been developed and utilised to various extents through construction of diversion structures and storage reservoirs for purposes of irrigation, drinking-water supply and industrial uses. These utilisations in turn produce *return flows* of varying extent; return flow being defined as the non-consumptive part of any diversion returned back to the river. Return flows to the stream from irrigation use in the basin are usually assumed to be 10% of the water diverted from the reservoir or diversion structure on the stream for irrigation. The return flows from diversions for domestic and industrial use is usually assumed as 80% of the use. The return flow to the stream from groundwater use is commonly ignored.

The natural flow in a given period at a site is obtained through water budgeting of observed flow, upstream utilisation, increase in storage, evaporation, other consumptive uses and return flows. The surface and groundwater components are usually treated separately.

- *CWC* (1983), on the basis of statistical analysis of available data and on rainfall–runoff relationships where flow data was meager or not available, estimated the total annual runoff of the river systems of India as 1881 km^3.
- The National Commission for Integrated water Resources Development (1999) used the then available estimates and data and assessed the total surface water resources of the country as 1952.87 km^3 (say 1953 km^3).

  It should be noted that the average annual natural (virgin) flow at the terminal point of a river is generally taken as the surface water resources of the basin. But this resource is available with a probability of about 50%, whereas it is customary to design irrigation projects with 75% dependability and domestic water supply projects for nearly 100% dependability. Obviously, the magnitude of water at higher values of dependability (say 75% and above) will be smaller than the average value.

The total catchment area of all rivers in India is approximately 3.05 million km^2. This can be considered to be made up of three classes of catchments:

1. Large catchments with basin area larger than 20,000 km^2.
2. Medium catchments with basin area between 20,000 and 2000 km^2.
3. Small catchments with area less than 2000 km^2.

Rao[9] has estimated that large catchments occupy nearly 85% of the country's total drainage area and produce nearly 85% of the annual runoff. The medium and minor catchments account for 7% and 8% of annual runoff volumes respectively. In the major river basins of the country, two mighty rivers, the Brahmaputra and the Ganga, together constitute 71.5% of the total yield in their class and contribute 61% of the country's river flow. Further, these two rivers rank 5 and 11 respectively in the list of the world's largest rivers, (Table 5.14). It is interesting to note that the fifteen rivers listed in Table 5.14 account for nearly 30% of the world's annual runoff.

Table 5.14 ***World's Fifteen Largest Rivers***

Sl. No.	River	Location	River mouth Discharge (Km^3/Yr)	Sediment Load (10^6 tonnes/Yr)
1	Amazon	South America	5330 ± 426	1160
2	Congo	Africa	1271 ± 130	43
3	Orinoco	South America	984 ± 112	150
4	Yangtze (Changjiang)	China	910 ± 133	480
5	Brahmaputra	India	613 ± 51	645
6	Yenisei	China	577 ± 42	5
7	Mississippi	USA	536 ± 130	210
8	Lena	Russia	526 ± 63	11
9	Parana	South America	476 ± 96	100
10	Ob-Irtysh	Russia	397 ± 61	16
11	Ganga	India	382 ± 76	405
12	Tocantins -Araguaia	Brazil	356 ± 64	75
13	Amur	Russia	312 ± 60	52
14	Mekong	SE Asia	292 ± 33	160
15	Irrawady	Myanmar	258 ± 26	260

The discharge in Col.4 indicates long term mean ± standard deviation.

[**Ref:** A. Dai and K.E. Trenberth, *New Estimates of Continental Discharge and Oceanic Freshwater Transport, 2003.]*
http://www.cgd.ucar.edu/cas/adai/papers/Dai_discharge_AMS03.pdf

According to an analysis of *CWC*, about 80% of average annual flow in the rivers of India is carried during monsoon months. This highlights the need for creating storages for proper utilisation of surface water resources of the country. Another interesting aspect of Indian rivers is that almost all the rivers flow through more than one state, highlighting the need for inter-state co-operation in the optimum development of water resources.

5.11.2 Utilisable Water Resources

Utilisable water resources mean the quantum of water withdrawable from its place of natural occurrence. Withdrawal of water from a river depends on topographic conditions and availability of land for the stated project. As a result of various limitations such as to topography, environmental consideration, non-availability of suitable locations and technological shortcomings, it will not be possible to utilise the entire surface water resources of the country. Further, surface water-storage structures, such as reservoirs, cause considerable loss by evaporation and percolation. Also, environmental considerations preclude total utilisation or diversion of surface water resources of a basin. From these considerations, it is necessary to estimate the optimum utilisable surface runoff of the country for planning purposes. Normally, the optimum utilisable surface runoff of a basin will be around 70% of the total surface runoff potential of the basin.

CWC in 1988 estimated the utilisable surface water resource of the country as 690.32 km^3. The National Commission for Integrated Water Resources Development[8] (1999) has adopted this value in preparing estimates of future water demand–supply scenarios up to the year 2050. Table 5.15 gives the basinwise distribution of utilisable surface water resource of the country.

Table 5.15 *Average Flow and Utilisable Surface Water Resource of Various Basins*

[Unit: km^3/Year] (Source: Ref. 8)

S. No.	River Basin	Surface water resources	Utilisable surface water resources
1.	Indus	73.31	46
2.	Ganga–Brahmaputra–Meghna Basin		
	2a Ganga sub-basin	525.02	250.0
	2b Brahmaputra sub-basin and	629.05	24.0
	2c Meghna (Barak) sub-basin	48.36	
3.	Subarnarekha	12.37	6.81
4.	Brahmani–Baitarani	28.48	18.30
5.	Mahanadi	66.88	49.99
6.	Godavari	110.54	76.30
7.	Krishna	69.81	58.00
8.	Pennar	6.86	6.32
9.	Cauvery	21.36	19.00
10.	Tapi	14.88	14.50

(Contd.)

Table 5.15 (*Contd.*)

11.	Narmada	45.64	34.50
12.	Mahi	11.02	3.10
13.	Sabarmati	3.81	1.93
14.	West flowing rivers of Kutchch and Saurashtra	15.10	14.96
15.	West flowing rivers south of Tapi	200.94	36.21
16.	East flowing rivers between Mahanadi and Godavari	17.08	
17.	East flowing rivers between Godavari and Krishna	1.81	13.11
18.	East flowing rivers between Krishna and Pennar	3.63	
19.	East flowing rivers between Pennar and Cauvery	9.98	16.73
20.	East flowing rivers south of Cauvery	6.48	
21.	Area North of Ladakh not draining into India	0	0
22.	Rivers draining into Bangladesh	8.57	0
23.	Rivers draining into Myanmar	22.43	0
24.	Drainage areas of Andaman, Nicobar and Lakshadweep islands	0	0
	Total	**1952.87**	**690.32**

In the computation of utilisable water resources as 690 km^3 it is assumed that adequate storage facility is available for balancing the monsoon flows into an average year round availability. The minimum storage required to achieve this is estimated as 460 km^3 against the present estimated total available storage capacity of 253 km^3. If more storage capacity could be developed carry-over from years of above normal rainfall to dry years would be possible. For comparison purposes, for about the same annual runoff the *USA* has storage of 700 km^3.

1. Utilisable Dynamic Groundwater Resources

The total replenishable groundwater resources of the country (dynamic) has been estimated by *CGWB* as 431.89 km^3/year and the utilisable dynamic groundwater potential as 396 km^3/year (details in Chapter 9, Section 9.15.3).

2. Water Available from Return Flows

Water used for a specific activity such as irrigation and domestic water supply includes consumptive and non-consumptive components. The non-consumptive component part of water use is returned back to hydrologic system either as surface flow or as addition to groundwater system or as soil moisture. However, due to economic and technological constraints and due to possibilities of diminished water quality, only a part of the return flow is recoverable for re-use. The utilisable return flow is an important component to be considered in the demand–supply analysis of utilisable water resources.

5.11.3 Total Water Requirement and Available Resources Scenario

1. Total Water Requirement for Different Uses

The estimated total water requirements, estimated by *NCIWRD*[8], for the two scenarios and for various sectors at three future horizons are shown in Table 5.16. Irrigation would continue to have the highest water requirement (about 68% of total water requirement), followed by domestic water including drinking and bovine needs.

Table 5.16 ***Water Requirement for Different Uses***

(Unit: Cubic Kilometer) [Source: Ref. 8]

Sl No.	Uses	Year 2010			Year 2025			Year 2050		
		Low	High	%	Low	High	%	Low	High	%
Surface Water										
1.	Irrigation	330	339	48	325	366	43	375	463	39
2.	Domestic	23	24	3	30	36	5	48	65	6
3.	Industries	26	26	4	47	47	6	57	57	5
4.	Power Inland	14	15	2	25	26	3	50	56	5
5.	Navigation Environment	7	7	1	10	10	1	15	15	1
6.	(Ecology) Evaporation	5	5	1	10	10	1	20	20	2
7.	Losses	42	42	6	50	50	6	76	76	6
	Total	**447**	**458**	**65**	**497**	**545**	**65**	**641**	**752**	**64**
Ground Water										
1.	Irrigation	213	218	31	236	245	29	253	344	29
2.	Domestic	19	19	2	25	26	3	42	46	4
3.	Indusrtries	11	11	1	20	20	2	24	24	2
4.	Power	4	4	1	6	7	1	13	14	1
	Total	**247**	**252**	**35**	**287**	**298**	**35**	**332**	**428**	**36**
	Grand Total	**694**	**710**	**100**	**784**	**843**	**100**	**973**	**1180**	**100**

2. Evaporation

In water-resources evaluation studies, it is common to adopt a percentage of the live capacity of a reservoir as evaporation losses. The *NCIWRD* has adopted a national average value of 15% of the live storage capacity of major projects and 25% of the live storage capacity of minor projects as evaporation losses in the country. The estimated evaporation losses from reservoirs are 42 km^3, 50 km^3 and 76 km^3 by the years 2010, 2025 and 2050 respectively.

3. Demand and Available Water Resources

Planning and allocation of water resources of the country are done keeping in view the national water policy. The National water Policy (1987, 2002) has provided the following priorities for allocation of water:

1. Drinking water,
2. Irrigation ,
3. Hydropower,
4. Ecology,
5. Agro-industries and non-agricultural industries, and
6. Navigation and other uses.

These priorities should be broadly followed. Further, priorities could be modified or added if warranted by area/region specific considerations.

A national level of demand assessment has been prepared by *NCIWRD*[8]. The summary of *NCIWRD* (1999) study relating to assessment of demand and available water resources is given in Table 5.17. The utilisable return flow is an important component to be considered in the demand–supply analysis of utilisable water resources. Estimated utilisable return flows of the country in surface and groundwater mode for different time horizons are shown in Table 5.17. It may be noted that the return flow contributes to an extent of nearly 20–25% in reducing the demand.

Table 5.17 ***Utilisable Water, Requirements and Return Flow***

(Quantity in Cubic Kilometre) [Source: Ref. 8]

Sl. No.	Particulars	Year 2010		Year 2025		Year 2050	
		Low Demand	High Demand	Low Demand	High Demand	Low Demand	High Demand
1	**Utilisable Water**						
	Surface Water	690	690	690	690	690	690
	Ground water	396	396	396	396	396	396
	Augmentation from canal						
	Irrigation	90	90	90	90	90	90
	Total	**996**	**996**	**996**	**996**	**996**	**996**
2	**Total Water Requirement**						
	Surface Water	447	458	497	545	641	752
	Groundwater	247	252	287	298	332	428
	Total	**694**	**710**	**784**	**843**	**973**	**1180**
3	**Return Flow**						
	Surface Water	52	52	70	74	91	104
	Groundwater	144	148	127	141	122	155
	Total	**196**	**200**	**197**	**215**	**213**	**259**
4	**Residual Utilisable water**						
	Surface Water	295	284	263	219	140	42
	Groundwater	203	202	146	149	96	33
	Total	**498**	**486**	**409**	**463**	**236**	**75**

While the table is self-explanatory, the following significant aspects may be noted:

(a) The available water resources of the country are adequate to meet the low demand scenario up to year 2050. However, at high demand scenario it barely meets the demand.

(b) Need for utmost efficiency in management of every aspect of water use, conservation of water resources and reducing the water demand to low demand scenario are highlighted.

5.12 *Glacier Resources of Indian Himalayas*

All major north Indian rivers owe their origin to the glaciers in the Himalayas. With the increasing concern about the global warming and its impact on the water resources of the country, there is an emergence of keen interest in the monitoring and study of the glaciers in Indian Himalayas. There are thousands of glaciers in the Himalayas and according to the latest inventory of the Geological Survey of India (*GSI*), there are 9575 glaciers in the Indian part of the Himalayas.

While the Himalayas cover a total area of about 8.45 Mkm^2, the Indian part of the Himalayas occupy only about 0.35 Mkm^2. This area lies at an elevation greater than 1000 m. All the three major north Indian river basins, viz., the Indus, Ganga and Brahmaputra basins are fed by the Himalayan glaciers. Salient features of the Indian Himalayan glaciers feeding the various Indian river basins are given in Table 5.18.

Table 5.18 ***Basin-wise details of Indian Himalayan Glaciers, (Ref. 12)***

Basin	Component Major Rivers	Number of Glaciers	Area (km^2)	Volume (km^3)
Indus	Ravi	172	193	8.0
	Chenab	1278	3059	206.2
	Jhelum	133	94	3.3
	Beas	277	599	36.9
	Satluj	926	1251	61.0
	Indus	1796	2165	104.6
	Shyok	2658	7106	601.7
	Kishenganga	222	174	5.9
	Gilgit	535	82	NA
	Sub Total	**7997**	**14724**	**1027.7**
Ganga	Yamuna	52	144	12.2
	Bhagirathi	238	755	67.0
	Alakananda	407	855	90.7
	Ghagra	271	729	43.8
	Sub Total	**968**	**2484**	**213.7**
Brahmaputra	Teesta	449	706	39.6
	Arunachal	161	223	10.0
	Sub Total	**610**	**929**	**49.6**
Total		**9575**	**18136**	**1291**

It is seen from Table 5.18 that the Indian glaciers cover an area of 18,136 km^2 and their volume content is estimated as 1291 km^3. This is a huge water bank responsible for sustenance of the great river basins; the Indus, Ganga and Brahmaputra. The Indus basin has the majority of Indian glaciers (7997 numbers) and contains nearly 80% of the volume of water. The glaciers that feed the Ganga basin are 968 in number and contain 16.6 % of waters of Indian glaciers.

The glaciers are the most vulnerable part of the hydrologic cycle affected by global warming. The impact of any change in the status of these glaciers on the river flows in the Indus, Ganga and Brahmaputra river systems is very substantial. In view of this, the study of melting of the Indian Himalayan glaciers is receiving considerable attention and detailed monitoring as a part of global warming studies related to India.

5.13 *Environmental Flows*

Development of surface-water resources in river basins for agriculture, urban use, hydropower, industrial use and such other uses beneficial to humans is increasing at a very fast pace all over the world. Recent awareness of environmental and ecological aspects of our planet has led to some startling revelations of over-exploitation of natural water resources. The flows of many rivers have been reduced or seasonally altered changing the size and frequency of floods. So also is the case with length and severity of droughts. The ecosystems of some of the rivers have been so severely and adversely impacted that they appear to be beyond redemption. It is estimated that over one half of the world's major river systems are presently adversely affected by flow regulation.

The growing concern over the riverine eco-system has led to an evolving movement to recognise that water-dependent ecosystems should not be jeopardised or compromised adversely in the name of development. A consequence of this awareness is the concept of environmental flows of a river. The World Bank defines the Environmental flows (*EF*) as: *The quality, quantity and timing of water flows required to maintain the components, functions, processes and resilience of aquatic eco systems which provide goods and services to people.* This concept of *EF* has been put succinctly in what is known as Brisbane declaration (2007) as: *Environmental flows describe the quantity, quality and timing of water flows required to sustain freshwater and estuarine ecosystem and human livelihoods and wellbeing that depend on these ecosystems.*

The river ecosystem include not just the flora and fauna that live in the river, but also that live in the flood plains and wetlands supported by the river floods. Further, it encompasses the groundwater dependent ecosystems that are supported by river seepage and estuaries. It is seen that the above definition (and the explanation) of *EF* has three important components: quantity, variability and quality. There is a linkage in all the three components that together constitutes the *EF*. The variability aspect underscores that *EF* stands for Environmental flows and cannot be represented by a single value of flow for a given river.

5.13.1 Evolution of *EF* Concept

In the early part of 1950s, there was a massive construction activity of dams world over for irrigation and hydropower development. At that time, practically, no serious attention was paid to upstream and downstream environmental impacts and ecological damages to the riverine eco system were literally ignored. However, when the consequences of such actions started becoming apparent in due course of time, concern about degradation of river eco system was expressed in different fora.

- Initially, it was thought that a *minimum flow* in the stream would serve the purpose of maintaining quality water in the river and ecological balance in the downstream stretch of the river. This concept, which came in to prominence in the 1970s, mandated that downstream of every impoundment and diversion structure should carry a certain minimum flow to be left unhindered for environmental, aesthetic, religious and cultural considerations. This flow was named variously as *Minimum flow, compensation flow, Reserved flow, Prescribed flow, in-stream flow and Environmental flow.* A host of methods was proposed to estimate this minimum flow and a value in the neighborhood of Q_{95} was considered as reasonable.
- The inadequacy of minimum flow to address the ecological concern adequately was recognised in early 1980s and the birth of the concept of environmental flows took place in that decade. Thus, the concept of *EF* is a relatively new concept that is hardly 3 decades old. This concept is being continuously reinforced and enlarged upon with inputs from various geographies and disciplines.
- In 1990s South Africa, *USA*, *UK* and *EU* have adopted various appropriate legislations to bring the *EF* in to the mainstream .
- The World Bank has adopted the *EF* as part of its mandatory component in their Integrated Water Resources Management (*IWRM*) activity related to all new projects.

5.13.2 Environmental Flow Assessment (*EFA*)

Environmental flow assessment (*EFA*) is the science of determining the quantity and quality of water required in a stream for its ecosystem conservation and resources protection. *EFA* tries to identify a environmental flow requirement (*EFR*) of a stream at a given location that can assure a reasonable healthy ecological condition of the river. Both spatial and temporal patterns of river flow are important. In a stream, high flows of different frequency are needed for breeding of birds, wetland flooding and riparian vegetation requirement. Moderate level flows are critical for migration of fish. Minimum (low) flows are important for water quality maintenance and for multipurpose uses of river by people and animals. In many locations the religious, cultural and aesthetic attributes of the river have to be preserved. Thus, the *EFR* is not only a minimum flow but also includes appropriate variability. A good EF regime mimics the salient variations of the stream to keep the river in desired level of health.

Considering the fact that most water resources development projects of a river have prime commercial and economical objectives, the *EFR* is a compromise between those developmental objectives and the maintenance of the river in a healthy and reasonable condition.

An environmental flows study is a relatively new discipline and as such there is a profusion of new ideas and methodologies for assessment of such flows. In a review of international Environmental Flows assessments (*EFA*), Tharme (2003) recorded

207 different methodologies within 44 countries. He has classified the different methodologies in to four categories as

- Hydrological index methods
- Hydraulic rating methods
- Habitat simulation methodologies
- Holistic methodologies

The details of the various methodologies are beyond the scope of this essay. However, a few salient features of these methods are presented in Table 5.19.

Table 5.19 *Major advantages and disadvantages of different methodologies of EFA*

Methodology	Duration of assessment (months)	Major advantages	Major disadvantages
Hydrological index method	0.5	Low cost, rapid to use.	Not truly site specific, Ecological links assumed.
Hydraulic rating method	2 – 4	Low cost, site specific	Ecological links assumed.
Habitat simulation	6 – 18	Ecological links included.	Extensive data collection; needs experts; high cost.
Holistic method	12 – 36	Covers most aspects.	Requires very large scientific expertize; very high cost.

It is seen that for quick assessment or for preliminary assessment either hydrologic index method or hydrologic rating method can be adopted. As an introduction to this growing field of *EFA*, one hydrologic rating method known as *Tennant's method* will be described briefly.

5.13.3 Tennant's Method

This method (also known sometimes as *Montana method*) was developed in *USA* in 1975. It was originally developed for protection of trout and is based on extensive field observations in mid-west of *USA*. Guidelines for flow management based on percentage of mean annual flow (*MAF*) rate have been suggested for a set of seven levels of eco-status of the river. The method has undergone many changes and is used for general *EFA* purposes in many parts of the world. In this method, the water year is considered in two halves, high flow season and low flow season. For each season, a certain flow, expressed as a percentage of mean annual flow rate is prescribed to achieve a desired level of eco-status. The method is based purely on hydrological data and on subjective assessments. It is suitable for initial studies and for low controversy situations. The basic guidelines of Tennant for *EFR* are given in Table 5.20.

Flow description of *good habitat* means that to achieve this level of eco status, the basin managers must provide 40% of *MAF* rate during half year of high flow (April through September) and 20% of *MAF* rate during the other half year of relatively low flow (October through March). Further, a flushing flow of 200% *MAF* rate for duration of 48 to 96 hours must be provided during high flow season.

Table 5.20 ***Environmental Flow Requirement by Tennant's Method***

Description of Flow	Flow to be released during	
	(*HFS*) April to September	(*LFS*) October to March
Flushing flow (from 48 to 96 hours)	200% *MAF* rate	Not applicable
Optimum range of flow	60% – 100% *MAF* rate	60% - 100 % *MAF* rate
Outstanding habitat	60% *MAF* rate	40% *MAF* rate
Excellent habitat	50% *MAF* rate	30% *MAF* rate
Good habitat	40% *MAF* rate	20% *MAF* rate
Fair and degrading habitat	30% *MAF* rate	10% *MAF* rate
Poor or minimum habitat	10% *MAF* rate	10% *MAF* rate
Severe degradation	< 10% *MAF* rate	<10% *MAF* rate
MAF rate = Mean Annual Flow rate.	*HFS* = High flow Season of 6 months	
LFS = Low Flow Season of 6 months		

Procedure

1. The mean annual flow (*MAF*) rate of the basin is obtained from the analysis of the available stream flow data. Monthly or 10 daily flow values of an average water year are then assembled.
2. All the goods and services provided by the river at the study stretch, below the proposed impoundment or diversion location, are listed and current status and desirable status are noted in terms of qualitative descriptive terms.
3. From the available data and due consideration of the other uses of the impounded/diverted water, a desirable level of eco-status of the river is arrived at and is described in terms of the 7 qualitative descriptors of Tennant.
4. Using Tenant's Table (Table 5.20), the monthly or 10-daily flow throughout the year required to be released in to the river is worked out. A flow hydrograph of the mean annual flow as well as the environmental flow requirement (*EFR*) are then prepared.

Example 5.14 illustrates the use of Tennant's method for calculating *EFR* of a river at a proposed reservoir.

Example 5.14 Mean monthly values of flow in a river downstream of a proposed reservoir are given. It is desired to keep the downstream ecosystem of the river in good habitat condition after the construction of the impoundment structure. Using Tennant's method, estimate the environmental flows required in the river to keep the desired ecological condition.

Month	June	July	Aug	Sept	Oct	Nov
Mean monthly flow (cumec.day)	8050	30150	32130	20600	15670	5800
Month	**Dec**	**Jan**	**Feb**	**March**	**April**	**May**
Mean monthly flow (cumec.day)	2010	1070	680	570	430	720

Solution

The calculation are shown in Table 5.21

Mean Annual daily flow = 117880/365 = 322.96 cumec.day

Mean Annual Flow rate = *MAF* rate = 322.96 cumecs

Flushing flow volume = 200% of *MAF* rate for 2 days = 2 × 322.96 = 645.92 cumec.day

For the *good habitat condition* desired:

- The high flow season (*HFS*) is from April to September and the *EF* is 40% of *MAF* rate.
  EFR volume in a month in *HFS* is calculated in Col. 4 as = Col. 2 × 322.96 × 0.4
- Similarly, The low flow season(*HFS*) is from October to March and the *EF* is 20% of *MAF* rate.
  in a *LFS* month, *EFR* volume = Col. 2 ×3.0274 ×0.20
- Flushing flow volume of 645.92 cumec.day is provided in August.
- $\text{Col. 8} = \dfrac{\text{Col.3}}{\text{Col.2}}$
- Col. 9 = 0.4 × 322.96 =129.2 cumecs if the month is in *HFS*, other wise
  = 0.2 × 322.96 = 64.6 cumecs.

For the month of August, flushing flow contributes an average rate of

$$= \frac{645.92}{31} = 20.84 \text{ cumecs.}$$

Columns 8 and 9 are used to plot the hydrograph of normal stream flow and *EFR*. The hydrographs are shown in Fig. 5.19

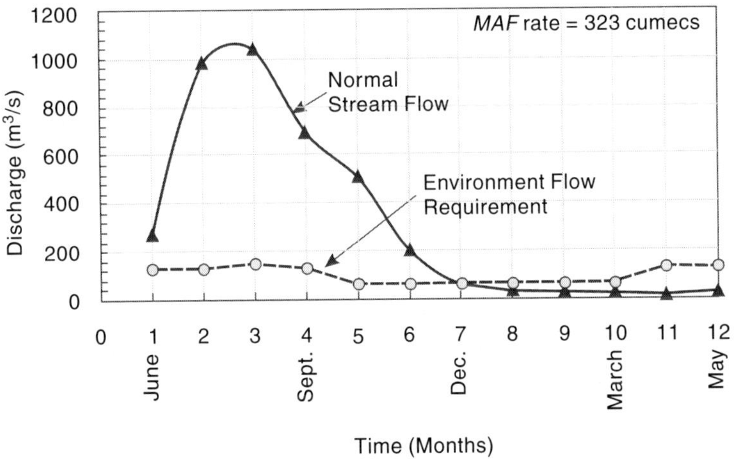

Fig. 5-21 ***Hydrographs of Normal Stream Flow and EFR – Example 5.14***

Table 5.21 ***Calculation of EFR by Tennants Method – Example 5.14***

Month	No. of days in the month	Mean monthly flow volume (cumec.day)	*EFR* (Cumec. day) in the month		Flushing flows in the month (cumec.day)	Total *EFR* volume in the month (cumec.day) [Col.4 + Col. 5+ Col.6]	Mean Monthly flow rate (cumecs)	Average *EFR* rate in the month (cumecs)
			In *HFS* @ 40% of MAF rate	In *LFS* @ 20% of MAF rate				
1	2	3	4	5	6	7	8	9
June	30	8050	3876			3876	268	129.2
July	31	30150	4005			4005	973	129.2
Aug	31	32130	4005		645.9	4651	1036	150.0
Sept	30	20600	3876			3876	687	129.2
Octo	31	15670		2002		2002	505	64.6
Nov	30	5800		1938		1938	193	64.6
Dec	31	2010		2002		2002	65	64.6
Jan	31	1070		2002		2002	35	64.6
Feb	28	680		1809		1809	24	64.6
March	31	570		2002		2002	18	64.6
April	30	430	3876			3876	14	129.2
May	31	720	4005			4005	23	129.2
Total	365	117880						

5.13.4 Status of *EF* Studies and Implementation in India

The concept of environmental flows has been slow in coming to India. However, there has been recognition of the need to provide its due share to ecology. The national water policy ranks *ecology* as fourth priority item in its priorities. Also, Item 6.3 of the same document states: "*In the planning , implementation and operation of a project, the preservation of the quality of environment and the ecological balance should be a primary consideration. The adverse impact on the environment, if any, should be minimized and should be offset by adequate compensatory measures. The project should, nevertheless, be sustainable.*"

Most pre-1980 literature related to water resources considered the water flowing to the sea as *wasted.* Around late 1990s, the concept of *minimum flow* was mooted in various fora and reports.

In terms of operational policies and action, probably the first major action was a Supreme Court judgment in 1999, which directed the Government to ensure a minimum flow of 10 m^3/s in Yamuna at Delhi for reasons of improving water quality. The Cauvery Water Dispute Tribunal, in its final order in 2007, states: "We reserve some quantity of water for (i) environmental protection, and (ii) inevitable escapage in to the sea as under:

(i) Quantity reserved for environmental protection = 10 TMC

(ii) Quantity determined for inevitable escapage in to the sea = 4 TMC."

The Krishna Water Dispute Tribunal (*KWDT*-II) in its report of 2010 states: "All the three states shall release in all 16 TMC of water for maintaining minimum *in-stream flow* and for environment and ecology".

In 2001, Government of India constituted a working group to advice the Water Quality Assessment Authority (*WQAA*) on the minimum flows in rivers to conserve the eco system. The major recommendation of the working group made in 2007 relating to the minimum flows are as follows:

- **For Himalayan Rivers**
  - Minimum flow to be not less than 2.5% of 75% dependable annual flow, Q_{75}
  - One flushing flow during monsoon with peak not less than 600% of Q_{75}.
- **For other rivers:**
  - Minimum flow in any ten daily periods to be not less than the observed ten daily flow with 99% exceedance. Where the ten daily flow data is not available this may be taken as 0.5% of $Q_{75.}$
  - One flushing flow during monsoon with a peak not less than 600% of Q_{75}.

Since the early 2000s, there have been considerable interest and research activities in the academic and research organisations in India on topics related to EFA. Refrences 14, 15 and 16 can be referred to for information on environmental flow related research work in Indian context.

REFERENCES

1. Central Water Commission, *Water Resources of India, CWC* Pub. No. 30/88, *CWC*, New Delhi, India, 1988.
2. Chow, V.T. (Ed.), *Handbook of Applied Hydrology,* McGraw-Hill, New York, *USA*, 1964.
3. Chow, V.T., Maidment, D.R. and Mays, L.W., *Applied Hydrology,* McGraw-Hill, Singapore, 1988.
4. *Jal Vigyan Sameeksha* (Hydrology Review), Pub. of High Level Tech. Com. on Hydrology, Nat. Inst. of Hydrology, Roorkee, India, Vol. I, No. 1, June 1986.
5. Linsley, R.K. et al, *Applied Hydrology*, Tata McGraw-Hill, New Delhi, India, 1979.
6. Linsley, R.K. et al, *Hydrology for Engineers,* SI Metric edition, McGraw-Hill Book Co., Singapore 1988.
7. Ministry of Agriculture, Govt. of India, *Handbook of Hydrology*, New Delhi, 1972.
8. Min. of Water Resources, *GOI, Report of The National Commission for Integrated Water Resources Development,* Vol. 1, New Delhi, Sept. 1999.
9. Rao, K.L., *India's Water Wealth,* Orient Longmans, New Delhi, India, 1975.
10. Ponce, V.M. and Hawkins, R.H., "Runoff Curve Number: Has it reached its maturity?, *J. of Hydrologic Engg., ASCE*, Vol. 1, No. 1, 1996.
11. Wagner, T.P., and R.K. Linsley, "Application of Stanford Watershed Model to an Indian Catchment", *Irrigation and Power*, J. of *CBIP* (India), Vol. 32, No. 4, Oct. 1975, pp. 465–475.

12. *CWC & NIH, Preliminary Consolidated Report on Effect of Climate Change on Water Resources,* Govt. of India, Min. of water resources, (*CWC & NIH*), New Delhi, June 2008.
[http://www.indiawaterportal.org/sites/indiawaterportal.org/files/Effect_of_Climate_Change_on_Water_Resources_June_2008.pdf]
13. Dyson, M etal., *Flow–The Essential of Environment Flows*, IUCN, Cambridge, U.K., 2003.
14. *GBP-EMP, Environmental Flow—State of the art with special reference to Rivers in the Ganga river basin*, Report No. 0122-*GBP-IIT-EFL*-SOA-01, *IIT* Kanpur, June 2011.
15. Smakhtin, V. U. and Anputhas, M. *An Assessment of Environmental Flow Requirements of Indian River Basins*. Research Report No. 107, *IWMI*, Sri Lanka, 2006.
16. *WQAA, Report of the Working Group to Advise Water Quality Assessment Authority (WQAA) on the Minimum Flows in the Rivers*, Central Water Commission, Ministry of Water Resources, Government of India, New Delhi, 2007.

REVISION QUESTIONS

5.1 List the factors affecting the seasonal and annual yield (annual runoff) of a catchments. Describe briefly the interaction of factors listed by you.

5.2 With the help of typical annual hydrographs describe the salient features of (i) Perennial, (ii) Intermittent, and (iii) Ephemeral streams.

5.3 Explain with the aid of a neat sketch the method of assigning the stream order numbers for a stream network in a catchment.

5.4 Write a brief note on the following related to a catchment: (i) Hypsometric curve, (ii) Drainage density, and (iii) compactness coefficient.

5.5 Explain briefly: (i) Water year (ii) Natural (Virgin) flow.

5.6 What is meant by 75% dependable yield of a catchment ? Indicate a procedure to estimate the same by using annual runoff volume time series.

5.7 Describe briefly the *SCS-CN* method of estimation of yield of a catchment through use of daily rainfall record.

5.8 Indicate a procedure to estimate the annual yield of a catchment by using Strange's tables.

5.9 Explain the procedure for calculating 75% dependable yield of a basin at a flow gauging station. List the essential data series required for this analysis.

5.10 What is watershed simulation? Explain briefly various stages in the simulation study.

5.11 What is a flow-duration curve? What information can be gathered from a study of the flow duration curve of a stream at a site?

5.12 With the aid neat sketches, describe a flow mass curve and explain how it could be used for the determination of
(a) the minimum storage needed to meet a constant demand
(b) the maximum constant maintainable demand from a given storage

5.13 What are the limitations of flow mass curve?

5.14 What is a residual mass curve? What is its use?

5.15 Describe briefly the sequent peak algorithm of calculating the reservoir capacity in a stream.

5.16 What is a hydrological drought? What are its components and their possible effects?

5.17 Write short notes on (i) Drought indices, (ii) Floodwater Harvesting, and (iii) Rooftop water harvesting.

5.18 List the measures that can be adopted to lessen the effects of droughts in a region.
5.19 Describe briefly the surface water resources of India.
5.20 Briefly explain the concept of environmental flows.
5.21 Describe Tennant's method of assessment of environmental flows of a stream at a chosen section.

PROBLEMS

5.1 Long-term observations at a streamflow-measuring station at the outlet of a catchment in a mountainous area gives a mean annual discharge of 65 m^3/s. An isohyetal map for the annual rainfall over the catchment gives the following areas closed by isohyets and the divide of the catchment:

Isohyet (cm)	Area (km^2)	Isohyet (cm)	Area (km^2)
140–135	50	120–115	600
135–130	300	115–110	400
130–125	450	110–105	200
125–120	700		

Calculate (a) the mean annual depth of rainfall over the catchment,
(b) the runoff coefficient.

5.2 A small stream with a catchment area of 70 km^2 was gauged at a location some distance downstream of a reservoir. The data of the mean monthly gauged flow, rainfall and upstream diversion are given. The regenerated flow reaching the stream upstream of the gauging station can be assumed to be constant at a value of 0.20 Mm^3/month. Obtain the rainfall runoff relation for this stream. What virgin flow can be expected for a monthly rainfall value or 15.5 cm?

Month	Monthly rainfall (cm)	Gauged monthly flow (Mm^3)	Upstream utilisation (Mm^3)
1.	5.2	1.09	0.60
2.	8.6	2.27	0.70
3.	7.1	1.95	0.70
4.	9.2	2.80	0.70
5.	11.0	3.25	0.70
6.	1.2	0.28	0.30
7.	10.5	2.90	0.70
8.	11.5	2.98	0.70
9.	14.0	3.80	0.70
10.	3.7	0.84	0.30
11.	1.6	0.28	0.30
12.	3.0	0.40	0.30

5.3 The following table shows the observed annual rainfall and the corresponding annual runoff for a small catchment. Develop the rainfall–runoff correlation equation for this catchment and find the correlation coefficient. What annual runoff can be expected from this catchment for an annual rainfall of 100 cm?

Year	1964	1965	1966	1967	1968	1969
Annual Rainfall (cm)	90.5	111.0	38.7	129.5	145.5	99.8
Annual Runoff (cm)	30.1	50.2	5.3	61.5	74.8	39.9
Year	**1970**	**1971**	**1972**	**1973**	**1974**	**1975**
Annual Rainfall (cm)	147.6	50.9	120.2	90.3	65.2	75.9
Annual Runoff (cm)	64.7	6.5	46.1	36.2	24.6	20.0

5.4 Flow measurement of river Netravati at Bantwal (catchment area = 3184 km^2) yielded the following annual flow volumes:

Year	Observed annual flow (Mm^3)	Year	Observed annual flow (Mm^3)
1970–71	15925	1980–81	16585
1971–72	14813	1981–82	14649
1972–73	11726	1982–83	10662
1973–74	11818	1983–84	11555
1974–75	12617	1984–85	10821
1975–76	15704	1985–86	9466
1976–77	8334	1986–87	9732
1977–78	12864		
1978–79	16195		
1979–80	10392		

The withdrawal upstream of the gauging station [for meeting irrigation, drinking water and industrial needs are 91 Mm^3 in 1970–71 and is found to increase linearly at a rate of 2 Mm^3/year. The annual evaporation loses from water bodies on the river can be assumed to be 4 Mm^3. Estimate the 75% dependable yield at Bantwal. If the catchment area at the mouth of the river is 3222 km^2, estimate the average yield for the whole basin.

5.5 The mean monthly rainfall and temperature of a catchment near Bangalore are given below. Estimate the annual runoff volume and the corresponding runoff coefficient by using Khosla's runoff formula.

Month	Jan	Feb	Mar	Apr	May	Jun	July	Aug	Sep	Oct	Nov	Dec
Temp (°C)	24	27	32	33	31	26	24	24	23	21	20	21
Rainfall (mm)	7	9	11	45	107	71	111	137	164	153	61	13

5.6 An irrigation tank has a catchment of 900 ha. Estimate, by using Strange's method, the monthly and total runoff volumes into the tank due to following monthly rainfall values.

Month	July	Aug	Sept	Oct
Monthly Rainfall (mm)	210	180	69	215

5.7 For a 500 ha watershed in South India with predominantly non-black cotton soil, the CN_{II} has been estimated as 68. (a) If the total rainfall in the past five days is 25 cm and the season is dormant season, estimate the runoff volume due to 80 mm of rainfall in a day? (b) What would be the runoff volume if the rainfall in the past five days were 35 mm?

5.8 Estimate the values of CN_I, CN_{II} and CN_{III} for a catchment with the following land use:

Land use	Soil group *C* (%)	Soil group *D* (%)	Total % area
Cultivated land (Paddy)	30	45	75
Scrub forest	6	4	10
Waste land	9	6	15

5.9 A 400 ha watershed has predominantly black cotton soil and its CN_{II} value is estimated as 73. Estimate the runoff volume due to two consecutive days of rainfall as follows:

Day	Day 1	Day 2
Rainfall (mm)	65	80

The AMC can be assumed to be Type III.

5.10 Compute the runoff volume due to a rainfall of 15 cm in a day on a 550 ha watershed. The hydrological soil groups are 50% of group *C* and 50% of group *D*, randomly distributed in the watershed. The land use is 55% cultivated with good-quality bunding and 45% waste land. Assume antecedent moisture condition of Type-III and use standard *SCS-CN* equations.

5.11 A watershed having an area 680 ha has a CN_{III} value of 77. Estimate the runoff volume due to 3 days of rainfall as below:

Day	Day 1	Day 2	Day 2
Rainfall (mm)	30	50	13

Assume the AMC at Day 1 to be of Type III. Use standard *SCS-CN* equations.

5.12 A watershed has the following land use:

(a) 400 ha of row crop with poor hydrologic condition

(b) 100 ha of good pasture land

The soil is of hydrologic soil group B. Estimate the runoff volume for the watershed under antecedent moisture category III when 2 days of consecutive rainfall of 100 mm and 90 mm occur. Use standard *SCS-CN* equations.

5.13 (a) Compute the runoff from a 2000 ha watershed due to 15 cm rainfall in a day. The watershed has 35% group *B* soil, 40% group *C* soil and 25% group *D* soil. The land use is 80% residential that is 65% impervious and 20% paved roads. Assume AMC II conditions.

(b) If the land were pasture land in poor condition prior to the development, what would have been the runoff volume under the same rainfall? What is the percentage increase in runoff volume due to urbanisation?

[**Note:** Use standard *SCS-CN* equations.]

5.14 Discharges in a river are considered in 10 class intervals. Three consecutive years of data of the discharge in the river are given below. Draw the flow-duration curve for the river and determine the 75% dependable flow.

Discharge range (m^3/s)	< 6	6.0–9.9	10 14.9	15–24.9	25–39	40–99	100–149	–150 249	250–349	>350
No. of occurrences	20	137	183	232	169	137	121	60	30	6

5.15 The average monthly inflow into a reservoir in a dry year is given below:

Month	Jun	Jul	Aug	Sep	Oct	Nov	Dec	Jan	Feb	Mar	Apr	May
Mean Monthly Flow (m^3/s)	20	60	200	300	200	150	100	80	60	40	30	25

If a uniform discharge at 90 m^3/s is desired from this reservoir what minimum storage capacity is required?

(*Hint:* Assume the next year to have similar flows as the present year.)

5.16 For the data given in Prob. 5.15, plot the flow mass curve and find:

(a) The minimum storage required to sustain a uniform demand of 70 m^3/s;

(b) If the reservoir capacity is 7500 cumec-day, estimate the maximum uniform rate of withdrawal possible from this reservoir.

5.17 The following table gives the monthly inflow and contemplated demand from a proposed reservoir. Estimate the minimum storage that is necessary to meet the demand

Month	Jan	Feb	Mar	Apr	May	Jun	July	Aug	Sept	Oct	Nov	Dec
Monthly inflow (Mm^3)	50	40	30	25	20	30	200	225	150	90	70	60
Monthly demand (Mm^3)	70	75	80	85	130	120	25	25	40	45	50	60

5.18 For the reservoir in Prob. 5.17 the mean monthly evaporation and rainfall are given below.

Month	Jan	Feb	Mar	Apr	May	Jun	July	Aug	Sept	Oct	Nov	Dec
Evaporation (cm)	6	8	13	17	22	22	14	11	13	12	7	5
Rainfall (cm)	1	0	0	0	0	19	43	39	22	6	2	1

If the average reservoir area can be assumed to be 30 km^2, estimate the change in the storage requirement necessitated by this additional data. Assume the runoff coefficient of the area flooded by the reservoir as equal to 0.4.

5.19 Following is the stream flow record of a stream and covers a critical 2 year period. What is the minimum size of the reservoir required on this stream to provide a constant downstream flow of 45 cumecs? Use Sequent Peak Algorithm.

Month (1st Year)	Monthly Discharge (ha.m)	Month (2nd Year)	Monthly Discharge (ha.m)
June	20	June	15
July	60	July	40
August	200	August	110
September	300	September	180
October	150	October	80
November	125	November	50
December	100	December	80
January	80	January	70
February	50	February	45
March	40	March	45
April	30	April	35
May	25	May	30

5.20 Solve Problem 5.18 using Sequent Peak Algorithm method.

5.21 An unregulated stream provides the following volumes through each successive 4-day period over a 40-day duration at a possible reservoir site. What would be the reservoir capacity needed to ensure maintaining the average flow over these 40 days, if the reservoir is full to start with? What is the average flow? What would be the approximate quantity of water wasted in spillage in this case?

Day	0	4	8	12	16	20	24	28	32	36	40
Runoff volume (Mm^3)	0	9.6	5.4	2.3	3.5	2.3	2.2	1.4	6.4	12.4	10.9

5.22 A reservoir is located in a region where the normal annual precipitation is 160 cm and the normal annual *US* class *A* pan evaporation is 200 cm. The average area of reservoir water surface is 75 km^2. If under conditions of 35% of the rainfall on the land occupied by the reservoir runoff into the stream, estimate the net annual increase or decrease in the stream flow as result of the reservoir. Assume evaporation pan coefficient = 0.70.

5.23 Mean monthly flows of a river downstream of a proposed reservoir are given. It is desired to keep the downstream ecosystem of the river in *fair habitat* condition after the construction of the reservoir. Using Tennant's method, estimate the environmental flows required in the river to keep the desired ecological condition.

Month	June	July	Aug	Sept	Oct	Nov	Dec	Jan	Feb	Mar	April	May
Mean monthly flow (Mm^3)	55	210	225	175	100	75	60	55	45	40	30	40

OBJECTIVE QUESTIONS

5.1 A mean annual runoff of 1 m^3/s from a catchment of area 31.54 km^2 represents an effective rainfall of
(a) 100 cm (b) 1.0 cm (e) 100 mm (d) 3.17 cm

5.2 Direct runoff is made up of
(a) surface runoff, prompt interflow and channel precipitation
(b) surface runoff, infiltration and evapotranspiration
(c) overland flow and infiltration
(d) rainfall and evaporation

5.3 A hydrograph is a plot of
(a) rainfall intensity against time (b) stream discharge against time
(c) cumulative rainfall against time (d) cumulative runoff against time

5.4 The term *base flow* denotes
(a) delayed groundwater flow reaching a stream
(b) delayed groundwater and snowmelt reaching a stream
(c) delayed groundwater and interflow
(d) the annual minimum flow in a stream

5.5 Virgin flow is
(a) the flow in the river downstream of a gauging station
(b) the flow in the river upstream of a gauging station
(c) the flow unaffected by works of man
(d) the flow that would exist in the stream if there were no abstractions to the precipitation

5.6 The water year in India starts from the first day of
(a) January (b) April (c) June (d) September

5.7 An ephemeral stream
(a) is one which always carries some flow
(b) does not have any base flow contribution
(c) is one which has limited contribution of groundwater in wet season
(d) is one which carries only snow-melt water

5.8 An intermittent stream
(a) has water table above the stream bed throughout the year
(b) has only flash flows in response to storms
(c) has flows in the stream during wet season due to contribution of groundwater.
(d) does not have any contribution of ground water at any time

5.9 Khosla's formula for monthly runoff R_m due to a monthly rainfall P_m is $R_m = P_m - L_m$ where L_m is
(a) a constant
(b) monthly loss and depends on the mean monthly catchment temperature
(c) a monthly loss coefficient depending on the antecedent precipitation index
(d) a monthly loss depending on the infiltration characteristics of the catchment

5.10 The flow-duration curve is a plot of
(a) accumulated flow against time
(b) discharge against time in chronological order
(c) the base flow against the percentage of times the flow is exceeded
(d) the stream discharge against the percentage of times the flow is equalled or exceeded

5.11 In a flow–mass curve study, the demand line drawn from a ridge in the curve did not interest the mass curve again. This represents that
(a) the reservoir was not full at the beginning
(b) the storage was not adequate
(c) the demand cannot be met by the inflow as the reservoir will not refill
(d) the reservoir is wasting water by spill

5.12 If in a flow–mass curve, a demand line drawn tangent to the lowest point in a valley of the curve does not intersect the mass curve at an earlier time period, it represents that
(a) the storage is inadequate
(b) the reservoir will not be full at the start of the dry period
(c) the reservoir is full at the beginning of the dry period
(d) the reservoir is wasting later by spill

5.13 The flow-mass curve is an integral curve of
(a) the hydrograph (b) the hyetograph
(c) the flow duration curve (d) the *S*-curve.

5.14 The total rainfall in a catchment of area 1200 km^2 during a 6-h storm is 16 cm while the surface runoff due to the storm is 1.2×10^8 m^3. The ϕ index is
(a) 0.1 cm/h (b) 1.0 cm/h
(c) 0.2 cm/h (d) cannot be estimated with the given data.

5.15 In India, a meteorological subdivision is considered to be affected by moderate drought if it receives a total seasonal rainfall which is
(a) less than 25% of normal value
(b) between 25% and 49% of normal value
(c) between 50% and 74% of normal value
(d) between 75% and 99% of normal value

5.16 An area is classified as a drought prone area if the probability P of occurrence of a drought is
(a) $0.4 < P \leq 1.0$ (b) $0.2 \leq P \leq 0.40$
(c) $0.1 \leq P < 0.20$ (d) $0.0 < P < 0.20$

5.17 In the standard *SCS-CN* method of modelling runoff due to daily rainfall, the maximum daily rainfall that would not produce runoff in a watershed with $CN = 50$ is about
(a) 65 mm (b) 35 mm (c) 50 mm (d) 25 mm

5.18 In the standard *SCS-CN* method, if $CN = 73$, the runoff volume for a one-day rainfall of 100 mm is about
(a) 38 mm (b) 2 mm (c) 56 mm (d) 81 mm

CHAPTER

6

HYDROGRAPHS

6.1 INTRODUCTION

While long-term runoff concerned with the estimation of yield was discussed in the previous chapter, the present chapter examines in detail the short-term runoff phenomenon. The storm hydrograph is the focal point of the present chapter.

Consider a concentrated storm producing a fairly uniform rainfall of duration, D over a catchment. After the initial losses and infiltration losses are met, the rainfall excess reaches the stream through overland and channel flows. In the process of translation a certain amount of storage is built up in the overland and channel-flow phases. This storage gradually depletes after the cessation of the rainfall. Thus, there is a time lag between the occurrence of rainfall in the basin and the time when that water passes the gauging station at the basin outlet. The runoff measured at the stream-gauging station will give a typical hydrograph as shown in Fig. 6.1. The duration of the rainfall is also marked in this figure to indicate the time lag in the rainfall and runoff. The hydrograph of this kind which results due to an isolated storm is typically single-peaked skew distribution of discharge and is known variously as *storm hydrograph*, *flood hydrograph* or simply *hydrograph.* It has three characteristic regions: (i) the rising limb AB, joining point A, the starting point of the rising curve and point B, the point of inflection, (ii) the crest segment BC between the two points of inflection with a peak P in between, (iii) the falling limb or *depletion curve* CD starting from the second point of inflection C.

The hydrograph is the response of a given catchment to a rainfall input. It consists of flow in all the three phases of runoff, viz. surface runoff, interflow and base flow, and embodies in itself the integrated effects of a wide variety of catchment and rainfall parameters having complex interactions. Thus, two different storms in a given catchment produce hydrographs differing from each other. Similarly, identical storms in two catchments produce hydrographs that are different. The interactions of various storms and catchments are in general extremely complex. If one examines the record of a large number of flood hydrographs of a stream, it will be found that many of them will have kinks, multiple peaks, etc. resulting in shapes much different from the simple single-peaked hydrograph of Fig. 6.1. These complex hydrographs are the result of storm and catchment peculiarities and their complex interactions. While it is theoretically possible to resolve a complex hydrograph into a set of simple hydrographs for purposes of hydrograph analysis, the requisite data of acceptable quality are seldom available. Hence, simple hydrographs resulting from isolated storms are preferred for hydrograph studies.

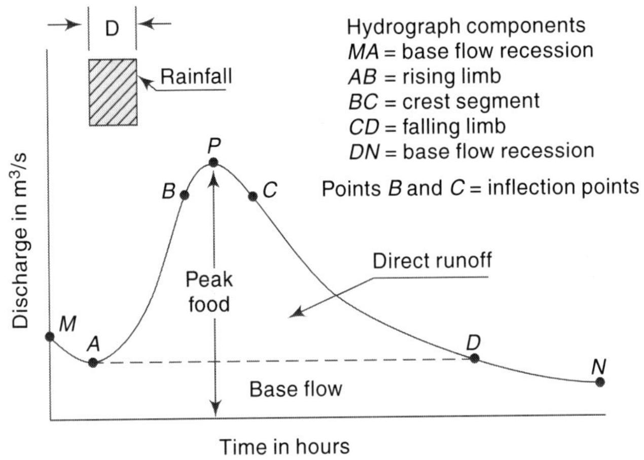

Fig. 6.1 ***Elements of a Flood Hydrograph***

6.2 *Factors Affecting Runoff Hydrograph*

The factors that affect the shape of the hydrograph can be broadly grouped into climatic factors and physiographic factors. Each of these two groups contains a host of factors and the important ones are listed in Table 6.1. Generally, the climatic factors control the rising limb and the recession limb is independent of storm characteristics, being determined by catchment characteristics only. Many of the factors listed in Table 6.1 are interdependent. Further, their effects are very varied and complicated. As such only important effects are listed below in qualitative terms only.

Table 6.1 ***Factors Affecting Runoff Hydrograph***

Physiographic factors	Climatic factors
1. Basin characterstics: (a) Shape (b) size (c) slope (d) nature of the valley (e) elevation (f) drainage density 2. Infiltration characteristics: (a) land use and cover (b) soil type and geological conditions (c) lakes, swamps and other storage 3. Channel characteristics: cross-section, roughness and storage capacity	1. Storm characteristics: precipitation, intensity, duration, magnitude and movement of storm. 2. Initial loss 3. Evapotranspiration

6.2.1 Shape of the Basin

The shape of the basin influences the time taken for water from the remote parts of the catchment to arrive at the outlet. Thus, the occurrence of the peak and hence the shape of the hydrograph are affected by the basin shape. Fan-shaped, i.e. nearly semicircular shaped catchments give high peak and narrow hydrographs while elongated catchments give broad and low-peaked hydrographs. Figure 6.2 shows schematically the hydrographs from three catchments having identical infiltration characteristics due to identical rainfall over the catchment. In catchment *A*, the hydrograph is skewed to the left, i.e. the peak occurs relatively quickly. In catchment *B*, the hydrograph is skewed to the right, the peak occurring with a relatively longer lag. Catchment *C* indicates the complex hydrograph produced by a composite shape.

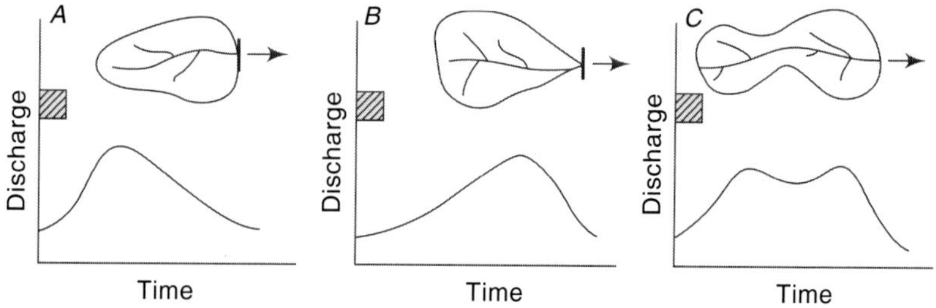

Fig. 6.2(a) ***Effect of Catchment Shape on the Hydrograph***

6.2.2 Size

Small basins behave different from the large ones in terms of the relative importance of various phases of the runoff phenomenon. In small catchments, the overland flow phase is predominant over the channel flow. Hence the land use and intensity of rainfall have important role on the peak flood. On large basins these effects are suppressed as the channel flow phase is more predominant. The peak discharge is found to vary as A^n where A is the catchment area and n is an exponent whose value is less than unity, being about 0.5. The time base of the hydrographs from larger basins will be larger than those of corresponding hydrographs from smaller basins. The duration of the surface runoff from the time of occurrence of the peak is proportional to A^m, where m is an exponent less than unity and is of the order of magnitude of 0.2.

6.2.3 Slope

The slope of the main stream controls the velocity of flow in the channel. As the recession limb of the hydrograph represents the depletion of storage, the stream channel slope will have a pronounced effect on this part of the hydrograph. Large stream slopes give rise to quicker depletion of storage and hence result in steeper recession limbs of hydrographs. This would obviously result in a smaller time base.

The basin slope is important in small catchments where the overland flow is relatively more important. In such cases the steeper slope of the catchment results in larger peak discharges.

6.2.4 Drainage Density

The drainage density is defined as the ratio of the total channel length to the total drainage area. A large drainage density creates situation conducive for quick disposal of runoff down the channels. This fast response is reflected in a pronounced peaked discharge. In basins with smaller drainage densities, the overland flow is predominant and the resulting hydrograph is squat with a slowly rising limb (Fig. 6.2b).

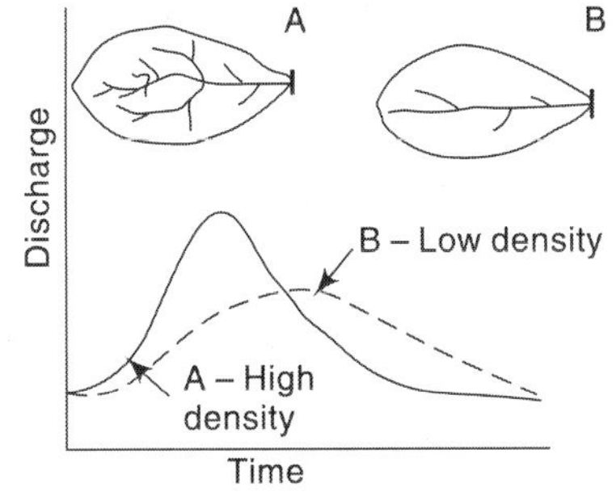

Fig. 6.2(b) ***Role of Drainage Density on the Hydrograph***

6.2.5 Land Use

Vegetation and forests increase the infiltration and storage capacities of the soils. Further, they cause considerable retardance to the overland flow. Thus, the vegetal cover reduces the peak flow. This effect is usually very pronounced in small catchments of area less than 150 km^2. Further, the effect of the vegetal cover is prominent in small storms. In general, for two catchments of equal area, other factors being identical, the peak discharge is higher for a catchment that has a lower density of forest cover. Of the various factors that control the peak discharge, probably the only factor that can be manipulated is land use and thus it represents the only practical means of exercising long-term natural control over the flood hydrograph of a catchment.

6.2.6 Urbanisation

Urbanisation of a rural watershed results in the following significant changes in the characteristics of the watershed;

(i) Urbanisation increases the amount of impervious surface areas such as roofing on built up areas, concrete and asphalt road surfaces, parking lots. This is a major factor that would inhibit infiltration and surface retention. Further, the amount of rainfall that would be available as surface runoff would be large. These effect of decrease in infiltration and initial losses and consequent increase in the effective (runoff producing) rainfall is shown schematically in Fig. 6.3(a). From this figure, it can clearly be seen as to how the decrease in infiltration due to urbanisation results in increase in runoff producing rain.

(ii) Due to urbanisation the natural drainage of the rural catchment will be removed or improved. Also, there will be a superposition of man-made artificial surface

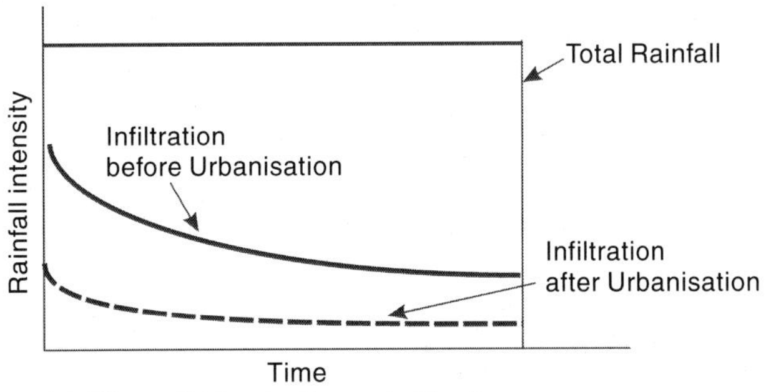

(a) Effect of Urbanisation on Infiltration

Fig. 6.3(a) ***Effect of Urbanisation***

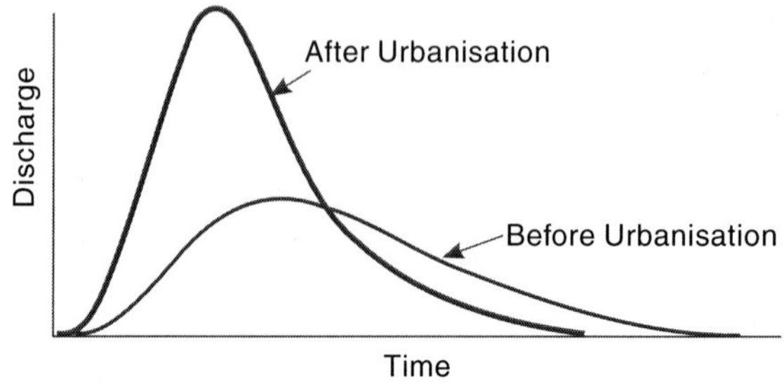

(b) Effect of Urbanisation on the Hydrograph

Fig. 6.3(b) ***Effect of Urbanisation***

drainage system consisting of impervious conduits. The relatively slow process of interflow to small streamlets which feed the large drainage will not exist any more.

The new drainage in urbanised catchment would start with rapid overland flow over relatively impervious surfaces of roads, pavements and parking lots to the nearest curb or drainage inlet. Then onwards the storm runoff in the drainage channel will move rapidly down the drainage system. In general, a unit of runoff would take the shortest path to the outlet. The net result of the changed drainage pattern is the rapid response of the system that would cause the hydrograph to have higher peak at a shorter time from the start and also rapid receding of the peak.

Thus, while the larger and higher impervious areas in the urban catchment causes more quantity of surface runoff, the drainage system of urban areas cause higher peaks at short time from start and a rapid recession of the peak. The change in the shape of the hydrograph due to urbanisation of a rural catchment is shown in Fig. 6.3(b) wherein the above characteristic features are clearly shown.

The increase in the volume of direct runoff due to urbanisation of the catchment can be estimated by using *SCS* method (Sec. 5.6). Example 5.7 illustrates the procedure through an illustrative example.

6.2.7 Climatic Factors

Among climatic factors the intensity, duration and direction of storm movement are the three important ones affecting the shape of a flood hydrograph. For a given duration, the peak and volume of the surface runoff are essentially proportional to the intensity of rainfall. This aspect is made use of in the unit hydrograph theory of estimating peak-flow hydrographs, as discussed in subsequent sections of this chapter. In very small catchments, the shape of the hydrograph can also be affected by the intensity.

The duration of storm of given intensity also has a direct proportional effect on the volume of runoff. The effect of duration is reflected in the rising limb and peak flow. Ideally, if a rainfall of given intensity i lasts sufficiently long enough, a state of equilibrium discharge proportional to iA is reached.

If the storm moves from upstream of the catchment to the downstream end, there will be a quicker concentration of flow at the basin outlet. This results in a peaked hydrograph. Conversely, if the storm movement is up the catchment, the resulting hydrograph will have a lower peak and longer time base. This effect is further accentuated by the shape of the catchment, with long and narrow catchments having hydrographs most sensitive to the storm-movement direction.

6.3 *Components of a Hydrograph*

As indicated earlier, the essential components of a hydrograph are (i) the rising limb, (ii) the crest segment, and (iii) the recession limb (Fig. 6.1). A few salient features of these components are described below.

6.3.1 Rising Limb

The rising limb of a hydrograph, also known as *concentration curve,* represents the increase in discharge due to the gradual building up of storage in channels and over the catchment surface. The initial losses and high infiltration losses during the early period of a storm cause the discharge to rise rather slowly in the initial periods. As the storm continues, more and more flow from distant parts reach the basin outlet. Simultaneously the infiltration losses also decrease with time. Thus, under a uniform storm over the catchment, the runoff increases rapidly with time. As indicated earlier, the basin and storm characteristics control the shape of the rising limb of a hydrograph.

6.3.2 Crest Segment

The crest segment is one of the most important parts of a hydrograph as it contains the peak flow. The peak flow occurs when the runoff from various parts of the catchment simultaneously contribute amounts to achieve the maximum amount of flow at the basin outlet. Generally, for large catchments, the peak flow occurs after the cessation of rainfall, the time interval from the centre of mass of rainfall to the peak being essentially controlled by basin and storm characteristics. Multiple-peaked complex hydrographs in a basin can occur when two or more storms occur in succession. Estimation of the peak flow and its occurrence, being important in flood-flow studies are dealt with in detail elsewhere in this book.

6.3.3 Recession Limb

The recession limb, which extends from the point of inflection at the end of the crest segment (point C in Fig. 6.1) to the commencement of the natural groundwater flow (point D in Fig. 6.1) represents the withdrawal of water from the storage built up in the basin during the earlier phases of the hydrograph. The starting point of the recession limb, i.e. the point of inflection represents the condition of maximum storage. Since the depletion of storage takes place after the cessation of rainfall, the shape of this part of the hydrograph is independent of storm characteristics and depends entirely on the basin characteristics.

The storage of water in the basin exists as (i) surface storage, which includes both surface detention and channel storage, (ii) interflow storage, and (iii) groundwater storage, i.e. base-flow storage. Barnes (1940) showed that the recession of a storage can be expressed as

$$Q_t = Q_0 K_r^t \tag{6.1}$$

in which Q_t is the discharge at a time t and Q_0 is the discharge at $t = 0$; K_r is a recession constant of value less than unity. Equation (6.1) can also be expressed in an alternative form of the exponential decay as

$$Q_t = Q_0 e^{-at} \tag{6.1a}$$

where $a = -\ln K_r$.

The recession constant K_r can be considered to be made up of three components to account for the three types of storages as

$$K_r = K_{rs} \cdot K_{ri} \cdot K_{rb}$$

where K_{rs} = recession constant for surface storage, K_{ri} = recession constant for interflow and K_{rb} = recession constant for base flow. Typically, the values of these recession constants, when time t is in days, are

$$K_{rs} = 0.05 \text{ to } 0.20 \qquad K_{ri} = 0.50 \text{ to } 0.85 \qquad K_{rb} = 0.85 \text{ to } 0.99$$

When the interflow is not significant, K_{ri} can be assumed to be unity.

If suffixes 1 and 2 denote the conditions at two time instances t_1 and t_2,

From Eq. (6.1)

$$\frac{Q_1}{Q_2} = K_r^{(t_1 - t_2)} \tag{6.2}$$

or from Eq. (6.1a)

$$\frac{Q_1}{Q_2} = e^{-a(t_1 - t_2)} \tag{6.2a}$$

Equation 6.1 (and also 6.1 a) plots as a straight line when plotted on a semi-log paper with discharge on the log–scale. The slope of this line represents the recession constant. Using this property and using Eq. 6.2 (or 6.2a) the value of K_r for a basin can be estimated by using observed recession data of a flood hydrograph. Example 6.1 explains the procedure in detail.

The storage S_t remaining at any time t is obtained as

$$S_t = \int_t^{\infty} Q_t\, dt = \int_t^{\infty} Q_0 e^{-at}\, dt = \frac{Q_t}{a} \tag{6.3}$$

Example 6.1 The recession limb of a flood hydrograph is given below. The time is indicated from the arrival of peak. Assuming the interflow component to be negligible, estimate the base flow and surface flow recession coefficients. Also, estimate the storage at the end of day-3.

Time from peak (day)	Discharge (m^3/s)	Time from Peak (day)	Discharge (m^3/s)
0.0	90	3.5	5.0
0.5	66	4.0	3.8
1.0	34	4.5	3.0
1.5	20	5.0	2.6
2.0	13	5.5	2.2
2.5	9.0	6.0	1.8
3.0	6.7	6.5	1.6
		7.0	1.5

Solution:

The data are plotted on a semi-log paper with discharge on the log-scale. The data points from t = 4.5 days to 7.0 days are seen to lie on straight line (line *AB* in Fig. 6.4). This indicates that the surface flow terminates at t = 4.5 days. The best fitting exponential curve for this straight-line portion (obtained by use of *MS* Excel) is

$$Q_t = 11.033e^{-0.2927t} \text{ with } r^2 = 0.9805.$$

The base flow recession coefficient K_{rb} is obtained as
$\ln K_{rb} = -0.2927$ and as such $K_{rb} = 0.746$.

[Alternatively, by using the graph, the value of K_{rb} could be obtained by selecting two points 1 and 2 on the straight line *AB* and using Eq. (6.2)].

The base flow recession curve is extended till $t \approx 1$ day as shown by the line *ABM* Fig. 6.4. The surface runoff depletion is obtained by subtracting the base flow from the given recession limb of the flood hydrograph. The computations are shown in the Table given below.

The surface flow values (Col. 4 of Table above) are plotted against time as shown in Fig. 6.4. It is seen that these points lie on a straight line, *XY*. The best fitting exponential curve for this straight-line portion *XY* (obtained by use of *MS* Excel) is

$$Q_t = 106.84e^{-1.3603t} \text{ with } r^2 = 0.9951$$

Time from peak (days)	Recession Limb of given flood hydrograph (m^3/s)	Base flow (Obtained by using K_{rb} = 0.746)	Surface runoff (m^3/s)
0.0	90.0		
0.5	66.0	10.455	55.545
1.0	34.0	7.945	26.055
1.5	20.0	6.581	13.419
2.0	13.0	5.613	7.387
2.5	9.0	4.862	4.138
3.0	6.7	4.249	2.451
3.5	5.0	3.730	1.270
4.0	3.8	3.281	0.519
4.5	3.0	2.884	
5.0	2.6	2.530	
5.5	2.2	2.209	
6.0	1.8	1.917	
6.5	1.6	1.647	
7.0	1.5	1.398	

The *surface flow recession coefficient* K_{rs} is obtained as
$\ln K_{rs} = -1.3603$ and as such $K_{rs} = 0.257$.

[Alternatively, by using the graph, the value of K_{rs} could be obtained by selecting two points 1 and 2 on the straight line *XY* and using Eq. (6.2)].

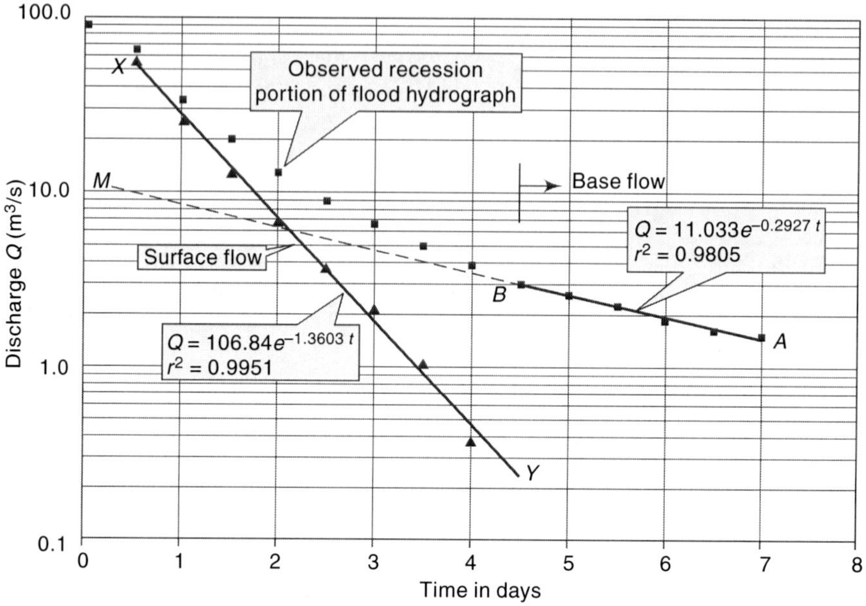

Fig. 6.4 ***Storage Recession Curve—Example 6.1***

The storage available at the end of day-3 is the sum of the storages in surface flow and groundwater recession modes and is given by

$$S_{t3} = \left(\frac{Q_{s3}}{-\ln K_{rs}} + \frac{Q_{b3}}{-\ln K_{rb}} \right)$$

For the surface flow recession using the best fit equation:

$$Q_{s3} = 106.84e^{-1.3603 \times 3} = 1.8048; \ -\ln K_{rs} = 1.3603$$

$$\frac{Q_{s3}}{-\ln K_{rs}} = \frac{1.8048}{1.3603} = 1.3267 \text{ cumec-days}$$

Similarly for the base flow recession:

$$Q_{b3} = 11.033e^{-0.2927 \times 3} = 4.585; \ -\ln K_{rb} = 0.2927$$

$$\frac{Q_{b3}}{-\ln K_{rb}} = \frac{4.585}{0.2927} = 15.665 \text{ cumec-days}$$

Hence, total storage at the end of 3 days $= S_{t3} = 1.3267 + 15.665$
$= 16.9917$ cumec. days $= 1.468 \text{ Mm}^3$

6.4 *Base Flow Separation*

In many hydrograph analyses, a relationship between the surface-flow hydrograph and the effective rainfall (i.e. rainfall minus losses) is sought to be established. The surface-flow hydrograph is obtained from the total storm hydrograph by separating the quick-response flow from the slow response runoff. It is usual to consider the interflow as a part of the surface flow in view of its quick responsc. Thus, only the base flow is to be deducted from the total storm hydrograph to obtain the surface flow hydrograph.

There are three methods of base-flow separation that are in common use.

6.4.1 Methods of Base-flow Separation

Method I—Straight-Line Method

In this method, the separation of the base flow is achieved by joining with a straight line the beginning of the surface runoff to a point on the recession limb representing the end of the direct runoff. In Fig. 6.5, point A represents the beginning of the direct runoff and it is usually easy to identify in view of the sharp change in the runoff rate at that point.

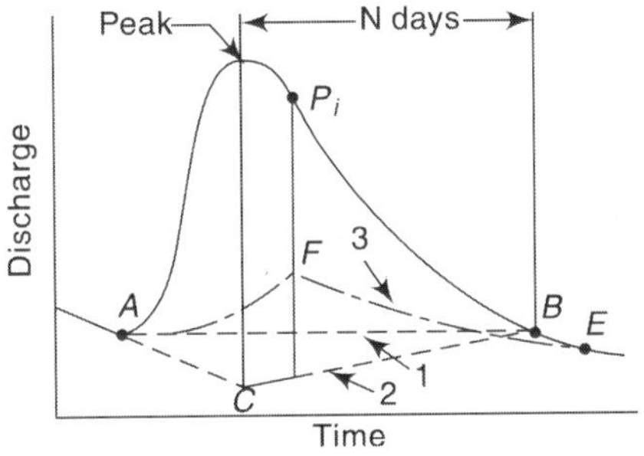

Fig. 6.5 ***Base Flow Seperation Methods***

Point B, marking the end of the direct runoff is rather difficult to locate exactly. An empirical equation for the time interval N (days) from the peak to the point B is

$$N = 0.83A^{0.2} \quad (6.4)$$

where A = drainage area in km^2 and N is in days. Points A and B are joined by a straight line to demarcate to the base flow and surface runoff. It should be realised that the value of N obtained as above is only approximate and the position of B should be decided by considering a number of hydrographs for the catchment. This method of base-flow separation is the simplest of all the three methods.

Method 2

In this method, the base flow curve existing prior to the commencement of the surface runoff is extended till it intersects the ordinate drawn at the peak (point C in Fig. 6.5). This point is joined to point B by a straight line. Segment AC and CB demarcate the base flow and surface runoff. This is probably the most widely used base-flow separation procedure.

Method 3

In this method, the base flow recession curve after the depletion of the flood water is extended backwards till it intersects the ordinate at the point of inflection (line EF in Fig. 6.5). Points A and F are joined by an arbitrary smooth curve. This method of base-flow separation is realistic in situations where the groundwater contributions are significant and reach the stream quickly.

It is seen that all the three methods of base-flow separation are rather arbitrary. The selection of anyone of them depends upon the local practice and successful predictions achieved in the past. The surface runoff hydrograph obtained after the base-flow separation is also known as *Direct Runoff Hydrograph* (*DRH*).

6.5 Effective Rainfall (ER)

Effective rainfall (also known as *Excess Rainfall*) (*ER*) is that part of the rainfall that becomes direct runoff at the outlet of the watershed. It is thus the total rainfall in a given duration from which abstractions such as infiltration and initial losses are subtracted. As such, *ER* could be defined as that rainfall that is neither retained on the land surface nor infiltrated into the soil.

For purposes of correlating *DRH* with the rainfall which produced the flow, the hyetograph of the rainfall is also pruned by deducting the losses. Figure 6.6 shows the hyetograph of a storm. The initial loss and infiltration losses are subtracted from it. The resulting hyetograph is known as *Effective Rainfall Hyetograph* (*ERH*). It is also known as *excess rainfall hyetograph*.

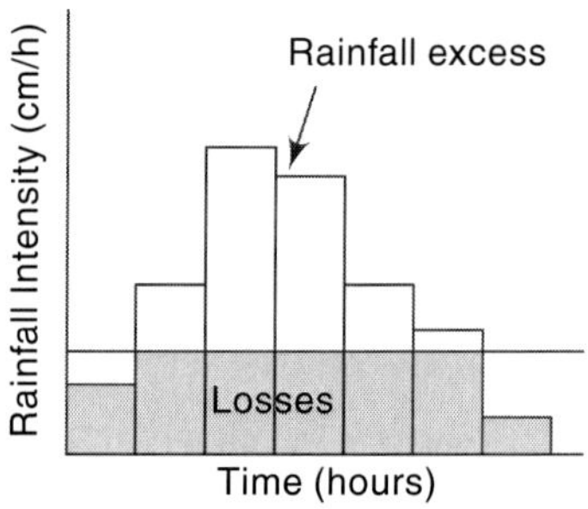

Fig. 6.6 ***Effective Rainfall Hyetograph (ERH)***

Both *DRH* and *ERH* represent the same total quantity but in different units. Since *ERH* is usually in cm/h plotted against time, the area of *ERH* multiplied by the catchment area gives the total volume of direct runoff which is the same as the area of *DRH*. The initial loss and infiltration losses are estimated based on the available data of the catchment.

Example 6.2 Rainfall of magnitude 3.8 cm and 2.8 cm occurring on two consecutive 4-h durations on a catchment of area 27 km^2 produced the following hydrograph of flow at the outlet of the catchment. Estimate the rainfall excess and ϕ index.

Time from start of rainfall (h)	–6	0	6	12	18	24	30	36	42	48	54	60	66
Observed flow (m^3/s)	6	5	13	26	21	16	12	9	7	5	5	4.5	4.5

Solution

The hydrograph is plotted to scale (Fig. 6.7). It is seen that the storm hydrograph has a base-flow component. For using the simple straight-line method of base-flow separation, by Eq. (6.4)

$$N = 0.83 \times (27)^{0.2} = 1.6 \text{ days} = 38.5 \text{ h}$$

However, by inspection, *DRH* starts at $t = 0$, has the peak at $t = 12$ h and ends at $t = 48$ h (which gives a value of $N = 48 - 12 = 36$ h). As $N = 36$ h appears to be more satisfactory than $N = 38.5$ h, in the present case *DRH* is assumed to exist from $t = 0$ to 48 h. A straight line base flow separation gives a constant value of 5 m^3/s for the base flow.

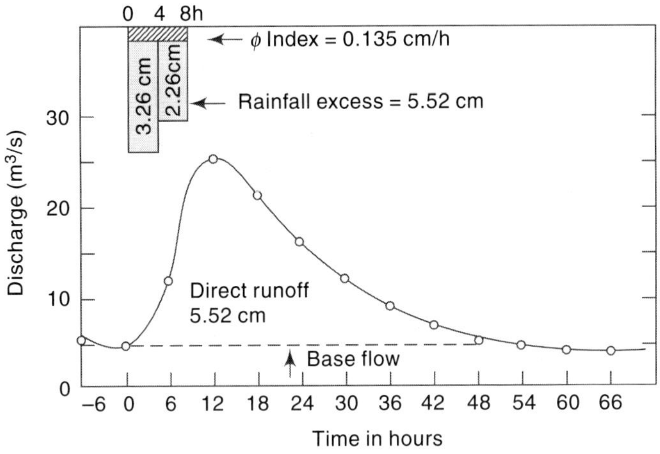

Fig. 6.7 ***Base Flow Separation—Example 6.2***

$$\text{Area of } DRH = (6 \times 60 \times 60)\left[\frac{1}{2}(8) + \frac{1}{2}(8+21) + \frac{1}{2}(21+16) + \frac{1}{2}(16+11)\right.$$

$$\left. + \frac{1}{2}(11+7) + \frac{1}{2}(7+4) + \frac{1}{2}(4+2) + \frac{1}{2}(2)\right]$$

$$= 3600 \times 6 \times (8 + 21 + 16 + 11 + 7 + 4 + 2) = 1.4904 \times 10^6 \text{ m}^3$$

= Total direct runoff due to storm

$$\text{Runoff depth} = \frac{\text{Runoff volume}}{\text{Catchment area}} = \frac{1.4904 \times 10^6}{27 \times 10^6} = 0.0552 \text{ m}$$

= 5.52 cm = rainfall excess

Total rainfall = 3.8 + 2.8 = 6.6 cm

Duration = 8 h

$$\phi \text{ index} = \frac{6.6 - 5.52}{8} = 0.135 \text{ cm/h}$$

Example 6.3

A storm over a catchment of area 5.0 km^2 had a duration of 14 hours. The mass curve of rainfall of the storm is as follows:

Time from start of storm (h)	0	2	4	6	8	10	12	14
Accumulated rainfall (cm)	0	0.6	2.8	5.2	6.6	7.5	9.2	9.6

If the ϕ index for the catchment is 0.4 cm/h, determine the effective rainfall hyetograph and the volume of direct runoff from the catchment due to the storm.

Solution:

First the depth of rainfall in a time interval Δt = 2 hours, in total duration of the storm is calculated, (Col. 4 of Table 6.2).

Table 6.2 ***Calculation for Example 6.3***

Time from start of storm, t (h)	Time interval Δt (h)	Accumulated rainfall in time t (cm)	Depth of rainfall in Δt (cm)	$\varphi\,\Delta t$ (cm)	ER (cm)	Intensity of ER (cm/h)
1	2	3	4	5	6	7
0	—	0	—	—	—	—
2	2	0.6	0.6	0.8	0	0
4	2	2.8	2.2	0.8	1.4	0.7
6	2	5.2	2.4	0.8	1.6	0.8
8	2	6.7	1.5	0.8	0.7	0.35
10	2	7.5	0.8	0.8	0	0
12	2	9.2	1.7	0.8	0.9	0.45
14	2	9.6	0.4	0.8	0	0

In a given time interval Δt, effective rainfall (ER) is given by

$$ER = (\text{actual depth of rainfall} - \phi\,\Delta t)$$

or $\quad ER = 0$, whichever is larger.

The calculations are shown in Table 6.2. For plotting the hyetograph, the intensity of effective rainfall is calculated in Col. 7.

The effective rainfall hyetograph is obtained by plotting ER intensity (Col. 7) against time from start of storm (Col. 1), and is shown in Fig. 6.8.

Total effective rainfall = Direct runoff due to storm = area of ER hyetograph = (0.7 + 0.8 + 0.35 + 0.45) × 2 = 4.6 m

Volume of Direct runoff =

$$\frac{4.6}{1000} \times 5.0 \times (1000)^2 = 23000\ \text{m}^3$$

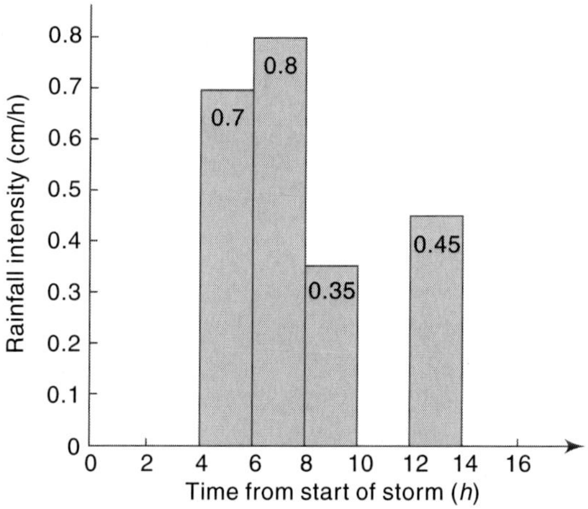

Fig. 6.8 ***ERH of Storm—Example 6.3***

6.6 *Unit Hydrograph*

6.6.1 Theory

The problem of predicting the flood hydrograph resulting from a known storm in a catchment has received considerable attention. A large number of methods are proposed to solve this problem and of them probably the most popular and widely used method is the *unit hydrograph method.* This method was first suggested by Sherman in 1932 and has undergone many refinements since then.

A *unit hydrograph* is defined as the hydrograph of direct runoff resulting from one unit depth (1 cm) of rainfall excess occurring uniformly over the basin and at a uniform rate for a specified duration (*D* hours). The term *unit* here refers to a unit depth of rainfall excess which is usually taken as 1 cm. The duration, being a very important characteristic, is used as a prefix to a specific unit hydrograph. Thus, one has a 6-h unit hydrograph, 12-h unit hydrograph, etc. and in general a *D-h* unit hydrograph applicable to a given catchment. The definition of a unit hydrograph implies the following:

- The unit hydrograph represents the lumped response of the catchment to a unit rainfall excess of *D*-h duration to produce a direct-runoff hydrograph. It relates only the direct runoff to the rainfall excess. Hence the volume of water contained in the unit hydrograph must be equal to the rainfall excess. As 1 cm depth of rainfall excess is considered, the area of the unit hydrograph is equal to a volume given by 1 cm over the catchment.
- The rainfall is considered to have an average intensity of *excess rainfall* (*ER*) of 1/*D* cm/h for the duration *D*-h of the storm.
- The distribution of the storm is considered to be uniform all over the catchment.

Figure 6.9 shows a typical 6-h unit hydrograph. Here, the duration of the rainfall excess is 6 h.

Area under the unit hydrograph = 12.92×10^6 m^3

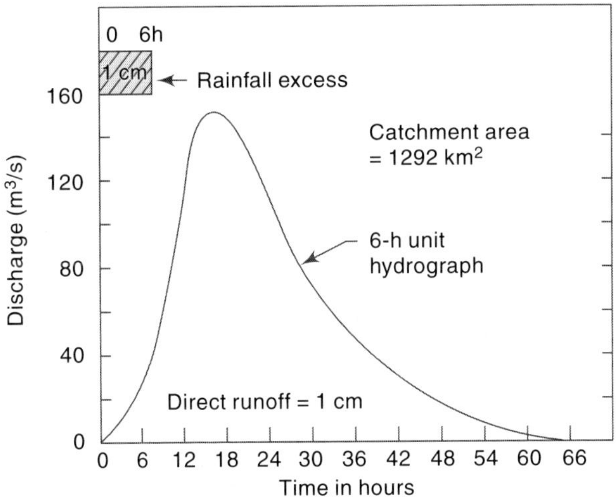

Fig. 6.9 ***Typical 6-h Unit Hydrograph***

Hence,

Catchment area of the basin = 1292 km^2

Two basic assumptions constitute the foundations for the unit-hydrograph theory. These are (i) the time invariance, and (ii) the linear response.

1. Time Invariance

This first basic assumption is that the direct-runoff response to a given effective rainfall in a catchment is time-invariant. This implies that the *DRH* for a given *ER* in a catchment is always the same irrespective of when it occurs.

2. Linear Response

The direct-runoff response to the rainfall excess is assumed to be linear. This is the most important assumption of the unit-hydrograph theory. Linear response means that if an input x_1 (t) causes an output y_1 (t) and an input x_2 (t) causes an output y_2 (t), then an input x_1 (t) + x_2 (t) gives an output y_1 (t) + y_2 (t). Consequently, if x_2 $(t) = r\, x_1$ (t), then y_2 $(t) = r\, y_1$ (t). Thus, if the rainfall excess in a duration *D* is *r* times the unit depth, the resulting *DRH* will have ordinates bearing ratio *r* to those of the corresponding *D*-h unit hydrograph. Since the area of the resulting *DRH* should increase by the ratio *r*, the base of the *DRH* will be the same as that of the unit hydrograph.

The assumption of linear response in a unit hydrograph enables the method of superposition to be used to derive *DRH*s. Accordingly, if two rainfall excess of *D*-h duration each occur consecutively, their combined effect is obtained by superposing the respective *DRH*s with due care being taken to account for the proper sequence of events. These aspects resulting from the assumption of linear response are made clearer in the following two illustrative examples.

Example 6.4 Given below are the ordinates of a 6-h unit hydrograph for a catchment. Calculate the ordinates of the DRH due to a rainfall excess of 3.5 cm occurring in 6 h.

Time (h)	0	3	6	9	12	15	18	24	30	36	42	48	54	60	69
UH ordinate (m³/s)	0	25	50	85	125	160	185	160	110	60	36	25	16	8	0

Solution

The desired ordinates of the *DRH* are obtained by multiplying the ordinates of the unit hydrograph by a factor of 3.5 as in Table 6.3. The resulting *DRH* as also the unit hydrograph are shown in Fig. 6.10 (a). Note that the time base of *DRH* is not changed and remains the same as that of the unit hydrograph. The intervals of coordinates of the unit hydrograph (shown in column 1) are not in any way related to the duration of the rainfall excess and can be any convenient value.

Table 6.3 ***Calculation of DRH Due to 3.5 ER—Example 6.4***

Time (h)	Ordinate of 6-h unit hydrograph (m³/s)	Ordinate of 3.5 cm *DRH* (m³/s)
1	2	3
0	0	0
3	25	87.5
6	50	175.0
9	85	297.5
12	125	437.5
15	160	560.0
18	185	647.5
24	160	560.0
30	110	385.0
36	60	210.0
42	36	126.0
48	25	87.5
54	16	56.0
60	8	28.0
69	0	0

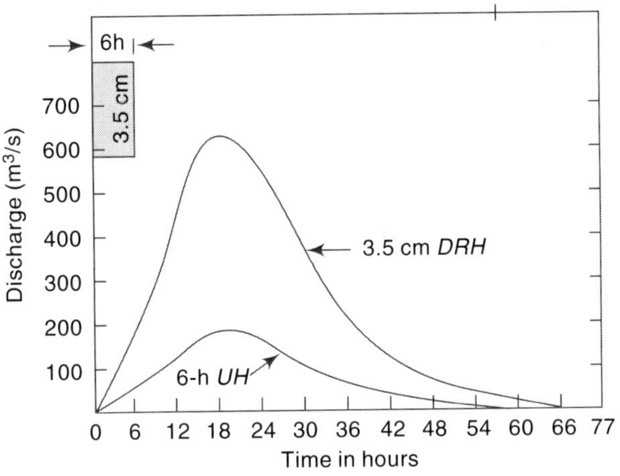

Fig. 6.10(a) ***3.5 cm DRH derived from 6-h Unit Hydrograph—Example 6.4***

Example 6.5 Two storms, each of 6-h duration and having rainfall excess values of 3.0 and 2.0 cm respectively, occur successively. The 2 cm *ER* rain follows the 3 cm rain. The 6 h unit hydrograph for the catchment is the same as given in Example 6.4. Calculate the resulting *DRH*.

Solution

First, the *DRH*s due to 3.0 and 2.0 cm *ER* are calculated, as in Example 6.3 by multiplying the ordinates of the unit hydrograph by 3 and 2 respectively. Noting that the 2 cm *DRH* occurs after the 3 cm *DRH*, the ordinates of the 2 cm *DRH* are lagged by 6 hrs as shown in column 4 of Table 6.4. Columns 3 and 4 give the proper sequence of the two *DRHs*. Using the method of superposition, the ordinates of the resulting *DRH* are obtained by combining the ordinates of the 3 and 2 cm *DRH*s at any instant. By this process the ordinates of the 5 cm *DRH* are obtained in column 5. Figure 6.10(b) shows the component 3 and 2 cm *DRH*s as well as the composite 5-cm *DRH* obtained by the method of superposition.

Table 6.4 ***Calculation of DRH by method of Superposition—Example 6.5***

Time (h)	Ordinate of 6 h *UH* (m^3/s)	Ordinate of 3 cm *DRH* (col. 2) × 3	Ordinate of 2 cm *DRH* (col. 2 lagged by 6 h) × 2	Ordinate of 5 cm *DRH* (col. 3 + col. 4) (m^3/s)	Remarks
1	2	3	4	5	6
0	0	0	0	0	
3	25	75	0	75	
6	50	150	0	150	
9	85	255	50	305	
12	125	375	100	475	
15	160	480	170	650	
18	185	555	250	805	

(Contd.)

(Contd.)

(21)	(172.5)	(517.5)	(320)	(837.5)	Interpolated value
24	160	480	370	850	
30	110	330	320	650	
36	60	180	220	400	
42	36	108	120	228	
48	25	75	72	147	
54	16	48	50	98	
60	8	24	32	56	
(66)	(2.7)	(8.1)	(16)	(24.1)	Interpolated value
69	0	0	(10.6)	(10.6)	Interpolated value
75	0	0	0	0	

[**Note:** 1. The entries in col. 4 are shifted by 6 h in time relative to col. 2.
2. Due to unequal time interval of ordinates a few entries have to be interpolated to complete the table. These interpolated values are shown in parentheses.]

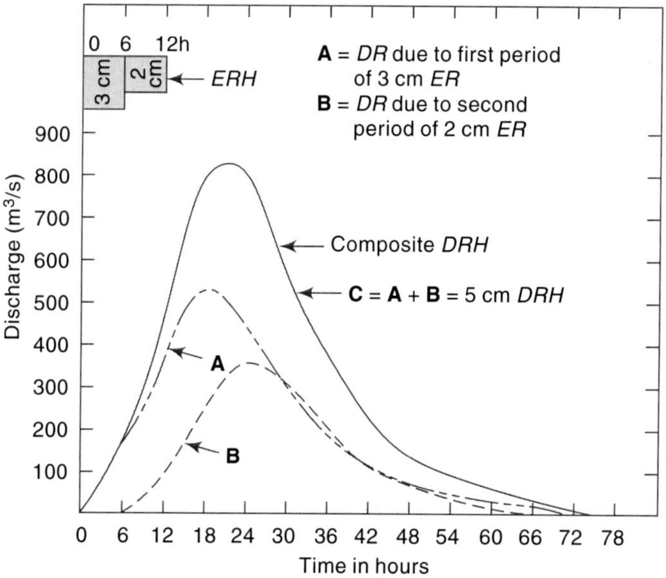

Fig. 6.10(b) ***Principle of Superposition—Example 6.5***

6.6.2 Application of Unit Hydrograph

Using the basic principles of the unit hydrograph, one can easily calculate the *DRH* in a catchment due to a given storm if an appropriate unit hydrograph was available. Let it be assumed that a *D-h* unit-hydrograph and the storm hyetograph are available. The initial losses and infiltration losses are estimated and deducted from the storm hyetograph to obtain the *ERH* (Sec. 6.5). The *ERH* is then divided into *M* blocks of

D-h duration each. The rainfall excess in each *D*-h duration is then operated upon the unit hydrograph successively to get the various *DRH* curves. The ordinates of these *DRH*s are suitably lagged to obtain the proper time sequence and are then collected and added at each time element to obtain the required net *DRH* due to the storm.

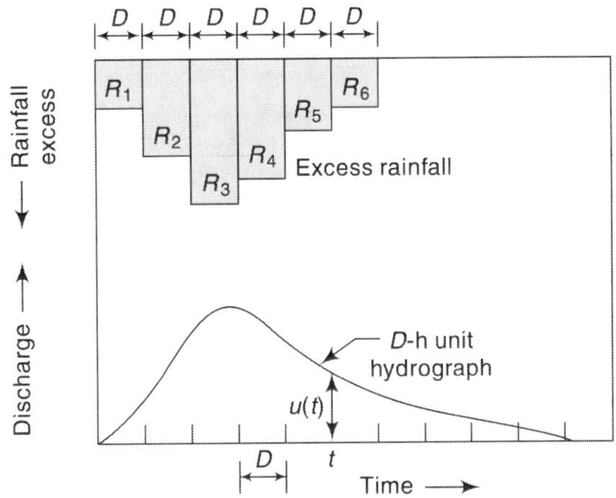

Fig. 6.11 ***DRH due to an ERH***

Consider Fig. 6.11 in which a sequence of *M* rainfall excess values R_1, R_2, …, R_i, … R_m each of duration *D*-h is shown. The line $u\,[t]$ is the ordinate of a *D*-h unit hydrograph at *t* h from the beginning.

The direct runoff due to R_1 at time *t* is

$$Q_1 = R_1 \cdot u[t]$$

The direct runoff due to R_2 at time $(t - D)$ is

$$Q_2 = R_2 \cdot u\,[t - D]$$

Similarly, $$Q_i = R_i \cdot u\,[t - (i - 1)\,D]$$

and $$Q_m = R_m \cdot u\,[t - (M - 1)\,D]$$

Thus, at any time *t*, the total direct runoff is

$$Q_t = \sum_{i=1}^{M} Q_i = \sum_{i=1}^{M} R_i \cdot u[t-(i-1)D] \qquad (6.5)$$

The arithmetic calculations of Eq. (6.5) are best performed in a tabular manner as indicated in Examples 6.5 and 6.6. After deriving the net *DRH*, the estimated base flow is then added to obtain the total flood hydrograph.

Digital computers are extremely useful in the calculations of flood hydrographs through the use of unit hydrograph. The electronic spread sheet (such as *MS* Excel) is ideally suited to perform the *DRH* calculations and to view the final *DRH* and flood hydrographs.

Example 6.6 The ordinates of a 6-hour unit hydrograph of a catchment is given below.

Time (h)	0	3	6	9	12	15	18	24	30	36	42	48
Ordinate of 6-h *UH*	0	25	50	85	125	160	185	160	110	60	36	25
Time (h)	54	60	69									
Ordinate of 6-h *UH*	16	8	0									

Derive the flood hydrograph in the catchment due to the storm given below:

Time from start of storm (h)	0	6	12	18
Accumulated rainfall (cm)	0	3.5	11.0	16.5

The storm loss rate (ϕ — index) for the catchment is estimated as 0.25 cm/h. The base flow can be assumed to be 15 m^3/s at the beginning and increasing by 2.0 m^3/s for every 12 hours till the end of the direct-runoff hydrograph.

Solution

The effective rainfall hyetograph is calculated as in the following table.

The direct runoff hydrograph is next calculated by the method of superposition as indicated in Table 6.5. The ordinates of the unit hydrograph are multiplied by the *ER* values successively. The second and third set of ordinates are advanced by 6 and 12 h respectively and the ordinates at a given time interval added. The base flow is then added to obtain the flood hydrograph shown in Col 8, Table 6.5.

Interval	1st 6 hours	2nd 6 hours	3rd 6 hours
Rainfall depth (cm)	3.5	(11.0 – 3.5) = 7.5	(16.5 – 11.0) = 5.5
Loss @ 0.25 cm/h for 6 h	1.5	1.5	1.5
Effective rainfall (cm)	2.0	6.0	4.0

Table 6.5 ***Calculation of Flood Hydrograph due to a known ERH*—Example 6.6**

Time	Ordinates of *UH*	*DRH* due to 2 cm *ER* Col. 2 × 2.0	*DRH* due to 6 cm *ER* Col. 2 × 6.0 (Advanced by 6 h)	*DRH* due to 4 cm *ER* Col. 2 × 4.0 (Advanced by 12 h)	Ordinates of final *DRH* (Col. 3 + 4 + 5)	Base flow (m^3/s)	Ordinates of flood hydrograph (m^3/s) (Col. 6 + 7)
1	2	3	4	5	6	7	8
0	0	0	0	0	0	15	15
3	25	50	0	0	50	15	65
6	50	100	0	0	100	15	115
9	85	170	150	0	320	15	335
12	125	250	300	0	550	17	567
15	160	320	510	100	930	17	947
18	185	370	750	200	1320	17	1337
(21)	(172.5)	(345)	960	340	1645	(17)	1662
24	160	320	1110	500	1930	19	1949
(27)	(135)	(270)	(1035)	640	1945	19	1964
30	110	220	960	740	1920	19	1939
36	60	120	660	640	1420	21	1441
42	36	72	360	440	872	21	893
48	25	50	216	240	506	23	529
54	16	32	150	144	326	23	349
60	8	16	96	100	212	25	237
66	(2.7)	(5.4)	48	64	117	25	142
69	0	0	—	—	—	—	—
72		0	16	32	48	27	75
75		0	0	—	—	—	—
78		0	0	(10.8)	(11)	27	49
81				0	0	27	27
84						27	27

[***Note:*** Due to the unequal time intervals of unit hydrograph ordinates, a few entries, indicated in parentheses have to be interpolated to complete the table.]

6.7 Derivation of Unit Hydrograph

6.7.1 From Isolated Storms

A number of isolated storm hydrographs caused by short spells of rainfall excess, each of approximately same duration [0.90 to 1.1 *D* h] are selected from a study of the continuously gauged runoff of the stream. For each of these storm hydrographs, the base flow is separated by adopting one of the methods indicated in Sec. 6.4.

The area under each *DRH* is evaluated and the volume of the direct runoff obtained is divided by the catchment area to obtain the depth of *ER*. The ordinates of the various *DRH*s are divided by the respective *ER* values to obtain the ordinates of the unit hydrograph.

Flood hydrographs used in the analysis should be selected to meet the following desirable features with respect to the storms responsible for them.

- The storms should be isolated storms occurring individually.
- The rainfall should be fairly uniform during the duration and should cover the entire catchment area.
- The duration of the rainfall should be 1/5 to 1/3 of the basin lag.
- The rainfall excess of the selected storm should be high. A range of *ER* values of 1.0 to 4.0 cm is sometimes preferred.

A number of unit hydrographs of a given duration are derived by the above method and then plotted on a common pair of axes as shown in Fig. 6.12. Due to the rainfall variations both in space and time and due to storm departures from the assumptions of the unit hydrograph theory, the various unit hydrographs thus developed will not be identical. It is a common practice to adopt a mean of such curves as the unit hydrograph of a given duration for the catchment. While deriving the mean curve, the average of peak flows and time to peaks are first calculated. Then a mean curve of best fit, judged by eye, is drawn through the averaged peak to close on an averaged base length. The volume of *DRH* is calculated and any departure from unity is corrected by adjusting

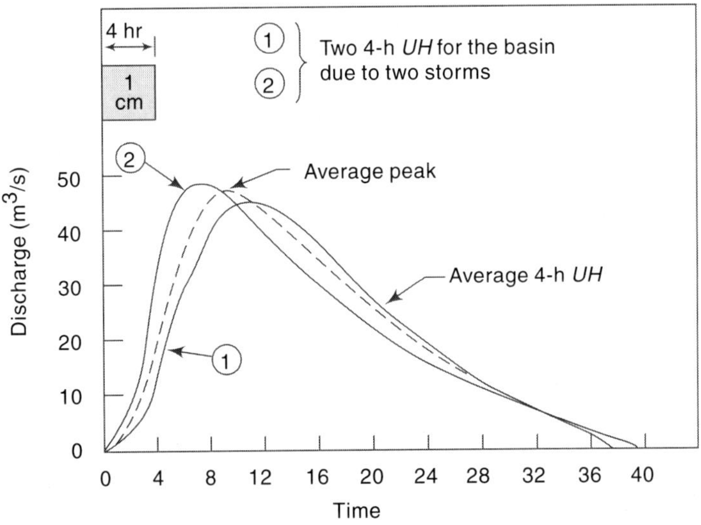

Fig. 6.12 ***Derivation of an Average Unit Hydrograph***

the value of the peak. The averaged *ERH* of unit depth is customarily drawn in the plot of the unit hydrograph to indicate the type and duration of rainfall causing the unit hydrograph.

By definition the rainfall excess is assumed to occur uniformly over the catchment during duration *D* of a unit hydrograph. An. ideal duration for a unit hydrograph is one wherein small fluctuations in the intensity of rainfall within this duration do not have any significant effect on the runoff. The catchment has a damping effect on the fluctuations of the rainfall intensity in the runoff-producing process and this damping is a function of the catchment area. This indicates that larger durations are admissible for larger catchments. By experience it is found that the duration of the unit hydrograph should not exceed 1/5 to 1/3 basin lag. For catchments of sizes larger than 250 km^2 the duration of 6 h is generally satisfactory.

Example 6.7 Following are the ordinates of a storm hydrograph of a river draining a catchment area of 423 km^2 due to a 6-h isolated storm. Derive the ordinates of a 6-h unit hydrograph for the catchment.

Time from start of storm (h)	–6	0	6	12	18	24	30	36	42	48
Discharge (m^3/s)	10	10	30	87.5	115.5	102.5	85.0	71.0	59.0	47.5
Time from start of storm (h)	54	60	66	72	78	84	90	96	102	
Discharge (m^3/s)	39.0	31.5	26.0	21.5	17.5	15.0	12.5	12.0	12.0	

Solution The flood hydrograph is plotted to scale (Fig. 6.13). Denoting the time from beginning of storm as *t*, by inspection of Fig. 6.12,

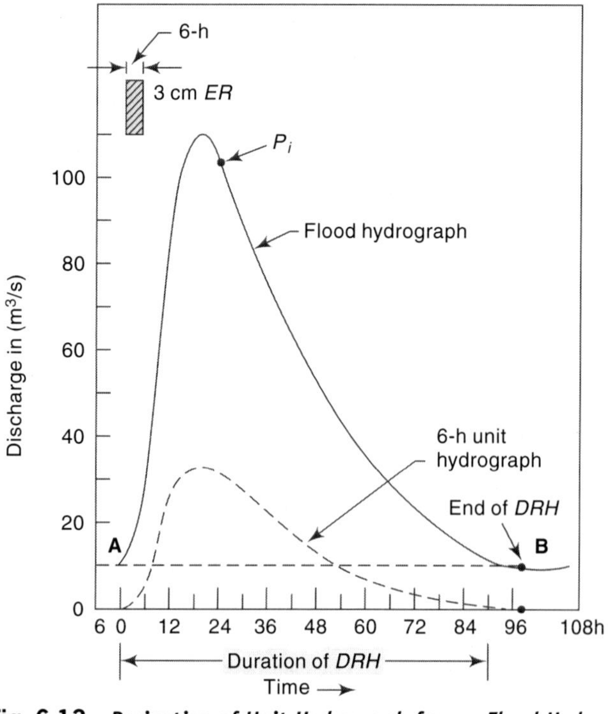

Fig. 6.13 ***Derivation of Unit Hydrograph from a Flood Hydrograph***

A = beginning of *DRH* $t = 0$
B = end of *DRH* $t = 90$ h
P_m = peak $t = 20$ h

Hence,

$$N = (90 - 20) = 70 \text{ h} = 2.91 \text{ days}$$

By Eq. (6.4),

$$N = 0.83\,(423)^{0.2} = 2.78 \text{ days}$$

However, $N = 2.91$ days is adopted for convenience. A straight line joining A and B is taken as the divide line for base-flow separation. The ordinates of *DRH* are obtained by subtracting the base flow from the ordinates of the storm hydrograph. The calculations are shown in Table 6.6.

$$\text{Volume of } DRH = 60 \times 60 \times 6 \times (\text{sum of } DRH \text{ ordinates})$$
$$= 60 \times 60 \times 6 \times 587 = 12.68 \text{ Mm}^3$$

$$\text{Drainage area} = 423 \text{ km}^2 = 423 \text{ Mm}^2$$

$$\text{Runoff depth} = ER \text{ depth} = \frac{12.68}{423} = 0.03 \text{ m} = 3 \text{ cm.}$$

The ordinates of *DRH* (col. 4) are divided by 3 to obtain the ordinates of the 6 h unit hydrograph (see Table 6.6).

Table 6.6 ***Calculation of the Ordinates of a 6 Unit Hydrograph*—Example 6.7**

Time from beginning of storm (h)	Ordinate of flood hydrograph (m^3/s)	Base Flow (m^3/s)	Ordinate of *DRH* (m^3/s)	Ordinate of 6-h unit hydrograph (Col. 4)/3
1	2	3	4	5
–6	10.0	10.0	0	0
0	10.0	10.0	0	0
6	30.0	10.0	20.0	6.7
12	87.5	10.5	77.0	25.7
18	111.5	10.5	101.0	33.7
24	102.5	10.5	101.0	33.7
30	85.0	11.0	74.0	24.7
36	71.0	11.0	60.0	20.0
42	59.0	11.0	48.0	16.0
48	47.5	11.5	36.0	12.0
54	39.0	11.5	27.5	9.2
60	31.5	11.5	20.0	
66	26.0	12.0	14.0	
72	21.5	12.0	9.5	
78	17.5	12.0	5.5	
84	15.0	12.5	2.5	
90	12.5	12.5	0	
96	12.0	12.0	0	
102	12.0	12.0	0	

Example 6.8 (a) The peak of flood hydrograph due to a 3 h duration isolated storm in a catchment is 270 m³/s. The total depth of rainfall is 5.9 cm. Assuming an average infiltration loss of 0.3 cm/h and a constant base flow of 20 m³/s, estimate the peak of the 3-h unit hydrograph (UH) of this catchment.

(b) If the area of the catchment is 567 km², determine the base width of the 3-h unit hydrograph by assuming it to be triangular in shape.

Solution

(a) Duration of rainfall excess = 3 h — Loss @ 0.3 cm/h for 3 h = 0.9 cm

Total depth of rainfall = 5.9 cm — Rainfall excess = 5.9–0.9 = 5.0 cm

Peak flow:

Peak of flood hydrograph = 270 m³/s — Peak of *DRH* = 250 m³/s

Base flow = 20 m³/s

$$\text{Peak of 3 h unit hydrograph} = \frac{\text{peak of } DRH}{\text{rainfall excess}} = \frac{250}{5.0} = 50 \text{ m}^3/\text{s}$$

(b) Let B = base width of the 3 h *UH* in hours.

Volume represented by the area of *UH* = Volume of 1 cm depth over the catchment

Area of *UH* = (Area of catchment × 1 cm)

$$\frac{1}{2} \times B \times 60 \times 60 \times 50 = 567 \times 10^6 \times \frac{1}{100}$$

$$B = \frac{567 \times 10^4}{9 \times 10^4} = 63 \text{ hours}$$

6.7.2 Unit Hydrograph from a Complex Storm

When suitable simple isolated storms are not available, data from complex storms of long duration will have to be used in unit-hydrograph derivation. The problem is to decompose a measured composite flood hydrograph into its component *DRH*s and base flow. A common unit hydrograph of appropriate duration is assumed to exist. This problem is thus the inverse of the derivation of flood hydrograph through use of Eq. (6.5). Consider a rainfall excess made up of three consecutive durations of *D*-h and *ER* values of R_1, R_2 and R_3. Figure 6.14 shows the *ERH*. By base flow separation of the resulting composite flood hydrograph, a composite *DRH* is obtained (Fig. 6.14). Let the ordinates of the composite *DRH* be drawn at a time interval of *D*-h. At various time intervals 1*D*, 2*D*, 3*D*, … from the start of the *ERH*, let the ordinates of the unit hydrograph be u_1, u_2, u_3, … and the ordinates of the composite *DRH* be $Q_1, Q_2, Q_3, \ldots,$

Then

$$Q_1 = R_1 u_1$$

$$Q_2 = R_1 u_2 + R_2 u_1$$

$$Q_3 = R_1 u_3 + R_2 u_2 + R_3 u_1$$

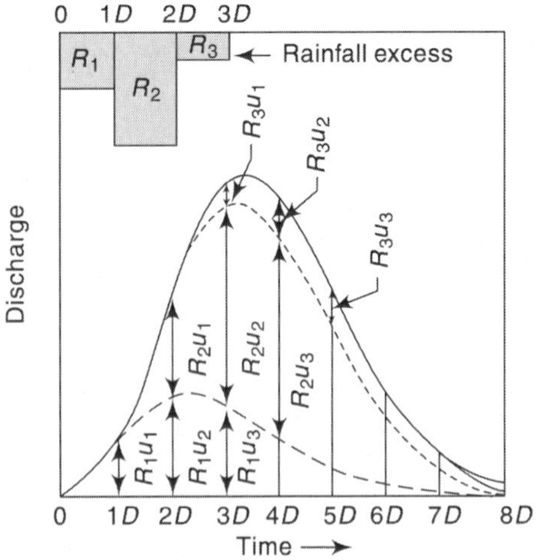

Fig. 6.14 ***Unit Hydrograph from a Complex Storm***

$$Q_4 = R_1 u_4 + R_2 u_3 + R_3 u_2$$
$$Q_5 = R_1 u_5 + R_2 u_4 + R_3 u_3$$
$$\ldots\ldots\ldots \quad (6.6)$$

so on.

From Eq. (6.6), the values of u_1, u_2, u_3, ... can be determined. However, this method suffers from the disadvantage that the errors propagate and increase as the calculations proceed. In the prescnce of errors the recession limb of the derived *D-h* unit hydrograph can contain oscillations and even negative values. Matrix methods with optimisation schemes are available for solving Eq. (6.6) in a digital computer.

6.8 *Unit Hydrographs of Different Durations*

Ideally, unit hydrographs are derived from simple isolated storms and if the duration of the various storms do not differ very much, say within a band of ± 20% *D*, they would all be grouped under one average duration of *D*-h. If in practical applications unit hydrographs of different durations are needed they are best derived from field data. Lack of adequate data normally precludes development of unit hydrographs covering a wide range of durations for a given catchment. Under such conditions a *D* hour unit hydrograph is used to develop unit hydrographs of differing durations *nD*. Two methods are available for this purpose.

- Method of superposition
- The *S*-curve

These are discussed below.

6.8.1 Method of Superposition

If a *D*-h unit hydrograph is available, and it is desired to develop a unit hydrograph of *nD* h, where *n* is an integer, it is easily accomplished by superposing *n* unit hydrographs with each graph separated from the previous on by *D*-h. Figure 6.15 shows three 4-h unit hydrographs *A*, *B* and *C*. Curve *B* begins 4 h after *A* and *C* begins 4-h, after *B*. Thus the combination of these three curves is a *DRH* of 3 cm due to an *ER* of 12-h duration. If the ordinates of this *DRH* are now divided by 3, one obtains a 12-h unit hydrograph. The calculations are easy if performed in a tabular form (Table 6.7).

Example 6.9 Given the ordinates of a 4-h unit hydrograph as below derive the ordinates of a 12-h unit hydrograph for the same catchment.

Time (h)	0	4	8	12	16	20	24	28	32	36	40	44
Ordinate of 4-h UH	0	20	80	130	150	130	90	52	27	15	5	0

Solution

The calculations are performed in a tabular form in Table 6.7. In this

Column 3 = ordinates of 4-h *UH* lagged by 4-h

Column 4 = ordinates of 4-h *UH* lagged by 8-h

Column 5 = ordinates of *DRH* representing 3 cm *ER* in 12-h

Column 6 = ordinates of 12-h *UH* = (Column 5)/3

The 12-h unit hydrograph is shown in Fig. 6.15.

6.8.2 The S-Curve

If it is desired to develop a unit hydrograph of duration mD, where m is a fraction, the method of superposition cannot be used. A different technique known as the S-curve method is adopted in such cases, and this method is applicable for rational values of m.

Table 6.7 ***Calculation of a 12-h Unit Hydrograph from a 4-h Unit Hydrograph—Example 6.9***

Time (h)	Ordinates of 4-h *UH* (m^3/s)			*DRH* of 3 cm in 12-h (m^3/s) (Col. 2+3+4)	Ordinate of 12-h *UH* (m^3/s) (Col. 5)/3
	A	*B* Lagged by 4-h	*C* Lagged by 8-h		
1	2	3	4	5	6
0	0	—	—	0	0
4	20	0	—	20	6.7
8	80	20	0	100	33.3
12	130	80	20	230	76.7
16	150	130	80	360	120.0
20	130	150	130	410	136.7
24	90	130	150	370	123.3
28	52	90	130	272	90.7
32	27	52	90	169	56.3
36	15	27	52	94	31.3
40	5	15	27	47	15.7
44	0	5	15	20	6.7
48		0	5	5	1.7
52			0	0	0

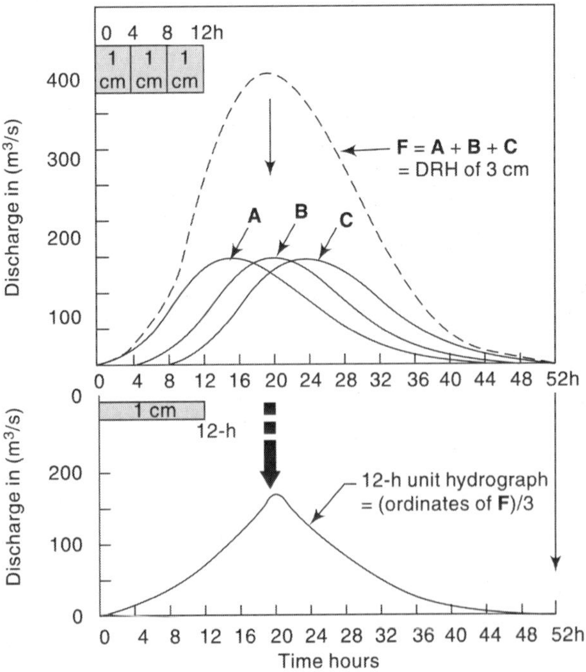

Fig. 6.15 ***Construction of a 12-h Unit Hydrograph from a 4-h Unit Hydrograph—Example 6.9***

The *S-curve*, also known as *S-hydrograph* is a hydrograph produced by a continuous effective rainfall at a constant rate for an infinite period. It is a curve obtained by summation of an infinite series of *D*-h unit hydrographs spaced *D*-h apart. Figure 6.16 shows such a series of *D*-h hydrograph arranged with their starting points *D*-h apart. At any given time the ordinates of the various curves occurring at that time coordinate are summed up to obtain ordinates of the *S*-curve. A smooth curve through these ordinates result in an *S*-shaped curve called *S*-curve.

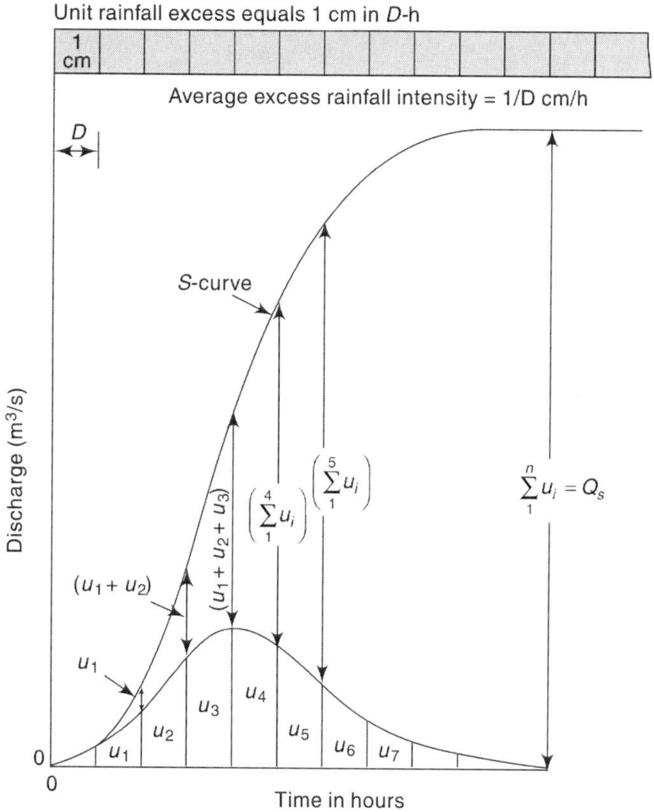

Fig. 6.16 ***S-curve***

This *S*-curve is due to a *D*-h unit hydrograph. It has an initial steep portion and reaches a maximum equilibrium discharge at a time equal to the time base of the first unit hydrograph. The average intensity of *ER* producing the *S*-curve is $1/D$ cm/h and the equilibrium discharge,

$$Q_s = \left(\frac{A}{D} \times 10^4\right) \text{m}^3/\text{h},$$

where A = area of the catchment in km^2 and D = duration in hours of *ER* of the unit hydrograph used in derivin the *S*-curve. Alternatively,

$$Q_s = 2.778 \frac{A}{D} \text{ m}^3/\text{s} \tag{6.7}$$

where A is the km^2 and D is in h. The quantity Q_s represents the maximum rate at which an *ER* intensity of $1/D$ cm/h can drain out of a catchment of area A. In actual construction of an S-curve, it is found that the curve oscillates in the top portion at around the equilibrium value due to magnification and accumulation of small errors in the hydrograph. When it occurs, an average smooth curve is drawn such that it reaches a value Q_s at the time base of the unit hydrograph.

[**Note:** It is desirable to designate the S-curve due to D-hour unit hydrograph as S_D-curve to give an indication that the average rainfall excess of the curve is $(1/D)$ cm/h. It is particularly advantageous when more than one S-curve is used as in such cases the curves would be designated as S_{D1}, S_{D2}, ... etc. to avoid possible confusion and mistakes.]

1. Construction of S–curve

By definition, an S-curve is obtained by adding a string of D-h unit hydrographs each lagged by D-hours from one another. Further, if T_b = base period of the unit hydrograph, addition of only T_b/D unit hydrographs are sufficient to obtain the S-curve. However, an easier procedure based on the basic property of the S-curve is available for the construction of S-curves.

i.e. $$U(t) = S(t) - S(t-D)$$

or $$S(t) = U(t) + S(t-D) \tag{6.8}$$

The term $S(t-D)$ could be called S-curve addition at time t so that

Ordinate of S-curve at any time t = Ordinate of D-h unit hydrograph at time t + S-curve addition at time t

Noting that for all $t \leq D$, $S(t-D) = 0$, Eq. (6.8) provides a simple recursive procedure for computation of S-curve ordinates. The procedure is explained in Example 6.10.

Example 6.10 Derive the S-curve for the 4-h unit hydrograph given below.

Time (h)	0	4	8	12	16	20	24	28
Ordinate of 4-h UH (m^3/s)	0	10	30	25	18	10	5	0

Solution

Computations are shown in Table 6.8. In this table, col. 2 shows the ordinates of the 4-h unit hydrograph. col. 3 gives the S-curve additions and col. 4 gives the ordinates of the S-curve. The sequence of entry in col. 3 is shown by arrows. Values of entries in col. 4 is obtained by using Eq. (6.8), i.e. by summing up of entries in col. 2 and col. 4 along each row.

Table 6.8 ***Construction of S-curve—Example 6.10***

Time in hours	Ordinate of 4-h *UH*	*S*-curve addition (m^3/s)	S_4-curve ordinate (m^3/s). (col. 2 + col. 3)
1	2	3	4
0	0		0
4	10	0	10
8	30	10	40
12	25	40	65
16	18	65	83
20	10	83	93
24	5	93	98
28	0	98	98

At $i = 4$ hours; Ordinate of 4-hUH = 10 m^3/s.

S-curve addition = ordinate of 4-h UH @{t = (4–4) = 0 hours} = 0

Hence S-curve ordinate Eq. (6.8) = 10 + 0 = 10 m^3/s,

At $t = 8$ hours; Ordinate of 4-hUH = 30 m^3/s.

S-curve addition = ordinate of 4-hUH @{t = (8–4) = 4 hours} = 10 m^3/s

Hence S-curve ordinate by Eq. (6.8) = 30 + 10 = 40 m^3/s.

At $t = 12$ hours; Ordinate of 4-hUH = 25 m^3/s.

S-curve addition = ordinate of 4-hUH @{t = (12–4) = 48 hours} = 40 m^3/s

Hence S-curve ordinate by Eq. (6.8) = 25 + 40 = 65 m^3/s.

This calculation is repeated for all time intervals till t = base width of UH = 28 hours. Plots of the 4-h UH and the derived S-curve are shown in Fig. 6.17.

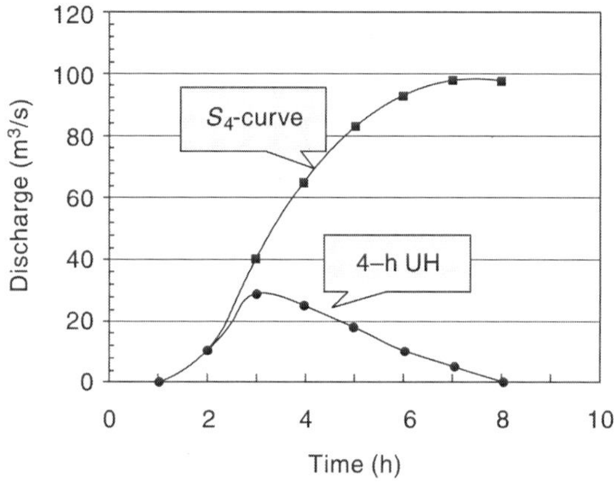

Fig. 6.17 ***Construction of S_4-curve—(Example 6.10)***

2. Derivation of T-hour Unit Hydrograph

Consider two D-h S-curves A and B displaced by T-h (Fig. 6.18). If the ordinates of B are subtracted from that of A, the resulting curve is a DRH produced by a rainfall excess

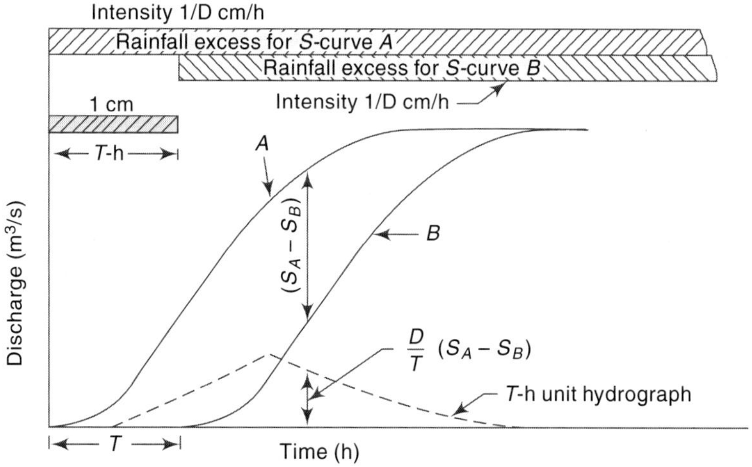

Fig. 6.18 ***Derivation of a T-h Unit Hydrograph by S-curve Lagging Method***

of duration T-h and magnitude $\left(\frac{1}{D} \times T\right)$ cm. Hence, if the ordinate differences of A and B, i.e. $(S_A - S_B)$ are divided by T/D, the resulting ordinates denote a hydrograph due to an *ER* of 1 cm and of duration T-h, i.e. a *T-h* unit hydrograph. The derivation of a T-h unit hydrograph as above can be achieved either by graphical means or by arithmetic computations in a tablar form as indicated in Example 6.11.

Example 6.11 Solve Example 6.9 by the *S*-Curve method.

Solution

Computations are shown in Table 6.9. Column 2 shows the ordinates of the 4-h unit hydrograph. Column 3 gives the *S*-curve additions and Column 4 the *S*-curve ordinates. The sequence of additions are shown by arrows. At t = 4 h, ordinate of the 4-h *UH* = ordinate of the *S*-curve. This value becomes the *S*-curve addition at $t = 2 \times 4 = 8$ h. At this t = 8 h, the ordinate of *UH* (80) + *S*-curve addition (20) = *S*-curve ordinate (100). The *S*-curve addition at 3 × 4 = 12 h is 100, and so on. Column 5 shows the *S*-curve lagged by 12 h. Column 6 gives the ordinate of *DRH* of (T/D) = 3 cm. Ordinates shown in Column 6 are divided by $(T/D = 3)$ to obtain the ordinates of the 2-h unit hydrograph shown in Column 7.

Table 6.9 ***Determination of a 12-h Unit Hydrograph by S-Curve Method—Example 6.11***

Time (h)	Ordinate of 4-h *UH* (m^3/s)	*S*-curve addition (m^3/s)	*S*-curve ordinate (m^3/s) (Col. 2 + Col. 3)	*S*-curve lagged by 12 h (m^3/s)	(Col. 4– Col. 5)	$\frac{\text{Col. 6}}{(12/4)}$ = 12-h *UH* ordinates (m^3/s)
1	2	3	4	5	6	7
0	0	—	0	—	0	0
4	20	0	20	—	20	6.7
8	80	20	100	—	100	33.3
12	130	100	230	0	230	76.7
16	150	230	380	20	360	120.0
20	130	380	510	100	410	136.7
24	90	510	600	230	370	123.3
28	52	600	652	380	272	90.7
32	27	652	679	510	169	56.3
36	15	679	694	600	94	31.3
40	5	694	699	652	47	15.7
44	0	699	699	679	20	6.7
48		699	699	694	5	1.7
52			699	699	0	0

Example 6.12 Ordinates of a 4-h unit hydrograph are given. Using this derive the ordinates of a 2-h unit hydrograph for the same catchment.

Time (h)	0	4	8	12	16	20	24	28	32	36	40	44
Ordinate or 4-h *UH* (m^3/s)	0	20	80	130	150	130	90	52	27	15	5	0

Solution

In this case, the time interval of the ordinates of the given unit hydrograph should be at least 2 h. As the given ordinates are at 4-h intervals, the unit-hydrograph is plotted and its ordinates at 2-h intervals determined. The ordinates are shown in column 2 of Table 6.10. The *S*-curve additions and *S*-curve ordinates are shown in columns 3 and 4 respectively. First, the *S*-curve ordinates corresponding to the time intervals equal to successive durations of the given unit hydrograph (in this case at 0, 4, 8, 12 … *h*) are determined by following the method of Example 6.11. Next, the ordinates at intermediate intervals (viz. at $t = 2, 6, 10, 14 \ldots$ h) are determined by having another series of *S*-curve additions. The sequence of these are shown by distinctive arrows in Table 6.9. To obtain a 2-h unit hydrograph the *S*-curve is lagged by 2 h (column 5) and this is subtracted from column 4 and the results listed in column 6. The ordinates in column 6 are now divided by $T/D = 2/4 = 0.5$, to obtain the required 2-h unit hydrograph ordinates, shown in column 7.

Table 6.10 ***Determination of 2-h Unit Hydrograph from a 4-h Unit Hydrograph—Example 6.12***

Time (h)	Ordinate of 4-h *UH* (m^3/s)	S_4-curve addition (m^3/s)	S_4-curve ordinate (Col. (2) + (3)) (m^3/s)	S_4-curve lagged by 2 h	(Col. (4) – Col. (5)) *DRH* of (2/4) = 0.5 cm	2-h *UH* ordinates Col. (6) / (2/4) (m^3/s)
1	2	3	4	5	6	7
0	0	—	0	—	0	0
2	8	—	8	0	8	16
4	20	0	20	8	12	24
6	43	8	51	20	31	62
8	80	20	100	51	49	98
10	110	51	161	100	61	122
12	130	100	230	161	69	138
14	146	161	307	230	77	154
16	150	230	380	307	73	146
18	142	307	449	380	69	138
20	130	380	510	449	61	122
22	112	449	561	510	51	102
24	90	510	600	561	39	78
26	70	561	631	600	31	62
28	52	600	652	631	21	42
30	38	631	669	652	17	34
32	27	652	679	669	10	20
34	20	669	689	679	10	(20)15
36	15	679	694	689	5	(10)10
38	10	689	699	694	5	(10)6
40	5	694	699	699	(0)	(0)3
42	2	699	701	699	(2)	(4)0
44	0	699	699	701	(–2)	(–4)0

Final adjusted values are given in col. 7.

Unadjusted values are given in parentheses.

The errors in interpolation of unit hydrograph ordinates often result in oscillation of *S*-curve at the equilibrium value. This results in the derived *T-h* unit hydrograph having an abnormal sequence of discharges (sometimes even negative values) at the tail end. This is adjusted by fairing the *S*-curve and also the resulting *T*-h unit-hydrograph by smooth curves. For example, in the present example the 2-h unit hydrograph ordinates at time > 36-h are rather abnormal. These values are shown in parentheses. The adjusted values are entered in column 7.

6.9 *Use and Limitations of Unit Hydrograph*

As the unit hydrographs establish a relationship between the *ERH* and *DRH* for a catchment, they are of immense value in the study of the hydrology of a catchment. They are of great use in (i) the development of flood hydrographs for extreme rainfall magnitudes for use in the design of hydraulic structures, (ii) extension of flood-flow records based on rainfall records, and (iii) development of flood forecasting and warning systems based on rainfall.

Unit hydrographs assume uniform distribution of rainfall over the catchment. Also, the intensity is assumed constant for the duration of the rainfall excess. In practice, these two conditions are never strictly satisfied. Non-uniform areal distribution and variation in intensity within a storm are very common. Under such conditions unit hydrographs can still be used if the areal distribution is consistent between different storms. However, the size of the catchment imposes an upper limit on the applicability of the unit hydrograph. This is because in very large basins the centre of the storm can vary from storm to storm and each can give different *DRH*s under otherwise identical situations. It is generally felt that about 5000 km^2 is the upper limit for unit-hydrograph use. Flood hydrographs in very large basins can be studied by dividing them into a number of smaller subbasins and developing *DRH*s by the unit-hydrograph method. These *DRH*s can then be routed through their respective channels to obtain the composite *DRH* at the basin outlet.

There is a lower limit also for the application of unit hydrographs. This limit is usually taken as about 200 ha. At this level of area, a number of factors affect the rainfall-runoff relationship and the unit hydrograph is not accurate enough for the prediction of *DRH*.

Other limitations to the use of unit hydrographs are the following:

- Precipitation must be from rainfall only. Snow-melt runoff cannot be satisfactory represented by unit hydrograph.
- The catchment should not have unusually large storages in terms of tanks, ponds, large flood-bank storages, etc. which affect the linear relationship between storage and discharge.
- If the precipitation is decidedly non-uniform, unit hydrographs cannot be expected to give good results.

In the use of unit hydrographs, very accurate reproduction of results should not be expected. Variations in the hydrograph base of as much as ±20% and in the peak discharge by ±10% are normally considered acceptable.

6.10 *Duration of the Unit Hydrograph*

The choice of the duration of the unit hydrograph depends on the rainfall records. If recording raingauge data are available any convenient time depending on the size of the basin can be used. The choice is not much if only daily rainfall records are available. A rough guide for the choice of duration D is that it should not exceed the least of (i) the time of rise, (ii) the basin lag, and (iii) the time of concentration. A value of D equal to about 1/4 of the basin lag is about the best choice. Generally, for basins with areas more than 1200 km^2 a duration D = 12 hours is preferred.

6.11 *Distribution Graph*

The distribution graph introduced by Bernard (1935) is a variation of the unit hydrograph. It is basically a D-h unit hydrograph with ordinates showing the percentage of the surface runoff occurring in successive periods of equal time intervals of D-h. The duration of the rainfall excess (D-h) is taken as the unit interval and distribution-graph ordinates are indicated at successive such unit intervals. Figure 6.19 shows a typical 4-h distribution graph. Note the ordinates plotted at 4-h intervals and the total area under the distribution graph adds up to 100%. The use of the distribution graph to generate a *DRH* for a known *ERH* is exactly the same as that of a unit hydrograph (Example 6.13). Distribution graphs are useful in comparing the runoff characteristics of different catchments.

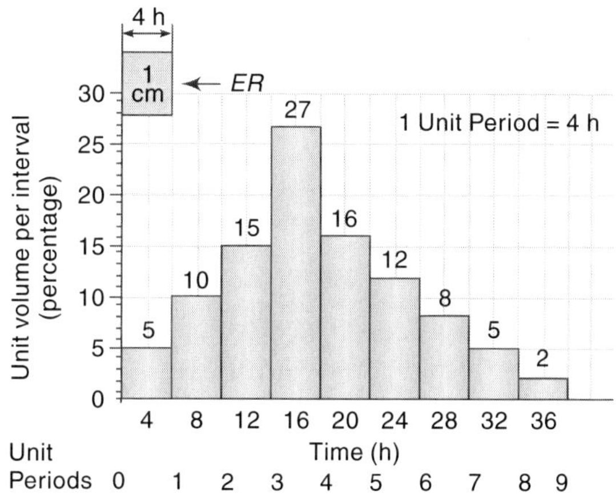

Fig. 6.19 *Four-hour Distribution Graph*

Example 6.13 A catchment of 200 hectares area has rainfalls of 7.5 cm, 2.0 cm and 5.0 cm in three consecutive days. The average ϕ index can be assumed to be 2.5 cm/day. Distribution-graph percentages of the surface runoff which extended over 6 days for every rainfall of 1-day duration are 5, 15, 40, 25, 10 and 5. Determine the ordinates of the discharge hydrograph by neglecting the base flow.

Solution

The calculations are performed in a tabular form in Table 6.11.

Table 6.11 *Calculation of DRH using Distribution Graph—Example 6.13*

Time interval (days)	Rain-fall (cm)	Infiltra-tion loss (cm)	Effective rainfall (cm)	Average distri-buition ratio (per-cent)	Distributed runoff for rain-fall excess of 5 cm	0	2.5 cm	Runoff cm	m³/s × 10⁻²
0–1	7.5	2.5	5.0	5	0.250	0		0.250	5.79
1–2	2.0	2.5	0	15	0.750	0	0	0.750	17.36
2–3	5.0	2.5	2.5	40	2.000	0	0.125	2.750	49.19

(Contd.)

(Table 6.11 Contd.)

3–4				25	1.250	0	0.375	2.125	37.62
4–5				10	0.500	0	1.000	1.625	34.72
5–6				5	0.250	0	0.625	1.500	20.25
6–7				0	0	0	0.250	0.875	5.79
7–8						0	0.125	0.250	2.89
8–9							0	0.125	0

[Runoff of 1 cm in 1 day = $\frac{200 \times 100 \times 100}{86400 \times 100}$ m^3/s for 1 day = 0.23148 m^3/s for 1 day]

(The runoff ordinates are plotted at the mid-points of the respective time intervals to obtain the *DRH*)

6.12 *Synthetic Unit Hydrograph*

6.12.1 Introduction

To develop unit hydrographs to a catchment, detailed information about the rainfall and the resulting flood hydrograph are needed. However, such information would be available only at a few locations and in a majority of catchments, especially those which are at remote locations, the data would normally be very scanty. In order to construct unit hydrographs for such areas, empirical equations of regional validity which relate the salient hydrograph characteristics to the basin characteristics are available. Unit hydrographs derived from such relationships are known as *synthetic-unit hydrographs*. A number of methods for developing synthetic-unit hydrographs are reported in literature. It should, however, be remembered that these methods being based on empirical correlations are applicable only to the specific regions in which they were developed and should not be considered as general relationships for use in all regions.

6.12.2 Snyder's Method

Snyder (1938), based on a study of a large number of catchments in the Appalachian Highlands of eastern United States developed a set of empirical equations for synthetic-unit hydrographs in those areas. These equations are in use in the *USA*, and with some modifications in many other countries, and constitute what is known as *Snyder's synthetic-unit hydrograph*.

The most important characteristic of a basin affecting a hydrograph due to a storm is *basin lag*. While actually basin lag (also known as *lag time*) is the time difference between the centroid of the input (rainfall excess) and the output (direct runoff hydrograph), because of the difficulty in determining the centroid of the direct runoff hydrograph (*DRH*) it is defined for practical purposes as the elapsed time between the centroid of rainfall excess and peak of *DRH*. Physically, lag time represents the mean time of travel of water from all parts of the watershed to the outlet during a given storm. Its value is determined essentially on the topographical features, such as the size, shape, stream density, length of main stream, slope, land use and land cover. The modified definition of basin time is very commonly adopted in the derivation of synthetic unit hydrographs for a given watershed.

The first of the Snyder's equation relates the basin lag t_p, defined as the time interval from the mid-point of rainfall excess to the peak of the unit hydrograph (Fig. 6.20), to the basin characteristics as

$$t_p = C_t(LL_{ca})^{0.3} \qquad (6.9)$$

where t_p = basin lag in hours

L = basin length measured along the water course from the basin divide to the gauging station in km

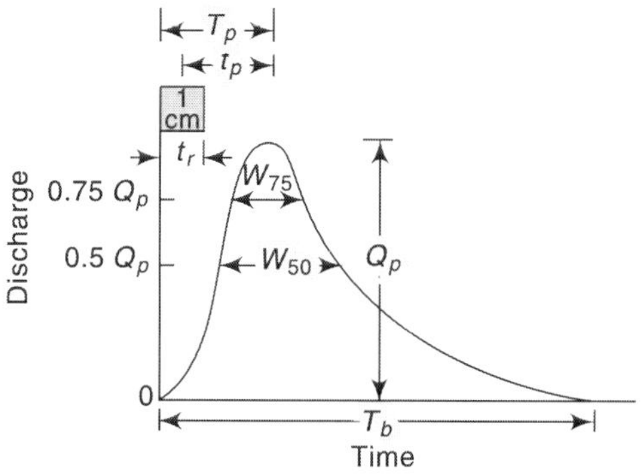

Fig. 6.20 ***Elements of a Synthetic Unit Hydrograph***

L_{ca} = distance along the main water course from the gauging station to a point opposite to the watershed centroid in km

C_t = a regional constant representing watershed slope and storage effects.

The value of C_t in Snyder's study ranged from 1.35 to 1.65. However, studies by many other investigators have shown that C_t depends upon the region under study and wide variations with the value of C_t ranging from 0.3 to 6.0 have been reported[6].

Linsley et al.[5] found that the basin lag t_p is better correlated with the catchment parameter $\left(\frac{LL_{ca}}{\sqrt{S}}\right)$ where S = basin slope. Hence, a modified form of Eq. (6.9) was suggested by them as

$$t_p = C_{tL}\left(\frac{LL_{ca}}{\sqrt{S}}\right)^n \qquad (6.10)$$

where C_{tL} and n are basin constants. For the basins in the *USA* studied by them, n was found to be equal to 0.38 and the values of C_{tL} were 1.715 for mountainous n drainage areas, 1.03 for foothill drainage areas and 0.50 for valley drainage areas.

Snyder adopted a standard duration t_r hours of effective rainfall given by

$$t_r = \frac{t_p}{5.5} \qquad (6.11)$$

The peak discharge Q_{ps} (m^3/s) of a unit hydrograph of *standard duration* t_r h is given by Snyder as

$$Q_{ps} = \frac{2.78\,C_p\,A}{t_p} \qquad (6.12)$$

where A = catchment area in km^2 and C_p = a regional constant. This equation is based on the assumption that the peak discharge is proportional to the average discharge of $\left(\frac{1\text{ cm} \times \text{catchment area}}{\text{duration of rainfall excess}}\right)$. The values of the coefficient C_p range from 0.56 to 0.69 for Snyder's study areas and is considered as an indication of the retention and storage capacity of the watershed. Like C_t, the values of C_p also vary quite considerably depending on the characteristics of the region and values of C_p in the range 0.31 to 0.93 have been reported.

If a non-standard rainfall duration t_R h is adopted, instead of the standard value t_r to derive a unit hydrograph the value of the basin lag is affected. The modified basin lag is give by

$$t_p' = t_p + \frac{t_R - t_r}{4} = \frac{21}{22} t_p + \frac{t_R}{4} \tag{6.13}$$

here t_p' = basin lag in hours for an effective duration of t_R h and t_p is as given by Eq. (6.9) or (6.10). The value of t_p' must be used instead of t_p in Eq. (6.11). Thus the peak discharge for a nonstandard *ER* of duration t_R is in m^3/s

$$Q_p = 2.78 C_p A / t_p' \tag{6.12a}$$

Note that when $t_R = t_r$

$$Q_p = Q_{ps}$$

The time base of a unit hydrograph (Fig. 6.20) is given by Synder as

$$T_b = 3 + \frac{t_p'}{8} \text{ days} = (72 + 3 t_p') \text{ hours} \tag{6.14}$$

where T_b = time base. While Eq. (6.14) gives reasonable estimates of T_b for large catchments, it may give excessively large values of the time base for small catchments. Taylor and Schwartz[1] recommend

$$T_b = 5 \left(t_p' + \frac{t_R}{2} \right) \text{ hours} \tag{6.15}$$

with t_b (given in h) taken as the next larger integer value divisible by t_R, i.e. T_b is about five times the time-to-peak.

To assist in the sketching of unit hydrographs, the widths of unit hydrographs at 50 and 75% of the peak (Fig. 6.20) have been found for *US* catchments by the US Army Corps of Engineers. These widths (in time units) are correlated to the peak discharge intensity and are given by

$$W_{50} = \frac{5.87}{q^{1.08}} \tag{6.16}$$

and

$$W_{75} = W_{50}/1.75 \tag{6.17}$$

where

W_{50} = width of unit hydrograph in h at 50% peak discharge

W_{75} = width of unit hydrograph in h at 75% peak discharge

$q = Q_p/A$ = peak discharge per unit catchment area in $m^3/s/km^2$

Since the coefficients C_t and C_p vary from region to region, in practical applications it is advisable that the value of these coefficients are determined from known unit hydrographs of a meteorologically homogeneous catchment and then used in the basin under study. This way Snyder's equations are of use in scaling the hydrograph information from one catchment to another similar catchment.

Example 6.14 Two catchments *A* and *B* are considered meteorologically similar. Their catchment characteristics are given below.

Catchment A	Catchment B
L= 30 km	L = 45 km
L_{ca}= 15 km	L_{ca} = 25 km
A = 250 km^2	A = 400 km^2

For catchment A, a 2-h unit hydrograph was developed and was found to have a peak discharge of 50 m^3/s. The time to peak from the beginning of the rainfall excess in this unit hydrograph was 9.0 h. Using Snyder's method, develop a unit hydrograph for catchment B.

Solution

For Catchment A:

$$t_R = 2.0 \text{ h}$$

Time to peak from beginning of ER

$$T_p = \frac{t_R}{2} + t_p' + = 9.0 \text{ h}$$

$\therefore$ $$t_p' = 8.0 \text{ h}$$

From Eq. (6.13),

$$t_p' = \frac{21}{22} t_p + \frac{t_R}{4} = \frac{21}{22} t_p + 0.5 = 8.0$$

$$t_p = \frac{7.5 \times 22}{21} = 7.857 \text{ h}$$

From Eq. (6.9),

$$t_p = C_t (L\, L_{ca})^{0.3} \qquad 7.857 = C_t (30 \times 15)^{0.3} \qquad C_t = 1.257$$

From Eq. (6.12a),

$$Q_p = 2.78\, C_p A / t_p' \qquad 50 = 2.78 \times C_p \times 250/8.0 \qquad C_p = 0.576$$

For Catchment B: Using the values of C_t = 1.257 and C_p = 0.576 in catchment B, the parameters of the synthetic-unit hydrograph for catchment B are determined. From Eq. (6.9),

$$t_p = 1.257\,(45 \times 25)^{0.3} = 10.34 \text{ h}$$

By Eq. (6.11),

$$t_r = \frac{10.34}{5.5} = 1.88 \text{ h}$$

Using t_R = 2.0 h, i.e. for a 2-h unit hydrograph, by Eq. (6.12),

$$t_p' = 10.34 \times \frac{21}{22} + \frac{2.0}{4} = 10.37 \text{ h}$$

By Eq. (6.12a),

$$Q_p = \frac{2.78 \times 0.576 \times 400}{10.37} = 61.77 \text{ m}^3/\text{s, say } 62 \text{ m}^3/\text{s}$$

From Eq. (6.16),

$$W_{50} = \frac{5.87}{(62/400)^{1.08}} = 44 \text{ h}$$

By Eq. (6.17),

$$W_{75} = \frac{44}{1.75} = 25 \text{ h}$$

Time base: From Eq. (6.14), $T_b = 72 + (3 \times 10.37) = 103$ h
From Eq. (6.14), $T_b = 5\,(10.37 + 10) \approx 58$ h

Considering the values of W_{50} and W_{75} and noting that the area of catchment *B* is rather small, $T_b \approx 58$ h is more appropriate in this case.

Finalising of Synthetic-Unit Hydrograph

After obtaining the values of Q_p, t_R, t_p', W_{75}, W_{50} and T_b from Snyder's equations, a tentative unit hydrograph is sketched and *S*-curve is then developed and plotted. As the ordinates of the unit hydrograph are tentative, the *S*-curve thus obtained will have kinks. These are then smoothened and a logical pattern of the *S*-curve is sketched. Using this *S*-curve t_R hour unit hydrograph is then derived back. Further, the area under the unit hydrograph is checked to see that it represents 1 cm of runoff. The procedure of adjustments through the *S*-curve is repeated till satisfactory results are obtained. It should be noted that out of the various parameters of the synthetic unit hydrograph the least accurate will be the time base T_b and this can be changed to meet other requirements.

6.12.3 SCS Dimensionless Unit Hydrograph

Dimensionless unit hydrographs based on a study of a large number of unit hydrographs are recommended by various agencies to facilitate construction of synthetic unit hydrographs. A typical dimensionless unit hydrograph developed by the *US* Soil Conservation Services (*SCS*) is shown in Fig. 6.21(a). In this, the ordinate is (Q/Q_p) which is the discharge Q expressed as a ratio to the peak discharge Q_p, and the abscissa is (t/T_p), which is the time t expressed as a ratio of the time to peak T_p. By definition, $Q/Q_p = 1.0$ when $t/T_p = 1.0$. The coordinates of the *SCS* dimensionless unit hydrograph is given in Table 6.12 for use in developing a synthetic unit hydrograph in place of Snyder's equations (6.14) through (6.17).

Table 6.12 ***Coordinates of SCS Dimensionless Unit Hydrograph***[4]

t/T_p	Q/Q_p	t/T_p	Q/Q_p	t/T_p	Q/Q_p
0.0	0.000	1.10	0.980	2.80	0.098
0.1	0.015	1.20	0.92	3.00	0.074
0.2	0.075	1.30	0.840	3.50	0.036
0.3	0.160	1.40	0.750	4.00	0.018
0.4	0.280	1.50	0.660	4.50	0.009
0.5	0.430	1.60	0.560	5.00	0.004
0.6	0.600	1.80	0.420		
0.7	0.770	2.00	0.320		
0.8	0.890	2.20	0.240		
0.9	0.970	2.40	0.180		
1.0	1.000	2.60	0.130		

6.12.4 SCS Triangular Unit Hydrograph

The value of Q_p and T_p may be estimated using a simplified model of a triangular unit hydrograph (Fig. 6.21(b)) suggested by *SCS*. This triangular unit hydrograph has the same percentage of volume on the rising side as the dimensionless unit hydrograph of Fig. 6.21(a).

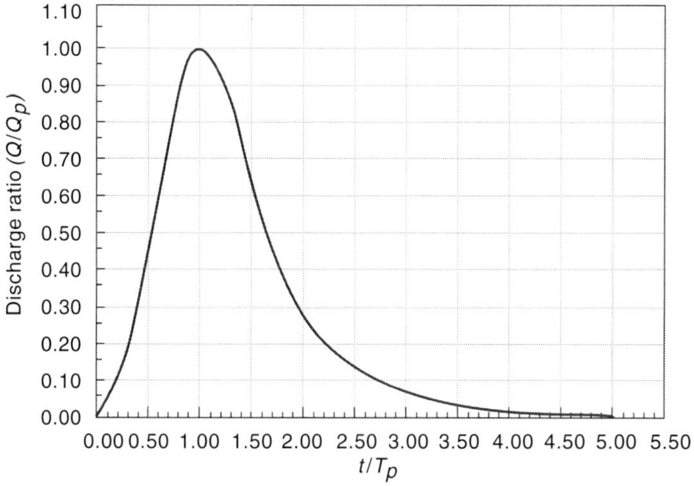

Fig. 6.21(a) ***Dimensionless SCS Unit Hydrograph***

In Fig. 6.21(b),

Q_p = peak discharge in m^3/s

t_r = duration of effective rainfall

T_p = time of rise = time to peak $= (t_r/2) + t_p$

t_p = lag time

T_b = base length

SCS suggests that the time of recession = $(T_b - T_p) = 1.67\ T_p$

Thus $\quad T_b = 2.67\ T_p$

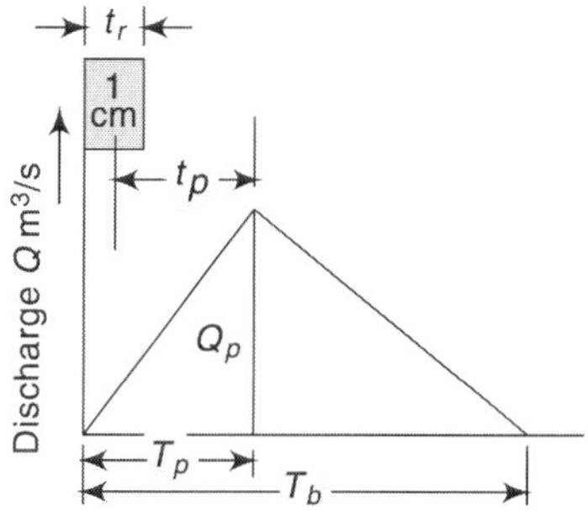

Fig. 6.21(b) ***SCS Triangular Unit Hydrograph***

Since the area under the unit hydrograph is equal to 1 cm,

If A = area of the watershed in km^2,

$$\frac{1}{2}Q_F \times (2.67T_p) \times (3600) = \frac{1}{100} \times A \times 10^6$$

$$Q_p = \frac{2A \times 10^4}{3600 \times 2.67T_p} = 2.08\frac{A}{T_p} \tag{6.18}$$

Further on the basis of a large number of small rural watersheds, *SCS* found that $t_p \approx 0.6\ t_c$, where t_c = time of concentration (described in detail in Sec. 7.2, Chapter 7).

Thus
$$T_p = \left(\frac{t_r}{2} + 0.6\,t_c\right) \tag{6.19}$$

The *SCS* triangular unit hydrograph is a popular method used in watershed development activities, especially in small watersheds.

Example 6.15 Develop a 30-minute *SCS* triangular unit hydrograph for a watershed of area 550 ha and time of concentration of 50 minutes.

Solution

$$A = 550 \text{ ha} = 5.5 \text{ km}^2$$

$$t_r = 30 \text{ min} = 0.50 \text{ h} \qquad t_c = 50 \text{ min} = 0.833 \text{ h}$$

lag time $t_p = 0.6\, t_c = 0.6 \times 0.833 = 0.50 \text{ h}$

$$T_p = \left(\frac{t_r}{2} + t_p\right) = 0.25 + 0.50 = 0.75 \text{ h}$$

$$Q = 2.08 \frac{A}{T_p} = 2.08 \times \frac{5.5}{0.75} = 15.25 \text{ m}^3\text{/s}$$

$$T_b = 2.67\, T_p = 2.67 \times 0.75 = 2.00 \text{ h}$$

The derived triangular unit hydrograph is shown in Fig. 6.22

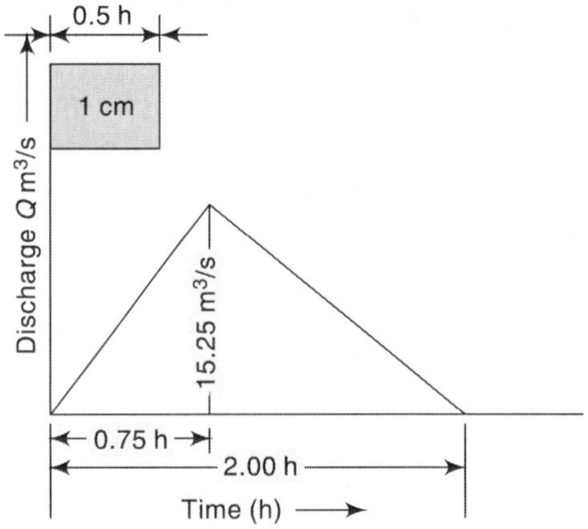

Fig. 6.22 ***Triangular Unit Hydrograph—Example 6.15***

Example 6.16 A 6-h unit hydrograph of a basin is triangular in shape and has a peak of 60 m^3/s. The peak occurs at 18 hours from the start and the base length of the unit hydrograph is 54 hours.

(a) What is the area of the catchment represented by the unit hydrograph?

(b) Calculate the equilibrium discharge of the S_6—curve of the basin.

(c) Calculate the flood hydrograph due to a rainfall excess of 2.5 cm in the first 6 hour period and 3.5 cm in the next 6 hour interval. The base flow can be assumed to be 15 m^3/s constant through out.

Solution

Figure 6.23 shows schematically the 6-h *UH* of the catchment.

(a) Volume of runoff due to 1 cm of rainfall excess = Area of *UH*

If A = area of the basin km^2

$$(A \times 10^6) \times \frac{1}{100} = \frac{1}{2} \times 54 \times (3600) \times 60$$

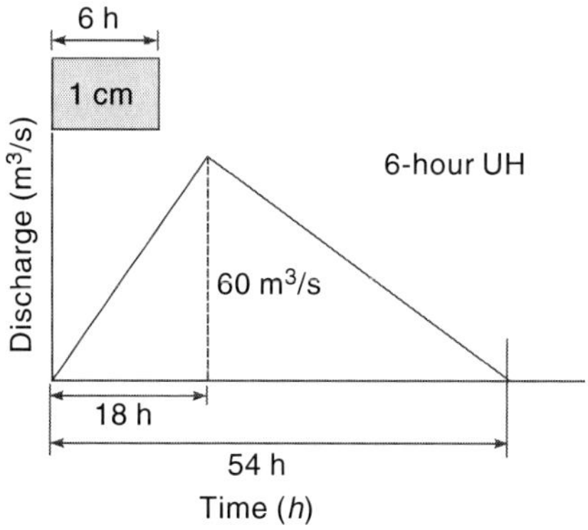

Fig. 6.23 ***6-hour triangular UH of Example 6.16***

$$A = \frac{5{,}832{,}000}{10^4} = 583.2 \text{ km}^2$$

(b) Q_s = equilibrium discharge of S_6 – curve by Eq. 6.7 = $\dfrac{2.778A}{D}$ where A is in km^2 and D is in hours. In the present case

$$Q_s = \frac{2.778A}{D} = \frac{2.778 \times 583.2}{6} = 270 \text{ m}^3/\text{s}$$

(c) Calculations of the flood hydrograph are shown in tabular form in Table 6.13.

Table 6.13 ***Calculation of Flood Hydrograph – Example 6.16***

1	2	3	4	5	6
Time (h)	6-h *UH* ordinates (m³/s)	Direct runoff due to ER=2.5 cm. [Col.2× 2.5]	Direct runoff due to ER = 3.5 cm. [Col.2× 3.5]	Base flow (m³/s)	Ordinates of Flood Hydrograph (m³/s) [cols. 3+4+5]
0	0	0	0	15	15
6	20	50	0	15	65
12	40	100	70	15	185
18	60	150	140	15	305
24	50	125	210	15	350
30	40	100	175	15	290
36	30	75	140	15	230
42	20	50	105	15	170
48	10	25	70	15	110
54	0	0	35	15	50
60			0	15	15

Example 6.17 The one-hour unit hydrograph of a small rural catchment is triangular in shape with a peak value of 3.6 m^3/s occurring at 3 hours from the start and a base time of 9 hours. Following urbanisation over a period of two decades the infiltration index φ has decreased from 0.7 cm/h to 0.4 cm/h. Also, the one-hour unit hydrograph has now a peak of 6.0 m^3/s at 1.5 hours from the start. The time base has shortened to 6 hours. If a design storm has intensities of 4 cm/h and 3.0 cm/h for two consecutive one-hour intervals, estimate the percentage increase in the peak storm runoff and in the volume of direct runoff due to urbanisation.

Solution

1. The effective rainfall hyetograph is first calculated for the cases of (a) before and (b) after urbanisation and is shown together with the respective 1-hour unit hydrograph in Fig. 6.24(a) and Fig. 6.24(b). The *ER* values are 3.3 cm and 2.3 cm in two consecutive 1-h intervals for before urbanisation situation and they are 3.6 cm and 2.6 cm in two consecutive one-hour time intervals for after urbanisation situation
2. The calculation of the *DRH* due to two consecutive 1-h *ER* values are shown in Table 6.14(a).Both the situations under considerations viz., before and after urbanisation cases are shown in the same table.
3. In Table 6.14(b), the results are summarized and percentage increase in the *DRH* volume and peak discharge are calculated.

Table 6.14(a) ***Calculation of DRH—Example 6.17***

Before Urbanisation Direct Runoff due to					After Urbanisation Direct Runoff due to				
Time (h)	1-h *UH* ord. (m^3/s)	*ER* = 3.3 cm {Col.2×3.3}	*ER* = 2.3 cm {Col.2×2.3} Advanced by 1-hour	Direct Runoff Hydrograph ord. (m^3/s)	Time (h)	1-h *UH* ord. (m^3/s)	*ER* = 3.6 cm {Col.7×3.6}	*ER* = 2.6 cm {Col.7×2.6} Advanced by 1-hour	Direct Runoff Hydrograph ord. (m^3/s)
1	2	3	4	5	6	7	8	9	10
0	0.00	0.00	0	0.00	0.0	0.00	0.00	0	0.00
1	1.20	3.96	0	3.96	0.5	2.00	7.20	0	7.20
2	2.40	7.92	2.76	10.68	1.0	4.00	14.40	0.00	14.40
3	3.60	11.88	5.52	17.40	1.5	6.00	21.60	5.20	26.80
4	3.00	9.90	8.28	**18.18**	2.0	5.33	19.20	10.40	29.60
5	2.40	7.92	6.9	14.82	2.5	4.67	16.80	15.60	**32.40**
6	1.80	5.94	5.52	11.46	3.0	4.00	14.40	13.87	28.27
7	1.20	3.96	4.14	8.10	3.5	3.33	12.00	12.13	24.13
8	0.60	1.98	2.76	4.74	4.0	2.67	9.60	10.40	20.00
9	0.00	0.00	1.38	1.38	4.5	2.00	7.20	8.67	15.87
10	0	0	0	0.00	5.0	1.33	4.80	6.93	11.73
Total of ordinates =				**90.72**	5.5	0.67	2.40	5.20	7.60
					6.0	0	0.00	3.47	3.47
					6.5	0	0	1.73	1.73
					7.0			0.00	0.00
					Total of ordinates =				**223.20**

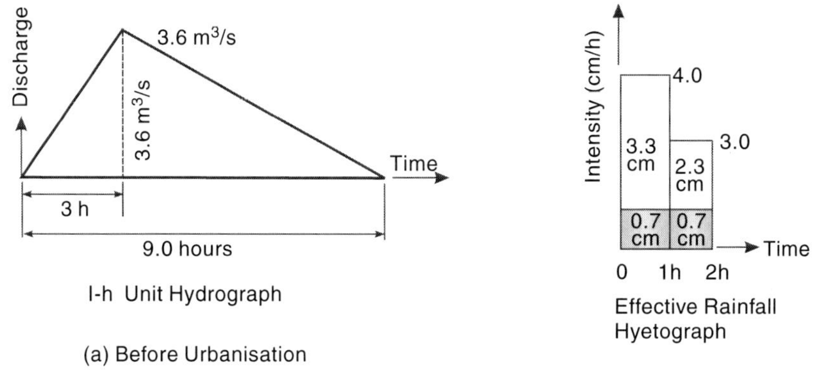

Fig. 6.24(a) ***ERH and 1-h UH before urbanisation***

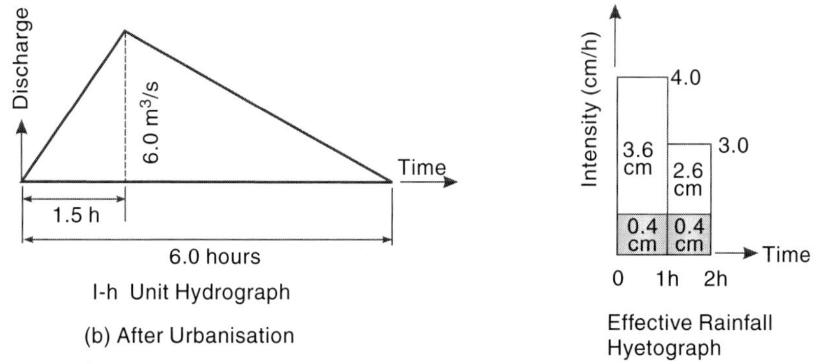

Fig. 6.24(b) ***ERH and 1-h UH after urbanisation***

Table 6.14(b) ***Summary of results—Example 6.17***

	Before urbanisation	After urbanisation
1. Total *DRH* volume	$= 90.72\ (m^3/s)(h)$ $= 90.72 \times (60 \times 60)$ $= 326592\ m^3$	$= 223.20\ (m^3/s)\ (0.5h)$ $= 223.20 \times 1800$ $= 401760\ m^3$
2. Increase in *DRH* volume	$= 401760\text{-}326592 = 75162\ m^3$	
3. Percentage increase in *DRH* volume	$= \frac{75168}{326592} \times 100 = 23.0$	
4. Peak discharges of *DRH*	$18.18\ m^3/s$	$32.40\ m^3/s$
5. Percentage increase in peak	$= \frac{(32.40 - 18.18)}{18.18} \times 100 = 78.2\%$	

6.12.5 The Indian Practice

Two approaches (short-term plan and long-term plan) were adopted by *CWC* to develop methodologies for estimation of design flood discharges applicable to small and medium catchments (25–1000 ha) of India.

Under the *short-term plan*, a quick method of estimating design flood peak has been developed[2] as follows:

The peak discharge of a D-h unit hydrograph Q_{pd} in m^3/s is

$$Q_{pd} = 1.79A^{3/4} \qquad \text{for } S_m > 0.0028 \tag{6.20}$$

and

$$Q_{pd} = 37.4A^{3/4}\, S_m^{2/3} \qquad \text{for } S_m < 0.0028 \tag{6.21}$$

where A = catchment area in km^2 and S_m = weighted mean slope given by

$$S_m = \left[\frac{L_{ca}}{(L_1/S_1)^{1/2} + (L_2/S_2)^{1/2} + \cdots + (L_n/S_n)^{1/2}}\right]^2 \tag{6.22}$$

in which L_{ca} = distance along the river from the gauging station to a point opposite to the centre of gravity of the area.

$L_1, L_2, \ldots L_n$ = length of main channel having slopes $S_1, S_2, \ldots S_n$ respectively, obtained from topographic maps.

The lag time in hours (i.e. time interval from the mid-point of the rainfall excess to the peak) of a 1-h unit hydrograph, t_{p1} is given by

$$t_{P1} = \frac{1.56}{[Q_{pd}/A]^{0.9}} \tag{6.23}$$

For design purposes the duration of rainfall excess in hours is taken as

$$D = 1.1\, t_{p1} \tag{6.24}$$

Equations (6.20) through (6.22) enable one to determine the duration and peak discharge of a design unit hydrograph. The time to peak has to be determined separately by using Eq. (6.9) or (6.10).

Under the *long-term plan*, a separate regional methodology has been developed by *CWC*. In this, the country is divided into 26 hydrometeorologically homogeneous subzones. For each subzone, a regional synthetic unit hydrograph has been developed. Detailed reports containing the synthetic unit hydrograph relations, details of the computation procedure and limitations of the method have been prepared, [e.g. *CWC* Reports No. CB/11/1985 and GP/10/1984 deal with flood estimation in Kaveri Basin (Sub-zone – 3i) and Middle Ganga Plains (Sub-zone – 1f) respectively.]

6.13 *Instantaneous Unit Hydrograph* (*IUH*)

6.13.1 Theory

The unit-hydrograph concept discussed in the preceding sections considered a D-h unit hydrograph. For a given catchment a number of unit hydrographs of different durations are possible. The shape of these different unit hydrographs depend upon the value of D. Figure 6.23 shows a typical variation of the shape of unit hydrographs for different values of D. As D is reduced, the intensity of rainfall excess being equal to $1/D$ increases and the unit hydrograph becomes more skewed. A finite unit hydrograph is indicated as the duration $D \rightarrow 0$. The limiting case of a unit hydrograph of zero duration is known as *instantaneous unit hydrograph* (*IUH*). Thus *IUH* is a fictitious, conceptual unit hydrograph which represents the surface runoff from the catchment due to an instantaneous precipitation of the rainfall excess volume of 1 cm. *IUH* is designated as $u(t)$ or sometimes as $u(0, t)$. It is a single-peaked hydrograph with a finite base width and its important properties can be listed as below:

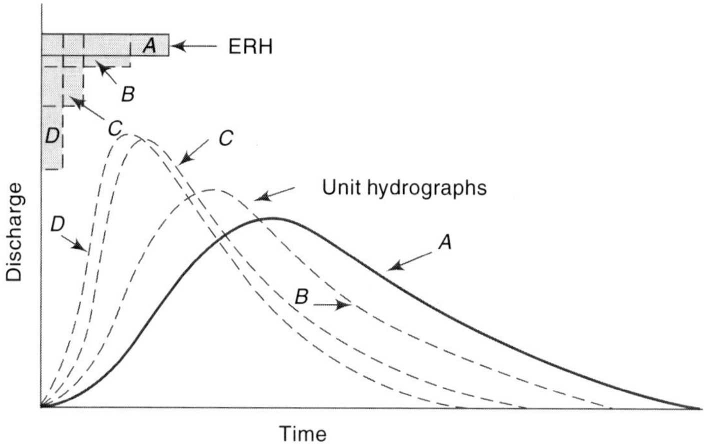

Fig. 6.25 ***Unit Hydrographs of Different Durations***

1. $0 \leq u\ (t) \leq$ a positive value, for $t > 0$;
2. $u(t) = 0$ for $t \leq 0$;
3. $u(t) \rightarrow = 0$ as $t \rightarrow \infty$;
4. $\int_0^\infty u\ (t)\ dt$ = unit depth over the catchment; and
5. time to the peak time to the centroid of the curve.

Consider an effective rainfall $I\ (\tau)$ of duration t_0 applied to a catchment as in Fig. 6.26. Each infinitesimal element of this *ERH* will operate on the *IUH* to produce a *DRH* whose discharge at time t is given by

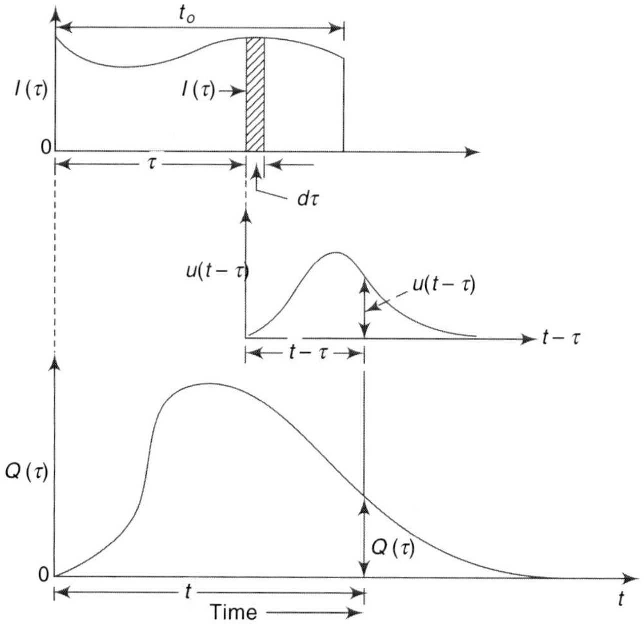

Fig. 6.26 ***Convolution of I (τ) and IUH***

$$Q(t) = \int_0^{t'} u(t - \tau)\, I(\tau)\, d\tau \qquad (6.25)$$

where $t' = t$ when $t < t_0$ and $t' = t_0$ when $t \geq t_0$

Equation (6.25) is called the *convolution integral* or *Duhamel integral.* The integral of Eq. (6.25) is essentially the same as the arithmetical computation of Eq. (6.5).

The main advantage of *IUH* is that it is independent of the duration of *ERH* and thus has one parameter less than a *D*-h unit hydrograph. This fact and the definition of *IUH* make it eminently suitable for theoretical analysis of rainfall excess-runoff relationship of a catchment. For a given catchment *IUH*, being independent of rainfall characteristics, is indicative of the catchment storage characteristics.

6.13.2 Derivation of IUH

Consider an *S*-curve, designated as S_1, derived from a *D*-h unit hydrograph. In this, the intensity of rainfall excess, $i = 1/D$ cm/h. Let S_2 be another *S*-curve of intensity i cm/h. If S_2 is separated from S_1 by a time interval dt and the ordinates are subtracted, a *DRH* due to a rainfall excess of duration dt and magnitude $i\,dt = dt/D$ h is obtained. A unit hydrograph of dt hours is obtained from this by dividing the above *DRH* by $i\,dt$. Thus, the dt-h unit hydrograph will have ordinates equal to $\left(\frac{S_2 - S_1}{i\,dt}\right)$. As dt is made smaller and smaller, i.e. as $dt \rightarrow 0$, an *IUH* results. Thus, for an *IUH*, the ordinate at any time is

$$u(t) = \lim_{dt \to 0}\left(\frac{S_2 - S_1}{i\,dt}\right) = \frac{1}{i}\frac{dS}{dt} \qquad (6.26)$$

If $i = 1$, then $u(t) = dS'/dt,$ (6.27)

where S' represents a *S*-curve of intensity 1 cm/h. Thus, the ordinate of an *IUH* at any time t is the slope of the *S*-curve of intensity 1 cm/h (i.e. *S*-curve derived from a unit hydrograph of 1-h duration) at the corresponding time. Equation (6.26) can be used in deriving *IUH* approximately.

*IUH*s can be derived in many other ways, notably by (i) harmonic analysis (ii) Laplace transform, and (iii) conceptual models. Details of these methods are beyond the scope of this book and can be obtained from Ref. 3. However, two simple models viz., Clark's model and Nash's model are described in Chapter 8 (Sections 8.8 and 8.9).

Derivation of D-hour Unit Hydrograph from *IUH*

For simple geometric forms of *IUH*, Eq. (6.25) can be used to derive a D-hour unit hydrograph. For complex shaped *IUH*s the numerical computation techniques used in deriving unit hydrographs of different durations (Sec. 6.7) can be adopted.

From Eq. 6.27, $dS' = u(t)\, dt$

Integrating between two points 1 and 2,

$$S_2' - S_1' = \int_{t_1}^{t_2} u(t)\, dt \qquad (6.28)$$

If $u(t)$ is essentially linear within the range 1–2 then for small values of $\Delta t = (t_2 - t_1)$, by taking

$$u(t) = \bar{u}(t) = \frac{1}{2}[u(t_1) + u(t_2)]$$

$$S_2' - S_1' = \frac{1}{2}[u(t_1) + u(t_2)](t_2 - t_1) \tag{6.29}$$

But $(S_2' - S_1')/(t_2 - t_1)$ = ordinate of a unit hydrograph of duration $D_1 = (t_2 - t_1)$. Thus, in general terms, for small values of D_1, the ordinates of a D_1-hour unit hydrograph are obtained by the equation

$$(D_1\text{-hour } UH)_t = \frac{1}{2}[(IUH)_t + (IUH)_{t-D_t}] \tag{6.30}$$

Thus, if two *IUH*s are lagged by D_1–hour where D_1 is small and their corresponding ordinates are summed up and divided by two, the resulting hydrograph will be a D_1-hour *UH*. After obtaining the ordinates of a *D*-hour unit hydrograph from Eq. (6.30), the ordinates of any *D*-hour *UH* can be obtained by the superposition method or *S*-curve method described in Sec. 6.8.2. From accuracy considerations, unless the limbs of *IUH* can be approximated as linear, it is desirable to confine D_1 to a value of 1-hour or less.

Example 6.18 The coordinates of the IUH of a catchment are given below. Derive the direct runoff hydrograph (*DRH*) for this catchment due to a storm of duration 4 hours and having a rainfall excess of 5 cm.

Time (hours)	0	1	2	3	4	5	6	7	8	9	10	11	12
IUH ordinate $u(t)$ (m^3/s)	0	8	35	50	47	40	31	23	15	10	6	3	0

Solution:

The calculations are performed in Table 6.15.

1. First, the ordinates of 1-h *UH* are derived by using Eq. (6.30)

   In Table 6.13, Col. 2 = ordinates of given *IUH* = $u(t)$

   Col. 3 = ordinates of *IUH* lagged by 1-h

   Col. 4 = $\frac{1}{2}$ (Col. 2 + Col. 3) = ordinates of 1-h *UH* by Eq. (6.30)

2. Using the 1-hour *UH*, the *S*-curve is obtained and lagging it by 4 hours the ordinates of 4-h *UH* are obtained.

   In Table 6.12, Col. 5 = *S*-curve additions

   Col. 6 = (Col. 4 + Col. 5) = *S*-curve ordinates

   Col. 7 = Col. 6 lagged by 4 hours = *S*-curve ordinates lagged by 4-h.

   Col. 8 = (Col. 6 – Col. 7) = Ordinates of a *DRH* due to 4 cm of *ER* in 4 hours.

   Col. 9 = (Col. 8)/4 = Ordinates of 4-hour *UH*

3. The required *DRH* ordinates due to 5.0 cm *ER* in 4 hours are obtained by multiplying the ordinates of 4-h *UH* by 5.0

   In Table 6.12, Col. 10 = (Col. 9) × 5.0 = ordinates of required *DRH*

   [***Note:*** Calculation of 4-hour *UH* directly by using D_1 = 4-h in Eq. (6.30) will lead to errors as the assumptions of linearity of $u(t)$ during D_1 may not be satisfied.]

Table 6.15 *Determination of DRH from IUH—Example 6.18*

1 Time (h)	2 $u(t)$ (m^3/s)	3 $u(t)$ lagged by 1 hour	4 Ordinate of 1-h *UH* (m^3/s)	5 S_1-Curve addition (m^3/s)	6 S_1-Curve ordinate (m^3/s)	7 S_1-Curve lagged by 4 hours (m^3/s)	8 *DRH* of 4 cm in 4 hours	9 Ordinate of 4-h *UH* (m^3/s)	10 *DRH* due to 5 cm *ER* in 4 hours (m^3/s)
			([2] + [3])/2		[4] + [5]		[6]–[7]	[8]/4	[9] × 5
0	0		0		0		0.0	0.00	0.00
1	8	0	4.0	0	4.0		4.0	1.00	5.00
2	35	8	21.5	4.0	25.5		25.5	6.38	31.88
3	50	35	42.5	25.5	68.0		68.0	17.00	85.00
4	47	50	48.5	68.0	116.5	0.0	116.5	29.13	145.63
5	40	47	43.5	116.5	160.0	4.0	156.0	39.00	195.00
6	31	40	35.5	160.0	195.5	25.5	170.0	42.50	212.50
7	23	31	27.0	195.5	222.5	68.0	154.5	38.63	193.13
8	15	23	19.0	222.5	241.5	116.5	125.0	31.25	156.25
9	10	15	12.5	241.5	254.0	160.0	94.0	23.50	117.50
10	6	10	8.0	254.0	262.0	195.5	66.5	16.63	83.13
11	3	6	4.5	262.0	266.5	222.5	44.0	11.00	55.00
12	0	3	1.5	266.5	268.0	241.5	26.5	6.63	33.13
13		0	0.0	268.0	268.0	254.0	14.0	3.50	17.50
14		0	0.0	268.0	268.0	262.0	6.0	1.50	7.50
15		0	0.0	268.0	268.0	266.5	1.5	0.38	1.88
16		0	0.0	268.0	268.0	268.0	0.0	0.00	0.00

REFERENCES

1. Butler, S. C., *Engineering Hydrology*, Prentice Hall Inc., *USA*, 1957.
2. Central Water Commission, 'Estimation of Design Flood Peak', *Report No.1*, Flood Estimation Directorate, *CWC*, New Delhi, India, 1973.
3. Chow, V. T., (Ed.), *Handbook of Applied Hydrology*, McGraw-Hill, New York, *USA*, 1964.
4. Gray, D. M., *Principles of Hydrology*, Water Inf. Center, Huntington, NY, *USA*, 1970.
5. Linsley, R. K. et al., *Hydrology for Engineers*, McGraw-Hill, New York, *USA*, 1958.
6. Sokolov, A. A. et al., *Flood Flow Computation*, The Unesco Press, Paris, France, 1976.

REVISION QUESTIONS

6.1 List the factors affecting a flood hydrograph. Discuss the role of these factors.

6.2 Describe the analysis of the recession limb of a flood hydrograph.

6.3 Explain the term Rainfall Excess (*ER*). How is *ERH* of a storm obtained?

6.4 Why is base flow separated from the flood hydrograph in the process of developing a unit hydrograph?

6.5 What is a unit hydrograph? List the assumptions involved in the unit hydrograph theory.

6.6 Describe briefly the procedure of preparing a *D*-hour unit hydrograph for a catchment.

6.7 Explain the procedure of using a unit hydrograph to develop the flood hydrograph due to a storm in a catchment.

6.8 Describe the *S*-curve method of developing a 6-h *UH* by using 12-h *UH* of the catchment.

6.9 Explain a procedure of deriving a synthetic unit hydrograph for a catchment by using Snyder's method.

6.10 What is an *IUH*? What are its characteristics?

6.11 Explain a procedure of deriving a *D*-h unit hydrograph from the *IUH* of the catchment.

6.12 Distinguish between

(a) Hyetograph and hydrograph (b) *D*-h *UH* and *IUH*

6.13 Describe the effect of urbanisation on runoff hydrograph of a catchment.

PROBLEMS

6.1 The flood hydrograph of a small stream is given below. Analyse the recession limb of the hydrograph and determine the recession coefficients. Neglect interflow.

Time (days)	Discharge (m^3/s)	Time (days)	Discharge (m^3/s)	Time (days)	Discharge (m^3/s)
0	155	2.0	9.0	4.0	1.9
0.5	70.0	2.5	5.5	5.0	1.4
1.0	38.0	3.0	3.5	6.0	1.2
1.5	19.0	3.5	2.5	7.0	1.1

Estimate the groundwater storage at the end of 7th day from the occurrence of peak.

6.2 On June 1, 1980 the discharge in a stream was measured as 80 m^3/s. Another measurement on June 21, 1980 yielded the stream discharge as 40 m^3/s. There was no rainfall in the catchment from April 15, 1980. Estimate the (a) recession coefficient,

(b) expected stream flow and groundwater storage available on July 10, 1980. Assume that there is no further rainfall in the catchment up to that date.

6.3 If $Q(t) = Q_0\, K^t$ describes the base flow recession in a stream, prove that the storage $S(t_1)$ left in the basin at any time for supplying base flow follows the linear reservoir model, viz. $S(t_1) = C\, Q\, (t_1)$, where C is a constant.
[*Hint:* Use the boundary condition: at $t = \infty$, $S_\infty = 0$ and $Q_\infty = 0$]

6.4 A 4 -hour storm occurs over an 80 km^2 watershed. The details of the catchment are as follows.

Sub Area (km^2)	ϕ-Index (mm/hour)	Hourly Rain (mm)			
		1st hour	2nd hour	3rd hour	4th hour
15	10	16	48	22	10
25	15	16	42	20	8
35	21	12	40	18	6
5	16	15	42	18	8

Calculate the runoff from the catchment and the hourly distribution of the effective rainfall for the whole catchment.

6.5 Given below are observed flows from a storm of 6-h duration on a stream with a catchment area of 500 km^2

Time (h)	0	6	12	18	24	30	36	42	48	54	60	66	72
Observed flow (m^3/s)	0	100	250	200	150	100	70	50	35	25	15	5	0

Assuming the base flow to be zero, derive the ordinates of the 6-h unit hydrograph.

6.6 A flood hydrograph of a river draining a catchment of 189 km^2 due to a 6 hour isolated storm is in the form of a triangle with a base of 66 hour and a peak ordinate of 30 m^3/s occurring at 10 hours from the start. Assuming zero base flow, develop the 6-hour unit hydrograph for this catchment.

6.7 The following are the ordinates of the hydrograph of flow from a catchment area of 770 km^2 due to a 6-h rainfall. Derive the ordinates of the 6-h unit hydrograph. Make suitable assumptions regarding the base flow.

Time from beginning of storm (h)	0	6	12	18	24	30	36	42	48	54	60	66	72
Discharge (m^3/s)	40	65	215	360	400	350	270	205	145	100	70	50	42

6.8 Analysis of the surface runoff records of a 1-day storm over a catchment yielded the following data:

Time (days)	0	1	2	3	4	5	6	7	8	9
Discharge (m^3/s)	20	63	151	133	90	63	44	29	20	20
Estimated base flow (m^3/s)	20	22	25	28	28	26	23	21	20	20

Determine the 24-h distribution graph percentages. If the catchment area is 600 km^2, determine the depth of rainfall excess.

6.9 The ordinates of a hydrograph of surface runoff resulting from 4.5 cm of rainfall excess of duration 8 h in a catchment are as follows:

Time (h)	0	5	13	21	28	32	35	41	45	55
Discharge (m³/s)	0	40	210	400	600	820	1150	1440	1510	1420
Time (h)	61	91	98	115	138					
Discharge (m³/s)	1190	650	520	290	0					

Determine the ordinates of the 8-h unit hydrograph for this catchment.

6.10 The peak of a flood hydrograph due to a 6-h storm is 470 m^3/s. The mean depth of rainfall is 8.0 cm. Assume an average infiltration loss of 0.25 cm/h and a constant base flow of 15 m^3/s and estimate the peak discharge of the 6-h unit hydrograph for this catchment.

6.11 Given the following data about a catchment of area 100 km^2, determine the volume of surface runoff and peak surface runoff discharge corresponding to a storm of 60 mm in 1 hour.

Time (h)	0	1	2	3	4	5
Rainfall (mm)	0	40	0	0	0	0
Runoff (m³/s)	300	300	1200	450	300	300

6.12 The ordinates of a 6-h unit hydrograph are given.

Time (h)	0	3	6	9	12	18	24	30	36	42	48	54	60	66
6-h *UH* ordinate (m²/s)	0	150	250	450	600	800	700	600	450	320	200	100	50	0

A storm had three successive 6-h intervals of rainfall magnitude of 3.0, 5.0 and 4.0 cm, respectively. Assuming a ϕ index of 0.20 cm/h and a base flow of 30 m^3/s, determine and plot the resulting hydrograph of flow.

6.13 The ordinates of a 6-h unit hydrograph are as given below:

Time (h)	0	6	12	18	24	30	36	42	48	54	60	66
ordinate of 6-h *UH* (m³/s)	0	20	60	150	120	90	66	50	32	20	10	0

If two storms, each of 1 cm rainfall excess and 6-h duration occur in succession, calculate the resulting hydrograph of flow. Assume base flow to be uniform at 10 m^3/s.

6.14 Using the 6-h unit hydrograph of Prob. 6.13 derive a 12-h unit hydrograph for the catchment.

6.15 The ordinates of the 2-h unit hydrograph of a basin are given:

Time (h)	0	2	4	6	8	10	12	14	16	18	20	22
2-h *UH* ordinate (m³/s)	0	25	100	160	190	170	110	70	30	20	6	0

Determine the ordinates of the S_2-curve hydrograph and using this S_2-curve, determine the ordinates of the 4-h unit hydrograph of the basin.

6.16 The 6-hour unit hydrograph of a catchment is triangular in shape with a base width of 64 hours and a peak ordinate of 30 m^3/s. Calculate the equilibrium discharge of the S_6-curve of the basin.

6.17 Ordinates of the one-hour unit hydrograph of a basin at one-hour intervals are 5, 8, 5, 3 and 1 m^3/s. Calculate the

(i) watershed area represented by this unit hydrograph. (ii) S_1-curve hydrograph.

(iii) 2-hour unit hydrograph for the catchment.

6.18 Using the ordinates of a 12-h unit hydrograph given below, compute the ordinates of the 6-h unit hydrograph of the basin.

Time (h)	Ordinate of 12-h *UH* (m^3/s)	Time (h)	Ordinate of 12-h *UH* (m^3/s)	Time (h)	Ordinate of 12-h *UH* (m^3/s)
0	0	54	130	108	17
6	10	60	114	114	12
12	37	66	99	120	8
18	76	72	84	126	6
24	111	78	71	132	3
30	136	84	58	138	2
36	150	90	46	144	0
42	153	96	35		
48	146	106	25		

[Note that the tail portion of the resulting 6-h *UH* needs fairing.]

6.19 The 3-h unit hydrograph for a basin has the following ordinates. Using the *S*-curve method, determine the 9-h unit hydrograph ordinates of the basin.

Time (h)	0	3	6	9	12	15	18	21	24	27	30
3-h *UH* ordinates (m^3/s)	0	12	75	132	180	210	183	156	135	144	96
Time (h)	**33**	**36**	**39**	**42**	**45**	**48**	**51**	**54**	**57**	**60**	
3-h *UH* ordinates (m^3/s)	87	66	54	42	33	24	18	12	6	6	

6.20 Using the given 6-h unit hydrograph derive the flood hydrograph due to the storm given below.

UH:

Time (h)	0	6	12	18	24	30	36	42	48	54	60	66
6-h *UH* ordinates (m^3/s)	0	20	60	150	120	90	66	50	32	20	10	0

Storm:

Time from beginning of the storm (h)	0	6	12	18
Accumulated rainfall (cm)	0	4	5	10

The φ index for the storm can be assumed to be 0.167 cm/h. Assume the base flow to be 20 m^3/s constant throughout.

6.21 The 6-hour unit hydrograph of a basin is triangular in shape with a peak of 100 m^3/s occurring at 24-h from the start. The base is 72-h.

(a) What is the area of the catchment represented by this unit hydrograph?

(b) Calculate the flood hydrograph due to a storm of rainfall excess of 2.0 cm during the first 6 hours and 4.0 cm during the second 6 hours interval. The base flow can be assumed to be 25 m^3/s constant throughout.

6.22 The 6-h unit hydrograph of a catchment of area 1000 km^2 can be approximated as a triangle with base of 69 h. Calculate the peak ordinate of this unit hydrograph.

6.23 The 4-h, distribution graph of a catchment of 50 km^2 area has the following ordinates:

Unit periods (4-h units)	1	2	3	4	5	6
Distribution (percentage)	5	20	40	20	10	5

If the catchment has rainfalls of 3.5. 2.2 and 1.8 cm in three consecutive 4-h periods, determine the resulting direct runoff hydrograph by assuming the φ-index for the storm as 0.25 cm/h.

6.24 The 6-h unit hydrograph of a catchment of area 259.2 km^2 is triangular in shape with a base width of 48 hours. The peak occurs at 12 h from the start. Derive the coordinates of the 6-h distribution graph for this catchment.

6.25 A 6-h unit hydrograph of a catchment of 500 km^2 area can be expressed as a triangle with base of 66 hours. (a) What is the peak ordinate of this unit hydrograph? Calculate the equilibrium discharge of the S_6–curve of the basin.

6.26 The following table gives the ordinates of a direct-runoff hydrograph resulting from two successive 3-h durations of rainfall excess values of 2.0 and 4.0 cm, respectively. Derive the 3-h unit hydrograph for the catchment.

Time (h)	0	3	6	9	12	15	18	21	24	27	30
Direct runoff (m^3/s)	0	120	480	660	460	260	160	100	50	20	0

6.27 Characteristics of two catchments *M* and *N* measured from a map are given below:

Item	Catchment *M*	Catchment *N*
L_{ca}	76 km	52 km
L	148 km	106 km
A	2718 km^2	1400 km^2

For the 6-h unit hydrograph in catchment *M*, the peak discharge is at 200 m^3/s and occurs at 37 h from the start of the rainfall excess. Assuming the catchments *M* and *N* are meteorologically similar, determine the elements of the 6-h synthetic unit hydrograph for catchment *N* by using Snyder's method.

6.28 A basin has an area of 400 km^2, and the following characteristics:

L = basin length = 35 km

L_{ca} = Length up to the centroid of the basin = 10 km

Snyder's coefficients: $C_t = 1.5$ and $C_p = 0.70$.

Develop synthetically the 3-h synthetic-unit hydrograph for this basin using Snyder's method.

6.29 Using the peak discharge and time to peak values of the unit hydrograph derived in Prob. 6.27, develop the full unit hydrograph by using the *SCS* dimensionless-unit hydrograph.

6.30 The rainfall excess of a storm is modelled as

$I(t) = 6$ cm/s for $0 \le t \le 4$ h

$I(t) = 0$ for $t \ge 4$ h

The corresponding direct runoff hydrograph is expressed in terms of depth over unit catchment area per hour (cm/h) as

$Q(t) = 6.0\, t$ cm/h for $0 \le t \le 4$ h

$Q(t) = 48 - 6.0\, t$ cm/h for $8 > t \geq 4$ h

$Q(t) = 0$ for $t > 8$

where t is in hours. Determine the (i) 4-h unit hydrograph of the catchment and corresponding *S*-curve of the catchment (ii) 3-h unit hydrogen of the catchment.

6.31 A 2-h unit hydrograph is given by

$U(t) = 0.5$ cm/h for $0 \leq t \leq 2$ h

$U(t) = 0$ for $t \geq 4$ h

(i) Determine the *S*-curve corresponding to the given 2-h *UH*

(ii) Using the *S*-curve developed above, determine the 4-h unit hydrograph

6.32 A 1-h unit hydrograph is rectangular in shape with a base of 3 hours and peak of 100 m^3/s. Develop the *DRH* due to an *ERH* given below:

Time since start (h)	1	2	3
Excess Rainfall (cm)	3	0	5

6.33 A 750 ha watershed has a time of concentration of 90 minutes.

(i) Derive the 15-minute unit hydrograph for this watershed by using *SCS* triangular unit hydrograph method.

(ii) What would be the *DRH* for a 15-minute storm having 4.0 cm of rainfall?

6.34 The *IUH* of a catchment is triangular in shape with a base of 36 h and peak of 20 m^3/s occurring at 8 hours from the start. Derive the 2-h unit hydrograph for this catchment.

6.35 The coordinates of the *IUH* of a catchment are as below:

Time (h)	0	1	2	3	4	5	6	8	10	12	14	16	18	20
Ordinates (m^3/s)	0	11	37	60	71	75	72	60	45	33	21	12	6	0

(a) What is the areal extent of the catchment?

(b) Derive the 3-hour unit hydrograph for this catchment.

OBJECTIVE QUESTIONS

6.1 The recession limb of a flood hydrograph can be expressed with positive values of coefficients, as $Q_t / Q_0 =$

(a) K_c^{at} (b) $a\,K_t^{-at}$ c) a^{-at} (d) e^{-at^2}

6.2 For a given storm, other factors remaining same,

(a) basins having low drainage density give smaller peaks in flood hydrographs

(b) basins with larger drainage densities give smaller flood peaks

(c) low drainage density basins give shorter time bases of hydrographs

(d) the flood peak is independent of the drainage density

6.3 Base-flow separation is performed

(a) on a unit hydrograph to get the direct-runoff hydrograph

(b) on a flood hydrograph to obtain the magnitude of effective rainfall

(c) on flood hydrographs to obtain the rainfall hyetograph

(d) on hydrographs of effluent streams only.

6.4 The shape of the recession limb of a hydrograph depends on

(a) basin as well as storm characteristics

(b) storm characteristics only

(c) basin characteristics only

(d) base flow only

6.5 A direct-runoff hydrograph due to a storm was found to be triangular in shape with a peak of 150 m^3/s, time from start of effective storm to peak of 24 h and a total time base of 72 h. The duration of the storm in this case was

(a) < 24 h (b) between 24 to 72 h
(c) 72 h (d) > 72 h

6.6 A unit hydrograph has one unit of

(a) peak discharge (b) rainfall duration
(c) direct runoff (d) the time base of direct runoff.

6.7 The basic assumptions of the unit-hydrograph theory are

(a) nonlinear response and time invariance
(b) time invariance and linear response
(c) linear response and linear time variance
(d) nonlinear time variance and linear response.

6.8 The *D*-hour unit hydrograph of a catchment may be obtained by dividing the ordinates of a single peak direct runoff hydrograph (*DRH*) due to a storm of *D* hour duration by the

(a) total runoff volume (in cm) (b) direct runoff volume (in cm)
(c) duration of *DRH* (d) total rainfall (in cm)

6.9 A storm hydrograph was due to 3 h of effective rainfall. It contained 6 cm of direct runoff. The ordinates of *DRH* of this storm

(a) when divided by 3 give the ordinates of a 6-h unit hydrograph
(b) when divided by 6 give the ordinates of a 3-h unit hydrograph
(c) when divided by 3 give the ordinates of a 3-h unit hydrograph
(d) when divided by 6 give the ordinates of a 6-h unit hydrograph.

6.10 A 3-hour storm over a watershed had an average depth of 27 mm. The resulting flood hydrograph was found to have a peak flow of 200 m^3/s and a base flow of 20 m^3/s. If the loss rate could be estimated as 0.3 cm/h, a 3-h unit hydrograph for this watershed will have a peak of

(a) 66.7 m^3/s (b) 100 m^3/s (c) 111.1 m^3/s (d) 33.3 m^3/s

6.11 A triangular *DRH* due to a storm has a time base of 80 hrs and a peak flow of 50 m^3/s occurring at 20 hours from the start. If the catchment area is 144 km^2, the rainfall excess in the storm was

(a) 20 cm (b) 7.2 cm (c) 5 cm (d) none of these.

6.12 The 12-hr unit hydrograph of a catchment is triangular in shape with a base width of 144 hours and a peak discharge value of 23 m^3/s. This unit hydrograph refers to a catchment of area

(a) 756 km^2 (b) 596 km^2 (c) 1000 km^2 (d) none of these.

6.13 The 6-h unit hydrograph of a catchment is triangular in shape with a base width of 64 h and peak ordinate of 20 m^3/s. If a 0.5 cm rainfall excess occurs in 6 h in that catchment, the resulting surface-runoff hydrograph will have

(a) a base of 128 h (b) a base of 32 h
(c) a peak of 40 m^3/s (d) a peak of 10 m^3/s

6.14 A 90 km^2 catchment has the 4-h unit hydrograph which can be approximated as a triangle. If the peak ordinate of this unit hydrograph is 10 m^3/s the time base is

(a) 120 h (b) 64 h (c) 50 h (d) none of these

6.15 A triangular *DRH* due to a 6-h storm in a catchment has a time base of 100 h and a peak flow of 40 m^3/s. The catchment area is 180 km^2. The 6-h unit hydrograph of this catchment will have a peak flow in m^3/s of

(a) 10 (b) 20 (c) 30 (d) none of these

6.16 A triangular direct runoff hydrograph due to an isolated storm has a time base of 60 h and a peak flow of 30 m^3/s that occurs at 20 h from the start. If the catchment area is 300 km^2, what is the rainfall excess in the storm?

(a) 108 mm (b) 216. mm (c) 10.8 mm (d) 8.3 mm

6.17 The ordinate of a 1-hour unit hydrograph at 1-hour intervals, starting from t = 0 are: 3, 5, 4, 2, 1 and 0 m^3/s. (i) The catchment area represented by this unit hydrograph, and (ii) the ordinate of 3-hour unit hydrograph for the catchment at t = 3 hour, are

(a) (1.0 km^2, 2.0 m^3/s) (b) (2.7 km^2, 3.3 m^3/s)
(c) (3.6 km^2, 3.7 m^3/s) (d) (5.4 km^2, 4.0 m^3/s)

6.18 A direct runoff hydrograph due to an isolated storm was triangular in shape with base of 80 h and peak of 200 m^3/s. If the catchment area is 1440 km^2, the effective rainfall of the storm is

(a) 2 cm (b) 5 cm (c) 10 cm (d) 20 cm

6.19 The 3-hour unit hydrograph U_1 of a catchment of area 250 km^2 is in the form of a triangle with peak discharge of 40 m^3/s. Another 3-hour unit hydrograph U_2 is also triangular in shape and has the same base width as U_1 but with a peak flow of 80 m^3/s. The catchment which U_2 refers to has an area of

(a) 125 km^2 (b) 250 km^2 (c) 1000 km^2 (d) 500 km^2

6.20 U_c is the 6-h unit hydrograph for a basin representing 1 cm of direct runoff and U_m is the direct runoff hydrograph for the same basin due to a rainfall excess of 1 mm in a storm of 6-hour duration.

(a) Ordinates of U_m are 1/10 the corresponding ordinates of U_c
(b) Base of U_m is 1/10 the base of U_c
(c) Ordinates of U_m are 10 times the corresponding ordinates of U_c
(d) Base of U_m is 10 times the base of U_c

6.21 A basin with an area of 756 km^2 has the 6-h unit hydrograph which could be approximated as a triangle with a base of 70 hours. The peak discharge of direct runoff hydrograph due to 5 cm of rainfall excess in 6 hours from that basin is

(a) 535 m^3/s (b) 60 m^3/s (c) 756 m^3/s (d) 300 m^3/s

6.22 The peak flow of a flood hydrograph caused by isolated storm was observed to be 120 m^3/s. The storm was of 6 hours duration and had a total rainfall of 7.5 cm. If the base flow and the φ-index are assumed to be 30 m^3/s and 0.25 cm/h respectively, the peak ordinate of the 6-h unit hydrograph of the catchment is

(a) 12.0 m^3/s (b) 15.0 m^3/s (c) 16.0 m^3/s (d) 20.0 m^3/s

6.23 The peak ordinate of 4-h unit hydrograph a basin is 80 m^3/s. An isolated storm of 4-hours duration in the basin was recorded to have a total rainfall of 7.0 cm. If it is assumed that the base flow and the φ-index are 20 m^3/s and 0.25 cm/h respectively, the peak of the flood discharge due to the storm could be estimated as

(a) 580 m^3/s (b) 360 m^3/s (c) 480 m^3/s (d) 500 m^3/s

6.24 The peak flow of a flood hydrograph caused by isolated storm was observed to be 100 m^3/s. The storm had a duration of 8.0 hours and the total depth of rainfall of 7.0 cm. The base flow and the φ-index were estimated as 20 m^3/s and 0.25 cm/h respectively. If in the above storm the total rainfall were 9.5 cm in the same duration of 8 hours, the flood peak would have been larger by

(a) 35.7% (b) 40% (c) 50% (d) 20%

6.25 For a catchment with an area of 360 km^2, the equilibrium discharge of the *S*-curve obtained by summation of 4-h unit hydrograph is

(a) 250 m^3/s (b) 90 m^3/s (c) 278 m^3/s (d) 360 m^3/s

6.26 For a catchment of area *A* an *S*-curve has been derived by using the *D*-hour unit hydrograph which has a time base *T*. In this *S*-curve

(a) the equilibrium discharge is independent of *D*
(b) the time at which the *S*-curve attains its maximum value is equal to *T*
(c) the time at which the *S*-curve attains its maximum value is equal to *D*
(d) the equilibrium discharge is independent of *A*

6.27 An *IUH* is a direct runoff hydrograph of

(a) of one cm magnitude due to rainfall excess of 1-h duration
(b) that occurs instantaneously due to a rainfall excess of 1-h duration
(c) of unit rainfall excess precipitating instantaneously over the catchment
(d) occurring at any instant in long duration

6.28 An instantaneous unit hydrograph is a hydrograph of

(a) unit duration and infinitely small rainfall excess
(b) infinitely small duration and of unit rainfall excess
(c) infinitely small duration and of unit rainfall excess of an infinitely small area
(d) unit rainfall excess on infinitely small area

CHAPTER 7

FLOODS

7.1 INTRODUCTION

A flood is an unusually high stage in a river, normally the level at which the river overflows its banks and inundates the adjoining area. The damages caused by floods in terms of loss of life, property and economic loss due to disruption of economic activity are all too well known. Thousands of crores of rupees are spent every year in flood control and flood forecasting. The hydrograph of extreme floods and stages corresponding to flood peaks provide valuable data for purposes of hydrologic design. Further, of the various characteristics of the flood hydrograph, probably the most important and widely used parameter is the flood peak. At a given location in a stream, flood peaks vary from year to year and their magnitude constitutes a hydrologic series which enable one to assign a frequency to a given flood-peak value. In the design of practically all hydraulic structures the peak flow that can be expected with an assigned frequency (say 1 in 100 years) is of primary importance to adequately proportion the structure to accommodate its effect. The design of bridges, culvert waterways and spillways for dams and estimation of scour at a hydraulic structure are some examples wherein flood-peak values are required.

At a stream site, a flood has two attributes: (i) magnitude of the peak (highest discharge rate), and (ii) stage of the peak (elevation of the water surface at the peak flow). The peak discharge rate is important in the design of the (i) spillways of dams and barrages, and (ii) capacities of bridge and culvert waterways. The stage is important in (i) estimating the extent of area inundated by the flood, (ii) deciding the elevation of bridge decks, and (iii) in deciding the minimum elevation of structures that are to be built on the flood plains.

7.2 *Rational Method*

Consider a rainfall of uniform intensity and very long duration occurring over a basin. The runoff rate gradually increases from zero to a constant value as indicated in Fig. 7.1. The runoff increases as more and more flow from remote areas of the catchment reach the outlet. Designating the time taken for a drop of water from

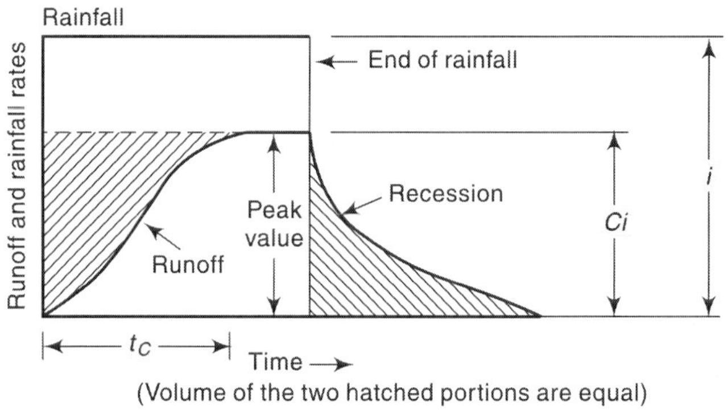

Fig. 7.1 ***Runoff Hydrograph due to Uniform Rainfall***

the farthest part of the catchment to reach the outlet as t_c = time of concentration, it is obvious that if the rainfall continues beyond t_c, the runoff will be constant and at the peak value. The peak value of the runoff is given by

$$Q_p = C\,A\,i;\ \text{for } t \geq t_c \tag{7.1}$$

where C = coefficient of runoff = (runoff/rainfall), A = area of the catchment and i = intensity of rainfall. This is the basic equation of the *rational method*. Using the commonly used units, Eq. (7.1) is written for field application as

$$Q_p = \frac{1}{3.6} C\,(i_{tc,p})A \tag{7.2}$$

where Q_p = peak discharge (m^3/s)

C = coefficient of runoff

$(i_{tc,p})$ = the mean intensity of precipitation (mm/h) for a duration equal to t_c and an exceedance probability P

A = drainage area in km^2

The use of this method to compute Q_p requires three parameters: t_c, $(i_{tc,p})$ and C.

7.2.1 Time of Concentration (t_c)

There are a number of empirical equations available for the estimation of the time of concentration. Two of these are described below.

1. US Practice

For small drainage basins, the time of concentration is assumed to be equal to the lag time of the peak flow. Thus,

$$t_c = t_p \text{ of Eq. (6.10)} = C_{tL}\left(\frac{LL_{ca}}{\sqrt{S}}\right)^n \tag{7.3}$$

where t_c = time of concentration in hours, C_{tL}, L, L_{ca}, n and S have the same meaning as in Eq. (6.10) of Chapter 6.

2. Kirpich Equation (1940)

This is the popularly used formula relating the time of concentration of the length of travel and slope of the catchment as

$$t_c = 0.01947\ L^{0.77}\ S^{-0.385} \tag{7.4}$$

where t_c = time of concentration (minutes)

L = maximum length of travel of water (m), and

S = slope of the catchment = $\Delta H/L$ in which

ΔH = difference in elevation between the most remote point on the catchment and the outlet.

For easy use, Eq. (7.4) is sometimes written as

$$t_c = 0.01947\ K_1^{0.77} \tag{7.4a}$$

where $K_1 = \sqrt{\dfrac{L^3}{\Delta H}}$

3. Rainfall Intensity ($i_{tc,\,p}$)

The rainfall intensity corresponding to a duration t_c and the desired probability of exceedance P, (i.e. return period $T = 1/P$) is found from the rainfall-frequency-duration relationship for the given catchment area (Chap. 2). This will usually be a relationship of the form of Eq. (2.15), viz.

$$i_{tc,\,p} = \frac{KT^x}{(t_c + a)^n}$$

in which the coefficients K, a, x and n are specific to a given area. Table 2.8 (preferably in its expanded form) could be used to estimate these coefficients to a specific catchment. In *USA* the peak discharges for purposes of urban area drainage are calculated by using $P = 0.05$ to 0.1. The recommended frequencies for various types of structures used in watershed development projects in India are as below:

Sl. No	Types of structure	Return Period (Years)
1	Storage and diversion dams having permanent spillways	50–100
2	Earth dams having natural spillways	25–50
3	Stock water dams	25
4	Small permanent masonry and vegetated waterways	10–15
5	Terrace outlets and vegetated waterways	10
6	Field diversions	15

7.2.2 Runoff Coefficient (C)

The coefficient C represents the integrated effect of the catchment losses and hence depends upon the nature of the surface, surface slope and rainfall intensity. The effect of rainfall intensity is not considered in the available tables of values of C. Some typical values of C are indicated in Table 7.1(a and b).

Equation (7.2) assumes a homogeneous catchment surface. If however, the catchment is non-homogeneous but can be divided into distinct sub-areas each having a different runoff coefficient, then the runoff from each sub-area is calculated separately and merged in proper time sequence. Sometimes, a non-homogeneous catchment may have component sub-areas distributed in such a complex manner that distinct sub-zones cannot be separated. In such cases, a weighted equivalent runoff coefficient C_e as below is used in place of C in Eq. (7.2).

$$C_e = \frac{\sum_1^N C_i A_i}{A} \tag{7.5}$$

where A_i = the areal extent of the sub area i having a runoff coefficient C_i and N = number of sub areas in the catchment.

Table 7.1(a) ***Value of the Runoff Coefficient C in Eq. (7.2)***

	Types of area	Value of C
A.	*Urban area* (P = 0.05 to 0.10)	
	Lawns:	
	Sandy-soil, flat, 2%	0.05–0.10
	Sandy soil, steep, 7%	0.15–0.20
	Heavy soil, average, 2.7%	0.18–0.22
	Residential areas:	
	Single family areas	0.30–0.50
	Multi-units, attached	0.60–0.75
	Industrial:	
	Light	0.50–0.80
	Heavy	0.60–0.90
	Streets	0.70–0.95
B.	*Agricultural Area*	
	Flat:	
	Tight clay; cultivated	0.50
	woodland	0.40
	Sandy loam; cultivated	0.20
	woodland	0.10
	Hilly:	
	Tight clay; cultivated	0.70
	woodland	0.60
	Sandy loam; cultivated	0.40
	woodland	0.30

Table 7.1(b) ***Values of C in Rational Formula for Watersheds with Agricultural and Forest Land Covers***

Sl. No	Vegetative cover	Slope (%)	Soil Texture		
			Sandy Loam	Clay and Silty Loam	Stiff Clay
1	Cultivated Land				
		0–5	0.30	0.50	0.60
		5–10	0.40	0.60	0.70
		10–30	0.52	0.72	0.82
2	Pasture Land				
		0–5	0.10	0.30	0.40
		5–10	0.16	0.36	0.55
		10–30	0.22	0.42	0.60
3	Forest Land				
		0–5	0.10	0.30	0.40
		5–10	0.25	0.35	0.50
		10–30	0.30	0.50	0.60

The rational formula is found to be suitable for peak-flow prediction in small catchments up to 50 km^2 in area. It finds considerable application in urban drainage designs and in the design of small culverts and bridges.

It should be noted that the word *rational* is rather a misnomer as the method involves the determination of parameters t_c and C in a subjective manner. Detailed description and the practice followed in using the rational method in various countries are given in detail in Ref. 7.

Example 7.1(a) An urban catchment has an area of 85 ha. The slope of the catchment is 0.006 and the maximum length of travel of water is 950 m. The maximum depth of rainfall with a 25-year return period is as below:

Duration (min)	5	10	20	30	40	60
Depth of rainfall (mm)	17	26	40	50	57	62

If a culvert for drainage at the outlet of this area is to be designed for a return period of 25 years, estimate the required peak-flow rate, by assuming the runoff coefficient as 0.3.

Solution

The time of concentration is obtained by the Kirpich formula [Eq.(7.4)] as

$$t_c = 0.01947 \times (950)^{0.77} \times (0.006)^{-0.385} = 27.4 \text{ minutes}$$

By interpolation,

Maximum depth of rainfall for 27.4-min durtion

$$= \frac{(50-40)}{10} \times 7.4 + 40 = 47.4 \text{ mm}$$

Average intensity $i_{tc,\,p} = \dfrac{47.4}{27.4} \times 60 = 103.8$ mm/h

By Eq. (7.2), $Q_p = \dfrac{0.30 \times 103.8 \times 0.85}{3.6} = 7.35 \text{ m}^3/\text{s}$

Example 7.1(b)

If in the urban area of Example 7.1(a), the land use of the area and the corresponding runoff coefficients are as given below, calculate the equivalent runoff coefficient.

Land use	Area (ha)	Runoff coefficient
Roads	8	0.70
Lawn	17	0.10
Residential area	50	0.30
Industrial area	10	0.80

Solution

Equivalent runoff coefficient $C_e = \dfrac{\sum_{1}^{N} C_i A_i}{A}$

$$C_e = \frac{[(0.7 \times 8) + (0.1 \times 17) + (0.3 \times 50) + (0.8 \times 10)]}{[8 + 17 + 50 + 10]}$$

$$= \frac{30.3}{85} = 0.36$$

Example 7.2

A 500 ha watershed has the land use/cover and corresponding runoff coefficient as given below:

Land use/cover	Area (ha)	Runoff coefficient
Forest	250	0.10
Pasture	50	0.11
Cultivated land	200	0.30

The maximum length of travel of water in the watershed is about 3000 m and the elevation difference between the highest and outlet points of the watershed is 25 m. The maximum intensity-duration-frequency relationship of the watershed is give by

$$i = \frac{6.311 T^{0.1523}}{(D + 0.50)^{0.945}}$$

water i = intensity in cm/h, T = Return period in years and D = duration of the rainfall in hours. Estimate the (i) 25-year peak runoff from the watershed and (ii) the 25-year peak runoff if the forest cover has decreased to 50 ha and the cultivated land has encroached upon the pasture and forest lands to have a total coverage of 450 ha.

Solution

Case 1: Equivalent runoff coefficient $C_e = \dfrac{\sum_1^N C_i A_i}{A}$

$$= \frac{[(0.10 \times 250) + (0.11 \times 50) + (0.30 \times 200)]}{500} = 0.181$$

By Eq. (7.4a) time of concentration $t_c = 0.01947\,(K_1)^{0.77}$ with $K_1 = \sqrt{\dfrac{L^3}{\Delta H}}$

Since $L = 3000$ m and $\Delta H = 25$ m $\quad K_1 = \sqrt{\dfrac{(3000)^3}{25}} = 32863$

$$t_c = 0.01947\,(32863)^{0.77} = 58.5 \text{ min} = 0.975 \text{ h}$$

Calculation of $i_{tc,\,p}$: Here duration $D = t_c = 0.975$ h. $T = 25$ years. Hence,

$$i = \frac{6.311(25)^{0.1523}}{(0.975 + 0.50)^{0.945}} = 10.304/1.447 = 7.123 \text{ cm/h} = 71.23 \text{ mm/h}$$

Peak flow by Eq. (7.2), $Q_p = (1/3.6)(C_e\, i\, A)$

$$= \frac{0.181 \times 71.23 \times (500/100)}{3.6} = 64.46 \text{ m}^3/\text{s}$$

Case 2: Here, equivalent $C = C_e = \dfrac{[(0.10 \times 50) + (0.30 \times 450)]}{500} = 0.28$

$$i = 71.23 \text{ mm/h and } A = 500 \text{ ha} = 5\ (\text{km})^2$$

$$Q_p = \frac{0.28 \times 71.23 \times (500/100)}{3.6} = 99.72 \text{ m}^3/\text{s}$$

7.3 *Empirical Formulae*

The empirical formulae used for the estimation of the flood peak are essentially regional formulae based on statistical correlation of the observed peak and important catchment properties. To simplify the form of the equation, only a few of the many parameters affecting the flood peak are used. For example, almost all formulae use the catchment area as a parameter affecting the flood peak and most of them neglect the flood frequency as a parameter. In view of these, the empirical formulae are applicable only in the region from which they were developed and when applied to other areas they can at best give approximate values.

7.3.1 Flood Peak-Area Relationships

By far the simplest of the empirical relationships are those which relate the flood peak to the drainage area. The maximum flood discharge Q_p from a catchment area A is given by these formulae as

$$Q_p = f(A)$$

While there are a vast number of formulae of this kind proposed for various parts of the world, only a few popular formulae used in various parts of India are given below.

1. Dickens Formula (1865)

$$Q_p = C_D A^{3/4} \tag{7.6}$$

where Q_p = maximum flood discharge (m^3/s) A = catchment area (km^2)

C_D = Dickens constant with value between 6 to 30

The following are some guidelines in selecting the value of C_D:

Region	Value of C_D
North-Indian plains	6
North-Indian hilly regions	11–14
Central India	14–28
Coastal Andhra and Orissa	22–28

For actual use, the local experience will be of aid in the proper selection of C_D. Dickens formula is used in the central and northern parts of the country.

2. Ryves Formula (1884)

$$Q_p = C_R A^{2/3} \tag{7.7}$$

where Q_p = maximum flood discharge (m^3/s) A = catchment area (km^2)

and C_R = Ryves coefficient

This formula originally developed for the Tamil Nadu region, is in use in Tamil Nadu and parts of Karnataka and Andhra Pradesh. The values of C_R recommended by Ryves for use are

C_R = 6.8 for areas within 80 km from the east coast

= 8.5 for areas which are 80–160 km from the east coast

= 10.2 for limited areas near hills

3. Inglis Formula (1930)

This formula is based on flood data of catchments in Western Ghats in Maharashtra. The flood peak Q_p in m^3/s is expressed as

$$Q_p = \frac{124\,A}{\sqrt{A + 10.4}} \tag{7.8}$$

where A is the catchment area in km^2.

Equation (7.8) with small modifications in the constant in the numerator (124) is in use Maharashtra for designs in small catchments.

7.3.2 Other Formulae

There are many such empirical formulae developed in various parts of the world. References 3 and 5 list many such formulae suggested for use in various parts of India as well as of the world.

There are some empirical formulae which relate the peak discharge to the basin area and also include the flood frequency. Fuller's formula (1914) derived for catchments in *USA* is a typical one of this kind and is given by

$$Q_{Tp} = C_f A^{0.8} (1 + 0.8 \log T) \qquad (7.9)$$

where Q_{Tp} = maximum 24-h flood with a frequency of T years in m^3/s, A = catchment area in km^2, C_f = a constant with values in the range 0.18 to 1.88.

Generally, in present times, the use of empirical flood formulae are not encouraged in all but minor works. Even there, extreme caution is advised in the use of empirical formulae.

7.3.3 Envelope Curves

In regions having same climatological characteristics, if the available flood data are meagre, the enveloping curve technique can be used to develop a relationship between the maximum flood flow and drainage area. In this method, the available flood peak data from a large number of catchments which do not significantly differ from each other in terms of meteorological and topographical characteristics are collected. The data are then plotted on a log-log paper as flood peak *vs* catchment area. This would result in a plot in which the data would be scattered. If an enveloping curve that would encompass all the plotted data points is drawn, it can be used to obtain maximum peak discharges for any given area. Envelop curves thus obtained are very useful in getting quick rough estimations of peak values. If equations are fitted to these enveloping curves, they provide empirical flood formulae of the type, $Q = f(A)$.

Regional envelope curves become obsolete in about three or more decades as additional high flood value data becomes available due to their occurrences in this intervening period. This necessitates frequent updating of the enveloping curve. Details of an enveloping curve of highest floods in the world (including Indian data) and its use are detailed in Sec. 7.10.

7.4 *Unit Hydrograph Method*

The unit hydrograph technique described in the previous chapter can be used to predict the peak-flood hydrograph if the rainfall producing the flood, infiltration characteristics of the catchment and the appropriate unit hydrograph are available. For design purposes, extreme rainfall situations are used to obtain the design storm, viz. the hydrograph of the rainfall excess causing extreme floods. The known or derived unit hydrograph of the catchment is then operated upon by the design storm to generate the desired flood hydrograph. Details about this use of unit hydrograph are given in Sec. 7.13.

7.5 *Flood Frequency Studies*

Hydrologic processes such as floods are exceedingly complex natural events. They are resultants of a number of component parameters and are therefore very difficult to model analytically. For example, the floods in a catchment depend upon the characteristics of the catchment, rainfall and antecedent conditions, each one of these factors in turn depend upon a host of constituent parameters. This makes the estimation of the flood peak a very complex problem leading to many different approaches. The empirical formulae and unit hydrograph methods presented in the previous sections are some of them. Another approach to the prediction of flood flows, and also applicable to other hydrologic processes such as rainfall etc. is the statistical method of frequency analysis.

The values of the annual maximum flood from a given catchment area for large number of successive years constitute a hydrologic data series called the *annual series.* The data are then arranged in decreasing order of magnitude and the probability P of each event being equalled to or exceeded (plotting position) is calculated by the plotting-position formula (Weibull formula)

$$P = \frac{m}{N+1} \tag{7.10}$$

where m = order number of the event and N = total number of events in the data. The recurrence interval, T (also called the *return period* or *frequency*) is calculated as

$$T = 1/P = \frac{N+1}{m} \tag{7.11}$$

The relationship between the excedence probability P and the probability of occurrence of the event occurring r number of times in n successive years is the same as described in Sec. 2.11. Thus, for example the probability of a random hydrologic event, say annual flood of a given magnitude in a given stream, r times in n successive years is given by

$$P_{r,\,n} = {}^nC_r P_r\, q^{n-r} = \frac{n!}{(n-r)!r!} P^r\, q^{n-r} \tag{7.12}$$

where $\quad q = 1 - P$

For example, consider an event (say, annual flood of a given magnitude in a stream) of exceedance probability P :

- The probability of the event occurring 3 times in n successive years is

$$P_{3,r} = {}^nC_3 P_3 q^{n-3} = \frac{n!}{(n-3)!3!} P^3\, q^{n-3}$$

- The probability of the event *not occurring at all* in n successive years is

$$P_{0,n} = {}^nC_0 P_0 q^n = q^n = (1 - P)^n$$

- The probability of the event occurring at least once in n successive years is $P_1 = (1 -$ Probability of the event not occurring at all$) = 1 - q^n = 1 - (1 - P)^n$.

7.5.1 Frequency Analysis

The terms *return period* and *recurrence interval* are used synonymously to denote the reciprocal of the annual probability of exceedance. In a flood series the variation of the flood magnitude with the corresponding return period T calculated by Eq. 7.11 forms the basic aspect of the flood-frequency analysis. If a flood series of length N is arranged in descending order and each term is assigned a rank number m indicating the number of times the event is equaled or exceeded, each flood event in the series can be assigned a return period T calculated by using Eq. 7.11. If the flood magnitude Q (in cumecs) is plotted in the Y-axis against the return period T (in years) on the X-axis, the resulting plot is called a *flood Frequency plot*. It is also known as *empirical probability distribution curve*. Depending upon the range of the parameters Q and T, the scales of these parameters can be arithmetic or logarithmic. However, it is common practice to have the scale for T as logarithmic. A best-fit curve (trend line) is drawn through the plotted points and the resulting curve of Q vs. T can be used for estimating the values of flood magnitudes for known recurrence period and vice versa for small return periods. They can be used confidently where limited interpolation is required. However, while large extrapolation is not desirable, very limited extrapolation can be adopted with due caution. The use of the procedure is illustrated through Example 7.3.

Example 7.3 Annual maximum recorded flood series of the river Koyna at Warunji, covering a period of 42 years, is given below. Construct the flood frequency plot on a semi-log scale and estimate the following:

(a) Flood magnitude of return period of (i)15 years, and (ii) 50 years

(b) Return period of a flood of 4000 cumecs magnitude.

Year	1967	1968	1969	1970	1971	1972	1973	1974	1975	1976
Flood Discharge, (m^3/s)	3830	757	1605	1492	1386	1523	1900	1583	1681	2780
Year	1977	1978	1979	1980	1981	1982	1983	1984	1985	1986
Flood Discharge, (m^3/s)	1610	1576	1317	1868	1262	1156	1493	1334	1054	1284
Year	1987	1988	1989	1990	1991	1992	1993	1994	1995	1996
Flood Discharge, (m^3/s)	1073	2590	1600	2530	1800	1238	1023	2675	772	1566
Year	1997	1998	1999	2000	2001	2002	2003	2004	2005	2006
Flood Discharge, (m^3/s)	3529	828	1397	677	623	830	627	2716	4641	4973
Year	2007	2008								
Flood Discharge, (m^3/s)	2243	1887								

Solution

The data is arranged in descending order and the rank number is assigned to each recorded event. The highest flood event has the rank $m = 1$. The exceedance interval

T is calculated for each event by using Eq. 7.11. In this case since the number of records $N = 42$, the recurrence interval, $T = \frac{N+1}{m} = \frac{43}{m}$. The ordered discharge values, corresponding m value and the recurrence interval T are shown in Table 7.2. A graph is plotted between the annual flood discharge as the ordinate (on arithmetic scale) and the return period T as the abscissa (on logarithmic scale), (Fig. 7.2). The best-fitting line (tread line) is drawn for the plotted points. The calculations and plotting can be advantageously be done in a spreadsheet (such as *MS* Excel). The equation of the trend line is obtained as

$$Q = 1132.8 \ln T + 685.53 .$$

Table 7.2, ***Calculation of Return Period, T—Example 7.3***

Rank (m)	Sorted Annual Flood Discharge (Cumecs)	Recurrence Interval (Years) $T = (43/m)$	Rank (m)	Sorted Annual Flood Discharge (Cumecs)	Recurrence Interval (Years) $T = (43/m)$
1	4973	43.00	22	1523	1.95
2	4641	21.50	23	1493	1.87
3	3830	14.33	24	1492	1.79
4	3529	10.75	25	1397	1.72
5	2780	8.60	26	1386	1.65
6	2716	7.17	27	1334	1.59
7	2675	6.14	28	1317	1.54
8	2590	5.38	29	1284	1.48
9	2530	4.78	30	1262	1.43
10	2243	4.30	31	1238	1.39
11	1900	3.91	32	1156	1.34
12	1887	3.58	33	1073	1.30
13	1868	3.31	34	1054	1.26
14	1800	3.07	35	1023	1.23
15	1681	2.87	36	830	1.19
16	1610	2.69	37	828	1.16
17	1605	2.53	38	772	1.13
18	1600	2.39	39	757	1.10
19	1583	2.26	40	677	1.08
20	1576	2.15	41	627	1.05
21	1566	2.05	42	623	1.02

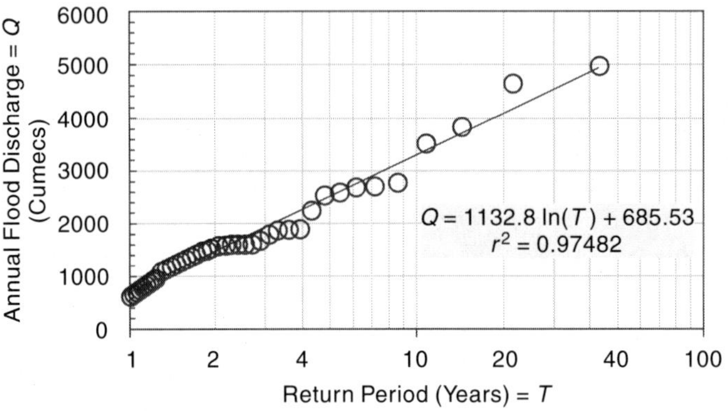

Fig. 7.2 ***Flood Frequency Plot—Example 7.3***

Using the trend line in the graph (or its equation), the required values of the parameters sought in the problem are obtained.

Flood magnitude of return period *(i)* $T = 15$ years, Q = 3753 cumecs.

(ii) T = *50* years, Q = 5117 cumecs

Return period of a flood of magnitude Q = 4000 cumecs, T = 18.7 years.

When large extrapolations of the return period T are involved, theoretical probability distributions have to be used. In frequency analysis of floods, the usual problem is to predict extreme flood events. Towards this, specific extreme-value distributions are assumed and the required statistical parameters calculated from the available data. Using these the flood magnitude for a specific return period is estimated.

Chow (1951) has shown that most frequency distribution functions applicable in hydrologic studies can be expressed by the following equation known as the *general equation of hydrologic frequency analysis*:

$$x_T = \bar{x} + K\sigma \tag{7.13}$$

where x_T = value of the variate X of a random hydrologic series with a return period T, $\bar{x}$ = mean of the variate σ = standard deviation of the variate, K = frequency factor which depends upon the return period, T and the assumed frequency distribution. Some of the commonly used frequency distribution functions for the predication of extreme flood values are:

1. Gumbel's extreme-value distribution
2. Log-Pearson Type III distribution
3. Log normal distribution

Only the first two distributions are dealt with in this book with emphasis on application. Further details and theoretical basis of these and other methods are available in Refs. 2, 3, 7 and 8.

7.6 *Gumbel's Method*

7.6.1 Basic Equations

This extreme value distribution was introduced by Gumbel (1941) and is commonly known as *Gumbel's distribution*. It is one of the most widely used probability distribution functions for extreme values in hydrologic and meteorologic studies for prediction of flood peaks, maximum rainfalls, maximum wind speed, etc.

Gumbel defined a flood as the largest of the 365 daily flows and the annual series of flood flows constitute a series of largest values of flows. According to his theory of extreme events, the probability of occurrence of an event equal to or larger than a value x_0 is

$$P(X \geq x_0) = 1 - e^{-e^{-y}} \tag{7.14}$$

in which y is a dimensionless variable given by

$$y = \alpha(x - a) \qquad a = \bar{x} - 0.45005\,\sigma_x \qquad \alpha = 1.2825/\sigma_x$$

Thus,

$$y = \frac{1.285(x - \bar{x})}{\sigma_x} + 0.577 \tag{7.15}$$

where $\bar{x}$ = mean and σ_x = standard deviation of the variate X. In practice it is the value of X for a given P that is required and as such Eq. (7.14) is transposed as

$$y_p = -\ln[-\ln(1 - P)] \tag{7.16}$$

Noting that the return period $T = 1/P$ and designating

y_T = the value of y, commonly called the reduced variate, for a given T

$$y_T = -\left[\ln . \ln \frac{T}{T-1}\right] \tag{7.17}$$

or

$$y_T = -\left[0.834 + 2.303 \log \log \frac{T}{T-1}\right] \tag{7.17a}$$

Now rearranging Eq. (7.15), the value of the variate X with a return period T is

$$x_T = \bar{x} + K\,\sigma_x \tag{7.18}$$

where

$$K = \frac{(y_T - 0.577)}{1.2825} \tag{7.19}$$

Note that Eq. (7.18) is of the same form as the general equation of hydrologic-frequency analysis (Eq. (7.13)). Further, Eqs. (7.18) and (7.19) constitute the basic Gumbel's equations and are applicable to an infinite sample size (i.e. $N \longrightarrow \infty$).

Since practical annual data series of extreme events such as floods, maximum rainfall depths, etc., all have finite lengths of record (Eq. (7.19)) is modified to account for finite N as given below for practical use.

7.6.2 Gumbel's Equation for Practical Use

Equation (7.18) giving the value of the variate X with a recurrence interval T is used as

$$x_T = \bar{x} + K\,\sigma_{n-1} \tag{7.20}$$

where σ_{n-1} = standard deviation of the sample of size $N = \sqrt{\dfrac{\Sigma(x - \bar{x})^2}{N-1}}$

K = frequency factor expressed as

$$K = \frac{y_T - \bar{y}_n}{S_n} \qquad (7.21)$$

in which y_T = reduced variate, a function of T and is given by

$$y_T = -\left[\ln . \ln \frac{T}{T-1}\right] \qquad (7.22)$$

or

$$y_T = \left[0.834 + 2.303 \log \log \frac{T}{T-1}\right]$$

$\bar{y}_n$ = reduced mean, a function of sample size N and is given in Table 7.3; for $N \to \infty$, $y_n \to 0.577$

S_n = reduced standard deviation, a function of sample size N and is given in Table 7.4; for $N \to \infty$, $S_n \to 1.2825$

These equations are used under the following procedure to estimate the flood magnitude corresponding to a given return period based on an annual flood series.

1. Assemble the discharge data and note the sample size N. Here, the annual flood value is the variate X. Find $\bar{x}$ and σ_{n-1} for the given data.
2. Using Tables 7.3 and 7.4 determine $\bar{y}_n$ and S_n appropriate to given N.
3. Find y_T for a given T by Eq. (7.22).
4. Find K by Eq. (7.21).
5. Determine the required x_T by Eq. (7.20).

The method is illustrated in Example 7.3.

To verify whether the given data follow the assumed Gumbel's distribution, the following procedure may be adopted. The value of x_T for some return periods $T < N$ are calculated by using Gumbel's formulae and plotted as x_T *vs* T on a convenient paper such as a semi-log, log-log or Gumbel probability paper. The use of Gumbel probability paper results in a straight line for x_T *vs* T plot. Gumbel's distribution has the property which gives $T = 2.33$ years for the average of the annual series when N is very large. Thus, the value of a flood with $T = 2.33$ years is called the *mean annual flood.* In graphical plots this gives a mandatory point through which the line showing variation of x_T with T must pass. For the given data, values of return periods (plotting positions) for various recorded values, x of the variate are obtained by the relation $T = (N + 1)/m$ and plotted on the graph described above. Figure 7.3 shows a good fit of observed data with the theoretical variation line indicating the applicability of Gumbel's distribution to the given data series. By extrapolation of the straight line x_T *vs* T, values of x_T for $T > N$ can be determined easily (Example 7.4).

7.6.3 Gumbel Probability Paper

The Gumbel probability paper is an aid for convenient graphical representation of Gumbel's distribution. It consists of an abscissa specially marked for various convenient values of the return period T. To construct the T scale on the abscissa, first construct an arithmetic scale of y_T values, say from –2 to +7, as in Fig. 7.3. For selected values of T, say 2, 10, 50, 100, 500 and 1000, find the values of y_T by Eq. (7.22) and mark off those positions on the abscissa. The T-scale is now ready for use as shown in Fig. 7.3.

Table 7.3 ***Reduced mean $\bar{y}_n$ in Gumbel's Extreme Value Distribution***

N = sample size

N	0	1	2	3	4	5	6	7	8	9
10	0.4952	0.4996	0.5035	0.5070	0.5100	0.5128	0.5157	0.5181	0.5202	0.5220
20	0.5236	0.5252	0.5268	0.5283	0.5296	0.5309	0.5320	0.5332	0.5343	0.5353
30	0.5362	0.5371	0.5380	0.5388	0.5396	0.5402	0.5410	0.5418	0.5424	0.5430
40	0.5436	0.5442	0.5448	0.5453	0.5458	0.5463	0.5468	0.5473	0.5477	0.5481
50	0.5485	0.5489	0.5493	0.5497	0.5501	0.5504	0.5508	0.5511	0.5515	0.5518
60	0.5521	0.5524	0.5527	0.5530	0.5533	0.5535	0.5538	0.5540	0.5543	0.5545
70	0.5548	0.5550	0.5552	0.5555	0.5557	0.5559	0.5561	0.5563	0.5565	0.5567
80	0.5569	0.5570	0.5572	0.5574	0.5576	0.5578	0.5580	0.5581	0.5583	0.5585
90	0.5586	0.5587	0.5589	0.5591	0.5592	0.5593	0.5595	0.5596	0.5598	0.5599
100	0.5600									

Table 7.4 ***Reduced Standard Deviation S_n in Gumbel's Extreme Value Distribution***

N = sample size

N	0	1	2	3	4	5	6	7	8	9
10	0.9496	0.9676	0.9833	0.9971	1.0095	1.0206	1.0316	1.0411	1.0493	1.0565
20	1.0628	1.0696	1.0754	1.0811	1.0864	1.0915	1.0961	1.1004	1.1047	1.1086
30	1.1124	1.1159	1.1193	1.1226	1.1255	1.1285	1.1313	1.1339	1.1363	1.1388
40	1.1413	1.1436	1.1458	1.1480	1.1499	1.1519	1.1538	1.1557	1.1574	1.1590
50	1.1607	1.1623	1.1638	1.1658	1.1667	1.1681	1.1696	1.1708	1.1721	1.1734
60	1.1747	1.1759	1.1770	1.1782	1.1793	1.1803	1.1814	1.1824	1.1834	1.1844
70	1.1854	1.1863	1.1873	1.1881	1.1890	1.1898	1.1906	1.1915	1.1923	1.1930
80	1.1938	1.1945	1.1953	1.1959	1.1967	1.1973	1.1980	1.1987	1.1994	1.2001
90	1.2007	1.2013	1.2020	1.2026	1.2032	1.2038	1.2044	1.2049	1.2055	1.2060
100	1.2065									

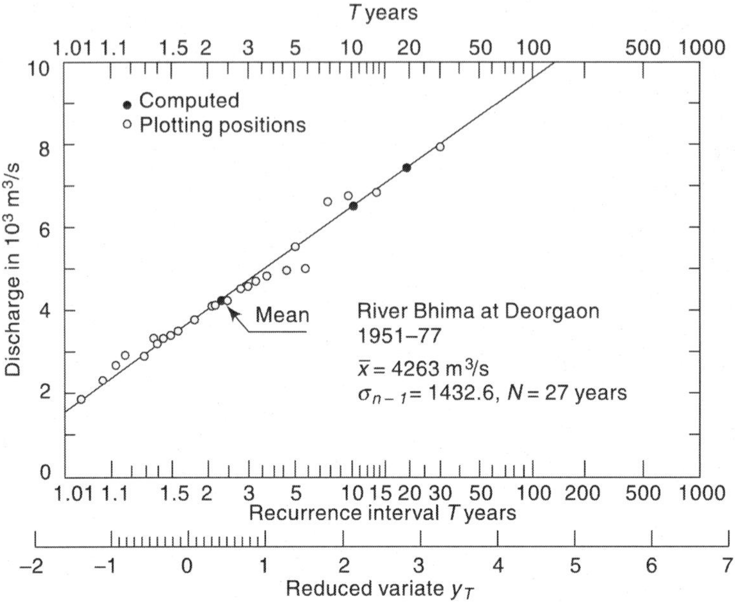

Fig. 7.3 ***Flood Probability Analysis by Gumbel's Distribution***

The ordinate of a Gumbel paper on which the value of the variate, x_T (flood discharge, maximum rainfall depth, etc.) are plotted may have either an arithmetic scale or logarithmic scale. Since by Eqs (7.18) and (7.19) x_T varies linearly with y_T, a Gumbel distribution will plot as a straight line on a Gumbel probability paper. This property can be used advantageously for graphical extrapolaton, wherever necessary.

Example 7.4 Annual maximum recorded floods in the river Bhima at Deorgaon, a tributary of the river Krishna, for the period 1951 to 1977 is given below. Verify whether the Gumbel extreme-value distribution fit the recorded values. Estimate the flood discharge with recurrence interval of (i) 100 years, and (ii) 150 years by graphical extrapolation.

Year	1951	1952	1953	1954	1955	1956	1957	1958	1959
Max. flood (m^3/s)	2947	3521	2399	4124	3496	2947	5060	4903	3757
Year	1960	1961	1962	1963	1964	1965	1966	1967	1968
Max. flood (m^3/s)	4798	4290	4652	5050	6900	4366	3380	7826	3320
Year	1969	1970	1971	1972	1973	1974	1975	1976	1977
Max. flood (m^3/s)	6599	3700	4175	2988	2709	3873	4593	6761	1971

Solution

The flood discharge values are arranged in descending order and the plotting position recurrence interval T_p for each discharge is obtained as

$$T_p = \frac{N+1}{m} = \frac{28}{m}$$

where m = order number. The discharge magnitude Q are plotted against the corresponding T_p on a Gumbel extreme probability paper (Fig. 7.3).

The statistics $\bar{x}$ and an σ_{n-1} for the series are next calculated and are shown in Table 7.5. Using these the discharge x_T for some chosen recurrence interval is calculated by using Gumbel's formulae [Eqs. (7.22), (7.21) and (7.20)].

Table 7.5 ***Calculation of T_p for Observed Data*—Example 7.4**

Order number m	Flood discharge x (m^3/s)	T_p (years)	Order number m	Flood discharge x (m^3/s)	T_p (years)
1	7826	28.00	15	3873	1.87
2	6900	14.00	16	3757	1.75
3	6761	9.33	17	3700	1.65
4	6599	7.00	18	3521	1.56
5	5060	5.60	19	3496	1.47
6	5050	4.67	20	3380	1.40
7	4903	4.00	21	3320	1.33
8	4798	3.50	22	2988	1.27
9	4652	3.11	23	2947	—
10	4593	2.80	24	2947	1.17
11	4366	2.55	25	2709	1.12
12	4290	2.33	26	2399	1.08
13	4175	2.15	27	1971	1.04
14	4124	2.00			

$N = 27$ years, $\bar{x} = 4263$ m^3/s, $\sigma_{n-1} = 1432.6$ m^3/s

From Tables 7.3 and 7.4, for $N = 27$, $y_n = 0.5332$ and $S_n = 1.1004$.

Choosing $T = 10$ years, by Eq. (7.22),

$$y_T = -[\ln . \ln(10/9)] = 2.25037$$

$$K = \frac{2.25307 - 0.5332}{1.1004} = 1.56$$

$$\bar{x}_T = 4263 + (1.56 \times 1432.6) = 6499 \text{ m}^3/\text{s}$$

Similarly, values of x_T are calculated for two more T values as shown below.

T years	x_T [obtained by Eq. (7.20)] (m^3/s)
5.0	5522
10.0	6499
20.0	7436

These values are shown in Fig. 7.3. It is seen that due to the property of the Gumbel's extreme probability paper these points lie on a straight line. A straight line is drawn through these points. It is seen that the observed data fit well with the theoretical Gumbel's extreme-value distribution.

[**Note:** In view of the linear relationship of the theoretical x_T and T on a Gumbel probability paper it is enough if only two values of T and the corresponding x_T are calculated. However, if Gumbel's probability paper is not available, a semi-log plot with log scale for T will have to be used and a large set of (x_T, T) values are needed to identify the theoretical curve.]

By extrapolation of the theoretical x_T *vs* T relationship, from Fig. 7.3,

At T = 100 years, $x_T = 9600$ m³/s

At T = 150 years, $x_T = 10{,}700$ m³/s

[By using Eqs (7.20) to (7.22), $x_{100} = 9558$ m³/s and $x_{150} = 10{,}088$ m³/s.]

Example 7.5

An analysis of annual flood series of a stream indicated the mean value and standard deviation of the flood series as 940 m³/s and 203 m³/s respectively. What is the magnitude of a flood of return period of 500 years in this stream? Assume that the annual flood series follow Gumbel's distribution and the sample size is very large.:

Solution

For a return period of $T = 500$ years, by Eq. 7.17,

$$y_T = -\left[\ln.\ln\left(\frac{T}{T-1}\right)\right] = -\left[\ln.\ln\left(\frac{500}{500-1}\right)\right] = 6.2136$$

$$\text{Frequency factor } K_T = \frac{y_T - 0.577}{1.2825} = \frac{6.2136 - 0.577}{1.2825} = 4.3950$$

$$x_T = \bar{x} + K_T\sigma_{n-1} = 940 + (4.395 \times 203) = 1832 \text{ m}^3/\text{s}$$

Example 7.6

The mean annual flood of a river is 600 m³/s and the standard deviation of the annual flood series is 150 m³/s. What is the probability of a flood of magnitude 1000 m³/s occurring in the river within next 5 years? Use Gumbel's method and assume the sample size to be very large.

Solution

$$\bar{x} = 600 \text{ m}^3/\text{s and } \sigma_{n-1} = 150 \text{ m}^3/\text{s}$$

$$x_T = \bar{x} + K_T\sigma_{n-1}$$

Inserting the given data, $1000 = 600 + K_T\,(150)$

$$K_T = 2.6667 = \frac{y_T - 0.577}{1.2825}$$

Hence, $y_T = 3.9970$

Also, by Eq. 7.17, $y_T = 3.9970 = -\left[\ln.\ln\frac{T}{T-1}\right]$

$$\frac{T}{T-1} = 1.01854$$

$T = 54.9$ years, say 55 years.

Probability of occurrence of a flood of magnitude 1000 $m^3/s = p = 1/55 = 0.0182$.

The probability of a flood of magnitude 1000 m^3/s occurring at least once in 5 years

$$= p_1 = 1 - (1 - p)^5 = 1 - (0.9818)^5$$
$$= 0.0877 = 11.4\%$$

Example 7.7 Flood-frequency computations for the river Chambal at Gandhisagar dam, by using Gumbel's method, yielded the following results.

Return period *T* (years)	Peak flood (m^3/s)
50	40,809
100	46,300

Estimate the flood magnitude in this river with a return period of 500 years.

Solution

By Eq. (7.20),

$$x_{100} = \bar{x} + K_{100}\,\sigma_{n-1} \qquad x_{50} = \bar{x} + K_{50}\,\sigma_{n-1}$$

$$(K_{100} - K_{50})\sigma_{n-1} = x_{100} - x_{50} = 46300 - 40809 = 5491$$

But $$K_T = \frac{y_T}{S_n} - \frac{\bar{y}_n}{S_n}$$

where S_n and $\bar{y}_n$ are constants for the given dat series.

$$\therefore \qquad (y_{100} - y_{50}) = \frac{\sigma_{n-1}}{S_n} = 5491$$

By Eq. (7.22)

$$y_{100} = -[\ln \cdot \ln (100/99)] = 4.60015$$

$$y_{50} = -[\ln \cdot \ln 50/99)] = 3.90194$$

$$\frac{\sigma_{n-1}}{S_n} = \frac{5491}{(4.60015 - 3.90194)} = 7864$$

For $T = 500$ years, by Eq. (7.22),

$$y_{500} = -\,[\ln \cdot \ln (500/499)]\;\; 6.21361$$

$$(y_{500} - y_{100})\frac{\sigma_{n-1}}{S_n} = x_{500} - x_{100}$$

$$(6.21361 - 4.60015) \times 7864 = x_{500} - 46300$$

$$x_{500} = 58988, \text{ say } 59{,}000 \text{ m}^3\text{/s}$$

Confidence Limits

Since the value of the variate for a given return period, x_T determined by Gumbel's method can have errors due to the limited sample data used, an estimate of the confidence limits of the estimate is desirable. The confidence interval indicates the limits about the calculated value between which the true value can be said to lie with a specific probability based on sampling errors only.

For a confidence probability c, the confidence interval of the variate x_T is bounded by values x_1 and x_2 given by[6]

$$x_{1/2} = x_T \pm f(c)\, S_e \tag{7.23}$$

where $f(c)$ = function of the confidence probability c determined by using the table of normal variates as

c in percent	50	68	80	90	95	99
$f(c)$	0.674	1.00	1.282	1.645	1.96	2.8

$$S_e = \text{probable error} = b\,\frac{\sigma_{n-1}}{\sqrt{N}} \tag{7.23a}$$

$$b = \sqrt{1 + 1.3K + 1.1K^2}$$

K = frequency factor given by Eq. (7.21)

σ_{n-1} = standard deviation of the sample

N = sample size.

It is seen that for a given sample and T, 80% confidence limits are twice as large as the 50% limits and 95% limits are thrice as large as 50% limits.

Example 7.8 Data covering a period of 92 years for the river Ganga at Raiwala yielded the mean and standard derivation of the annual flood series as 6437 and 2951 m^3/s respectively. Using Gumbel's method estimate the flood discharge with a return period of 500 years. What are the (a) 95% and (b) 80% confidence limits for this estimate.

Solution

From Tale 7.3 for $N = 92$ years, $\bar{y}_n = 0.5589$ and $S_n = 1.2020$ from Table 7.4.

$$Y_{500} = -[\ln \cdot \ln (500/499)] = 6.21361$$

$$K_{500} = \frac{6.21361 - 0.5589}{1.2020} = 4.7044$$

$$x_{500} = 6437 + (4.7044 \times 2951) = 20320 \text{ m}^3\text{/s}$$

From Eq. (7.33a)

$$b = \sqrt{1 + 1.3(4.7044) + 1.1(4.7044)^2} = 5.61$$

$$S_e \text{ probable error} = 5.61 \times \frac{2951}{\sqrt{92}} = 1726$$

(a) For 95% confidence probability $f(c) = 1.96$ and by Eq. (7.23)

$$x_{1/2} = 20320 \pm (1.96 \times 1726) \quad x_1 = 23703 \text{ m}^3\text{/s and } x_2 = 16937 \text{ m}^3\text{/s}$$

Thus estimated discharge of 20320 m^3/s has a 95% probability of lying between 23700 and 16940 m^3/s

(b) For 80% confidence probability, $f(c) = 1.282$ and by Eq. (7.23)

$x_{1/2} = 20320 \pm (1.282 \times 1726)$ giving $x_1 = 22533 \text{ m}^3/\text{s}$ and $x_2 = 18107 \text{ m}^3/\text{s}$

The estimated discharge of 20320 m^3/s has a 80% probability of lying between 22530 and 18110 m^3/s.

For the data of Example 7.8, the values of x_T for different values of T are calculated and shown plotted on a Gumbel probability paper in Fig. 7.4. This variation is marked as "fitted line" in the figure. Also shown in this plot are the 95 and 80% confidence limits for various values of T. It is seen that as the confidence probability increases, the confidence interval also increases. Further, an increase in the return period T causes the confidence band to spread. Theoretical work by Alexeev (1961) has shown that for Gumbel's distribution the coefficient of skew $C_s \rightarrow 1.14$ for very low values of N. Thus the Gumbel's distribution will give erroneous results if the sample has a value of C_s very much different from 1.14.

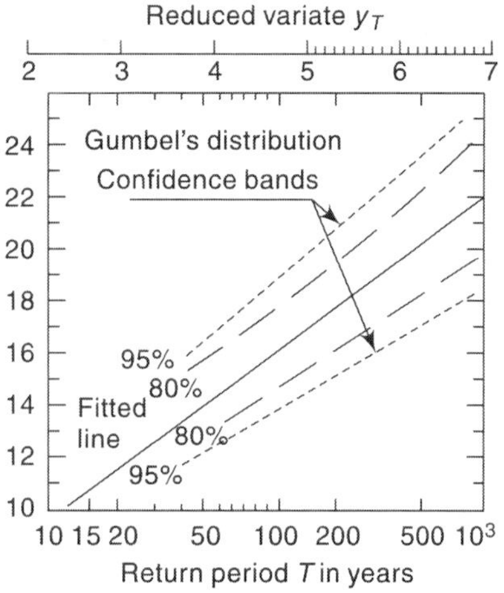

Fig. 7.4 ***Confidence Bands for Gumbels Distribution—Example 7.8***

7.7 Log-Pearson Type III Distribution

This distribution is extensively used in *USA* for projects sponsored by the *US* Government. In this the variate is first transformed into logarithmic form (base 10) and the transformed data is then analysed. If X is the variate of a random hydrologic series, then the series of Z variates where

$$z = \log x \tag{7.24}$$

are first obtained. For this Z series, for any recurrence interval T, Eq. (7.13) gives

$$z_T = \bar{z} + K_z \sigma_z \tag{7.25}$$

where K_z = a frequency factor which is a function of recurrence interval T and the coefficient of skew C_s,

σ_z = standard deviation of the Z variate sample

$$= \sqrt{\Sigma(z-\bar{z})^2/(N-1)} \tag{7.25a}$$

and C_s = coefficient of skew of variate Z

$$= \frac{N\Sigma(z-\bar{z})^3}{(N-1)(N-2)(\sigma_z)^3} \tag{7.25b}$$

$\bar{z}$ = mean of the z values

N = sample size = number of years of record

The variations of $K_z = f(C_s, T)$ is given in Table 7.6.

After finding z_T by Eq. (7.25), the corresponding value of x_T is obtained by Eq. (7.24) as

$$x_T = \text{antilog } (z_T) \quad (7.26)$$

Sometimes, the coefficient of skew C_s, is adjusted to account for the size of the sample by using the following relation proposed by Hazen 1930).

$$\hat{C}_s = C_s\left(\frac{1+8.5}{N}\right) \quad (7.27)$$

where $\hat{C}_s$ = adjusted coefficient of skew. However, the standard procedure for use of log-Pearson Type III distribution adopted by *U.S.* Water Resources Council does not include this adjustment for skew.

When the skew is zero, i.e. $C_s = 0$, the log-Pearson Type III distribution reduces to *log normal distribution.* The log-normal distribution plots as a straight line on logarithmic probability paper.

Table 7.6 ***$K_z = F(C_s, T)$ for Use in Log-Pearson Type III Distribution***

Coefficient of skew, C_s	Recurrence interval T in years						
	2	10	25	50	100	200	1000
3.0	–0.396	1.180	2.278	3.152	4.051	4.970	7.250
2.5	–0.360	1.250	2.262	3.048	3.845	4.652	6.600
2.2	–0.330	1.284	2.240	2.970	3.705	4.444	6.200
2.0	–0.307	1.302	2.219	2.912	3.605	4.298	5.910
1.8	–0.282	1.318	2.193	2.848	3.499	4.147	5.660
1.6	–0.254	1.329	2.163	2.780	3.388	3.990	5.390
1.4	–0.225	1.337	2.128	2.706	3.271	3.828	5.110
1.2	–0.195	1.340	2.087	2.626	3.149	3.661	4.820
1.0	–0.164	1.340	2.043	2.542	3.022	3.489	4.540
0.9	–0.148	1.339	2.018	2.498	2.957	3.401	4.395
0.8	–0.132	1.336	1.998	2.453	2.891	3.312	4.250
0.7	–0.116	1.333	1.967	2.407	2.824	3.223	4.105
0.6	–0.099	1.328	1.939	2.359	2.755	3.132	3.960
0.5	–0.083	1.323	1.910	2.311	2.686	3.041	3.815
0.4	–0.066	1.317	1.880	2.261	2.615	2.949	3.670
0.3	–0.050	1.309	1.849	2.211	2.544	2.856	3.525
0.2	–0.033	1.301	1.818	2.159	2.472	2.763	3.380
0.1	–0.017	1.292	1.785	2.107	2.400	2.670	3.235
0.0	**0.000**	**1.282**	**1.751**	**2.054**	**2.326**	**2.576**	**3.090**
–0.1	0.017	1.270	1.716	2.000	2.252	2.482	2.950
–0.2	0.033	1.258	1.680	1.945	2.178	2.388	2.810
–0.3	0.050	1.245	1.643	1.890	2.104	2.294	2.675
–0.4	0.066	1.231	1.606	1.834	2.029	2.201	2.540

Contd

Table 7.6 *Contd*

–0.5	0.083	1.216	1.567	1.777	1.955	2.108	2.400
–0.6	0.099	1.200	1.528	1.720	1.880	2.016	2.275
–0.7	0.116	1.183	1.488	1.663	1.806	1.926	2.150
–0.8	0.132	1.166	1.448	1.606	1.733	1.837	2.035
–0.9	0.148	1.147	1.407	1.549	1.660	1.749	1.910
–1.0	0.164	1.128	1.366	1.492	1.588	1.664	1.880
–1.4	0.225	1.041	1.198	1.270	1.318	1.351	1.465
–1.8	0.282	0.945	1.035	1.069	1.087	1.097	1.130
–2.2	0.330	0.844	0.888	0.900	0.905	0.907	0.910
–3.0	0.396	0.660	0.666	0.666	0.667	0.667	0.668

[***Note:*** $C_s = 0$ corresponds to log-normal distribution]

Example 7.9 For the annual flood series data of the river Bhima given in Example 7.4, estimate the flood discharge for a return period of (a) 100 years (b) 200 years, and (c) 1000 years by using log-Pearson Type III distribution.

Solution

The variate $z = \log x$ is first calculated for all the discharges (Table 7.7). Then the statistics $\overline{Z}$, σ_z and C_s are calculated from Table 7.7 to obtain

Table 7.7 ***Variate Z—Example 7.9***

Year	Flood x (m³/s)	$z = \log x$	Year	Flood x (m³/s)	$z = \log x$
1951	2947	3.4694	1965	4366	3.6401
1952	3521	3.5467	1966	3380	3.5289
1953	2399	3.3800	1967	7826	3.8935
1954	4124	3.6153	1968	3320	3.5211
1955	3496	3.5436	1969	6599	3.8195
1956	2947	3.4694	1970	3700	3.5682
1957	5060	3.7042	1971	4175	3.6207
1958	4903	3.6905	1972	2988	3.4754
1959	3751	3.5748	1973	2709	3.4328
1960	4798	3.6811	1974	3873	3.5880
1961	4290	3.6325	1975	4593	3.6621
1962	4652	3.6676	1976	6761	3.8300
1963	5050	3.7033	1977	1971	3.2947
1964	690	38388			

$$\sigma_z = 0.1427 \qquad C_s = \frac{27 \times 0.0030}{(26)(25)(0.1427)^3}$$

$$\overline{Z} = 3.6071 \qquad C_s = 0.043$$

The flood discharge for a given T is calculated as below. Here, values of K_z for given T and $C_s = 0.04$ are read from Table 7.6.

T (years)	$\bar{Z} = 3.6071$	$\sigma_z = 0.1427$	$C_s = 0.043$	x_T = antilog z_T (m^3/s)
	K_z (from Table 7.6) (for $C_s = 0.043$)	$K_z \sigma_z$	$Z_T = \bar{Z} + K_z \sigma_z$	
100	2.358	0.3365	3.9436	8782
200	2.616	0.3733	3.9804	9559
1000	3.152	0.4498	4.0569	11400

Example 7.10 For the annual flood series data analysed in Example 7.9 estimate the flood discharge for a return period of (a) 100 years, (b) 200 years, and (c) 1000 years by using log-normal distribution. Compare the results with those of Example 7.9.

Solution

Log-normal distribution is a special case of log-Pearson type III distribution with $C_s = 0$. Thus in this case C_s is taken as zero. The other statistics are $\bar{z} = 3.6071$ and $\sigma_z = 0.1427$ as calculated in Example 7.9.

The value of K for a given return period T and $C_s = 0$ is read from Table 7.6. The estimation of the required flood discharge is done as shown below.

T (years)	$z = 3.6071$	$\sigma_z = 0.1427$	$C_s = 0$	x_T = antilog z_T (m^3/s)
	K_z (from Table 7.6)	$K_z \sigma_z$	$Z_T = Z + K_z \sigma_z$	
100	2.326	0.3319	3.9390	8690
200	2.576	0.3676	3.9747	9434
1000	3.090	0.4409	4.0480	11170

On comparing the estimated x_T with the corresponding values in Example 7.9, it is seen that the inclusion of the positive coefficient of skew ($C_s = 0.047$) in log-Pearson type III method gives higher values than those obtained by the log-normal distribution method. However, as the value of C_s is small, the difference in the corresponding values of x_T by the two methods is not appreciable.

[**Note:** If the coefficient of skew is negative, the log-Pearson type III method gives consistently lower values than those obtained by the log-normal distribution method.]

Example 7.11 Annual flood flow series of a river were analysed and was found to follow log normal distribution. The frequency analysis of the data yielded the following results:

Return Period T in years	Peak Flood (m^3/s)
100	12500
200	15000

The following table of the variation of frequency factor with the return period in log-normal distribution is available:

Return Period (*T*)	50	100	200	1000
Frequency factor (K_z)	2.054	2.326	2.576	3.090

Estimate the flood magnitude in the river with a return period of 1000 years.

Solution

Let x = Random variate (Flood discharge) in the problem

The transformed variate $z = \log x$ is distributed log-normally.

For the variate z at any return period T

$$z_T = \bar{z} + K_z \sigma_z$$

For the given data:

At T = 100 years: $x_{100} = 12500$, $z_{100} = \log 12500 = 4.0969$ and

$K_{100} = 2.326$

Hence, $\bar{z} + 2.326\, \sigma_z = 4.0969$ (i)

At T = 200 years: $x_{200} = 15000$, $z_{200} = \log 15000 = 4.1761$ and

$K_{200} = 2.576$

Hence, $\bar{z} + 2.576\, \sigma_z = 4.1761$ (ii)

From Eqs.(1) and (ii); $0.25\, \sigma_z = 0.0792$ and $\sigma_z = 0.3168$

Substituting for σ_z in Eq.(ii); $\bar{z} = 4.1761 - (2.576 \times 0.3168) = 3.3600$

When T = 1000 years, $K_{1000} = 3.090$

$$z_{1000} = \bar{z} + 3.090\, \sigma_z = 3.3600 + (3.090 \times 0.3168) = 4.3389$$

Flood with a return period of 1000 years = x_{1000} = antilog 4.3389 = 21,823 m^3/s

7.8 *Partial Duration Series*

In the annual hydrologic data series of floods, only one maximum value of flood per year is selected as the data point. It is likely that in some catchments there are more than one independent floods in a year and many of these may be of appreciably high magnitude. To enable all the large flood peaks to be considered for analysis, a flood magnitude larger than an arbitrary selected base value are included in the analysis. Such a data series is called *partial-duration series.*

In using the partial-duration series, it is necessary to establish that all events considered are independent. Hence the partial-duration series is adopted mostly for rainfall analysis where the conditions of independency of events are easy to establish. Its use in flood studies is rather rare. The recurrence interval of an event obtained by annual series (T_A) and by the partial duration series (T_p) are related by

$$T_P = \frac{1}{\ln T_A - \ln(T_A - 1)} \tag{7.28}$$

From this, it can be seen that the difference between T_A and T_P is significant for $T_A < 10$ years and that for $T_A > 20$, the difference is negligibly small.

7.9 *Regional Flood-Frequency Analysis*

When the available data at a catchment is too short to conduct frequency analysis, a regional analysis is adopted. In this, a hydrologically homogeneous region from the statistical point of view is considered. Available long-time data from neighbouring catchments are tested for homogeneity and a group of stations satisfying the test are identified. This group of stations constitutes a region and all the station data of this region are pooled and analysed as a group to find the frequency characteristics of the region. The mean annual flood Q_{ma}, which corresponds to a recurrence interval of 2.33 years is used for non-dimensionalising the results. The variation of Q_{ma} with drainage area and the variation of Q_T/Q_{ma} with T where Q_T is the discharge for any T are the basic plots prepared in this analysis. Details of the method are available in Ref. 2.

7.10 *Extremes of Extremes—Envelope Curve of Highest Floods in the World*

The concept of envelope curve has been applied to develop relation for highest floods that have occurred anywhere in the world. Highest flood in a river basin is defined as the highest recorded flood in that river basin in its history. A collection of highest floods of different basins in a large region forms a set that could be analysed by envelope method. Highest floods, which by its definition are floods with very low probability of exceedance, in a large region or world over have some general characteristics. Using the discharge intensity, defined as discharge per unit catchment area, the following broad characteristics are identified:

- Large unit discharges are found in the rivers of moist tropics
- Only those basins that have full exposure to tropical cyclones, monsoon winds, or other heavy rain bearing wind systems have higher unit discharge. Typical examples are the Brahmaputra, Ganga, Yangtze, Mekong and Huangue rivers. Also, Amazon and Orinaco river basins belong to this category
- Correspondingly, several large river basins like the Congo, Niger, Nile, Zambezi, Murray, Colorado and Mississippi do not produce high unit discharge because of their geographic locations and meteorological conditions prevailing over them.

Thus, for analysis, the highest flood flow and catchment area data of river basins across the globe are plotted on a log–log scale plot and an enveloping curve drawn to envelope the data points. The curve represents the physical upper limit of flood from a catchment of known area. The meteorological conditions producing such high floods are homogeneous in a general sense of being extreme combinations. Further, the data itself has been sorted out by meteorological factors and there is no need for further classification. The envelope curve represents floods of limiting state of highest floods and is indicative of maximum possible flood from a catchment of known size.

These curves are of very low probability of exceedance; return period being of the order of 1000 years.

Rakhecha (Ref. 9) has analysed the available data of highest floods in Indian river basins and also the available highest flood data of world river basins and has noted the following significant points:

- The highest recorded floods in some river basins in India are comparable to highest reported floods in other parts of the world for drainage basins larger than 1000 km^2.
- Two floods recorded in India in recent past, viz., (a) flood of 6th Sept. 1970 on the Narmada river at Garudeshwar, and (b) flood of 11th Aug. 1979 on the Machhu river (in Kutch, Gujarat), are found to be record breaking events in the world.

Data on highest floods in 32 Indian basins and 40 other river basins spread world over available in Ref. 9 is shown plotted in Fig. 7.5 on a log-log plot. It is seen that the data of Indian rivers are remarkably comparable with the highest flood discharges recorded in other parts of the world for areas greater than 1000 km^2. For catchment areas less than 1000 km^2, the Indian data is consistently smaller than the world highest flood data. The enveloping line covering all the plotted points except for one outlier, could be expressed as

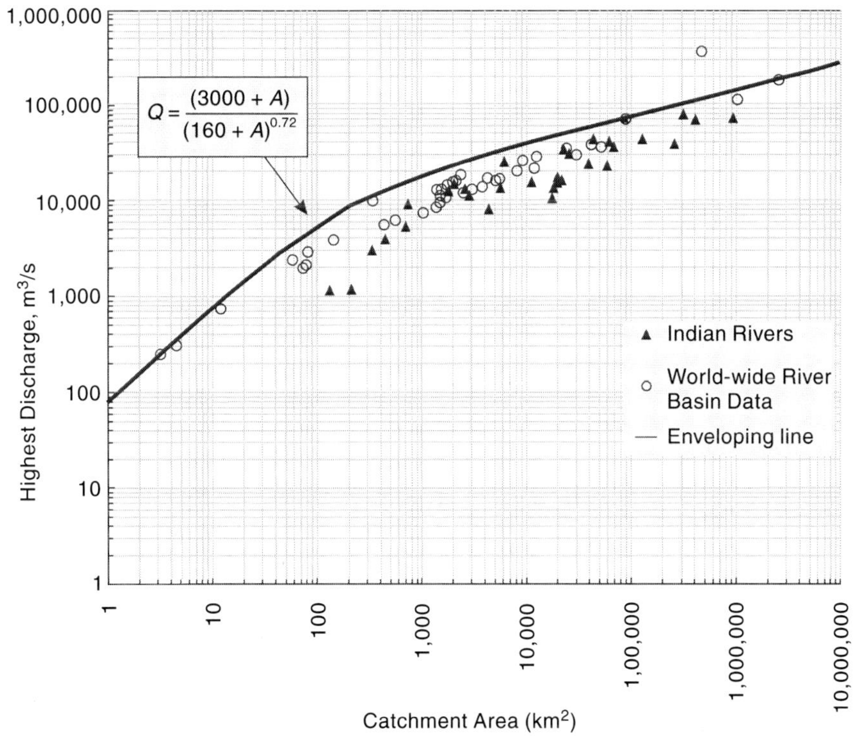

Fig. 7.5 ***Enveloping Line for Worldwide River Basin Data on Highest Floods (Data from Ref. 9)***

$$Q = \frac{(3000 + A)}{(160 + A)^{0.72}} \tag{7.29}$$

where Q = Highest flood discharge in m^3/s and A = catchment area in km^2.

The highest flood discharge calculated by the enveloping curve (or by Eq. 7.29 of world data on river basins (which includes Indian data) are extremely useful in (i) examining the safety of existing major dams and hydraulic structures, (ii) for investigating the estimates of possible maximum floods (*PMF*) for any new or ongoing dam projects, especially those with inadequate on-site data.

7.11 *Data for Frequency Studies*

The flood-frequency analysis described in the previous sections is a direct means of estimating the desired flood based upon the available flood flow data of the catchment. The results of the frequency analysis depend upon the length of data. The minimum number of years of record required to obtain satisfactory estimates depends upon the variability of data and hence on the physical and climatological characteristics of the basin. Generally, a minimum of 30 years of data is considered as essential. Smaller lengths of records are also used when it is unavoidable. However, frequency analysis should not be adopted if the length of records is less than 10 years.

In the frequency analysis of time series, such as of annual floods, annual yields and of precipitation, sometimes one comes across very long (say of the order of 100 years) times series. In such cases it is necessary to test the series for *homogeneity* to ascertain that there is no significant difference in the causative hydrological processes over the span of the time series. A time series is called time-homogeneous (also known *stationary*) if identical events under consideration in the series are likely to occur at all times. Departure from time homogeneity is reflected either in trend or periodicity or persistence of the variable over time. Potential non-homogeneity region, (if any), could be detected by (i) mass curve, or (ii) by moving mean of the variable. Statistical tests like *F-test* for equality of variances and *t-test* for significance of differences of means are adopted to identify non-homogeneous region/s in the series. Only the contiguous homogeneous region of the series covering the recent past is to be adopted for frequency analysis. However, it is prudent to test all time series, whether long or short, for time-homogeneity before proceeding with the frequency analysis. Thus the cardinal rule with the data of time series would be that the data should be reliable and homogeneous.

Flood-frequency studies are most reliable in climates that are uniform from year to year. In such cases a relatively short record gives a reliable picture of the frequency distribution.

7.12 *Design Flood*

In the design of hydraulic structures, it is not practical from economic considerations to provide for the safety of the structure and the system at the maximum possible flood in the catchment. Small structures such as culverts and storm drainages can be designed for less severe floods as the consequences of a higher than design flood may not be very serious. It can cause temporary inconvenience like the disruption of traffic and very rarely severe property damage and loss of life. On the other hand, storage structures such as dams demand greater attention to the magnitude of floods used in the design. The failure of these structures causes large loss of life and great property damage on the downstream of the structure. From this it is apparent that the type, importance of the structure and economic development of the surrounding area dictate the design criteria for choosing the flood magnitude. This section highlights the procedures adopted in selecting the flood magnitude for the design of some hydraulic structures.

The following definitions are first noted.

- **Design Flood** Flood adopted for the design of a structure.
- **Spillway Design Flood** Design flood used for the specific purpose of designing the spillway of a storage structure. This term is frequently used to denote the maximum discharge that can be passed in a hydraulic structure without any damage or serious threat to the stability of the structure.
- **Standard Project Flood (*SPF*)** The flood that would result from a severe combination of meteorological and hydrological factors that are reasonably applicable to the region. Extremely rare combinations of factors are excluded.
- **Probable Maximum Flood (*PMF*)** The extreme flood that is physically possible in a region as a result of severemost combinations, including rare combinations of meteorological and hydrological factors.

The *PMF* is used in situations where the failure of the structure would result in loss of life and catastrophic damage and as such complete security from potential floods is sought. On the other hand, *SPF* is often used where the failure of a structure would cause less severe damages. Typically, the *SPF* is about 40 to 60% of the PMF for the same drainage basin. The criteria used for selecting the design flood for various hydraulic structures vary from one country to another. Table 7.8 gives a brief summary of the guidelines adopted by *CWC* India, to select design floods.

Table 7.8 ***Guidelines for Selecting Design Floods (CWC, India)***[1]

S. No.	Structure	Recommended design flood
1.	Spillways for major and medium projects with storages more than 60 Mm^3	(a) *PMF* determined by unit hydrograph and probable maximum precipitation (*PMP*) (b) If (a) is not applicable or possible flood-frequency method with $T = 1000$ years
2.	Permanent barrage and minor dams with capacity less than 60 Mm^3	(a) *SPF* determined by unit hydrograph and standard project storm (*SPS*) which is usually the largest recorded storm in the region (b) Flood with a return period of 100 years. (a) or (b) whichever gives higher value.

Contd

Table 7.8 *Contd*

3.	Pickup weirs	Flood with a return period of 100 or 50 years depending on the importance of the project.
4.	Aqueducts (a) Waterway (b) Foundations and free board	 Flood with $T = 50$ years Flood with $T = 100$ years
5.	Project with very scanty or inadequate data	Empirical formulae

The Indian Standard Guidelines for Design of Floods for Dams

"IS : 11223—1985 : Guidelines for fixing spillway capacity" (Ref. 4) is currently used in India for selection of design floods for dams. In these guidelines, dams are classified according to size by using the hydraulic head and the gross storage behind the dam. The hydraulic head is defined as the difference between the maximum water level on the upstream and the normal annual average flood level on the downstream. The classification is shown in Table 7.9. The overall size classifications for dams would be greater of that indicated by either of the two parameters. For example, a dam with a gross storage of 5 Mm^3 and hydraulic head of 15 m would be classified as *intermediate* size dam.

The inflow design flood (*IDF*) for safety of the dam is taken for each class of dam as given in Table 7.9.

Table 7.9 ***Size Classification and Inflow Design floods for Dams***

Class	Gross Storage (Mm^3)	Hydraulic Head (m)	Inflow Design Flood for Safety
Small	0.5 to 10.0	7.5 to 12.0	100-year flood
Intermediate	10.0 to 60.0	12.0 to 30.0	Standard project Flood (*SPF*)
Large	> 60.0	>30.0	Probable Maximum Flood (*PMF*)

7.13 *Design Storm*

To estimate the design flood for a project by the use of a unit hydrograph, one needs the design storm. This can be the storm-producing probable maximum precipitation (*PMP*) for deriving *PMF* or a standard project storm (*SPS*) for *SPF* calculations. The computations are performed by experienced hydrometeorologists by using meteorological data. Various methods ranging from highly sophisticated hydrometeorological methods to simple analysis of past rainfall data are in use depending on the availability of reliable relevant data and expertise.

The following is a brief outline of a procedure followed in India:

- The duration of the critical rainfall is first selected. This will be the basin lag if the flood peak is of interest. If the flood volume is of prime interest, the duration of the longest storm experienced in the basin is selected.
- Past major storms in the region which conceivably could have occurred in the basin under study are selected. *DAD* analysis is performed and the enveloping

curve representing maximum depth–duration relation for the study basin obtained.

- Rainfall depths for convenient time intervals (e.g. 6 h) are scaled from the enveloping curve. These increments are to be arranged to get a critical sequence which produces the maximum flood peak when applied to the relevant unit hydrograph of the basin.

  The critical sequence of rainfall increments can be obtained by trial and error. Alternatively, increments of precipitation are first arranged in a table of relevant unit hydrograph ordinates, such that (i) the maximum rainfall increment is against the maximum unit hydrograph ordinate, (ii) the second highest rainfall increment is against the second largest unit hydrograph ordinate, and so on, and (iii) the sequence of rainfall increments arranged above is now reversed, with the last item first and first item last. The new sequence gives the design storm (Example 7.11).
- The design storm is then combined with hydrologic abstractions most conducive to high runoff, viz. low initial loss and lowest infiltration rate to get the hyetograph of rainfall excess to operate upon the unit hydrograph.

  Further details about the above procedure and other methods for computing design storms are available in Ref. 7. Reference 1 gives details of the estimation of the design flood peak by unit hydrographs for small drainage basins of areas from 25–500 km^2.

Example 7.12 The ordinates of cumulative rainfall from the enveloping maximum depth—duration curve for a basin are given below. Also, given are the ordinates of a 6-h unit hydrograph. Design the critical sequence of rainfall excesses by taking the φ index to be 0.15 cm/h.

Solution:

The critical storm and rainfall excesses are calculated in a tabular form in Table 7.10.

Time from start (h)	0	6	12	18	24	30	36	42	48	54	60
Cumulative rainfall (cm)	0	15	24.1	30	34	37	39	40.5	41.3		
6-h *UH* ordinate (m^3/s)	0	20	54	98	126	146	154	152	138	122	106
Time from start (h)	66	72	78	84	90	96	102	108	114	129	132
6-h *UH* ordinate (m^3/s)	92	79	64	52	40	30	20	14	10	6	0

Table 7.10 *Calculation of Critical Storm*—**Example 7.11**

Time (h)	Cumulative rainfall (cm)	6-h incremental rainfall (cm)	Ordinate of 6-h *UH* (m^3/s)	First arrangement of rainfall increment	Design sequence of rainfall increment	Infiltration loss (cm)	Rainfall excess of design storm (cm)
1	2	3	4	5	6	7	8
0	0		0		0	0	0
6	15.0	15.0	20		1.5	0.9	0.6
12	24.1	9.1	54		2.0	0.9	1.1
18	30.0	5.9	98	0.8	4.0	0.9	3.1
24	34.0	4.0	126	3.0	9.1	0.9	8.2
30	37.0	3.0	146	5.9	15.0	0.9	14.1
36	39.0	2.0	154	15.0	5.9	0.9	5.0
42	40.5	1.5	152	9.1	3.0	0.9	2.1
48	41.3	0.8	138	4.0	0.8	0.9	0
54			122	2.0			
60			106	1.5			
66			92				
72			79				
78			64				
84			52				
90			40				
96			30				
102			20				
108			14				
114			10				
120			6				
132			0				

1. (Column 6 is reversed sequence of column 5)
2. Infiltration loss = 0.15 cm/h = 0.9 cm/6 h

7.14 *Risk, Reliability and Safety Factor*

7.14.1 Risk and Reliability

The designer of a hydraulic structure always faces a nagging doubt about the risk of failure of his structure. This is because the estimation of the hydrologic design values (such as the design flood discharge and the river stage during the design flood) involve a natural or inbuilt uncertainty and as such a hydrological risk of failure. As an example, consider a weir with an expected life of 50 years and designed for a flood magnitude of return period T = 100 years. This weir may fail if a flood magnitude greater than the design flood occurs within the life period (50 years) of the weir.

The probability of occurrence of an event ($x \geq x_T$) at least once over period of n successive year is called the risk, $\bar{R}$. Thus, the risk is given by $R = 1 -$ (probability of non-occurrence of the event $x \geq x_T$ in n years)

$$R = 1 - (1 - P)^n$$

$$= 1 - \left(1 - \frac{1}{T}\right)^n \tag{7.30}$$

where P = probability $P\ (x \geq x_T) = \dfrac{1}{T}$

T = return period

The reliability R_e, is defined as

$$R_e = 1 - \bar{R} = \left(1 - \frac{1}{T}\right)^n \tag{7.31}$$

It can be seen that the return period for which a structure should be designed depends upon the acceptable level of risk. In practice, the acceptable risk is governed by economic and policy considerations.

7.14.2 Safety Factor

In addition to the hydrological uncertainty, as mentioned above, a water resource development project will have many other uncertainties. These may arise out of structural, constructional, operational and cnvironmental causes as well as from non-technological considerations such as economic, sociological and political causes. As such, any water resource development project will have a safety factor for a given hydrological parameter M as defined below

$$\text{Safty factor (for the parameter } M) = (S\,F)_m = \frac{\text{Actual value of the parameter } M \text{ adopted in the design of the project}}{\text{Value of the parameter } M \text{ obtained from hydrological considerations only}}$$

$$= \frac{C_{am}}{C_{hm}} \tag{7.32}$$

The parameter M includes such items as flood discharge magnitude, maximum river stage, reservoir capacity and free board. The difference ($C_{am} - C_{hm}$) is known as *safety margin.*

The concepts of risk, reliability and safety factor form the building block of the emerging field of reliability based design.

Example 7.13 A bridge has an expected life of 25 years and is designed for a flood magnitude of return period 100 years. (a) What is the risk of this hydrologic design? (b) If a 10% risk is acceptable, what return period will have to be adopted?

Solution

(a) The risk $\bar{R} = 1 - \left(1 - \frac{1}{T}\right)^n$

Here $n = 25$ years and $T = 100$ years

$$\bar{R} = 1 - \left(1 - \frac{1}{100}\right)^{25} = 0.222$$

Hence the inbuilt risk in this design is 22.2%

(b) If $\bar{R} = 10\% = 0.10$ $\qquad 0.10 = 1 - \left(1 - \frac{1}{T}\right)^{25}$

$\left(1 - \frac{1}{T}\right)^{25} = 0.90$ and $T = 238$ years = say 240 years.

Hence to get 10% acceptable risk, the bridge will have to be designed for a flood of return period $T = 240$ years.

Example 7.14 Analysis of annual flood series of a river yielded a sample mean of 1000 m^3/s and standard deviation of 500 m^3/s. Estimate the design flood of a structure on this river to provide 90% assurance that the structure will not fail in the next 50 years. Use Gumbel's method an assume the sample size to be very large.

Solution

$$\bar{x} = 100 \text{ m}^3/\text{s} \text{ and } \sigma_{n-1} = 500 \text{ m}^3/\text{s}$$

Reliability $\quad R_e = 0.90 = \left(1 - \frac{1}{T}\right)^{50}$

$$1 - \frac{1}{T} = (0.90)^{1/50} = 0.997895$$

$$T = 475 \text{ years} \quad x_T = \bar{x} + K\sigma_{n-1} \qquad K = \frac{y_T - 0.577}{1.2825}$$

Also, $\quad y_T = -\left[\ln . \ln \frac{475}{(475-1)}\right] = 6.16226$

$$K = \frac{6.16226 - 0.577}{1.2825} = 4.355$$

$$x_T = 1000 + (4.355) \times 500 = 3177 \text{ m}^3/\text{s}$$

Example 7.15 Annual flood data of the river Narmada at Garudeshwar covering the period 1948 to 1979 yielded for the annual flood discharges a mean of 29,600 m^3/s and a standard deviation of 14,860 m^3/s. For a proposed bridge on this river near this site it is decided to have an acceptable risk of 10% in its expected life of 50 years. (a) Estimate the flood discharge by Gumbel's method for use in the design of this structure. (b) If the actual flood value adopted in the design is 125,000 m^3/s what are the safety factor and safety margin relating to maximum flood discharge?

Solution

Risk $\bar{R} = 0.10$

Life period of the structure $n = 50$ years

Hence $\bar{R} = 0.10 = 1 - \left(1 - \frac{1}{T}\right)^{50}$

$$\left(1 - \frac{1}{T}\right) = (1 - 0.10)^{1/50} = 0.997895$$

$$T = 475 \text{ years}$$

Gumbel's method is now used to estimate the flood magnitude for this return period of $T = 475$ years.

Record length N = 1948 to 1979 = 32 years

From Tables 7.3 and 7.4, $\bar{y}_n = 0.5380$ and $S_n = 1.1193$

$$y_T = -\left[\ln.\ln\frac{T}{T-1}\right] = -\left[\ln.\ln\frac{475}{(475-1)}\right] = 6.16226$$

$$K = \frac{y_T - \bar{y}_n}{S_n} = \frac{(6.16226 - 0.5380)}{1.1193} = 5.0248$$

$$x_T = \bar{x}_T + K\,\sigma_{n-1}$$

$$= 29600 + (5.0248 \times 14860) = 104268$$

say = 105,000 m^3/s = hydrological design flood magnitude

Actual flood magnitude adopted in the project is = 125,000 m^3/s

Safety factor = $(SF)_{\text{flood}}$ = 125,000/105,000 = 1.19

Safety margin of flood magnitude = 125,000 – 105,000 = 20,000 m^3/s

REFERENCES

1. Central Water Commission, India, "Estimation of design flood peak", Flood Estimation Directorate, *Report No.* 1/73, New Delhi, 1973.
2. Chow. V.T., *Handbook of Applied Hydrology*, McGraw-Hill, New York, NY, 1964.
3. Gray, D.M., *Principles of Hydrology,* Water Inf. Centre, Huntington, NY, 1970.
4. Indian Bureau of Standards, "*Guidelines for fixing Spillway Capacity*", IS: 11223 – 1985.
5. Khushalani, K.B. and M. Khushalani, *Irrigation Practice and Design.* Vol. 1, Oxford & *IBH*, New Delhi, 1971.
6. Nemec, J., *Engineering Hydrology,* Tata McGraw-Hill, New Delhi, 1973.

7. Sokolov, A.A., S.E. Rantz, and M. Roche, *Flood Computation— Methods Compiled from World Experience,* The *UNESCO* Press, Paris, 1976.
8. *UNO/WMO*, "Assessment of magnitude and frequency of flood flows", United Nations Pub., *Water sources Series No. 30*, 1967.
9. Rakhecha, P. R., *Highest floods in India*, *IAHS* Pub. No. 271, 2002.
10. Rodier, I. A., and Roche, M., World catalogue of maximum Observed Floods, ISHS Pub. No., 143, 1984.
11. O' Connor, J. E., and Costa, J. E., *The World's largest Floods, Past and Present; Their Causes and Magnitude*, USGS Circular 1254, *USA*, 2004.

REVISION QUESTIONS

7.1 Explain the rational method of computing the peak discharge of a small catchment. Where is this method commonly used and what are its merits and demerits?

7.2 Discuss the factors affecting the runoff coefficient C in rational formula.

7.3 What do you understand by time of concentration of a catchment? Describe briefly methods of estimation of the time of concentration.

7.4 What is the importance of time of concentration of a catchment in the estimation of flood by rational formula?

7.5 Annual flood series having N consecutive entries are available for a catchment. Describe a procedure to verify whether the data follow Gumbel's distribution.

7.6 Write a brief note on frequency factor and its estimation in Gumbel's method.

7.7 If the annual flood series data for a catchment are available for N consecutive years, explain a procedure to determine a flood discharge with a return period of T, (where $T > N$), by using

(a) Log-Pearson type III distribution, and (b) Log-normal distribution.

7.8 What are the limitations of flood frequency studies?

7.9 Explain briefly the following terms:

(a) Design flood (b) Standard project flood
(c) Probable maximum flood (d) Design storm

7.10 What are the recommended design floods for

(a) Spillways of dams (b) Terrace outlets and vegetated waterways
(c) Field diversions (d) Permanent barrages
(e) Waterway for aqueducts

7.11 Explain briefly the following terms:

(a) Risk (b) Reliability (c) Safety margin

7.12 Write a note on the use of envelope curve of extreme flood peaks indicting its uses.

PROBLEMS

7.1 A catchment of area 120 ha has a time of concentration of 30 min and runoff coefficient of 0.3. If a storm of duration 45 min results in 3.0 cm of rain over the catchment estimate the resulting peak flow rate.

7.2 Information on the 50-year storm is given below.

Duration (minutes)	15	30	45	60	180
Rainfall (mm)	40	60	75	100	120

A culvert has to drain 200 ha of land with a maximum length of travel of 1.25 km. The general slope of the catchment is 0.001 and its runoff coefficient is 0.20. Estimate the peak flow by the rational method for designing the culvert for a 50-year flood.

7.3 A basin is divided by 1-h isochrones into four sub-areas of size 200, 250, 350 and 170 hectares from the upstream end of the outlet respectively. A rainfall event of 5-h duration with intensities of 1.7 cm/h for the first 2 h and 1.25 cm/h for the next 3 h occurs uniformly over the basin. Assuming a constant runoff coefficient of 0.5, estimate the peak rate of runoff.

(**Note:** An *isochrone* is a line on the catchment map joining points having equal time of travel of surface runoff. See Sec. 8.8.)

7.4 An urban catchment of area 3.0 km^2 consists of 52% of paved areas, 20% parks, 18% multi-unit residential area. The remaining land use can be classified as light industrial area. The catchment is essentially flat and has sandy soil. If the time of concentration is 50 minutes, estimate the peak flow due to a design storm of depth 85 mm in 50 minutes.

7.5 In estimating the peak discharge of a river at a location X the catchment area was divided into four parts A, B, C and D. The time of concentration and area for different parts are as follows:

Part	Time of Concentration	Area (in ha)
A	One Hour	600
B	Two Hours	750
C	Three Hours	1000
D	Four Hours	1200

Records of a rainstorm lasting for four hours as observed and the runoff factors during different hours are as follows:

Time (in hours)		Rainfall (mm)	Runoff factor
From	To		
0	1	25.0	0.50
1	2	50.0	0.70
2	3	50.0	0.80
3	4	23.5	0.85

Calculate the maximum flow to be expected at X in m^3/s assuming a constant base flow of 42.5 m^3/s.

7.6 A catchment area has a time of concentration of 20 minutes and an area of 20 ha. Estimate the peak discharge corresponding to return period of 25 yrs. Assume a runoff coefficient of 0.25. The intensity-duration-frequency for the storm in the area can be expressed by $i = KT^x/(D + a)^n$, where i = intensity in cm/h, T = retum period in years, and D = duration of storm in hours, with coefficients $K = 6.93$, $x = 0.189$, $a = 0.50$, $n = 0.878$.

7.7 A 100 ha watershed has the following characteristics:

i. Maximum length of travel of water in the catchment = 3500 m
ii. Difference in elevation between the most remote point on the catchment and the outlet = 65 m
iii. Land use/cover details:

Land use/cover	Area (ha)	Runoff coefficient
Forest	30	0.25
Pasture	10	0.16
Cuiltivated land	60	0.40

The maximum intensity-duration-frequency relationship for the watershed is given by

$$i = \frac{3.97\,T^{0.165}}{(D+0.15)^{0.733}}$$

where i = intensity in cm/h, T = Return period in years and D = duration of rainfall in hours. Estimate the 25-year peak runoff from the watershed that can be expected at the outlet of the watershed.

7.8 A rectangular paved area 150 m × 450 m has a longitudinal drain along one of its longer edges. The time of concentration for the area is estimated to be 30 minutes and consists of 25 minutes for over land flow across the pavement to the drain and 5 minutes for the maximum time from the upstream end of the drain to the outlet at the other end.

(a) Construct the isochrones at 5 minutes interval for this area.

(b) A rainfall of 7 cm/h occurs on this plot for D minutes and stops abruptly. Assuming a runoff coefficient of 0.8 sketch idealized outflow hydrographs for D = 5 and 40 minutes.

7.9 A rectangular parking lot is 150 m wide and 300 m long. The time of overland flow across the pavement to the longitudinal gutter along the centre is 20 minutes and the estimated total time of concentration to the downstream end of the gutter is 25 minutes. The coefficient of runoff is 0.92. If a rainfall of intensity 6 cm/h falls on the lot for 10 minutes and stops abruptly determine the peak rate of flow.

7.10 A flood of 4000 m^3/s in a certain river has a return period of 40 years. (a) What is its probability of exceedance? (b) What is the probability that a flood of 4000 m^3/s or greater magnitude may occur in the next 20 years? (c) What is the probability of occurrence of a flood of magnitude less than 4000 m^3/s?

7.11 Complete the following:

(a) Probability of a 10 year flood occurring at least once in the next 5 years is __________

(b) Probability that a flood of magnitude equal to or greater than the 20 year flood will not occur in the next 20 years is __________

(c) Probability of a flood equal to or greater than a 50 year flood occurring next year is __________

(d) Probability of a flood equal to or greater than a 50 year flood occurring three times in the next 10 years is __________

(e) Probability of a flood equal to or greater than a 50 year flood occurring at least once in next 50 years is __________

7.12 A table showing the variation of the frequency factor K in the Gumbel's extreme value distribution with the sample size N and return period T is often given in books. The following is an incomplete listing of K for T = 1000 years. Complete the table.

Sample size, *N*	**25**	**30**	**35**	**40**	**45**	**50**	**55**	**60**	**65**	**70**
Value of $K(T, N)$ for T =1000 years	5.842	5.727	—	5.576	—	5.478	—	—	—	5.359

7.13 The following table gives the observed annual flood values in the river Krishna at Karad Estimate the flood peaks with return periods of 50, 100 and 1000 years by using:
(a) Gumbel's extreme value distribution, (b) log-Pearson type III distribution, and
(c) log-normal distribution

Annual Flood Discharge in the river Krishna at Karad

Water Year Beginning	Flood discharge (m^3/s)	Water Year Beginning	Flood discharge (m^3/s)	Water Year Beginning	Flood discharge (m^3/s)
1965	4760	1980	2848	1995	1356
1966	2673	1981	1942	1996	2944
1967	4482	1982	1261	1997	5954
1968	1527	1983	1931	1998	1541
1969	4267	1984	1797	1999	2111
1970	2428	1985	1393	2000	774
1971	2156	1986	1801	2001	911
1972	2190	1987	1323	2002	1121
1973	3051	1988	3205	2003	937
1974	3101	1989	2504	2004	4163
1975	3231	1990	4361	2005	6312
1976	4562	1991	3150	2006	6708
1977	3422	1992	2127	2007	3868
1978	2187	1993	1812	2008	2884
1979	2435	1994	3915		

7.14 A hydraulic structure on a stream has been designed for a discharge of 350 m^3/s. If the available flood data on the stream is for 20 years and the mean and standard deviation for annual flood series are 121 and 60 m^3/s respectively, calculate the return period for the design flood by using Gumbel's method.

7.15 For a data of maximum-recorded annual floods of a river the mean and the standard deviation are 4200 m^3/s and 1705 m^3/s respectively. Using Gumbel's extreme value distribution, estimate the return period of a design flood of 9500 m^3/s. Assume an infinite sample size.

7.16 For a river, the estimated flood peaks by using log-normal distribution are as follows:

Return Period T in years	Peak Flood (m^3/s)
50	450
100	600

What flood discharge in this river will have a return period of 1000 years?
The variation of frequency factor with the return period in log-normal distribution is as follows:

Return Period (T)	50	100	200	1000
Frequency factor (K_z)	2.054	2.326	2.576	3.090

7.17 The flood data of a river was analysed for the prediction of extreme values by Log-Pearson Type III distribution. Using the variate $z = \log Q$, where Q = flood discharge in the river, it was found that $\bar{z} = 2.510$, $\sigma_z = 0.162$ and coefficient of skew $C_s = 0.70$. (a) Estimate the flood discharges with return periods of 50, 100, 200 and 1000 years in this river. (b) What would be the corresponding flood discharge if log-normal distribution was used?

7.18 The frequency analysis of flood data of a river by using Log Pearson Type III distribution yielded the following data:
Coefficient of Skewness = 0.4

Return Period T in years	Peak Flood (m^3/s)
50	10,000
200	15,000

Given the following data regarding the variation of the frequency factor K with the return period T for $C_s = 0.4$, estimate the flood magnitude in the river with a return period of 1000 yrs.

Return Period (T)	50	200	1000
Frequency factor (K)	2.261	2.949	3.670

7.19 A river has 40 years of annual flood flow record. The discharge values are in m^3/s. The logarithms to base 10 of these discharge values show a mean value of 3.2736, standard deviation of 0.3037 and a coefficient of skewness of 0.07. Calculate the 50 year return period annual flood discharge by,
(a) Log-normal distribution, and
(b) Log-Pearson type III distribution.

7.20 The following data give flood-data statistics of two rivers in UP:

S. No.	River	Length of records (years)	Mean annual flood (m^3/s)	σ_{n-1}
1	Ganga at Raiwala	92	6437	2951
2	Yamuna at Tajewala	54	5627	3360

(a) Estimate the 100 and 1000 year floods for these two rivers by using Gumbel's method.
(b) What are the 95% confidential intervals for the predicted values?

7.21 For a river, the estimated flood peaks for two return periods by the use of Gumbel's method are as follows:

Return Period (years)	Peak Flood (m^3/s)
100	435
50	395

What flood discharge in this river will have a return period of 1000 years?

7.22 Using 30 years data and Gumbel's method, the flood magnitudes, for return periods of 100 and 50 years for a river are found to be 1200 and 1060 m^3/s respectively.
(a) Determine the mean and standard deviation of the data used, and
(b) Estimate the magnitude of a flood with a return period of 500 years.

7.23 The ordinates of a mass curve of rainfall from a severe storm in a catchment is given. Ordinates of a 12-h unit hydrograph applicable to the catchment are also given. Using the given mass curve, develop a design storm to estimate the design flood for the catchment. Taking the ϕ index as 0.15 cm/h, estimate the resulting flood hydrograph. Assume the base flow to be 50 m^3/s.

Time (h)	0	12	24	36	48	60	72	84	96	108	120	132
Cumulative rainfall (cm)	0	10.2	30.5	34.0	36.0							
12-h UH ordinate (m^3/s)	0	32	96	130	126	98	75	50	30	15	7	0

7.24 A 6-hour unit hydrograph is in the form of a triangle with a peak of 50 m^3/s at 24 hours from start. The base is 54 hours. The ordinates of a mass curve of rainfall from a severe storm in the catchment is as below:

Time (h)	0	6	12	18	24
Cumulative Rainfall (cm)	0	5	12	15	17.6

Using this data, develop a design storm and estimate the design flood for the catchment. Assume ϕ index = 0.10 cm/h and the base flow = 20 m^3/s.

7.25 A water-resources project has an expected life of 20 years. (a) For an acceptable risk of 5% against the design flood, what design return period is to be adopted? (b) If the above return period is adopted and the life of the structure can be enhanced to 50 years, what is the new risk value?

7.26 A factory is proposed to be located on the edge of the 50-year flood plain of a river. If the design life of the factory is 25 years, what is the reliability that it will not be flooded during its design life?

7.27 A spillway has a design life of 20 years. Estimate the required return period of a flood if the acceptable risk of failure of the spillway is 10% (a) in any year, and (b) over its design life.

7.28 Show that if the life of a project n has a very large value, the risk of failure is 0.632 when the design period is equal to the life of the project, n.

(*Hint:* Show that $\left(1-\dfrac{1}{n}\right)^n = e^{-1}$ for large values of n)

7.29 The regression analysis of a 30-year flood data at a point on a river yielded sample mean of 1200 m^3/s and standard deviation of 650 m^3/s. For what discharge would you design the structure to provide 95% assurance that the structure would not fail in the next 50 years? Use Gumbel's method. The value of the mean and standard deviation of the reduced variate for $N = 30$ are 0.53622 and 1.11238 respectively.

7.30 Analysis of the annual flood peak data of river Damodar at Rhondia, covering a period of 21 years yielded a mean of 8520 m^3/s and a standard deviation of 3900 m^3/s. A proposed water control project on this river near this location is to have an expected life of 40 years. Policy decision of the project allows an acceptable reliability of 85%.

(a) Using Gumbel's method recommend the flood discharge for this project.

(b) If a safety factor for flood magnitude of 1.3 is desired, what discharge is to be adored? What would be the corresponding safety margin?

OBJECTIVE QUESTIONS

7.1 A catchment consists of 30% area with runoff coefficient of 0.4 with remaining 70% area with runoff coefficient of 0.60. The equivalent runoff coefficient of the catchment will be be

(a) 0.48 (b) 0.54 (c) 0.63 (d) 0.76

7.2 The land use of an area and he corresponding runoff coefficient are as follows:

Sl. No.	Land use	Area (ha)	Runoff coefficient
1	Roads	10	0.70
2	Lawn	20	0.10
3	Residential area	50	0.30
4	Industrial area	20	0.80

What is the equivalent coefficient?
(a) 0.15 (b) 0.36 (c) 0.40 (d) 0.5

7.3 A culvert is designed for a peak flow Q_p on the basis of the rational formula. If a storm of the same intensity as used in the design but of duration twice larger occurs the resulting peak discharge will be
(a) Q_p (b) $2\,Q_p$ (c) $Q_p/2$ (d) Q_p^2

7.4 A watershed of area 90 ha has a runoff coefficient of 0.4. A storm of duration larger than the time of concentration of the watershed and of intensity 4.5 cm/h creates a peak discharge of
(a) 11.3 m^3/s (b) 0.45 m^3/s (c) 450 m^3/s (d) 4.5 m^3/s

7.5 A rectangular parking lot, with direction of overland flow parallel to the larger side, has a time of concentration of 25 minutes. For the purpose of design of drainage, four rainfall patterns as below are to be considered.
A = 35 mm/h for 15 minutes, B = 45 mm/h for 10 minutes,
C = 10 mm/h for 60 minutes, D = 15 mm/h for 25 minutes,
The greatest peak rate of runoff is expected in the storm
(a) A (b) B (c) C (d) D

7.6 For an annual flood series arranged in decreasing order of magnitude, the return period for a magnitude listed at position m in a total of N entries, by Weibull formula is
(a) m/N (b) $m/(N + 1)$ (c) $(N + 1)/m$ (d) $N/(m + 1)$.

7.7 The probability that a hundred year flood may not occur at all during the 50 year life of a project is
(a) 0.395 (b) 0.001 (c) 0.605 (d) 0.133

7.8 The probability of a flood, equal to or greater than 1000 year flood, occurring next year is
(a) 0.0001 (b) 0.001 (c) 0.386 (d) 0.632

7.9 The probability of a flood equal to or greater than 50 year flood, occurring at least one in next 50 years is
(a) 0.02 (b) 0.636 (c) 0.364 (d) 1.0

7.10 The general equation for hydrological frequency analysis states that x_T = value of a variate with a return period of T years is given by x_T =
(a) $\bar{x} - K\sigma$ (b) $x/K\sigma$ (c) $K\sigma$ (d) $x + K\sigma$

7.11 For a return period of 1000 years the Gumbel's reduced variate y_T is
(a) 6.907 (b) 4.001 (c) 5.386 (d) 6.632

7.12 An annual flood series contains 100 years of flood data. For a return period of 200 years the Gumbel's reduced variate can be taken as
(a) 5.296 (b) –4.600 (c) 1.2835 (d) 0.517

7.13 To estimate the flood magnitude with a return period of T years by the Log–Pearson Type III method, the following data pertaining to annual flood series is sufficient
(a) mean, standard deviation and coefficient of skew of discharge data
(b) mean and standard deviation of the log of discharge and the number of years of data
(c) mean, standard deviation and coefficient of skew of log of discharge data
(d) mean and standard deviation of the log of discharges

7.14 If the recurrence interval of an event is T_A in annual series and T_p in partial duration series then
(a) T_A is always smaller than T_p
(b) difference between T_A and T_p is negligible for $T_A < 5$ years
(c) difference between T_A and T_p is negligible for $T_A > 10$ years
(d) difference between T_A and T_p is not negligible till $T_A > 100$ years

7.15 The term mean annual flood denotes
(a) mean floods in partial-duration series
(b) mean of annual flood flow series
(c) a flood with a recurrence interval of 2.33 years
(d) a flood with a recurrence interval of N/2 years, where N = number of years of record.

7.16 The use of unit hydrographs for estimating floods is generally limited to catchments of size less than
(a) 5000 km^2 (b) 500 km^2 (c) 10^6 km^2 (d) 5000 ha

7.17 The probable maximum flood is
(a) the standard project flood of an extremely large river
(b) a flood adopted in the design of all kinds of spillways
(c) an extremely large but physically possible flood in the region
(d) the maximum possible flood that can occur anywhere in the country

7.18 The standard project flood is
(a) Smaller than probable maximum flood in the region
(b) The same as the design flood used for all small hydraulic structures
(c) Larger than the probable maximum flood by a factor implying factor of safety
(d) The same as the probable maximum flood

7.19 A hydraulic structure has been designed for a 50 year flood. The probability that exactly one flood of the design capacity will occur in the 75 year life of the structure is
(a) 0.02 (b) 0.220 (c) 0.336 (d) 0.780

7.20 The return period that a designer must use in the estimation of a flood for a hydraulic structure, if he is willing to accept 20% risk that a flood of that or higher magnitude will occur in the next 10 years is
(a) 95 years (b) 75 years (c) 45 years (d) 25 years

7.21 A hydraulic structure with a life of 30-years is designed for a 30 year flood. The risk of failure of the structure during its life is
(a) 0.033 (b) 0.638 (c) 0.362 (d) 1.00

7.22 A bridge is designed for a 50-year flood. The probability that only one flood of the design capacity or higher will occur in the 75 years life of the bridge is
(a) 0.020 (b) 0.220 (c) 0.786 (d) 0.336

7.23 The hydraulic risk of a 100-years flood occurring during the 2- year service life of a project is
(a) 9.8% (b) 99% (c) 19.9% (d) 1.99%

CHAPTER 8

FLOOD ROUTING

8.1 INTRODUCTION

The flood hydrograph discussed in Chap. 6 is in fact a wave. The stage and discharge hydrographs represent the passage of waves of the river depth and discharge respectively. As this wave moves down the river, the shape of the wave gets modified due to various factors, such as channel storage, resistance, lateral addition or withdrawal of flows, etc. When a flood wave passes through a reservoir, its peak is attenuated and the time base is enlarged due to the effect of storage. Flood waves passing down a river have their peaks attenuated due to friction if there is no lateral inflow. The addition of lateral inflows can cause a reduction of attenuation or even amplification of a flood wave. The study of the basic aspects of these changes in a flood wave passing through a channel system forms the subject matter of this chapter.

Flood routing is the technique of determining the flood hydrograph at a section of a river by utilising the data of flood flow at one or more upstream sections. The hydrologic analysis of problems such as flood forecasting, flood protection, reservoir design and spillway design invariably include flood routing. In these applications two broad categories of routing can be recognised. These are

1. Reservoir routing,
2. Channel routing.

In *reservoir routing,* the effect of a flood wave entering a reservoir is studied. Knowing the volume-elevation characteristic of the reservoir and the outflow-elevation relationship for the spillways and other outlet structures in the reservoir, the effect of a flood wave entering the reservoir is studied to predict the variations of reservoir elevation and outflow discharge with time. This form of reservoir routing is essential (i) in the design of the capacity of spillways and other reservoir outlet structures, and (ii) in the location and sizing of the capacity of reservoirs to meet specific requirements.

In *channel routing,* the change in the shape of a hydrograph as it travels down a channel is studied. By considering a channel reach and an input hydrograph at the upstream end, this form of routing aims to predict the flood hydrograph at various sections of the reach. Information on the flood-peak attenuation and the duration of high-water levels obtained by channel routing is of utmost importance in flood-forecasting operations and flood-protection works.

A variety of routing methods are available and they can be broadly classified into two categories as: (i) hydrologic routing, and (ii) hydraulic routing. Hydrologic routing methods employ essentially the equation of continuity. Hydraulic methods, on the other hand, employ the continuity equation together with the equation of motion of unsteady flow. The basic differential equations used in the hydraulic routing, known as St. Venant equations afford a better description of unsteady flow than hydrologic methods.

8.2 *Basic Equations*

The passage of a flood hydrograph through a reservoir or a channel reach is an unsteady-flow phenomenon. It is classified in open-channel hydraulics as gradually varied unsteady flow. The equation of continuity used in all hydrologic routing as the primary equation states that the difference between the inflow and outflow rate is equal to the rate of change of storage, i.e.

$$I - Q = \frac{dS}{dt} \tag{8.1}$$

where I = inflow rate, Q = outflow rate and S = storage. Alternatively, in a small time interval Δt the difference between the total inflow volume and total outflow volume in a reach is equal to the change in storage in that reach

$$\bar{I}\,\Delta t - \bar{Q}\,\Delta t = \Delta S \tag{8.2}$$

where $\bar{I}$ = average inflow in time Δt, $\bar{Q}$ = average outflow in time Δt and ΔS = change in storage. By taking $\bar{I} = (I_1 + I_2)/2$, $\bar{Q} = (Q_1 + Q_2)/2$ and $\Delta S = S_2 - S_1$ with suffixes 1 and 2 to denote the beginning and end of time interval Δt, Eq. (8.2) is written as

$$\left(\frac{I_1 + I_2}{2}\right)\Delta t - \left(\frac{Q_1 + Q_2}{2}\right)\Delta t = S_2 - S_1 \tag{8.3}$$

The time interval Δt should be sufficiently short so that the inflow and outflow hydrographs can be assumed to be straight lines in that time interval. Further, Δt must be shorter than the time of transit of the flood wave through the reach.

In the differential form the *equation of continuity* for unsteady flow in a reach with no lateral flow is give by

$$\frac{\partial Q}{\partial x} + T\frac{\partial y}{\partial t} = 0 \tag{8.4}$$

where T = top width of the section and y = depth of flow.

The *equation of motion* for a flood wave is derived from the application of the momentum equation as

$$\frac{\partial y}{\partial x} + \frac{V}{g}\frac{\partial V}{\partial x} + \frac{1}{g}\frac{\partial V}{\partial t} = S_0 - S_f \tag{8.5}$$

where V = velocity of flow at any section, S_0 = channel bed slope and S_f = slope of the energy line. The continuity equation [Eq. (8.4)] and the equation of motion [Eq. (8.5)] are believed to have been first developed by A.J.C. Barrè de Saint Venant (1871) and are commonly known as St. Venant equations. Hydraulic-flood routing involves the numerical solution of St.Venant equations. Details about these equations, such as their derivations and various forms are available in Ref. 9.

8.3 *Hydrologic Storage Routing (Level Pool Routing)*

A flood wave $I(t)$ enters a reservoir provided with an outlet such as a spillway. The outflow is a function of the reservoir elevation (h) only, i.e. $Q = Q(h)$. The storage in the reservoir is a function of the reservoir elevation, $S = S(h)$. Further, due to the passage of the flood wave through the reservoir, the water level in the reservoir changes with time, $h = h(t)$ and hence the storage and discharge change with time (Fig. 8.1). It is required to find the variation of S, h and Q with time, i.e. find $S = S(t)$, $Q = Q(t)$ and $h = h(t)$ given $I = I(t)$.

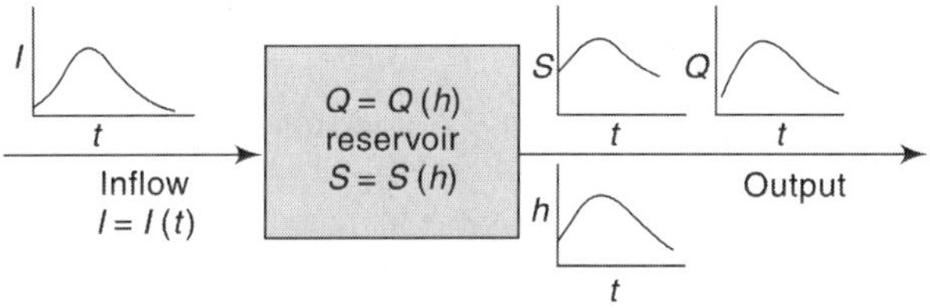

Fig. 8.1 ***Storage Routing (Schematic)***

If an uncontrolled spillway is provided in a reservoir, typically

$$Q = \frac{2}{3} C_d \sqrt{2g}\, L_e H^{3/2} = Q(h)$$

where H = head over the spillway, L_e = effective length of the spillway crest and C_d = coefficient of discharge. Similarly, for other forms of outlets, such as gated spillways, sluice gates, etc. other relations for $Q(h)$ will be available.

For reservoir routing, the following data have to be known: Graphs or Tables showing

- Storage volume *vs* elevation for the reservoir;
- Water-surface elevation *vs* outflow and hence storage *vs* outflow discharge;
- Inflow hydrograph, $I = I(t)$; and
- Initial values of S, I and Q at time $t = 0$.

There are a variety of methods available for routing of floods through a reservoir. All of them use Eq. (8.2) but in various rearranged manners. As the horizontal water surface is assumed in the reservoir, the storage routing is also known as *Level Pool Routing*.

Two commonly used semi-graphical methods and a numerical method are described below.

8.3.1 Modified Pul's Method

Equation (8.3) is rearranged as

$$\left(\frac{I_1 + I_2}{2}\right)\Delta t + \left(S_1 - \frac{Q_1 \Delta t}{2}\right) = \left(S_2 + \frac{Q_2 \Delta t}{2}\right) \tag{8.6}$$

At the starting of flood routing, the initial storage and outflow discharges are known. In Eq. (8.6) all the terms in the left-hand side are known at the beginning of a time step Δt. Hence, the value of the function $\left(S_2 + \frac{Q_2 \Delta t}{2}\right)$ at the end of the time step is calculated by Eq. (8.6). Since the relation $S = S(h)$ and $Q = Q(h)$ are known, $\left(S + \frac{Q\Delta t}{2}\right)_2$ will enable one to determine the reservoir elevation and hence the discharge at the end of the time step. The procedure is repeated to cover the full inflow hydrograph.

For practical use in hand computation, the following semigraphical method is very convenient.

1. From the known storage-elevation and discharge-elevation data, prepare a curve of $\left(S + \frac{Q\Delta t}{2}\right)$ *vs* elevation (Fig. 8.2). Here Δt is any chosen interval, approximately 20 to 40% of the time of rise of the inflow hydrograph.
2. On the same plot prepare a curve of outflow discharge *vs* elevation (Fig. 8.2).

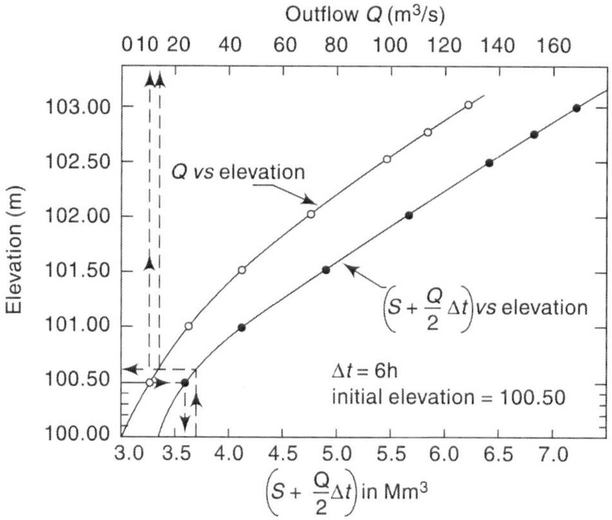

Fig. 8.2 ***Modified Pul's Method of Storage Routing***

3. The storage, elevation and outflow discharge at the starting of routing are known. For the first time inteval Δt, $\left(\frac{I_1 + I_2}{2}\right)\Delta t$ and $\left(S_1 + \frac{Q_1 \Delta t}{2}\right)$ are known and hence by Eq. (8.6)the term $\left(S_2 + \frac{Q_2 \Delta t}{2}\right)$ is determined.

4. The water-surface elevation corresponding to $\left(S_2 + \frac{Q_2\,\Delta t}{2}\right)$ is found by using the plot of step (1). The outflow discharge Q_2 at the end of the time step Δt is found from plot of step (2).

5. Deducting $Q_2\Delta t$ from $\left(S_2 + \frac{Q_2\,\Delta t}{2}\right)$ gives $\left(S - \frac{Q\Delta t}{2}\right)_1$ for the beginning of the next time step.

6. The procedure is repeated till the entire inflow hydrograph is routed.

Example 8.1 A reservoir has the following elevation, discharge and storage relationships:

Elevation (m)	Storage (10^6 m^3)	Outflow discharge (m^3/s)
100.00	3.350	0
100.50	3.472	10
101.00	3.380	26
101.50	4.383	46
102.00	4.882	72
102.50	5.370	100
102.75	5.527	116
103.00	5.856	130

When the reservoir level was at 100.50 m, the following flood hydrograph entered the reservoir.

Time (h)	0	6	12	18	24	30	36	42	48	54	60	66	72
Discharge (m^3/s)	10	20	55	80	73	58	46	36	55	20	15	13	11

Route the flood and obtain (i) the outflow hydrograph, and (ii) the reservoir elevation vs time curve during the passage of the flood wave.

Solution

A time interval Δt = 6 h is chosen. From the available data the elevation-discharge – $\left(S + \frac{Q\Delta t}{2}\right)$ table is prepared.

$$\Delta t = 6 \times 60 \times 60 = 0.0216 \times 10^6 \text{ s} = 0.0216 \text{ Ms}$$

Elevation (m)	100.00	100.50	101.00	101.50	102.00	102.50	102.75	103.00
Discharge Q (m^3/s)	0	10	26	46	72	100	116	130
$\left(S + \frac{Q\Delta t}{2}\right)$ (Mm^3)	3.35	3.58	4.16	4.88	5.66	6.45	6.78	7.26

Graph of Q *vs* elevation and $\left(S+\frac{Q\Delta t}{2}\right)$ *vs* elevation is prepared (Fig. 8.2). At the start of routing, elevation = 100.50 m, Q = 10.0 m^3/s, and $\left(S-\frac{Q\Delta t}{2}\right)$ = 3.362 Mm^3. Starting from this value of $\left(S-\frac{Q\Delta t}{2}\right)$, Eq. (8.6) is used to get $\left(S+\frac{Q\Delta t}{2}\right)$ at the end of first time step of 6 h as

$$\left(S+\frac{Q\Delta t}{2}\right)_2 = (I_1+I_2)\frac{\Delta t}{2}+\left(S-\frac{Q\Delta t}{2}\right)_1$$

$$= (10+20)\times\frac{0.0216}{2} + (3.362) = 3.686 \text{ Mm}^3.$$

Looking up in Fig. 8.2, the water-surface elevation corresponding to $\left(S+\frac{Q\Delta t}{2}\right)$ = 3.686 Mm^3 is 100.62 m and the corresponding outflow discharge Q is 13 m^3/s. For the next step, Initial value of $\left(S-\frac{Q\Delta t}{2}\right) = \left(S+\frac{Q\Delta t}{2}\right)$ of the previous step – $Q\,\Delta t$

$$= (3.686 - 13 \times 0.0216) = 3.405 \text{ Mm}^3$$

The process is repeated for the entire duration of the inflow hydrograph in a tabular form as shown in Table 8.1.

Using the data in columns 1, 8 and 7, the outflow hydrograph (Fig. 8.3) and a graph showing the variation of reservoir elevation with time (Fig. 8.4) are prepared.

Sometimes a graph of $\left(S-\frac{Q\Delta t}{2}\right)$ *vs* elevation prepared from known data is plotted in Fig. 8.2 to aid in calculating the items in column 5. Note that the calculations are sequential in nature and any error at any stage is carried forward. The accuracy of the method depends upon the value of Δt; smaller values of Δt give greater accuracy.

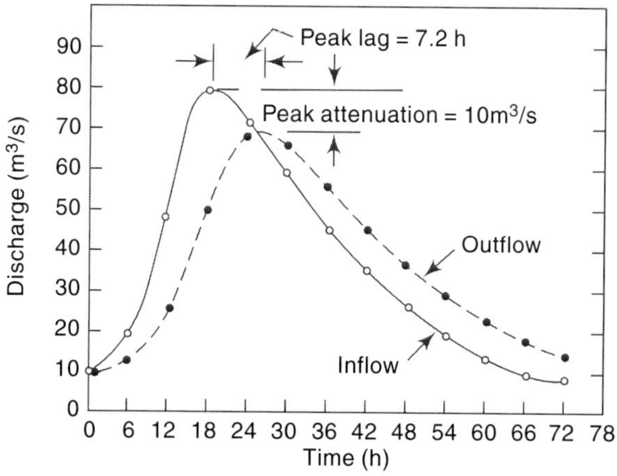

Fig. 8.3 ***Variation of Inflow and Outflow Discharges—Example 8.1***

Table 8.1 ***Flood Routing through a Reservoir—Modified Pul's Method—Example 8.1***

$\Delta t = 6$ h $= 0.0216$ Ms, $\bar{I} = (I_1 + I_2)/2$

Time (h)	Inflow I (m³/s)	$\bar{I}$ (m³/s)	$\bar{I} \cdot \Delta t$ (Mm³)	$S - \frac{\Delta t Q}{2}$ (Mm³)	$S + \frac{\Delta t Q}{2}$ (Mm³)	Elevation (m)	Q (m³/s)
1	2	3	4	5	6	7	8
0	10					100.50	10
		15.00	0.324	3.362	3.636		
6	20					100.62	13
		37.50	0.810	3.405	4.215		
12	55					101.04	27
		67.50	1.458	3.632	5.090		
18	80					101.64	53
		76.50	1.652	3.945	5.597		
24	73					101.96	69
		65.50	1.415	4.107	5.522		
30	58					101.91	66
		52.00	1.123	4.096	5.219		
36	46					101.72	57
		41.00	0.886	3.988	4.874		
42	36					101.48	48
		31.75	0.686	3.902	4.588		
48	27.5					101.30	37
		23.75	0.513	3.789	4.302		
54	20					100.10	25
		17.50	0.378	3.676	4.054		
60	15					100.93	23
		14.00	0.302	3.557	3.859		
66	13					100.77	18
		12.00	0.259	3.470	3.729		
72	11					100.65	14
				3.427			

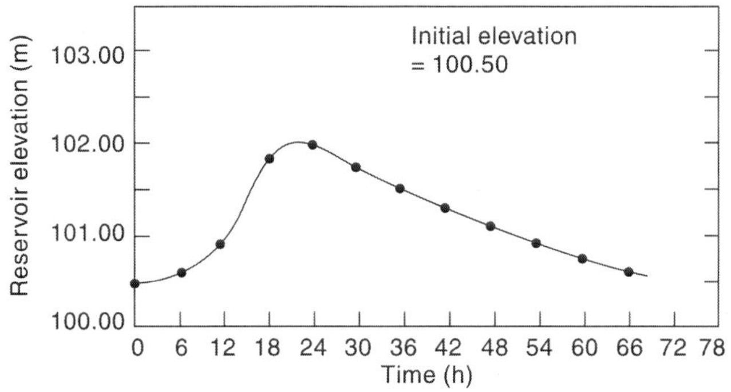

Fig. 8.4 ***Variation of Reservoir Elevation with Time—Example 8.1***

8.3.2 Goodrich Method

Another popular method of hydrologic reservoir routing, known as Goodrich method, utilises Eq. (8.3) rearranged as

$$I_1 + I_2 - Q_1 - Q_2 = \frac{2S_2}{\Delta t} - \frac{2S_1}{\Delta t}$$

where suffixes 1 and 2 stand for the values at the beginning and end of a time step Δt respectively. Collecting the known and initial values together,

$$(I_1 + I_2) + \left(\frac{2S_1}{\Delta t} - Q_1\right) = \left(\frac{2S_2}{\Delta t} + Q_2\right) \tag{8.7}$$

For a given time step, the left-hand side of Eq. (8.7) is known and the term $\left(\frac{2S}{\Delta t} + Q\right)_2$ is determined by using Eq. (8.7). From the known storage-elevation-discharge data, the function $\left(\frac{2S}{\Delta t} + Q\right)_2$ is established as a function of elevation. Hence, the discharge, elevation and storage at the end of the time step are obtained. For the next time step,

$$\left[\left(\frac{2S}{\Delta t} + Q\right)_2 - 2Q_2\right] \text{ of the previous time step}$$

$$= \left(\frac{2S}{\Delta t} - Q\right)_1 \text{ for use as the initial values}$$

The procedure is illustrated in Example 8.2.

Example 8.2 Route the following flood hydrograph through the reservoir of Example 8.1 by the Goodrich method:

Time (h)	0	6	12	18	24	30	36	42	48	54	60	66
Inflow (m^3/s)	10	30	85	140	125	96	75	60	46	35	25	20

The initial conditions are: when $t = 0$, the reservoir elevation is 100.60 m.

Solution

A time increment $\Delta t = 6$ h $= 0.0216$ Ms is chosen. Using the known storage-elevation-discharge data, the following table is prepared.

A graph depicting Q *vs* elevation and $\left(\frac{2S}{\Delta t}+Q\right)$ *vs* elevation is prepared from this data (Fig. 8.5).

Elevation (m)	100.00	100.50	101. 00	101.50	102.00	102.50	102.75	103.00
Outflow Q (m^3/s)	0	10	26	46	72	100	116	130
$\left(\frac{2S}{\Delta t}+Q\right)$ **(m^3/s)**	310.2	331.5	385.3	451.8	524.0	597.2	627.8	672.2

At $t = 0$, Elevation = 100.60 m, from Fig. 8.5, $Q = 12$ m^3/s, and

$$\left(\frac{2S}{\Delta t}+Q\right) = 340 \text{ m}^3/\text{s}$$

$$\left(\frac{2S}{\Delta t}-Q\right)_1 = 340 - 24 = 316 \text{ m}^3/\text{s}$$

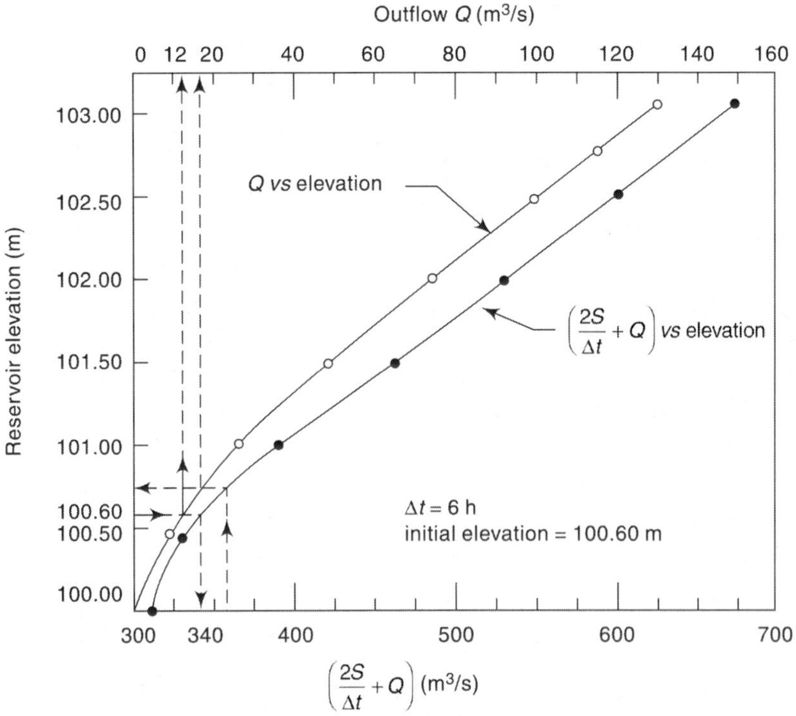

Fig. 8.5 ***Goodrich Method of Storage Routing—Example 8.2***

For the first time interval of 6 h,

$$I_1 = 10,\ I_2 = 30,\ Q_1 = 12, \text{ and}$$

$$\left(\frac{2S}{\Delta t} + Q\right)_2 = (10 + 30) + 316 = 356 \text{ m}^3\text{/s}$$

From Fig. 8.5, the reservoir elevation for this $\left(\frac{2S}{\Delta t} + Q\right)_2$ is 100.74 m

For the next time increment,

$$\left(\frac{2S}{\Delta t} - Q\right)_1 = 356 - 2 \times 17 = 322 \text{ m}^3\text{/s}$$

The procedure is repeated in a tabular form (Table 8.2) till the entire flood is routed.

Using the data in columns 1, 7 and 8, the outflow hydrograph and a graph showing the variation of reservoir elevation with time (Fig. 8.6) are plotted.

In this method also, the accuracy depends upon the value of Δt chosen; smaller values of Δt give greater accuracy.

Table 8.2 ***Reservoir Routing—Goodrich Method—Example 8.2***

$\Delta t = 6.0 \text{ h} = 0.0216 \text{ Ms}$

Time (h)	I (m^3/s)	$(I_1 + I_2)$	$\left(\frac{2S}{\Delta t}+Q\right)$ (m^3/s)	$\left(\frac{2S}{\Delta t}+Q\right)$ (m^3/s)	Elevation (m)	Discharge Q (m^3/s)
1	2	3	4	5	6	7
0	10			(340)	100.6	12
		40	316	356		
6	30				100.74	17
		115	322	437		
12	85				101.38	40
		225	357	582		
18	140				102.50	95
		265	392	657		
24	125				102.92	127
		221	403	624		
30	96				102.70	112
		171	400	571		
36	75				102.32	90
		135	391	526		
42	60				102.02	73
		106	380	486		
48	46				101.74	57
		81	372	453		
54	35				101.51	46
		60	361	421		
60	25				101.28	37
		45	347	392		
66	20				101.02	27
			335			

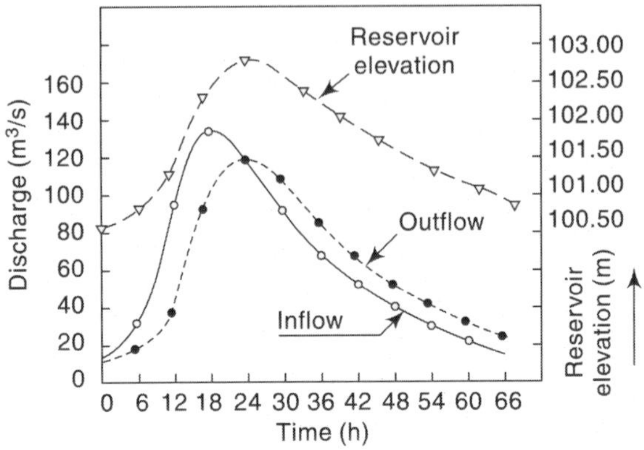

Fig. 8.6 ***Results of Reservoir Routing—Example 8.2***

8.3.3 Standard Fourth-Order Runge-Kutta Method (*SRK*)

The Pul's method and Goodrich method of level pool routing are essentially semi-graphical methods. While they can be used for writing programs for use in a computer, a more efficient computation procedure can be achieved by use of any of the Runge-Kutta methods. The standard fourth-order Runge-Kutta method (*SRK*) is the most accurate one.

Designating

S = storage at a water surface elevation H in the reservoir = $S(H)$

A = area of the reservoir at elevation H = function of $H = A(H)$

Q = outflow from the reservoir = function of $H = Q(H)$

$$dS = A(H) \cdot dH \tag{8.8}$$

By continuity equation,

$$\frac{dS}{dt} = I(t) - Q(H = A(H)\ \frac{dH}{dt}$$

$$\frac{dH}{dt} = \frac{I(t) - Q(H)}{A(H)} = \text{Function of } (t, H) = F(t, H) \tag{8.9}$$

If the routing is conducted from the initial condition, (at $t = t_0$ and $I = I_0$; $Q = Q_0$, $H = H_0$, $S = S_0$) in time steps Δt, the water surface elevation H at $(i + 1)$th step is given in *SRK* method as

$$H_{i+1} = H_i + \frac{1}{6}(K_1 + 2K_2 + 2K_3 + K_4)\,\Delta t \tag{8.10}$$

where $K_1 = F(t_i, H_i)$

$$K_2 = F\left(t_i + \frac{\Delta t}{2}, H_i + \frac{1}{2}K_1\,\Delta t\right)$$

$$K_3 = F\left(t_i + \frac{\Delta t}{2}, H_i + \frac{1}{2}K_2\,\Delta t\right)$$

$$K_4 = F(t_i + \Delta t, H_i + K_3\,\Delta t)$$

In Eq. (8.10), the suffix i denotes the values at the ith step, and suffix $(i + 1)$ denotes the values at the $(i + 1)$th step. At $i = 1$ the initial conditions I_0, Q_0, S_0 and H_0 prevail. Starting from the known initial conditions and knowing Q *vs* H and A *vs* H relationships, a given hydrograph $I = I(t)$ is routed by selecting a time step Δt. At any time $t = (t_0 + i\,\Delta t)$, the value of H_i is known and the coefficients K_1, K_2, K_3, K_4 are determined by repeated appropriate evaluation of the function $F(t, H)$. It is seen that the SRK method directly determines H_{i+1} by four evaluations of the function $F(t, H)$.

Knowing the values of H at various time intervals, i.e. $H = H(t)$, the other variables $Q(H)$ and $S(H)$ can be calculated to complete the routing operation.

Developing a computer program for level pool routing by using SRK is indeed very simple.

8.4 *Attenuation*

Figures 8.3 and 8.6 show the typical result of routing a flood hydrograph through a reservoir. Owing to the storage effect, the peak of the outflow hydrograph will be smaller than that of the inflow hydrograph. This reduction in the peak value is called *attenuation*. Further, the peak of the outflow occurs after the peak of the inflow; the time difference between the two peaks is known as *lag*. The attenuation and lag of a flood hydrograph at a reservoir are two very important aspects of a reservoir operating under a flood-control criterion. By judicious management of the initial reservoir level at the time of arrival of a critical flood, considerable attenuating of the floods can be achieved. The storage capacity of the reservoir and the characteristics of spillways and other outlets controls the lag and attenuation of an inflow hydrograph.

In Figs. 8.3 and 8.6 in the rising part of the outflow curve where the inflow curve is higher than the outflow curve, the area between the two curves indicate the accumulation of flow as storage. In the falling part of the outflow curve, the outflow curve is higher than the inflow curve and the area between the two indicate depletion from the storage. When the outflow from a storage reservoir is uncontrolled, as in a freely operating spillway, the peak of the outflow hydrograph will occur at the point of intersection of the inflow and outflow hydrographs (Figs. 8.3 and 8.6), as proved in Example 8.3.

Example 8.3 Show that in the level pool routing the peak of the outflow hydrograph must intersect the inflow hydrograph.

Solution

S = a function of water surface elevation in the reservoir = $S(H)$

$$\frac{dS}{dt} = A\frac{dH}{dt}$$

where A = area of the reservoir at elevation H.

Outflow Q = function of $H = Q(H)$

At peak outflow $\quad \dfrac{dQ}{dt} = 0$, hence $\dfrac{dS}{dt} = 0$

Also when $\dfrac{dH}{dt} = 0, \dfrac{dS}{dt} = 0$

By continuity equation $I - Q = \dfrac{dS}{dt}$

When $\dfrac{dS}{dt} = 0, I = Q$

Hence, when the peak outflow occurs, $I = Q$ and thus the peak of the outflow hydrograph must intersect the inflow hydrograph (Figs. 8.3 and 8.6).

8.5 *Hydrologic Channel Routing*

8.5.1 Basics

In reservoir routing presented in the previous sections, the storage was a unique function of the outflow discharge, $S = f(Q)$. However, in channel routing the storage is a function of both outflow and inflow discharges and hence a different routing method is needed. The flow in a river during a flood belongs to the category of gradually varied unsteady flow. The water surface in a channel reach is not only not parallel to the channel bottom but also varies with time (Fig. 8.7). Considering a channel reach having a flood flow, the total volume in storage can be considered under two categories as

1. Prism storage
2. Wedge storage

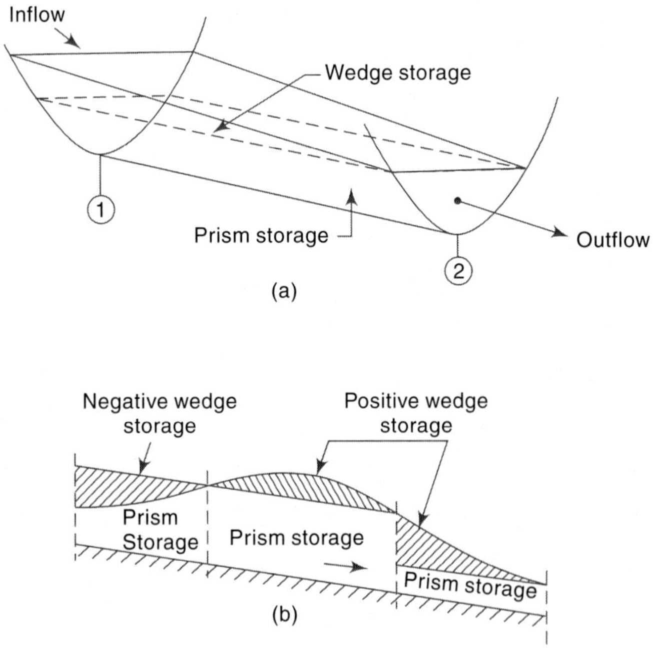

Fig. 8.7 ***Storage in a Channel Reach***

1. Prism Storage

It is the volume that would exist if the uniform flow occurred at the downstream depth, i.e. the volume formed by an imaginary plane parallel to the channel bottom drawn at the outflow section water surface.

2. Wedge Storage

It is the wedgelike volume formed between the actual water surface profile and the top surface of the prism storage.

At a fixed depth at a downstream section of a river reach, the prism storage is constant while the wedge storage changes from a positive value at an advancing flood to a negative value during a receding flood. The prism storage S_p is similar to a reservoir and can be expressed as a function of the outflow discharge, $S_p = f(Q)$. The wedge storage can be accounted for by expressing it as $S_w = f(I)$. The total storage in the channel reach can then be expressed as

$$S = K[x\, I^m + (1 - x)Q^m] \tag{8.11}$$

where K and x are coefficients and m = a constant exponent. It has been found that the value of m varies from 0.6 for rectangular channels to a value of about 1.0 for natural channels.

8.5.2 Muskingum Equation

Using $m = 1.0$, Eq. (8.11) reduces to a linear relationship for S in terms of I and Q as

$$S = K\,[x\, I + (1 - x)\, Q] \tag{8.12}$$

and this relationship is known as the *Muskingum equation*. In this, the parameter x is known as *weighting factor* and takes a value between 0 and 0.5. When $x = 0$, obviously the storage is a function of discharge only and Eq. (8.12) reduces to

$$S = KQ \tag{8.13}$$

Such a storage is known as *linear storage* or *linear reservoir*. When $x = 0.5$ both the inflow and outflow are equally important in determining the storage.

The coefficient K is known as *storage-time constant* and has the dimensions of time. It is approximately equal to the time of travel of a flood wave through the channel reach.

1. Estimation of *K* and *x*

Figure 8.8 shows a typical inflow and outflow hydrograph through a channel reach. Note that the outflow peak does not occur at the point of intersection of the inflow and outflow hydrographs. Using the continuity equation [Eq. (8.3)],

$$(I_1 + I_2)\,\frac{\Delta t}{2} - (Q_1 + Q_2)\,\frac{\Delta t}{2} = \Delta S$$

the increment in storage at any time t and time element Δt can be calculated. Summation of the various incremental storage values enable one to find the channel storage S *vs* time t relationship (Fig. 8.8).

If an inflow and outflow hydrograph set is available for a given reach, values of S at various time intervals can be determined by the above technique. By choosing a trial value of x, values of S at any time t are plotted against the corresponding $[x\,I + (1 - x)\,Q]$ values. If the value of x is chosen correctly, a straight-line relationship as given by Eq. (8.12) will result. However, if an incorrect value of x is used, the plotted

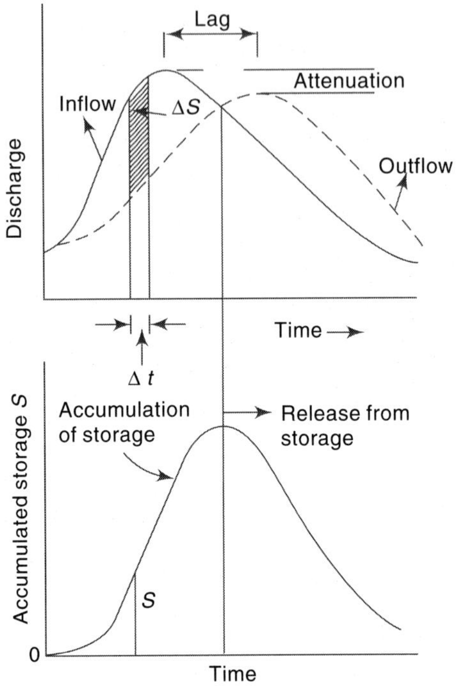

Fig. 8.8 ***Hydrographs and Storage in Channel Routing***

points will trace a looping curve. By trial and error, a value of x is so chosen that the data very nearly describe a straight line (Fig 8.9). The inverse slope of this straight line will give the value of K.

Normally, for natural channels, the value of x lies between 0 to 0.3. For a given reach, the values of x and K are assumed to be constant.

Example 8.4 The following inflow and outflow hydrographs were observed in a river reach. Estimate the values of K and x applicable to this reach for use in the Muskingum equation.

Time (h)	0	6	12	18	24	30	36	42	48	54	60	66
Inflow (m^3/s)	5	20	50	50	32	22	15	10	7	5	5	5
Outflow (m^3/s)	5	6	12	29	38	35	29	23	17	13	9	7

Solution

Using a time increment Δt = 6 h, the calculations are performed in a tabular manner as in Table 8.3. The incremental storage ΔS and S are calculated in columns 6 and 7 respectively. It is advantageous to use the units [(m^3/s).h] for storage terms.

As a first trial, x = 0.30 is selected and the value of $[x\,I + (1 - x)\,Q]$ evaluated (column 8) and plotted against S in Fig. 8.9. Since a looped curve is obtained, further trials are per-formed with x = 0.35 and 0.25. It is seen from Fig. 8.9 that for x = 0.25 the data very nearly describe a straight line and as such x = 0.25 is taken as the appropriate value for the reach.

From Fig. 8.9, K = 13.3 h

Table 8.3 *Determination of K and x—Example 8.4*

$\Delta t = 6$h Storage in (m^3/s) . h

Time (h)	I (m^3/s)	Q (m^3/s)	($I - Q$)	Average ($I - Q$)	ΔS = Col. 5 × Δt (m^3/s . h)	$S = \Sigma \Delta S$ (m^3/s . h)	[$x\,I + (1 - x)\,Q$] (m^3/s) $x = 0.35$	$x = 0.30$	$x = 0.25$
1	2	3	4	5	6	7	8	9	10
0	5	5	0			0	5.0	5.0	5.0
				7.0	42				
6	20	6	14			42	10.9	10.2	9.5
				26.0	156				
12	50	12	38			198	25.3	23.4	21.5
				29.5	177				
18	50	29	21			375	36.4	35.3	34.3
				7.5	45				
24	32	38	–6			420	35.9	36.2	36.5
				–9.5	–57				
30	22	35	–13			363	30.5	31.1	31.8
				–13.5	–81				
36	15	29	–14			282	24.1	24.8	25.5
				–13.5	–81				
42	10	23	–13			201	18.5	19.1	19.8
				–11.5	–69				
48	7	17	–10			132	13.5	14.0	14.5
				–9.0	–54				
54	5	13	–8			78	10.2	10.6	11.0
				–6.0	–36				
60	5	9	–4			42	7.6	7.8	8.0
				–3.0	–18				
66	5	7	–2			24	6.3	6.4	6.5

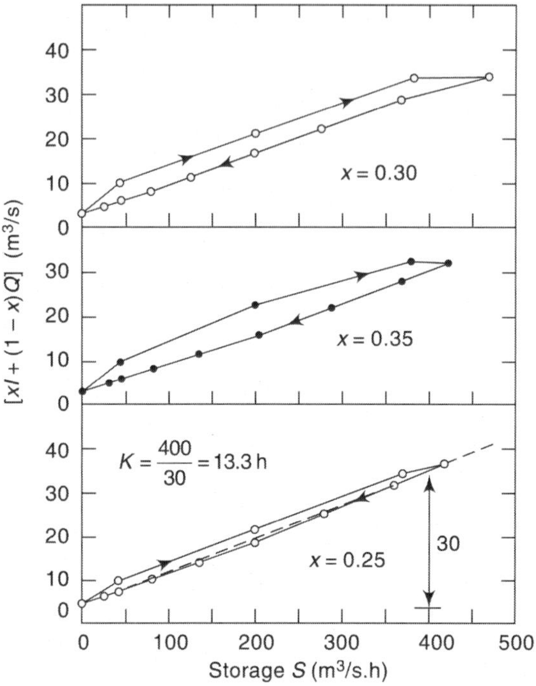

Fig. 8.9 ***Determination of K and x for a channel reach—Example 8.4***

2. Muskingum Method of Routing

For a given channel reach by selecting a routing interval Δt and using the Muskingum equation, the change in storage is

$$S_2 - S_1 = K[x\,(I_2 - I_1) + (1 - x)\,(Q_2 - Q_1)] \tag{8.14}$$

where suffixes 1 and 2 refer to the conditions before and after the time interval Δt. The continuity equation for the reach is

$$S_2 - S_1 = \left(\frac{I_2 + I_1}{2}\right)\Delta t - \left(\frac{Q_2 + Q_1}{2}\right)\Delta t \tag{8.15}$$

From Eqs (8.14) and (8.15), Q_2 is evaluated as

$$Q_2 = C_0\,I_2 + C_1\,I_1 + C_2\,Q_1 \tag{8.16}$$

where

$$C_0 = \frac{-Kx + 0.5\,\Delta t}{K - Kx + 0.5\,\Delta t} \tag{8.16a}$$

$$C_1 = \frac{Kx + 0.5\,\Delta t}{K - Kx + 0.5\,\Delta t} \tag{8.16b}$$

$$C_2 = \frac{K - Kx - 0.5\,\Delta t}{K - Kx + 0.5\,\Delta t} \tag{8.16c}$$

Note that $C_0 + C_1 + C_2 = 1.0$, Eq. (8.16) can be written in a general form for the nth time step as

$$Q_n = C_0\,I_n + C_1\,I_{n-1} + C_2\,Q_{n-1} \tag{8.16A}$$

Equation (8.16) is known as *Muskingum Routing Equation* and provides a simple linear equation for channel routing. It has been found that for best results the routing interval Δt should be so chosen that $K > \Delta t > 2Kx$. If $\Delta t < 2Kx$, the coefficient C_0 will be negative. Generally, negative values of coefficients are avoided by choosing appropriate values of Δt.

To use the Muskingum equation to route a given inflow hydrograph through a reach, the values of K and x for the reach and the value of the outflow, Q_1, from the reach at the start are needed. The procedure is indeed simple.

- Knowing K and x, select an appropriate value of Δt
- Calculate C_0, C_1 and C_2.
- Starting from the initial conditions I_1, Q_1 and known I_2 at the end of the first time step Δt calculate Q_2 by Eq. (8.16).
- The outflow calculated in step (c) becomes the known initial outflow for the next time step. Repeat the calculations for the entire inflow hydrograph.

The calculations are best done row by row in a tabular form. Example 8.5 illustrates the computation procedure. Spreadsheet (such as *MS* Excel) is ideally suited to perform the routing calculations and to view the inflow and outflow hydrographs.

Example 8.5 Route the following flood hydrograph through a river reach for which $K = 12.0$ h and $x = 0.20$. At the start of the inflow flood, the outflow discharge is 10 m³/s.

Time (h)	0	6	12	18	24	30	36	42	48	54
Inflow (m³/s)	10	20	50	60	55	45	35	27	20	15

Solution

Since $K = 12$ h and $2Kx = 2 \times 12 \times 0.2 = 4.8$ h, Δt should be such that 12 h > Δt > 4.8 h. In the present case $\Delta t = 6$ h is selected to suit the given inflow hydrograph ordinate interval.

Using Eqs. (8. 16-a, b and c) the coefficients C_0, C_1 and C_2 are calculated as

$$C_0 = \frac{-12 \times 0.20 + 0.5 \times 6}{12 - 12 \times 0.2 + 0.5 \times 6} = \frac{0.6}{12.6} = 0.048$$

$$C_1 = \frac{12 \times 0.2 + 0.5 \times 6}{12.6} = 0.429$$

$$C_2 = \frac{12 - 12 \times 0.2 - 0.5 \times 6}{12.6} = 0.523$$

For the first time interval, 0 to 6 h,

$$I_1 = 10.0 \qquad C_1 I_1 = 4.29$$
$$I_2 = 20.0 \qquad C_0 I_2 = 0.96$$
$$Q_1 = 10.0 \qquad C_2 Q_1 = 5.23$$

From Eq. (8.16) $Q_2 = C_0 I_2 + C_1 I_1 + C_2 Q_1 = 10.48$ m³/s

For the next time step, 6 to 12 h, $Q_1 = 10.48$ m³/s. The procedure is repeated for the entire duration of the inflow hydrograph. The computations are done in a tabular form as shown in Table 8.4. By plotting the inflow and outflow hydrographs the attenuation and peak lag are found to be 10 m³/s and 12 h respectively.

Table 8.4 ***Muskingum Method of Routing—Example 8.5***

$\Delta t = 6$ h

Time (h)	I (m^3/s)	$0.048\,I_2$	$0.429\,I_1$	$0.523\,Q_1$	Q (m^3/s)
1	2	3	4	5	6
0	10				10.00
		0.96	4.29	5.23	
6	20				10.48
		2.40	8.58	5.48	
12	50				16.46
		2.88	21.45	8.61	
18	60				32.94
		2.64	25.74	17.23	
24	55				45.61
		2.16	23.60	23.85	
30	45				49.61
		1.68	19.30	25.95	
36	35				46.93
		1.30	15.02	24.55	
42	27				40.87
		0.96	11.58	21.38	
48	20				33.92
		0.72	8.58	17.74	
54	15				27.0

Alternative Form of Eq. (8. 16)

Equations (8.14) and (8.15) can be combined in an alternative form of the routing equation as

$$Q_2 = Q_1 + B_1\,(I_1 - Q_1) + B_2(I_2 - I_1) \tag{8.17}$$

where $B_1 = \dfrac{\Delta t}{K(1-x) + 0.5\,\Delta t}$ $\qquad B_2 = \dfrac{0.5\,\Delta t - K_x}{K(1-x) + 0.5\,\Delta t}$

The use of Eq. (8.17) is essentially the same as that of Eq. (8.16).

8.6 *Hydraulic Method of Flood Routing*

The hydraulic method of flood routing is essentially a solution of the basic St. Venant equations [Eqs (8.4) and (8.5)]. These equations are simultaneous, quasi-linear, first-order, partial differential equations of the hyperbolic type and are not amenable to general analytical solutions. Only for highly simplified cases can one obtain the analytical solution of these equations. The development of modern, high-speed digital computers during the past two decades has given rise to the evolution of many sophisticated numerical techniques. The various numerical methods for solving St. Venant equations can be broadly classified into two categories:

1. Approximate methods
2. Complete numerical methods

8.6.1 Approximate Methods

These are based on the equation of continuity only or on a drastically curtailed equation of motion. The hydrological method of storage routing and Muskingum channel routing discussed earlier belong to this category.

Other methods in this category are *diffusion analogy and kinematic wave* models.

8.6.2 Complete Numerical Methods

These are the essence of the hydraulic method of routing and are classified into many categories as mentioned below:

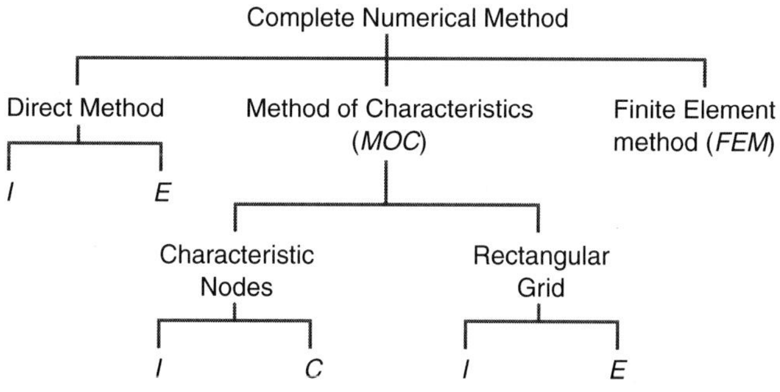

I = Implicit method, E = Explicit method

In the direct method, the partial derivatives are replaced by finite differences and the resulting algebraic equations are then solved. In the method of characteristics (*MOC*) St. Venant equations are converted into a pair of ordinary differential equations (i.e. characteristic forms) and then solved by finite difference techniques. In the finite element method (*FEM*) the system is divided into a number of elements and partial differential equations are integrated at the nodal points of the elements.

The numerical schemes are further classified into explicit and implicit methods. In the explicit method the algebraic equations are linear and the dependent variables are extracted explicitly at the end of each time step. In the implicit method, the dependent variables occur implicitly and the equations are nonlinear. Each of these two methods have a host of finite-differencing schemes to choose from. Details of hydraulic flood routing and a bibliography of relevant literature are available in Refs. 6, 8 and 9.

A large number of software, both in commercial and in common domain, are available for flood-flow analysis in natural channels. Among these, the *HEC-RAS* of *US* Army Corps of Engineers, *FLDWAV* of *US* National Weather Service , *MIKE-II* of *DIH*, Denmark and *FLO-2D* of *USA* are very popular. *HEC-RAS* is available with user manual for download by individuals free of charge, ***(http://www.hec.usace.army.mil/software/hec-ras/hecras-features.html)***. *HEC-RAS* stands for Hydrologic Engineering Center's River Analysis System. It contains four one-dimensional analysis components for (i) steady-flow water-surface profile computations, (ii) flood-flow simulations, (iii) mobile boundary sediment transport computations, and (iv) water quality analysis.

8.7 *Routing in Conceptual Hydrograph Development*

Even though the routing of floods through a reservoir or channel discussed in the previous section were developed for field use, they have found another important use in the conceptual studies of hydrographs. The routing through a reservoir which gives attenuation and channel routing which gives translation to an input hydrograph are treated as two basic modifying operators. The following two fictitious items are used in the studies for development of synthetic hydrographs through conceptual models:

1. **Linear Reservoir** A reservoir in which the storage is directly proportional to the discharge ($S = KQ$). This element is used to provide attenuation to a flood wave.

2. **Linear Channel** A fictitious channel in which the time required to translate a discharge Q through a given reach is constant. An inflow hydrograph passes through such a channel with only translation and no attenuation.

Conceptual modelling for *IUH* development has undergone rapid progress since the first work by Zoch (1937). Detailed reviews of various contributions to this field are available in Refs. 2 and 4 and the details are beyond the scope of this book. However, a simple method, viz., Clark's method (1945), which utilises the Muskingum method of routing through a linear reservoir is indicated below as a typical example of the use of routing in conceptual models. Nash's model which uses routing through a cascade of linear reservoirs is also presented, in Sec. 8.9, as another example of a conceptual model.

8.8 *Clark's Method For IUH*

Clark's method, also known as *time–area histogram* method, aims at developing an *IUH* due to an instantaneous rainfall excess over a catchment. It is assumed that the rainfall excess first undergoes pure translation and then attenuation. The translation is achieved by a travel time–area histogram and the attenuation by routing the results of the above through a linear reservoir at the catchment outlet.

8.8.1 Time–Area Curve

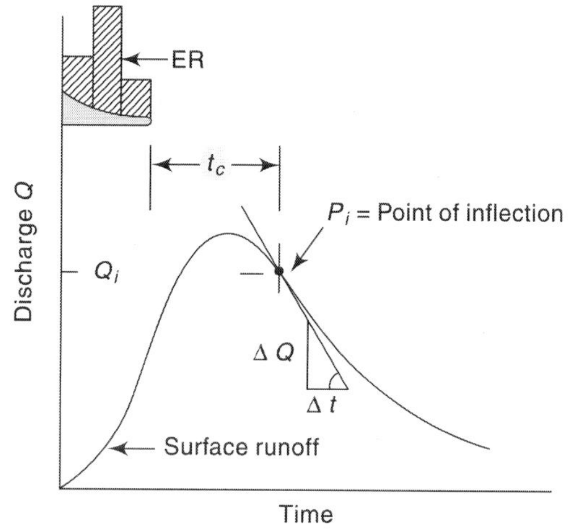

Fig. 8.10 ***Surface Runoff of a Catchment***

Time here refers to the time of concentration. As defined earlier in Sec. 7.2, the time of concentration t_c is the time required for a unit volume of water from the farthest point of catchment to reach the outlet. It represents the maximum time of translation of the surface runoff of the catchment. In gauged areas, the time interval between the end of the rainfall excess and the point of inflection of the resulting surface runoff (Fig. 8.10) provides a good way of estimating t_c from known rainfall-runoff data. In ungauged areas the empirical formulae Eq. (7.3) or (7.4) can be used to estimate t_c.

The total catchment area drains into the outlet in t_c hours. If points on the area having equal time of travel, (say t_1 *h* where $t_1 < t_c$), are considered and located on a map of the catchment, a line joining them is called an *Isochrone* (or *runoff iso-chrone*). Figure (8.11) shows a catchment being divided into *N* (= 8) subareas by isochrones having an equal time interval. To assist in drawing isochrones, the longest water course is chosen and its profile plotted as elevation *vs* distance from the outlet; the distance is then divided into *N* parts and the elevations of the subparts measured on the profile transferred to the contour map of the catchment.

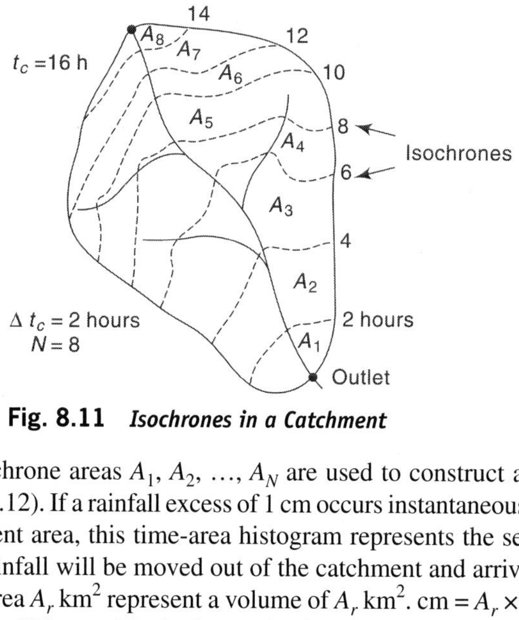

Fig. 8.11 ***Isochrones in a Catchment***

The inter-isochrone areas A_1, A_2, ..., A_N are used to construct a travel time-area histogram (Fig. 8.12). If a rainfall excess of 1 cm occurs instantaneously and uniformly over the catchment area, this time-area histogram represents the sequence in which the volume of rainfall will be moved out of the catchment and arrive at the outlet. In Fig. 8.12, a subarea A_r km^2 represent a volume of A_r km^2. cm = $A_r \times 10^4$ (m^3) moving out in time $\Delta t_c = t_c/N$ hours. The hydrograph of outflow obtained by this figure while properly accounting for the sequence of arrival of flows, do not provide for the storage properties of the catchment. To overcome this deficiency, Clark assumed a linear reservoir to be hypothetically available at the outlet to provide the requisite attenuation.

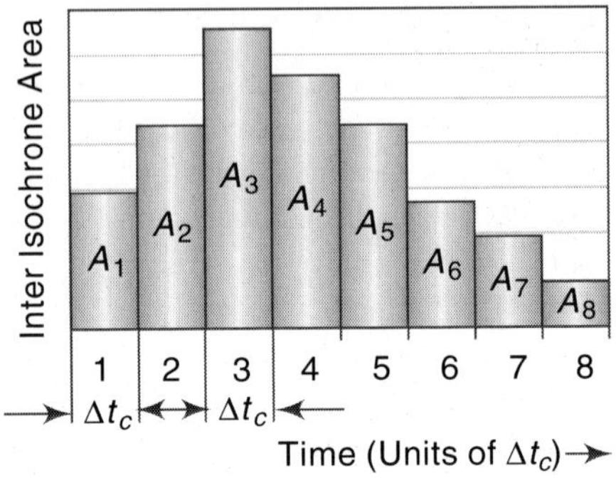

Fig. 8.12 ***Time-area Histogram***

8.8.2 Routing

The linear reservoir at the outlet is assumed to be described by $S = KQ$, where K is the storage time constant. The value of K can be estimated by considering the point of inflection P_i of a surface runoff hydrograph (Fig. 8.10). At this point, the inflow into the channel has ceased and beyond this point the flow is entirely due to withdrawal from the channel storage. The contiunity equation

$$I - Q = \frac{dS}{dt}$$

becomes $$-Q = \frac{dS}{dt} = K\frac{dQ}{dt} \qquad \text{(by Eq. 8.13)}$$

Hence, $$K = -Q_i/(dQ/dt)_i \qquad (8.\ 18)$$

where suffix i refers to the point of inflection, and K can be estimated from a known surface runoff hydrograph of the catchment as shown in Fig. 8.10. The constant K can also be estimated from the data on the recession limb of a hydrograph (Sec. 6.3).

Knowing K of the linear reservoir, the inflows at various times are routed by the Muskingum method. Note that since a linear reservoir is used $x = 0$ in Eq. (8.12). The inflow rate between an inter-isochrone area A_r km^2 with time interval Δt_c (h) is

$$I = \frac{A_r \times 10^4}{3600\ \Delta t_c} = 2.78\frac{A_r}{\Delta t_c} \quad (\text{m}^3/\text{s})$$

The Muskingum routing equation would now be by Eq. (8.16),

$$Q_2 = C_0 I_2 + C_1 I_1 + C_2 Q_1 \qquad (8.19)$$

where $$C_0 = (0.5\ \Delta t_c)/(K + 0.5\ \Delta t_c)$$

$$C_1 = (0.5\ \Delta t_c)/(K + 0.5\ \Delta t_c)$$

$$C_2 = (K - 0.5\ \Delta t_c)/(K + 0.5\ \Delta t_c)$$

i.e. $C_0 = C_1$. Also since the inflows are derived from the histogram $I_1 = I_2$ for each interval. Thus, Eq. (8.19) becomes

$$Q_2 = 2\ C_1 I_1 + C_2 Q_1 \qquad (8.20)$$

Routing of the time-area histogram by Eq. (8.20) gives the ordinates of IUH for the catchment. Using this *IUH* any other *D*-h unit hydrograph can be derived.

Example 8.6 A drainage basin has the following characteristics:

Area = 110 km^2, time of concentration = 18 h, storage constant = 12 h and inter-isochrone area distribution as below:

Travel time t (h)	0–2	2–4	4–6	6–8	8–10	10–12	12–14	14–16	16–18
Inter-Isochrone area (km^2)	3	9	20	22	16	18	10	8	4

Determine the IUH for this catchment.

Solution

$$K = 12 \text{ h}, \qquad t_c = 18\text{h}, \qquad \Delta t_c = 2 \text{ h}$$

$$C_1 = \frac{0.5 \times 2}{12 + 0.5 \times 2} = 0.077$$

$$C_2 = \frac{12 - 0.5 \times 2}{12 + 0.5 \times 2} = 0.846$$

Equation (8.20) becomes $Q_2 = 0.154 I_1 + 0.846\, Q_1$ = Ordinate of IUH

At $t = 0, \quad Q_1 = 0$

$$I_1 = 2.78\, A_r/2 = 1.39\, A_r \text{ m}^3/\text{s}$$

The calculations are shown in Table 8.5.

Table 8.5 ***Calculations of IUH—Clark's Method—Example 8.6***

Time (h)	Area A_r (km^2)	I (m^3/s)	0.154 I_1	0.846 Q_1	Ordinate of IUH (m^3/s)
1	2	3	4	5	6
0	0	0	0	0	0
2	3	4.17	0.64	0	0.64
4	9	12.51	1.93	0.54	2.47
6	20	27.80	4.28	2.09	6.37
8	22	30.58	4.71	5.39	10.10
10	16	22.24	3.42	8.54	11.96
12	18	25.02	3.85	10.12	13.97
14	10	13.90	2.14	11.82	13.96
16	8	11.12	1.71	11.81	13.52
18	4	5.56	0.86	11.44	12.30
20	0	0	0	10.40	10.40
22				8.80	8.80
24				7.45	7.45
26				6.30	6.30
28				5.30	5.30
				⋮	⋮
				so on	so on

8.9 Nash's Conceptual Model

Nash[7] (1957) proposed the following conceptual model of a catchment to develop an equation for *IUH*. The catchment is assumed to be made up of a series of n identical linear reservoirs each having the same storage constant K. The first reservoir receives a unit volume equal to 1 cm of effective rain from the catchment instantaneously. This inflow is routed through the first reservoir to get the outflow hydrograph. The outflow from the first reservoir is considered as the input to the second; the outflow from the second reservoir is the input to the third and so on for all the n reservoirs. The conceptual cascade of reservoirs as above and the shape of the outflow hydrographs from each reservoir of the cascade is shown in Fig. 8.13. The outflow hydrograph from the nth reservoir is taken as the *IUH* of the catchment

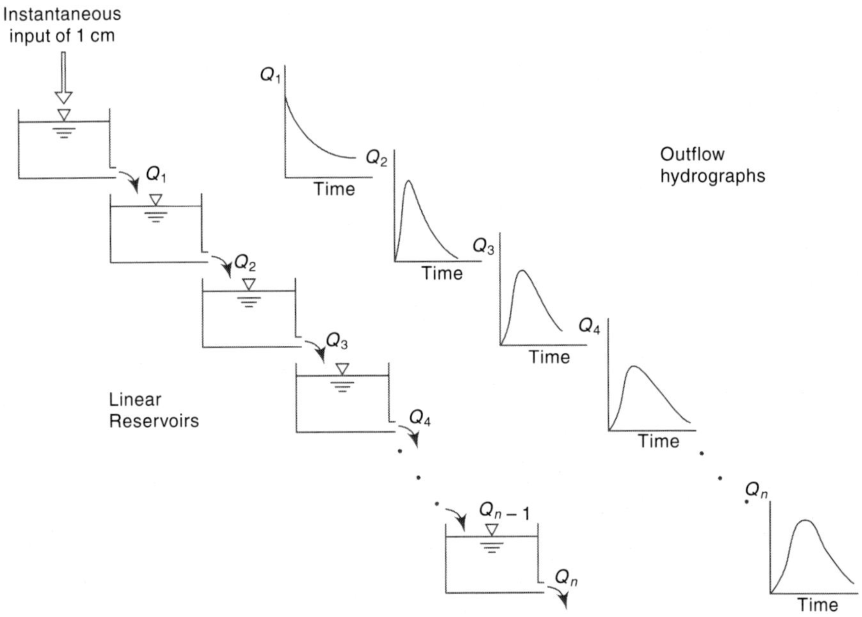

Fig. 8.13 ***Nash Model: Cascade of Linear Reservoirs***

From the equation of continuity (8.1), $I - Q = \dfrac{dS}{dt}$ (8.1)

For a linear reservoir $S = K\,Q$ and hence $\dfrac{dS}{dt} = K\,\dfrac{dQ}{dt}$ (8.21)

Subtituting in Eq. (8.1) and rearranging,

$$K\,\frac{dQ}{dt} + Q = I \tag{8.22}$$

and the solution of this differential equation, where Q and I are functions of time t, is

$$Q = \frac{1}{K} e^{-t/K} \int e^{t/K} I \, dt \tag{8.23}$$

Now for the first reservoir, the input is applied instantaneously Hence, for $t > 0$, $I = 0$. Also at $t = 0$, $\int I \, dt$ = instantaneous volume inflow = 1 cm of effective rain. Hence for the first reservoir Eq. (8.23) becomes,

$$Q_1 = \frac{1}{K} e^{-t/K} \tag{8.24}$$

For the second reservoir $Q_2 = \frac{1}{K} e^{-t/K} \int e^{t/K} I \, dt$

Here I = input = Q_1 given by Eq. (8.24). Thus,

$$Q_2 = \frac{1}{K} e^{-t/K} \int e^{t/K} \frac{1}{K} e^{-t/K} \, dt = \frac{1}{K^2} t e^{-t/k} \tag{8.25}$$

For the third reservoir in Eq. (8.23)

$$I = Q_2 \text{ and } Q_3 \text{ is obtained as } Q_3 = \frac{1}{2} \frac{1}{K^3} t^2 e^{-t/k} \tag{8.26}$$

Similarly, for the hydrograph of outflow from the nth reservoir Q_n is obtained as

$$Q_n = \frac{1}{(n-1)! K^n} t^{n-1} e^{-t/K} \tag{8.27}$$

As the outflow from the *n*th reservoir was caused by 1 cm of excess rainfall falling instantaneously over the catchment Eq. (8.27) describes the *IUH* of the catchment. Using the notation $u(t)$ to represent the ordinate of the *IUH*, Eq. (8.27) to represent the *IUH* of a catchment is written as

$$u(t) = \frac{1}{(n-1)! K^n} t^{n-1} e^{-t/K} \tag{8.28}$$

Here, if t is in hours, $u(t)$ will have the dimensions of cm/h; K and n are constants for the catchment to be determined by effective rainfall and flood hydrograph characteristics of the catchment.

It should be remembered that Eq. (8.28) is based on a conceptual model and as such if n for a catchment happens to be a fraction, it is still alright. To enable $(n-1)!$ to be determined both for integer and fractional values of n, the gamma function $\Gamma(n)$ is use to replace $(n-1)!$ so that

$$u(t) = \frac{1}{K\Gamma(n)} (t/K)^{n-1} e^{-t/K} \tag{8.29}$$

When n is an integer, $\Gamma(n) = (n-1)!$ which can be evaluated easily. However, when n is not an integer, the value of $\Gamma(n)$ is obtained from Gamma Tables[10] (Table 8.6).

Table 8.6 ***Gamma Function Γ (n)***

n	$\Gamma(n)$	n	$\Gamma(n)$	n	$\Gamma(n)$
1.00	1.000000	1.34	0.892216	1.68	0.905001
1.02	0.988844	1.36	0.890185	1.70	0.908639
1.04	0.978438	1.38	0.888537	1.72	0.912581
1.06	0.968744	1.40	0.887264	1.74	0.916826
1.08	0.959725	1.42	0.886356	1.76	0.921375
1.10	0.951351	1.44	0.885805	1.78	0.926227
1.12	0.943590	1.46	0.885604	1.80	0.931384
1.14	0.936416	1.48	0.885747	1.82	0.936845
1.16	0.929803	1.50	0.886227	1.84	0.942612
1.18	0.923728	1.52	0.887039	1.86	0.948687
1.20	0.918169	1.54	0.888178	1.88	0.955071
1.22	0.913106	1.56	0.889639	1.90	0.961766
1.24	0.908521	1.58	0.891420	1.92	0.968774
1.26	0.904397	1.60	0.893515	1.94	0.976099
1.28	0.900718	1.62	0.895924	1.96	0.983743
1.30	0.897471	1.64	0.898642	1.98	0.991708
1.32	0.894640	1.66	0.901668	2.00	1.000000

Note: Use the relation $\Gamma(n + 1) = n\,\Gamma(n)$ to evaluate $\Gamma(n)$ for any n.

EXAMPLE: (a) To find Γ (0.6) : $\Gamma(1.6) = \Gamma\,(0.6 + 1) = 0.6\,\Gamma\,(0.6)$

$$\text{Thus}\quad \Gamma\,(0.6) = \frac{\Gamma(1.6)}{0.6} = \frac{0.8935}{0.6} = 1.489$$

(b) To find Γ (4.7) : $\Gamma(4.7) = \Gamma(3.7+1) = 3.7\,\Gamma(3.7)$

$$= 3.7 \times 2.7\,\Gamma(2.7) = 3.7 \times 2.7 \times 1.7 \times \Gamma(1.7)$$

$$= 3.7 \times 2.7 \times 1.7 \times 0.9086 = 15.431$$

Determination of *n* and *K* of Nash's Model

From the property of the *IUH* given by Eq. (8.28), it can be shown that the first moment of the *IUH* about the origin $t = 0$ is given by

$$M_1 = nK \tag{8.30}$$

Also the second moment of the IUH about the origin $t = 0$ is given by

$$M_2 = n\,(n + 1)\,K^2 \tag{8.31}$$

Using these properties the values of n and K for a catchment can be determined adequately if the *ERH* and a corresponding *DRH* are available. If

M_{Q1} = first moment of the *DRH* about the time origin divided by the total direct runoff, and

M_{I1} = first moment of the *ERH* about the time origin divided by the total effective rainfall,

then,
$$M_{Q1} - M_{I1} = nK \tag{8.32}$$

Further, if

M_{Q2} = second moment of *DRH* about the time origin divided by total direct runoff, and

M_{I2} = second moment of *ERH* about the time origin divided by total excess rainfall,

then,
$$M_{Q2} - M_{I2} = n\,(n + 1)\,K^2 + 2nK\,M_{I1} \tag{8.33}$$

Knowing M_{I1}, M_{I2}, M_{Q1} and M_{Q2}, values of K and n for a given catchment can be calculated by Eqs. (8.32) and (8.33).

Example 8.7 illustrates the method of determining n and K of the Nash's model. Example 8.8 describes the computation of *IUH* and a *D*-hour *UH* when the values of n and K are known.

Example 8.7 For a catchment the effective rainfall hyetograph of an isolated storm and the corresponding direct runoff hydrograph is given below. Determine the coefficients *n* and *K* of Nash model IUH.

Coordinates of ERH:

Time from start of storm (h)	Effective rainfall intensity (cm/s)
0 to 1.0	4.3
1.0 to 2.0	3.2
2.0 to 3.0	2.4
3.0 to 4.0	1.8

Coordinates of DRH:

Time from start of storm (h)	Direct runoff (m^3/s)	Time from start of storm (h)	Direct runoff (m^3/s)
0	0	9	32.7
1	6.5	10	23.8
2	15.4	11	16.4
3	43.1	12	9.6
4	58.1	13	6.8
5	68.2	14	3.2
6	63.1	15	1.5
7	52.7	16	0
8	41.9		

Solution

The *ERH* is shown in Fig. 8.14(a) as a histogram. Each block has the total rainfall in a time interval of 1 hour marked on it.

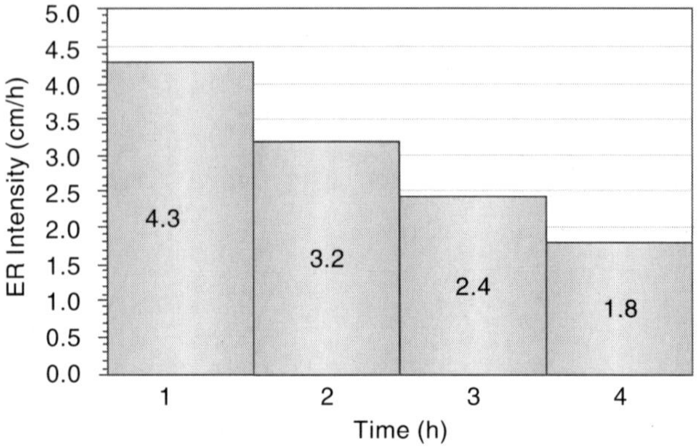

Fig. 8.14(a) ***Excess Rainfall Hyetograph of Example 8.7***

M_{I1} = first moment of the *ERH* about the time origin divided by the total rainfall excess.

$$M_{I1} = \frac{\sum(\text{Incremental area of } ERH \times \text{moment arm})}{\text{total area of } ERH}$$

M_{I2} = second moment of the *ERH* about the time original divided by the total rainfall excess.

$$= \left\{\frac{1}{\text{total area of } ERH}\right\}\left\{\begin{array}{l}\sum[\text{incremental area} \times (\text{moment arm})^2] \\ +\sum[\text{second moment of the incremental} \\ \quad \text{area about its own centroid}]\end{array}\right\}$$

The calculations of M_{I1} and M_{I2} are shown in Table 8.7(a)

Table 8.7(a) ***Calculation of M_{I1} and M_{I2}—Example 8.7***

1 Time (h)	2 Excess rainfall in Δt (cm)	3 Interval Δt (h)	4 Incre. area	5 moment arm	6 First moment	7 Second moment part (a)	8 Second moment part (b)
0	0	0	0	0	0	0	0
1	4.3	1	4.3	0.5	2.15	1.08	0.358
2	3.2	1	3.2	1.5	4.8	7.20	0.267
3	2.4	1	2.4	2.5	6.0	15.00	0.200
4	1.8	1	1.8	3.5	6.3	22.05	0.150
Sum			**11.7**		**19.25**	**45.325**	**0.975**

In Table 8.7(a)

Col. 6 = first moment of the incremental area about the origin = (Col. 4 × Col. 5)

Col. 7 = Col. 4 × (Col. 5)^2

Col. 8 = second moment of the incremental area about its own centroid

$$= \frac{1}{12} \times (\Delta t)^3 \,(ER) = \frac{1}{12} \times (\text{Col. } 3)^3 \times (\text{Col. } 2)$$

From the data of Table 8.7(a):

$$M_{I1} = (\text{sum of Col. 6})/(\text{sum of Col. 4}) = 19.25/11.7 = 1.645$$

$$M_{I2} = \{(\text{sum of Col. 7}) + (\text{sum of Co1. 8})\}/(\text{sum of Col. 4})$$

$$= (45.325 + 0.975)\}/11.75 = 3.957$$

The *DRH* is shown plotted in Fig. 8.14(b). A time interval of Δt = 1 hour is chosen and considering the average *DR* in this interval the *DRH* is taken to be made up of large number of rectangular blocks.

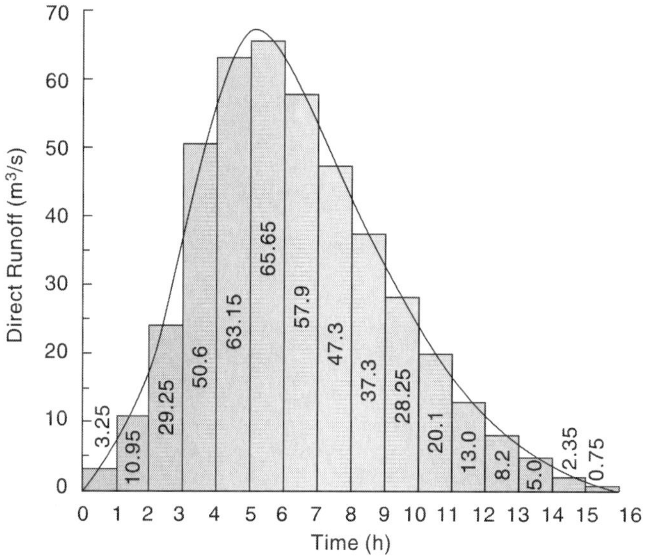

Fig. 8.14(b) ***Direct Runoff Hydrograph of Example 8.7***

For the DRH

M_{Q1} = first moment of the *DRH* about the time original divided by the total direct runoff

$$= \frac{\sum(\text{Incremental area of } DRH \times \text{moment arm})}{\text{total area of } DRH}$$

M_{Q2} = second moment of the *DRH* about the time original divided by the total direct runoff

$$= \left\{\frac{1}{\text{total area of } DRH}\right\}\left\{\begin{array}{l}\sum[\text{incremental area} \times (\text{moment arm})^2] \\ +\sum[\text{second moment of the incremental} \\ \quad \text{area about its own centroid}]\end{array}\right\}$$

The calculations of M_{Q1} and M_{Q2} are shown in Table 8.7(b).

Table 8.7(b) ***Calculation of M_{Q1} and M_{Q2}—Example 8.7***

1	2	3	4	5	6	7	8	9
Time (h)	Ord. of *DRH* (m³/s)	Average *DR* rate in Δt (m³/s)	Interval Δt (*h*)	Increment area	Moment arm	First Moment	Second Moment part (a)	Second Moment part (b)
0	0	0.00	0	0.00	0	0.00	0.00	0.00
1	6.5	3.25	1	3.25	0.5	1.63	0.81	0.27
2	15.4	10.95	1	10.95	1.5	16.43	24.64	0.91
3	43.1	29.25	1	29.25	2.5	73.13	182.81	2.44
4	58.1	50.60	1	50.60	3.5	177.10	619.85	4.22
5	68.2	63.15	1	63.15	4.5	284.18	1278.79	5.26
6	63.1	65.65	1	65.65	5.5	361.08	1985.91	5.47
7	52.7	57.90	1	57.90	6.5	376.35	2446.28	4.83
8	41.9	47.30	1	47.30	7.5	354.75	2660.63	3.94
9	32.7	37.30	1	37.30	8.5	317.05	2694.93	3.11
10	23.8	28.25	1	28.25	9.5	268.38	2549.56	2.35
11	16.4	20.10	1	20.10	10.5	211.05	2216.03	1.68
12	9.6	13.00	1	13.00	11.5	149.50	1719.25	1.08
13	6.8	8.20	1	8.20	12.5	102.50	1281.25	0.68
14	3.2	5.00	1	5.00	13.5	67.50	911.25	0.42
15	1.5	2.35	1	2.35	14.5	34.08	494.09	0.20
16	0	0.75	1	0.75	15.5	11.63	180.19	0.06
Sum				443.00		2806.30	21246.25	36.92

In Table 8.7(b):

Col. 7 = first moment of the incremental area of *DRH* about the origin

= (Col. 4 × Col. 5)

Col. 8 = Col. 5 × (Col. 6)^2

Col. 9 = second moment of the incremental area about its own centroid

$$= \frac{1}{12} \times (\text{Col. } 4)^3 \times (\text{Col. } 3)$$

From the data of Table 8.7(b):

M_{Q1} = (sum of Col. 7)/(sum of Col. 5) = 2806.3/443 = 6.33

M_{Q2} = {(sum of Col. 8) + (sum of Col. 9)}/(sum of Col. 5)

= (21246.25 + 36.92)/443 = 48.04

[**Note** that in the calculation of M_{I2} and M_{Q2}, for small values of Δt the second term in the bracket, *viz.* second moment part (b) = Σ [second moment of incremental area about its own centroid], is relatively small in comparison with the first term [part(a)] and can be neglected without serious error.]

By Eq. (8.30), $n\,K = M_{Q1} - M_{I1} = 6.335 - 1.645 = 4.690$

By Eq. (8.31), $M_{Q2} - M_{I2} = n\,(n+1)\,K^2 + 2\,n\,K\,M_{I1} = (nK)^2 + (nK)\,K + 2(nK)\,M_{I1}$

Substituting for nK, M_{Q2}, M_{I2} and M_{I1}

$$48.04 - 3.96 = (4.69)^2 + (4.69)\,K + 2\,(4.69)\,(1.645)$$

$$K = 6.654/4.69 = 1.42 \text{ hours}$$

$$n = nK/K = 4.69/1.42 = 3.30$$

Example 8.8 For a catchment of 300 km^2 area the values of the Nash model coefficients are found to have values of $n = 4.5$ and $K = 3.3$ hours. Determine the ordinates of (a) *IUH* and (b) 3-h unit hydrograph of the catchment.

Solution

The ordinates of *IUH* by Nash model are given by

$$u(t) = \frac{1}{K\Gamma(n)}(t/K)^{n-1}\,e^{(t/K)}$$

In the present case $n = 4.5$, $K = 3.3$ hours and $u(t)$ is in cm/h.

$$\Gamma(n) = \Gamma(4.5) = 3.5\,\Gamma(3.5) = 3.5 \times 2.5\,\Gamma(2.5)$$

$$= 3.5 \times 2.5 \times 1.5 \times \Gamma(1.5)$$

From Table 8.6, $\Gamma(1.5) = 0.886227$

Hence $\Gamma(4.5) = 3.5 \times 2.5 \times 1.5 \times 0.886227 = 11.632$

$$u(t) = \frac{1}{3.3 \times 11.632}(t/3.5)^{3.5}\,e^{-(t/3.3)} = 0.02605\,(t/3.3)^{3.5}\,e^{-(t/3.3)}$$

Values of $u(t)$ for various values of t are calculated as shown in Table 8.8. An interval of one hour is chosen. In Table 8.8, Col. 3 gives the ordinates of $u(t)$ in cm/h. Multiplying these values by $(2.78 \times A)$ where A = area of the catchment in km^2 gives the values of $u(t)$ in m^3/s, (Col. 4).

Thus Col. 4 = (Col. 3) × 2.78 × 300 = (Col. 3) × 834

Col. 5 is the ordinate of $u(t)$ [i.e. Col. 4] lagged by one hour

Col. 6 = (Col. 4 + Col. 6)/2 = ordinate of 1-h *UH* by Eq. (6.26)

The *S*-curve technique is used to derive the 3-h *UH* from the 1-h *UH* obtained in Col. 6.

Col. 7 = S_1–curve addition.

Col. 8 = S_1–curve ordinates

Col. 9 = S_1–curve ordinates lagged by 3 hours

Col. 10 = (Col. 8 – Col. 9) = ordinates of a *DRH* of 3 cm occurring in 3 hours.

Col. 11= (Col. 10)/3 = ordinates of 3-h *UH*

Table 8.8 ***Calculation of 3-Hour UH by Nash Method—Example 8.8***

$K = 3.3\ h$ $n = 4.5$ $\Gamma(n) = 11.632$ Area of the catchment = 300 km^2

1	2	3	4	5	6	7	8	9	10	11
Time *t* in hours	(*t/K*)	*u*(*t*) (cm/h)	*u*(*t*) (m^3/s)	*u*(*t*) lagged by 1 hour	1-h *UH* (m^3/s)	S_1 – Curve addition	Ordinate of S_1 – Curve	3-h lagged S_1 – Curve	*DRH* of 3 cm in 3 hours (m^3/s)	Ord. of 3-h *UH* (m^3/s)
0	0.000	0.0000	0.000		0.000	0.000	0.000		0.000	0.00
1	0.303	0.0003	0.246	0.000	0.123	0.000	0.123		0.123	0.40
2	0.606	0.0025	2.054	0.246	1.150	0.123	1.273		1.273	0.42
3	0.909	0.0075	6.271	2.054	4.162	1.273	5.435	0.000	5.435	1.81
4	1.212	0.0152	12.676	6.271	9.473	5.435	14.909	0.123	14.786	4.93
5	1.515	0.0245	20.444	12.676	16.560	14.909	31.469	1.273	30.196	10.07
6	1.818	0.0343	28.583	20.444	24.513	31.469	55.982	5.435	50.547	16.85
7	2.121	0.0434	36.209	28.583	32.396	55.982	88.378	14.909	73.469	24.49
8	2.424	0.0512	42.676	36.209	39.442	88.378	127.820	31.469	96.351	32.12
9	2.727	0.0571	47.600	42.676	45.138	127.820	172.958	55.982	116.975	38.99
10	3.030	0.0610	50.834	47.600	49.217	172.958	222.175	88.378	133.797	44.60
11	3.333	0.0628	52.411	50.834	51.623	222.175	273.798	127.820	145.978	48.66
12	3.636	0.0629	52.490	52.411	52.451	273.798	326.248	172.958	153.291	51.10
13	3.939	0.0615	51.303	52.490	51.897	326.248	378.145	222.175	155.970	51.99
14	4.242	0.0589	49.112	51.303	50.207	378.145	428.353	273.798	154.555	51.52
15	4.545	0.0554	46.180	49.112	47.646	428.353	475.998	326.248	149.750	49.92
16	4.848	0.0513	42.751	46.180	44.466	475.998	520.464	378.145	142.319	47.44
17	5.152	0.0468	39.039	42.751	40.895	520.464	561.359	428.353	133.006	44.34
18	5.455	0.0422	35.219	39.039	37.129	561.359	598.488	475.998	122.489	40.83
19	5.758	0.0377	31.431	35.219	33.325	598.488	631.812	520.464	111.348	37.12
20	6.061	0.0333	27.779	31.431	29.605	631.812	661.417	561.359	100.058	33.35
21	6.364	0.0292	24.337	27.779	26.058	661.417	687.475	598.488	88.988	29.66
22	6.667	0.0254	21.153	24.337	22.745	687.475	710.221	631.812	78.408	26.14
23	6.970	0.0219	18.253	21.153	19.703	710.221	729.924	661.417	68.507	22.84
24	7.273	0.0188	15.647	18.253	16.950	729.924	746.875	687.475	59.399	19.80
25	7.576	0.0160	13.332	15.647	14.489	746.875	761.364	710.221	51.143	17.05
26	7.879	0.0135	11.295	13.332	12.313	761.364	773.677	729.924	43.753	14.58
27	8.182	0.0114	9.520	11.295	10.408	773.677	784.085	746.875	37.210	12.40
28	8.485	0.0096	7.986	9.520	8.753	784.085	792.838	761.364	31.474	10.49
29	8.788	0.0080	6.669	7.986	7.328	792.838	800.166	773.677	26.488	8.83
30	9.091	0.0067	5.546	6.669	6.108	800.166	806.273	784.085	22.188	7.40
31	9.394	0.0055	4.594	5.546	5.070	806.273	811.344	792.838	18.505	6.17
32	9.697	0.0045	3.792	4.594	4.193	811.344	815.537	800.166	15.371	5.12
33	10.000	0.0037	3.119	3.792	3.456	815.537	818.992	806.273	12.719	4.24
34	10.303	0.0031	2.558	3.119	2.838	818.992	821.831	811.344	10.487	3.50
35	10.606	0.0025	2.091	2.558	2.324	821.831	824.155	815.537	8.618	2.87
36	10.909	0.0020	1.704	2.091	1.897	824.155	826.052	818.992	7.060	2.35
37	11.212	0.0017	1.385	1.704	1.545	826.052	827.597	821.831	5.766	1.92
38	11.515	0.0013	1.123	1.385	1.254	827.597	828.851	824.155	4.696	1.57
39	11.818	0.0011	0.909	1.123	1.016	828.851	829.867	826.052	3.815	1.27
40	12.121	0.0009	0.733	0.909	0.821	829.867	830.688	827.597	3.091	1.03

8.10 *Flood Control*

The term *flood control* is commonly used to denote all the measures adopted to reduce damages to life and property by floods. Currently, many people prefer to use the term *flood management* instead of *flood control* as it reflects the activity more realistically. As there is always a possibility, however remote it may be, of an extremely large flood occurring in a river the complete control of the flood to a level of zero loss is neither physically possible nor economically feasible. The flood control measures that are in use can be classified as

1. Structural Measures
   - Storage and detention reservoirs
   - Levees (flood embankments)
   - Flood ways (new channels)
   - Channel improvement
   - Watershed management
2. Non-structural Methods
   - Flood plain zoning
   - Flood forecast/warning
   - Evacuation and relocation
   - Flood insurance

8.10.1 Structural Methods

1. Storage Reservoirs

Storage reservoirs offer one of the most reliable and effective methods of flood control. Ideally, in this method, a part of the storage in the reservoir is kept apart to absorb the incoming flood. Further, the stored water is released in a controlled way over an extended time so that downstream channels do not get flooded. Figure 8.15 shows an ideal operating plan of a flood control reservoir. As most of the present-day storage reservoirs have multipurpose commitments, the manipulation of reservoir levels to satisfy many conflicting demands is a very difficult and complicated task. It so happens that many storage reservoirs while reducing the floods and flood damages do not always aim at achieving optimum benefits in the flood-control aspect. To achieve complete flood control in the entire length of the river, a large number of reservoirs at strategic locations in the catchment will be necessary.

The Hirakud and Damodar Valley Corporation (*DVC*) reservoirs are examples of major reservoirs in the country which have specific volumes earmarked for flood absorption.

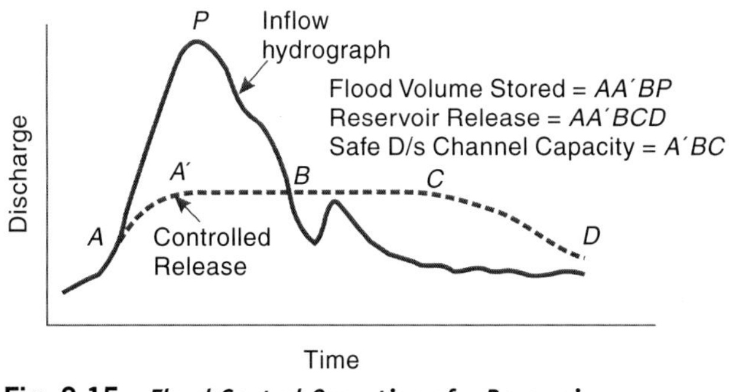

Fig. 8.15 ***Flood Control Operation of a Reservoir***

2. Detention Reservoirs

A detention reservoir consists of an obstruction to a river with an uncontrolled outlet. These are essentially small structures and operate to reduce the flood peak by providing temporary storage and by restriction of the outflow rate. These structures are not common in India.

3. Levees

Levees, also known as *dikes or flood embankments,* are earthen banks constructed parallel to the course of the river to confine it to a fixed course and limited cross-sectional width. The heights of levees will be higher than the design flood level with sufficient free board. The confinement of the river to a fixed path frees large tracts of land from inundation and consequent damage (Fig. 8.16).

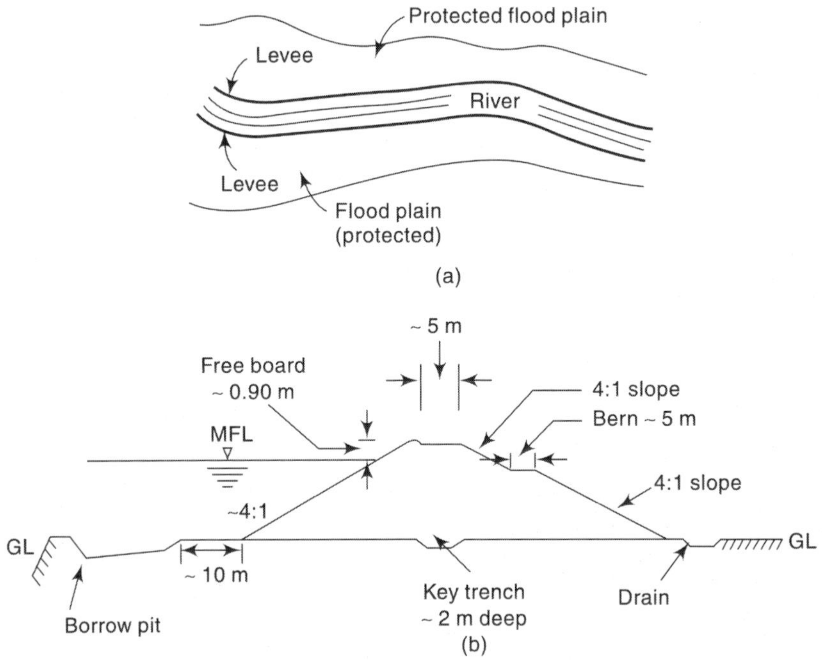

Fig. 8.16 ***A typical levee: (a) Plan (schematic) (b) Cross section***

Levees are one of the oldest and most common methods of flood-protection works adopted in the world. Also, they are probably the cheapest of structural flood-control measures. While the protection offered by a levee against food damage is obvious, what is not often appreciated is the potential damage in the event of a levee failure. The levees, being earth embankments, require considerable care and maintenance. In the event of being overtopped, they fail and the damage caused can be enormous. In fact, the sense of protection offered by a levee encourages economic activity along the embankment and if the levee is overtopped the loss would be more than what would have been if there were no levees. Confinement of flood banks of a river by levees to a narrower space leads to higher flood levels for a given discharge. Further, if the bed levels of the river also rise, as they do in aggrading rivers, the top of the levees have to be raised at frequent time intervals to keep up its safety margin.

The design of a levee is a major task in which costs and economic benefits have to be considered. The cross-section of a levee will have to be designed like an earth dam for complete safety against all kinds of saturation and drawdown possibilities. In many instances, especially in locations where important structures and industries are to be protected, the water side face of levees are protected by stone or concrete revetment. Regular maintenance and contingency arrangements to fight floods are absolutely necessary to keep the levees functional.

Masonry structures used to confine the river in a manner similar to levees are known as *flood walls*. These are used to protect important structures against floods, especially where the land is at a premium.

4. Floodways

Floodways are natural channels into which a part of the flood will be diverted during high stages. A floodway can be a natural or human-made channel and its location is controlled essentially by the topography. Generally, wherever they are feasible, floodways offer an economical alternative to other structural flood-control measures. To reduce the level of the river Jhelum at Srinagar, a supplementary channel has been constructed to act as a floodway with a capacity of 300 m^3/s. This channel is located 5 km upstream of Srinagar city and has its outfall in lake Wullar. In Andhra Pradesh, a floodway has been constructed to transfer a part of the flood waters of the river Budamaru to river Krishna to prevent flood damages to the urban areas lying on the downstream reaches of the river Budamaru.

5. Channel Improvement

The works under this category involve

- Widening or deepening of the channel to increase the cross-sectional area
- Reduction of the channel roughness, by clearing of vegetation from the channel perimeter
- Short circuiting of meander loops by cutoff channels, leading to increased slopes

All these three methods are essentially short-term measures and require continued maintenance.

6. Watershed Management

Watershed management and land treatment in the catchment aims at cutting down and delaying the runoff before it gets into the river. Watershed management measures include developing the vegetative and soil cover in conjunction with land treatment works like Nalabunds, check dams, contour bunding, zing terraces, etc. These measures are towards improvement of water infiltration capacity of the soil and reduction of soil erosion. These treatments cause increased infiltration, greater evapotranspiration and reduction in soil erosion; all leading to moderation of the peak flows and increasing of dry weather flows. Watershed treatment is nowadays an integral part of flood management. It is believed that while small and medium floods are reduced by watershed management measures, the magnitude of extreme floods are unlikely to be affected by these measures.

8.10.3 Non-Structural Methods

The flood management strategy has to include the philosophy of *living with the floods.* The following non-structural measures encompass this aspect.

1. Flood Plain Zoning

When the river discharges are very high, it is to be expected that the river will overflow its banks and spill into flood plains. In view of the increasing pressure of population, this basic aspects of the river are disregarded and there are greater encroachment of flood plains by humans leading to distress.

Flood-plain management identifies the flood-prone areas of a river and regulates the land use to restrict the damage due to floods. The locations and extent of areas likely to be affected by floods of different return periods are identified and development plans of these areas are prepared in such a manner that the resulting damages due to floods are within acceptable limits of risk. Figure 8.17 shows a conceptual zoning of a flood-prone area.

Zone	Flood Return Period	Example of Uses
1	100 Years	Residential houses, Offices, Factories, etc.
2	25 Years	Parks
3	Frequent	No construction/Encroachments

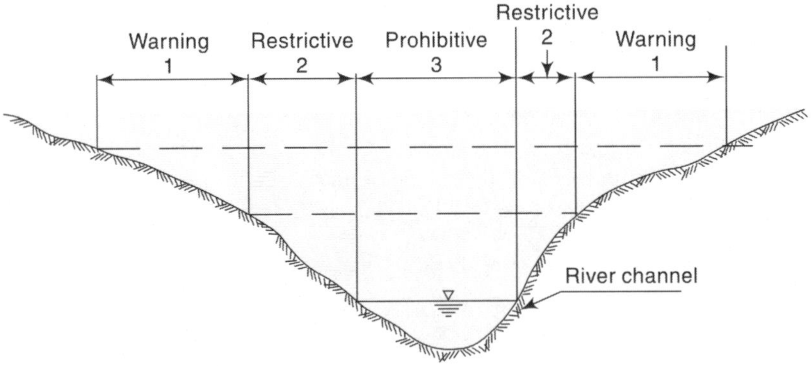

Fig. 8.17 ***Conceptual Zoning of a Flood Plain***

2. Flood Forecasting and Warning

Forecasting of floods sufficiently in advance enables a warning to be given to the people likely to be affected and further enables civil authorities to take appropriate precautionary measures. It thus forms a very important and relatively inexpensive non-structural flood management measure. However, it must be realised that a flood warning is meaningful only if it is given sufficiently in advance. Further, erroneous warnings will cause the populace to lose confidence and faith in the system. Thus, the dual requirements of reliability and advance notice are the essential ingredients of a flood-forecasting system.

The flood forecasting techniques can be broadly divided into three categories:

(a) Short range forecasts (b) Medium range forecasts

(c) Long range forecasts.

(a) Short-Range Forecasts In this, the river stages at successive stations on a river are correlated with hydrological parameters, such as rainfall over the local area, antecedent precipitation index, and variation of the stage at the upstream base point during the travel time of a flood. This method can give advance warning of 12-40 hours for floods. The flood forecasting used for the metropolitan city of Delhi is based on this technique.

(b) Medium-Range Forecasts In this method, rainfall–runoff relationships are used to predict flood levels with warning of 2–5 days. Coaxial graphical correlations of runoff, with rainfall and other parameters like the time of the year, storm duration and antecedent wetness have been developed to a high stage of refinement by the *US* Weather Bureau.

(c) Long-Range Forecasts Using radars and meteorological satellite data, advance information about critical storm-producing weather systems, their rain potential and time of occurrence of the event are predicted well in advance.

3. Evacuation and Relocation

Evacuation of communities along with their live stocks and other valuables in the chronic flood-affected areas and relocation of them in nearby safer locations is an area specific measure of flood management. This would be considered as non-structural measure when this activity is a temporary measure confined to high floods. However, permanent shifting of communities to safer locations would be termed as *structural measure*. Raising the elevations of buildings and public utility installations above normal flood levels is termed as *flood proofing* and is sometimes adopted in coastal areas subjected to severe cyclones.

4. Flood Insurance

Flood insurance provides a mechanism for spreading the loss over large numbers of individuals and thus modifies the impact of loss burden. Further, it helps, though indirectly, flood plain zoning, flood forecasting and disaster preparedness activities.

8.11 *Flood Control in India*

In India, the Himalayan rivers account for nearly 60% of the flood damage in the country. Floods in these rivers occur during monsoon months and usually in the months of August or September. The damages caused by floods are multi-dimensional and are difficult to estimate. The website of *CWC* [***www.cwc.nic.in***] contains data relating to damages due to floods since 1953.

During 1953–2004, the average number of human lives and cattle lost due to floods in the country were 1597 and 94,000 respectively. It is estimated that annually, on an average about 40 M ha of land is liable to flooding and of this about 14 M ha have some kind of flood-control measure. At the beginning of the current millennium, in the country, as a part of flood control measure there were about 15800 km of levees and about 32000 km of drainage channels affording protection from floods.

On an average, about 7.5 M ha land is affected by floods annually. Out of this, about 3.56 M ha are lands under crops. Similarly, annually about 3.345 lakhs of people are affected and about 12.15 lakhs houses are damaged by floods. On an average, about 60 to 80% of flood damages occur in the states of U.P., Bihar, West Bengal, Assam and Orissa.

Flood forecasting is handled by *CWC* in close collaboration with the IMD which lends meteorological data support. Nine flood Met offices with the aid of recording raingauges provide daily synoptic situations, actual rainfall amounts and rainfall forecasts to *CWC*. The *CWC* has 157 flood-forecasting stations, of which 132 stations are for river stage forecast and 25 for inflow forecast, situated in various basins to provide a forecasting service.

A national programme for flood management was launched in 1954 and an amount of 3165 crores was spent till 1992. The tenth plan (2002-2007) had an expenditure of 4468 crores for flood management. These figures highlight the seriousness of the flood problem and the efforts made towards mitigating flood damages. The experience gained in the flood control measures in the country are embodies in the report of the Rashtriya Barh Ayog (*RBA*) (National Flood Commission) submitted in March 1980. This report, containing a large number of recommendation on all aspects of flood control forms the basis for the evolution of the present national policy on floods. According to the national water policy, while structural flood control measures will continue to be necessary, the emphasis should be on non-structural methods so as to reduce recurring expenditure on flood relief.

8.12 *Global Warming and its Impact on Water Resources of India*

In common usage, the term *global warming* is used to refer to the warming that occurs as a result of increased emissions of green house gases (*GHG*) from human activities. The *US* Environmental protection agency defines global warming as follows: *Global warming is an average increase in the temperature of the atmosphere, which can contribute to changes in global climate patterns.* Global warming can occur from a variety of causes, both natural and human induced.

1. Greenhouse Effect

The earth's atmosphere contains many gases. Short-wave radiation from the sun passes through the atmosphere; while doing so, a part of it is absorbed, part of it is reflected and the remaining passes on to the surface of the earth. The earth's surface emits long-wave radiation towards the outer space. Some gases, known as greenhouse gases (*GHGs)* absorb some of this long-wave radiation and re-radiate it back to the earth. This phenomenon, known as greenhouse effect, modifies the heat balance of the earth. The most abundant naturally occurring *GHG* is water vapor. Other common *CHGs* include: carbon dioxide, nitrous oxide, methane, ozone and Chlorofluorocarbons (*CFCs*). These are increasingly being contributed by human activities in the planet.

Further, excess of *CFCs* in the atmosphere affects the ozone layer that has a beneficial role of reflecting incoming short wave radiation. Thus, the increasing concentrations of *GHG* gradually increase the average temperature on the earth with consequent impacts on different aspects of the hydrologic cycle.

2. Evidences of Climatic Change

While the causes and possibilities of global warming were suspected as back as the 19th century, the scientists in the 1950s and 60s expressed real serious studies and concerns. Since then there has been a spate of studies and reports in various international fora on global warming. The Intergovernmental panel on climate change (*IPCC*) (2007) lists the following evidences of climate change:

- Over the last century, the global average temperature has risen by 0.74°C ± 0.18°C. Fig. 1-18 shows the observed rise in mean global temperature since 1860. The ordinate is (Global mean temperature in °C—Average temperature during the period 1900-1930). The trend in the rise in temperature is unambiguous.
- Change in the precipitation in the middle and high latitudes of the northern hemisphere.
- Rise in the global mean sea level in the range 1.0–2.0 mm/year in the 20th century.
- Decreasing snow cover and land ice
- Behaviour of *ENSO* has been unusual since 1970 compared to previous 100 years.

Scientific studies of *GHG* emissions globally have confirmed that humans are contributing substantially to observed global climate changes. Further, there is a general consensus among policy makers and scientists that global emissions have to be cut in a determined manner to contain the effects global warming.

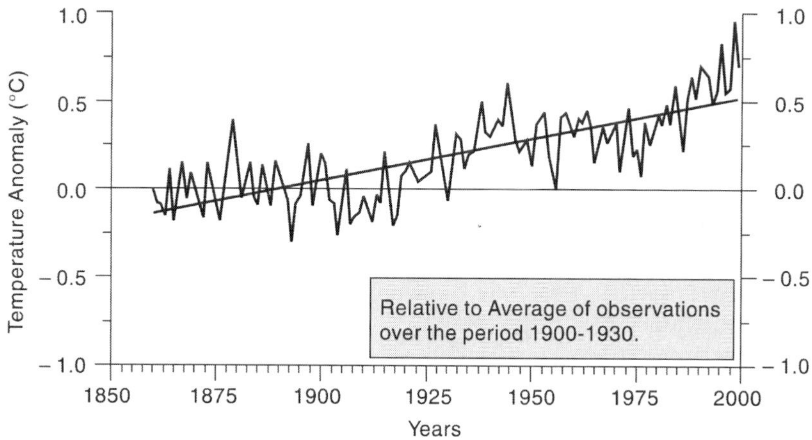

Fig. 8.18 ***Observed Rise in Global Mean Temperature (Ref. 13.)***

3. Science of Climate Change

Since last two decades, a very large number of serious scientific studies have been initiated to study all the relevant aspects of global warming. *Global Circulation Models* (*GCM*) are currently the most credible tools available to simulate the response of the global climate system to increasing *GHG* concentrations. Different future scenarios of *GHG* emissions are considered and for each of them the climate variables (like temperature, precipitation, wind speed and pressure) are estimated on a global scale. The spatial scales on which the *GCM* operate are very coarse being of the order of 3.75° latitude and longitude. This is roughly about 400 km scale in the tropical latitudes. While the *GCMs* can give global picture for a given *GHG* scenario, they are not suitable to identify the impact of global warming on a regional basis. The methodology to bring down the results of *GSM* studies to the level of a region (e.g., a river basin) a technique known as *downsizing* is used. There are several techniques of downsizing. This is a very fast developing specialty field and currently techniques are available for reasonably good downscaling. Ref. 11 gives a good account of available downscaling techniques.

4. Impact of Climate Change on the Hydrology of a Region

The following hydrologic implications can be expected in a region as a result of climate change effects.

Sl. No	Climate Change Effect	Hydrologic Implications in the Region
1	Increase in temperature	• Increase in evapotranspiration • Effect on water quality
2	Change in precipitation pattern	• Impact on stream flow quantity • Impact on intensity, frequency and magnitude of floods and draughts
3	Rise in sea level	• Inundation of land in coastal areas • Inland intrusion of saline waters
4	Melting of glaciers and ice caps	• Impact on flow of glacier-fed rivers

5. Current Predictions of Impact of Global Warming on Water Resources of India

A large number of studies have been initiated in India to study the impact of global warming on the hydrology of the country. In view of the differing scenarios, assumptions and data used many of the results on specific regional studies, at best, are of indicative in nature only. It should be remembered that that any predictions due to climate change is at a time horizon of 80 to 100 years and subject to a type of adaptation scenario used in the studies. As such, they are meant to develop strategies for adaptations to face the problem. A few significant results reported in *IPCC* (2007) are summarised below as an indication of possible impact on water resources of the country. *IPCC* (2007) predicts that

- the global temperature will rise by 2.0 to 4.5°C by the end of the century with an increase of 2.7°C – 4.3°C over India by 2080.
- the rainfall over the sub-continent will increase by 6 – 8% by 2100
- the mean sea level would rise by about 88 cm by 2100.

Regional climatic modeling applied to India using *IPCC*—Scenario *A*2 and *B*2 depict the following significant changes:(Ref. 12):

- A 20% rise in all India summer monsoon rainfall and a further rise are projected for all states except Punjab, Rajasthan and Tamil Nadu, which show a slight decrease.
- Extreme rise in maximum and minimum temperatures is expected and similarly extreme precipitation is projected over West coast and West-Central India.
- The rise in the temperatures causes a rise in the evapotranspiration (*ET*) from a basin. Thus, in some situations a rise in precipitation does not necessarily mean a rise in the runoff; as the corresponding rise in *ET* would have offset any gain in runoff due to increase in rainfall.

  The impact of climate change is varied across various river basins in the country. In 12 river basins reported in Ref. 12, seven rivers show increase in runoff and remaining five show decrease in runoff. Rivers Cauvery, Brahmini, Mahanadi, Narmada, Tapi, Godavari and Ganga show an increase in runoff. Rivers Krishna, Mahi, Pennar, Sabarmati and Luni show decreasing runoff due to climatic change.
- Further, the likely decrease in the runoff of Luni and Sabarmati rivers are very severe, being to an extent of 2/3 of the existing average runoff. This may lead to very severe drought conditions.

The above predictions are for a particular set of *GHG* emission and control scenarios and present a probable outcome at the end of the century. It is generally felt that it is highly likely The negative impacts of climate change on freshwater systems outweigh its benefits, with runoff declining in most streams and rivers.

In spite of their limitations the results of studies on climate change are important to the policy makers to initiate action towards developing appropriate adaptation strategies to cope with the eventual scenarios of heightened water stress.

REFERENCES

1. Butler, S.S., "Point slope approach for reservoir flood routing", *J. of Hyd. Div., Proc. ASCE*, Oct. 1982, pp 1102–1113.
2. Chow, V.T., *Handbook of Applied Hydrology*, McGraw-Hill, New York, NY, 1964.
3. Chow, V.T., Maidment, D.R. and Mays, L.W., *Applied Hydrology,* McGraw-Hill Book Co., Singapore, 1988.
4. Kraijenhoff Van der lam, D.A., "Rainfall-runoff relations and computational model", Ch. 15 in *Drainage Principles and Applications*, Vol. II, Pub. No. 16, Int. Inst. for Land Reclamation and Improvement, Wageningen, The Netherlands, 1973.
5. Linsley, R.K. et al., *Applied Hydrology*, Tata McGraw-Hill, New Delhi, India, 1975.
6. Mahmood, K. and Yevjevich V. (Ed.) *Unsteady Flow in Open Channels*, Vols.1 and 2, Water Resources Pub., Fort Collins, Colo, USA, 1975.
7. Nash, J.E., "The form of instantaneous unit hydrograph", IASH, Pub. No. 45, Vol. 3–4, pp 114–121,1957.
8. Streeter, V.L. and Wylie, E.B., *Hydraulic Transients*, McGraw-Hill, New York, NY, 1967.
9. Subramanya, K., *Flow in Open Channels*, 3rd Ed. Tata McGraw-Hill, New Delhi, India, 2009.
10. Pierce, B.O. and Foster, R.M., *A Short Table of Integrals*, Oxford, IBH New Delhi, India, 1963.

11. Ghosh, S., and Mishra, C, *Assessing Hydrological Impacts of Climate Change: Modelling techniques and challenges*, The Open Hydrology Journal, Vol. 4, pp. 115–121, 2010.
12. *MOEF, State of Environment Report, India-2009*, Government of India, Ministry of Environment and Forest, New Delhi, 2009.
13. *CWC & NIH, Preliminary Consolidated Report on Effect of Climate Change on Water Resource*, Govt. of India, Ministry of Water Resources, Central water Commission and National Institute of Hydrology, New Delhi, 2008.

REVISION QUESTIONS

8.1 Distinguish between:
(a) Hydraulic and hydrologic method of flood routing
(b) Hydrologic storage routing and hydrologic channel routing
(c) Prism storage and wedge storage

8.2 What are the basic equations used for flood routing by
(a) Hydrologic method (b) Hydraulic method

8.3 Define the problem of level pool routing. Describe a commonly used method of reservoir routing.

8.4 Describe a numerical method of hydrologic reservoir routing.

8.5 What is the basic premise in the Muskingum method of flood routing? Describe a procedure for estimating the values of the Muskingum coefficients K and x for a stream reach.

8.6 Describe the Muskingum method of routing an inflow hydrograph through a channel reach. Assume the values of the coefficients K and x for the reach are known.

8.7 Explain briefly
(a) Isochrone (b) Time of concentration
(c) Linear reservoir (d) Linear channel

8.8 Explain briefly the basic principles involved in the development of *IUH* by
(a) Clark's method (b) Nash's model.

8.9 Describe the various structural methods adopted for management of floods.

8.10 Describe the various non-structural measures of flood management.

8.11 Describe the problem of floods and their management with special reference to Indian scene.

8.12 What is greenhouse effect? List common greenhouse gases involved in global climate change.

8.13 What are the important evidences available for global warming? List major impacts of global warming on the hydrology of a region.

PROBLEMS

8.1 The storage, elevation and outflow data of a reservoir are given below:

Elevation (m)	Storage 10^6 m^3	Outflow discharge (m^3/s)
299.50	4.8	0
300.20	5.5	0
300.70	6.0	15

Contd.

Contd.

301.20	6.6	40
301.70	7.2	75
302.20	7.9	115
302.70	8.8	160

The spillway crest is at elevation 300.20 m. The following flood flow is expected into the reservoir.

Time (h)	0	3	6	9	12	15	18	21	24	27
Discharge (m^3/s)	10	20	52	60	53	43	32	22	16	10

If the reservoir surface is at elevation 300.00 m at the commencement of the inflow, route the flood to obtain (a) the outflow hydrograph, and (b) the reservoir elevation *vs* time curve.

8.2 Solve Prob. 8.1 if the reservoir elevation at the start of the inflow hydrograph is at 301.50 m.

8.3 A small reservoir has the following storage elevation relationship.

Elevation (m)	55.00	58.00	60.00	61.00	62.00	63.00
Storage (10^3 m^3)	250	650	1000	1250	1500	1800

A spillway provided with its crest at 60.00 m elevation has the discharge relationship $Q = 15\ H^{3/2}$, where H = head of water over the spillway crest. When the reservoir elevation is at 58.00 m a flood as given below enters the reservoir. Route the flood and determine the maximum reservoir elevation, peak outflow and attenuation of the flood peak.

Time (h)	0	6	12	15	18	24	30	36	42
Inflow (m^3/s)	5	20	40	60	50	32	22	15	10

8.4 The storage-elevation-discharge characteristic of a reservoir is as follows:

Elevation (m)	100.00	100.50	101.00
Discharge (m^3/s)	12	18	25
Storage (10^3 m^3)	400	450	550

When the reservoir elevation is at 101.00 m the inflow is at a constant rate of 10 m^3/s. Find the time taken for the water surface to drop to the 100.00 m elevation.

8.5 A small reservoir has a spillway crest at 200.00 m elevation. Above this elevation, the storage and outflow from the reservoir can be expressed as

Storage: $S = 36000 + 18000\ y\ (m^3)$
Outflow: $Q = 10\ y\ (m^3/s)$
where y = height of the reservoir level above the spillway crest in m.

Route an inflow flood hydrograph which can be approximated by a triangle as

$I = 0$ at $t = 0$ h
$I = 30\ m^3/s$ at $t = 6$ h (peak flow)
$I = 0$ at $t = 26$ h (end of inflow).

Assume the reservoir elevation as 200.00 m at $t = 0$ h.
Use a time step of 2 h.

8.6 A detention reservoir was found to have a linear storage discharge relationship, $Q = KQ$

(a) Show that the storage routing equation of an inflow hydrograph through this reservoir is $Q_2 = C_1\bar{I}_1 + C_2Q_2$ where C_1 and C_2 are constants and $\bar{I}_1 = (I_1 + I_2)/2$. Determine the values of C_1 and C_2 in terms of K and the routing time step Δt.

(b) If $K = 4.0$ h and $\Delta t = 2$ h, route the following inflow hydrograph through this reservoir. Assume the initial condition that at $t = 0$, $I_1 = Q_1 = 0$.

Time (h)	0	2	4	6	8	10	12	14	16	18
Inflow (m^3/s)	0	20	60	100	80	60	40	30	20	10

8.7 Observed values of inflow and outflow hydrographs at the ends of a reach in a river are given below. Determine the best values of K and x for use in the Muskingum method of flood routing.

Time (h)	0	6	12	18	24	30	36	42	48	54	60	66
Inflow (m^3/s)	20	80	210	240	215	170	130	90	60	40	28	16
Outflow (m^3/s)	20	20	50	150	200	210	185	155	120	85	55	23

8.8 Route the following flood through a reach for which $K = 22$ h and $x = 0.25$. Plot the inflow and outflow hydrographs and determine the peak lag and attenuation. At $t = 0$ the outflow discharge is 40 m^3/s.

Time (h)	0	12	24	36	48	60	72	84	96	108	120	132	144
Inflow (m^3/s)	40	65	165	250	240	205	170	130	115	85	70	60	54

8.9 The storage in the reach of a stream has been studied. The values of x and K in Muskingum equation have been identified as 0.28 and 1.6 days. If the inflow hydrograph to the reach is as given below, compute the outflow hydrograph. Assume the outflow from the reach at $t = 0$ as 3.5 m^3/s.

Time (h)	0	6	12	18	24	30
Inflow (m^3/s)	35	55	92	130	160	140

8.10 Route the following flood hydrograph through a river reach for which Muskingum coefficient $K = 8$ h and $x = 0.25$.

Time (h)	0	4	8	12	16	20	12	24	28
Inflow (m^3/s)	8	16	30	30	25	20	40	15	10

The initial outflow discharge from the reach is 8.0 m^3/s.

8.11 A stream has a uniform flow of 10 m^3/s. A flood in which the discharge increases linearly from 10 m^3/s to a peak of 70 m^3/s in 6 h and then decreases linearly to a value of 10 m^3/s in 24 h from the peak arrives at a reach. Route the flood through the reach in which $K = 10$ h and $x = 0$

8.12 A drainage basin has area = 137 km^2, storage constant $K = 9.5$ h and time of concentration = 7 h. The following inter-isochrone area distribution data are available:

Time (h)	0–1	1–2	2–3	3–4	4–5	5–6	6–7
Inter-isochrone area (km^2)	10	38	20	45	32	10	2

Determine (a) the *IUH*, and (b) the 1-h unit hydrograph for the catchment.

8.13 Solve Prob. 8.11 $K = 10$ h and $x = 0.5$. Determine the peak lag and attenuation and compare with the corresponding values of Prob. 8.11.

8.14 Show that the reservoir routing equation for a linear reservoir is

$$\frac{dQ}{dt} + \alpha Q = \alpha I$$

where α is a constant. Obtain the outflow from such a reservoir due to an inflow $I = I_0 + \beta t$ occurring from $t = 0$ to t_0 with the boundary condition $Q = 0$ at $t = 0$.

8.15 Given that $n = 4.0$ and $K = 6.0$ are the appropriate values of the coefficients in the Nash model for *IUH* of a catchment, determine the ordinates of *IUH* in cm/h at 3 hours interval. If the catchment area is 500 km^2, determine the ordinates of the *IUH* in m^3/s.

8.16 Show that in the *IUH* obtained by using the Nash model the peak flow occurs at a time

$$t_p = K(n-1)$$

and the magnitude of the peak flow is

$$u(t)_p = \frac{1}{K\Gamma(n)} e^{(1-n)} (n-1)^{n-1}$$

8.17 For a sub-basin in lower Godavari catchment, with an area of 250 km^2 the following values of Nash model coefficients were found appropriate: $n = 3.3$ and $K = 1.69$ h. Determine the co-ordinates of (a) *IUH* at 1-h interval, and (b) 1-hour *UH* at 1-h interval.

8.18 For a catchment *X* of area 100 km^2, an *ERH* of an isolated storm and its corresponding *DRH* were analysed to determine the first and second moments relative to the total area of the respective curves and the following values were obtained:

(1) (First moment of the curve)/(total area of the curve):

ERH = 11.0 h   *DRH* = 25.0 h

(2) (Second moment of the curve)/(total area of the curve):

ERH = 170 h^2 *DRH* = 730 h^2

Determine the *IUH* with ordinates at 2 hour interval for catchment *X* by using Nash model.

8.19 For a catchment, the effective rainfall hyetograph due to an isolated storm is given in Table 8.9(a). The direct runoff hydrograph resulting from the above storm is given in Table 8.9(b). Determine the values of Nash model *IUH* coefficients *n* and *K* for the above catchment.

Table 8.9(a) ***ERH***

Time (h)	0 to 6	6 to 12	12 to 18	18 to 24
ERH ordinates (cm/s)	4.3	2.8	3.9	2.7

Table 8.9(b) ***DRH***

Time (h)	*DR* m^3/s	Time (h)	*DR* m^3/s
0	0	36	160
6	20	42	75
12	140	48	30
18	368	54	10
24	380	60	0
30	280		

OBJECTIVE QUESTIONS

8.1 The hydrologic flood-routing methods use
(a) equation of continuity only (b) both momentum and continuity equations
(c) energy equation only (d) equation of motion only

8.2 The hydraulic methods of flood routing use
(a) equation of continuity only
(b) both the equation of motion and equation of continuity
(c) energy equation only
(b) equation of motion only

8.3 The St. Venant equations for unsteady open-channel flow are
(a) continuity and momentum equations
(b) momentum equation in two different forms
(c) momentum and energy equations
(d) energy and continuity equations

8.4 The prism storage in a river reach during the passage of a flood wave is
(a) a constant (b) a function of inflow and outflow
(c) function of inflow only (d) function of outflow only

8.5 The wedge storage in a river reach during the passage of a flood wave is
(a) a constant (b) negative during rising phase
(c) positive during rising phase (d) positive during falling phase

8.6 In routing a flood through a reach, the point of intersection of inflow and outflow hydrographs coincides with the peak of outflow hydrograph
(a) in all cases of flood routing
(b) when the inflow is into a reservoir with an uncontrolled outlet
(c) in channel routing only
(d) in all cases of reservoir routing

8.7 Which of the following is a proper reservoir-outing equation?

(a) $\frac{1}{2}(I_1 - I_2)\,\Delta t + \left(S_1 + \frac{Q_1\,\Delta t}{2}\right) = \left(S_2 - \frac{Q_2\,\Delta t}{2}\right)$

(b) $(I_1 + I_2)\,\Delta t + \left(\frac{2\,S_1}{\Delta t} - Q_1\right) = \left(\frac{2\,S_2}{\Delta t} + Q_2\right)$

(c) $\frac{1}{2}(I_1 + I_2)\,\Delta t + \left(S_2 - \frac{Q_2\,\Delta t}{2}\right) = \left(S_1 + \frac{Q_1\,\Delta t}{2}\right)$

(d) $(I_1 + I_2) + \left(\frac{2\,S_1}{\Delta t} - Q_1\right) = \left(\frac{2\,S_2}{\Delta t} + Q_2\right)$

8.8 A flood wave with a known inflow hydrograph is routed through a large uncontrolled reservoir. The outflow hydrograph will have
(a) attenuated peak with reduced time base
(b) attenuated peak with increase time base
(c) increased peak with increased time base
(d) increased peak with reduced time base

8.9 The Muskingum method of flood routing is a
(a) form of reservoir routing method
(b) hydraulic routing method

(c) complete numerical solution of St. Venant equations
(d) hydrologic channel-routing method.

8.10 The Muskingum method of flood routing assumes the storage S is related to inflow rate I and outflow rate Q of a reach as $S =$

(a) $K[xI - (1 - x)Q]$ (b) $K[xQ + (1 - x)I]$
(c) $K[xI + (1 - x)Q]$ (d) $Kx[I - (1 - x)Q]$

8.11 The Muskingum method of flood routing gives $Q_2 = C_0I_2 + C_1I_1 + C_2Q_1$. The coefficients in this equation will have values such that

(a) $C_0 + C_1 = C_2$ (b) $C_0 - C_1 - C_2 = 1$
(c) $C_0 + C_1 + C_2 = 0$ (d) $C_0 + C_1 + C_2 = 1$.

8.12 The Muskingum channel routing equation is written for the outflow from the reach Q in terms of the inflow I and coefficients C_0, C_1 and C_2 as

(a) $Q_2 = C_0 I_0 + C_1Q_1 + C_2I_2$ (b) $Q_2 = C_0I_2 + C_1I_1 + C_2Q_2$
(c) $Q_2 = C_0I_0 + C_1I_1 + C_2I_2$ (d) $Q_2 = C_0Q_0 + C_1Q_1 + C_2I_2$

8.13 In the Muskingum method of channel routing, the routing equation is written as $Q_2 = C_0I_2 + C_1I_1 + C_2Q_1$, If the coefficients $K = 12$ h and $x = 0.15$ and the time step for routing $\Delta t = 4$ h, the coefficient C_0 is

(a) 0.016 (b) 0.048 (c) 0.328 (d) 0.656

8.14 In the Muskingum method of channel routing the weighing factor x can have a value

(a) between –0.5 to 0.5 (b) between 0.0 to 0.5
(c) between 0.0 to 1.0 (d) between –1.0 to +1.0

8.15 In the Muskingum method of channel routing if $x = 0.5$, it represents an outflow hydrograph

(a) that has reduced peak
(b) with an amplified peak
(c) that is exactly the same as the inflow hydrograph
(d) with a peak which is exactly half of the inflow peak

8.16 If the storage S, inflow rate I and outflow rate Q for a river reach is written as

$$S = K [x I^n + (1 - x) Q^n]$$

(a) $n = 0$ represents storage routing through a reservoir
(b) $n = 1$ represents the Muskingum method
(c) $n = 0$ represents the Muskingum method
(d) $n = 0$ represents a linear channel

8.17 A linear reservoir is one in which the

(a) volume varies linearly with elevation
(b) storage varies linearly with the outflow rate
(c) storage varies linearly with time
(d) storage varies linearly with the inflow rate

8.18 An isochrone is a line on the basin map

(a) joining raingauge stations with equal rainfall duration
(b) joining points having equal standard time
(c) connecting points having equal time of travel of the surface runoff to the catchment outlet
(d) that connects points of equal rainfall depth in a given time interval

8.19 In the Nash model for *IUH* given by

$$u(t) = \frac{1}{K\Gamma(n)} (t/K)^{n-1} (e)^{-t/K}$$

the usual units of $u(t)$, n and K are, respectively;

(a) cm/h, h, h (b) h^{-1}, h, h
(c) h^{-1}, dimensionless number, h (d) cm/h, dimensionless number, h

8.20 The peak ordinate of the *IUH* of a catchment was obtained from Nash model as 0.03 cm/h. If the area of the catchment is 550 km^2 the value of the peak ordinate in m^3/s is
(a) 16.5 (b) 45.83 (c) 30.78 (d) 183.3

8.21 If the Gamma function Γ (1.5) = 0.886, the value of Γ (0.5) is
(a) 0.5907 (b) 1.329 (c) –0.886 (d) 1.772

8.22 In the Nash model for IUH, if M_{I1} = the first moment of ERH about the time origin divided by the total effective rainfall and M_{Q1} = the first moment of DRH about the time origin divided by the total direct runoff then
(a) $M_{Q1} - M_{I1} = nK$ (b) $M_{I1} - M_{Q1} = nK^2$
(c) $M_{Q1} - M_{I1} = n\,(n+1)\,K$ (d) $M_{I1} - M_{Q1} = 2$ nK

CHAPTER 9

GROUNDWATER

9.1 *INTRODUCTION*

In the previous chapters various aspects of surface water hydrology that deal with surface runoff have been discussed. Study of subsurface flow is equally important since about 30% of the world's freshwater resources exist in the form of groundwater. Further, the subsurface water forms a critical input for the sustenance of life and vegetation in arid zones. Due to its importance as a significant source of water supply, various aspects of groundwater dealing with the exploration, development and utilisation have been extensively studied by workers from different disciplines, such as geology, geophysics, geochemistry, agricultural engineering, fluid mechanics and civil engineering and excellent treatises are available (Ref. 1, 2 and 4 through 10). This chapter confines itself to only an elementary treatment of the subject of groundwater as a part of engineering hydrology.

9.1.1 Sources of Groundwater

Groundwater is derived from precipitation and recharge from surface water. It is the water that has infiltrated into the earth directly from precipitation, recharge from streams and other natural water bodies and artificial recharge due to action of man. Infiltration and further downward percolation from sources like rain, melting of snow and ice, rivers and streams, lakes, reservoirs, canals and other watercourses are the usual main sources that contribute to the groundwater of a region.

9.2 *Forms of Subsurface Water*

Water in the soil mantle is called subsurface water and is considered in two zones (Fig. 9.1):

1. Saturated zone
2. Aeration zone

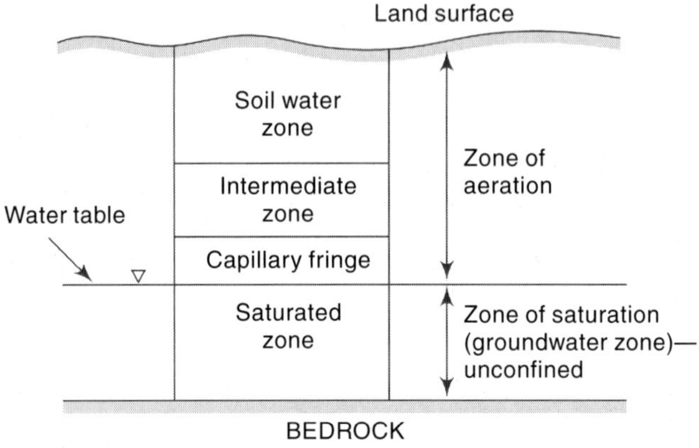

Fig. 9.1 ***Classification of Subsurface Water***

9.2.1 Saturated Zone

This zone, also known as groundwater zone, is the space in which all the pores of the soil are filled with water. The water table forms its upper limit and marks a free surface, i.e. a surface having atmospheric pressure.

9.2.2 Zone of Aeration

In this zone, the soil pores are only partially filled with water. The space between the land surface and the water table marks the extent of this zone. Any flow of water in this zone is in unsaturated soil matrix condition. The zone of aeration has three subzones.

- **Soil Water Zone** This lies close to the ground surface in the major root band of the vegetation from which the water is transported to the atmosphere by evapotranspiration.
- **Capillary Fringe** In this subzone, the water is held by capillary action. This subzone extends from the water table upwards to the limit of capillary rise.
- **Intermediate Zone** This subzone lies between the soil water zone and the capillary fringe. The thickness of the zone of aeration and its constituent subzones depend upon the soil texture and moisture content and vary from region to region. The soil moisture in the zone of aeration is of importance in agricultural practice and irrigation engineering. The present chapter is concerned with the saturated zone only.

9.2.3 Infiltration and Percolation

Infiltration and percolation are both processes that move water in the pores of the soil. The main difference between these two processes is that infiltration occurs closer to the surface of the soil. Infiltration denotes the entry of rainwater or ponded water in to the earth surface. The interface is the essence of infiltration process. The descending motion of the infiltered water through soil and rock is called *percolation*. Infiltration delivers water from the surface into the soil and plant rooting zone while percolation moves it through the soil profile to replenish groundwater supplies or become part of the sub-surface runoff process. Thus, the percolation process represents the flow of water from unsaturated zone to the saturated zone.

9.3 *Saturated Formation*

All earth materials, from soils to rocks have pore spaces. Although these pores are completely saturated with water below the water table, from the groundwater utilization aspect only such material through which water moves easily and hence can be extracted with ease are significant. On this basis, the saturated formations are classified into four categories:

1. Aquifer 2. Aquitard, 3. Aquiclude 4. Aquifuge

1. Aquifer

An *aquifer* is a saturated formation of earth material which not only stores water but yields it in sufficient quantity. Thus, an aquifer transmits water relatively easily due to its high permeability. Unconsolidated deposits of sand and gravel form good aquifers.

2. Aquitard

It is a formation through which only seepage is possible and thus the yield is insignificant compared to an aquifer. It is partly permeable. A sandy clay unit is an example of aquitard. Through an aquitard appreciable quantities of water may leak to an aquifer below it.

3. Aquiclude

It is a geological formation which is essentially impermeable to the flow of water. It may be considered as closed to water movement even though it may contain large amounts of water due to its high porosity. Clay is an example of an aquiclude.

4. Aquifuge

It is a geological formation which is neither porous nor permeable. There are no interconnected openings and hence it cannot transmit water. Massive compact rock without any fractures is an aquifuge.

The definitions of aquifer, aquitard and aquiclude as above are relative. A formation which may be considered as an aquifer at a place where water is at a premium (e.g. arid zones) may be classified as an aquitard or even aquiclude in an area where plenty of water is available.

The availability of groundwater from an aquifer at a place depends upon the rates of withdrawal and replenishment *(recharge)*. Aquifers play the roles of both a transmission conduit and a storage. Aquifers are classified as unconfined aquifers and confined aquifers on the basis of their occurrence and field situation. An *unconfined aquifer* (also known as *water table aquifer)* is one in which a free water surface, i.e. a water table exists (Fig. 9.2). Only the saturated zone of this aquifer is of importance in groundwater studies. Recharge of this aquifer takes place through infiltration of precipitation from the ground surface. A well driven into an unconfined aquifer will indicate a static water level corresponding to the water table level atthat location.

A *confined aquifer,* also known as *artesian aquifer,* is an aquifer which is confined between two impervious beds such as aquicludes or aquifuges (Fig. 9.2). Recharge of this aquifer takes place only in the area where it is exposed at the ground surface. The water in the confined aquifer will be under pressure and hence the piezometric level will be much higher than the top level of the aquifer. At some locations: the

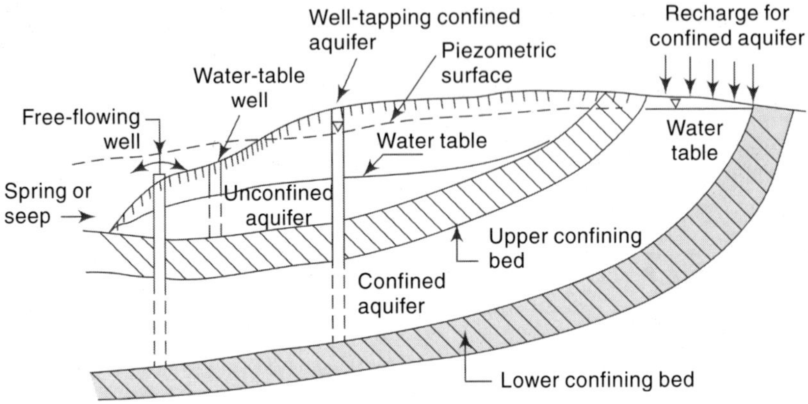

Fig. 9.2 ***Confined and Unconfined Acquifers***

piezometric level can attain a level higher than the land surface and a well driven into the aquifer at such a location will flow freely without the aid of any pump. In fact, the term *artesian* is derived from the fact that a large number of such freeflow wells were found in Artois, a former province in north France. Instances of free-flowing wells having as much as a 50 m head at the ground surface are reported.

A confined aquifer is called a *leaky aquifer* if either or both of its confining beds are aquitards.

9.3.1 Water Table

A water table is the free water surface in an unconfined aquifer. The static level of a well penetrating an unconfined aquifer indicates the level of the water table at that point. The water table is constantly in motion adjusting its surface to achieve a balance between the recharge and outflow from the subsurface storage. Fluctuations in the water level in a dug well during various seasons of the year, lowering of the groundwater table in a region due to heavy pumping of the wells and the rise in the water table of an irrigated area with poor drainage, are some common examples of the fluctuation of the water table. In a general sense, the water table follows the topographic features of the surface. If the water table intersects the land surface the groundwater comes out to the surface in the form of *springs* or *seepage.*

Sometimes a lens or localised patch of impervious stratum can occur inside an unconfined aquifer in such a way that it retains a water table above the general water table (Fig. 9.3). Such a water table retained around the impervious material is known as *perched water table.* Usually, the perched water table is of limited extent and the yield from such a situation is very small. In groundwater exploration, a perched water table is quite often confused with a general water table.

The position of the water table relative to the water level in a stream determines whether the stream contributes water to the groundwater storage or the other way about. If the bed of the stream is below the groundwater table, during periods of low flows in the stream, the water surface may go down below the general water table elevation and the groundwater contributes to the flow in the stream. Such streams which receive groundwater flow are called *effluent streams* (gaining streams) (Fig. 9.4 (a)). Perennial rivers and streams are of this kind. If, however, the water table is below the bed of the

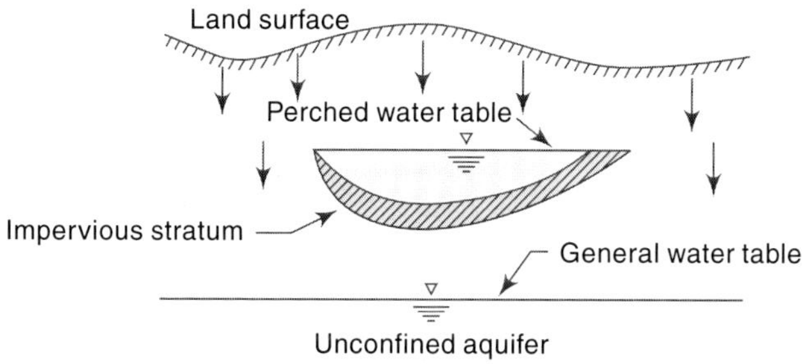

Fig. 9.3 ***Perched Water Table***

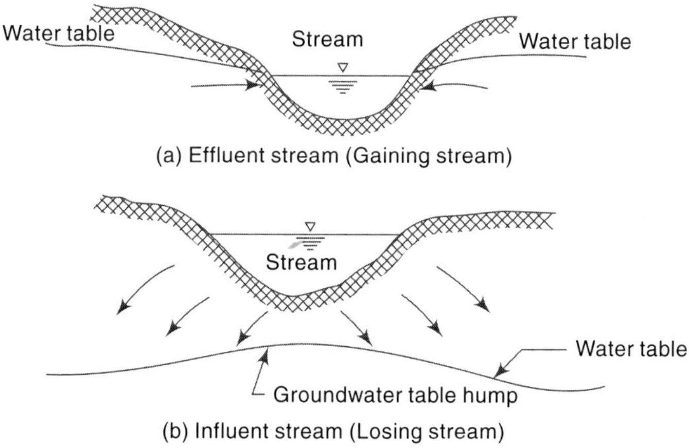

Fig. 9.4 ***Effluent and Influent Streams***

stream, the stream-water percolates to the groundwater storage and a hump is formed in the groundwater table (Fig. 9.4 (b)). Such streams which contribute to the groundwater are known as *influent streams* (losing streams). Intermittent rivers and streams which go dry during long periods of dry spell (i.e. no rain periods) are of this kind.

9.4 *Aquifer Properties*

The important properties of an aquifer are its capacity to release the water held in its pores and its ability to transmit the flow easily. These properties essentially depend upon the composition of the aquifer.

9.4.1 Porosity

The amount of pore space per unit volume of the aquifer material is called *porosity*. It is expressed as

$$n = \frac{V_v}{V_0} \tag{9.1}$$

where n = porosity, V_v = volume of voids and V_0 = volume of the porous medium. In an unconsolidated material the size distribution, packing and shape of particles determine the porosity. In hard rocks the porosity is dependent on the extent, spacing and the pattern of fracturing or on the nature of solution channels. In qualitative terms porosity greater than 20% is considered as large, between 5 and 20% as medium and less than 5% as small.

9.4.2 Specific Yield

While porosity gives a measure of the water-storage capability of a formation, not all the water held in the pores is available for extraction by pumping or draining by gravity. The pores hold back some water by molecular attraction and surface tension. The actual volume of water that can be extracted by the force of gravity from a unit volume of aquifer material is known as the *specific yield*, S_y. The fraction of water held back in the aquifer is known as *specific retention*, S_r. Thus porosity

$$n = S_y + S_r \tag{9.2}$$

The representative values of porosity and specific yield of some common earth materials are given in Table 9.1.

Table 9.1 ***Porosity and Specific Yield of Selected Formations***

Formation	Porosity, %	Specific yield, %
Clay	45–55	1–10
Sand	35–40	10–30
Gravel	30–40	15–30
Sand stone	10–20	5–15
Shale	1–10	0.5–5
Lime stone	1–10	0.5–5

It is seen from Table 9.1 that although both clay and sand have high porosity the specific yield of clay is very small compared to that of sand.

9.4.3 Darcy's Law

In 1856, Henry Darcy, a French hydraulic engineer, on the basis of his experimental findings proposed a law relating the velocity of flow in a porous medium. This law, known as *Darcy's law*, can be expressed as

$$V = Ki \tag{9.3}$$

where V = apparent velocity of seepage = Q/A in which Q = discharge and A = cross-sectional area of the porous medium. V is sometimes also known as discharge velocity.

$i = -\frac{dh}{ds}$ = hydraulic gradient, in which h = piezometric head and s = distance measured in the general flow direction; the negative sign emphasizes that the piezometric head drops in the direction of flow. K = a coefficient, called *coefficient of permeability*, having the units of velocity.

The discharge Q can be expressed as

$$Q = K\,i\,A \tag{9.3a}$$

$$= KA\left(-\frac{\Delta H}{\Delta s}\right)$$

where $(-\Delta H)$ is the drop in the hydraulic grade line in a length Δs of the porous medium.

Darcy's law is a particular case of the general viscous fluid flow. It has been shown valid for laminar flows only. For practical purposes, the limit of the validity of Darcy's law can be taken as Reynolds number of value unity, i.e.

$$Re = \frac{Vd_a}{\nu} = 1 \tag{9.4}$$

where Re = Reynolds number

d_a = representative particle size, usually $d_a = d_{10}$ where d_{10} represents a size such that 10% of the aquifer material is of smaller size. Often, the mean grain size is used as d_a.

ν = kinematic viscosity of water

Except for flow in fissures and caverns, to a large extent groundwater flow in nature obeys Darcy's law. Further, there is no known lower limit for the applicability of Darcy's law.

It may be noted that the *apparent velocity* V used in Darcy's law is not the actual velocity of flow through the pores. Owing to irregular pore geometry the actual velocity of flow varies from point to point and the *bulk pore velocity* (v_a) which represents the actual speed of travel of water in the porous media is expressed as

$$v_a = \frac{V}{n} \tag{9.5}$$

where n = porosity. The bulk pore velocity v_a is the velocity that is obtained by tracking a tracer added to the groundwater.

9.4.4 Coefficient of Permeability

The coefficient of permeability, also designated as *hydraulic conductivity,* reflects the combined effects of the porous medium and fluid properties. From an analogy of laminar flow through a conduit (*Hagen-Poiseuille flow*) the coefficient of permeability K can be expressed as

$$K = Cd_m^2 \frac{\gamma}{\mu} \tag{9.6}$$

where d_m = mean particle size of the porous medium, $\gamma = \rho g$ = unit weight of fluid, ρ = density of the fluid, g = acceleration due to gravity, μ = dynamic viscosity of the fluid and C = a shape factor which depends on the porosity, packing, shape of grains and grain-size distribution of the porous medium. Thus, for a given porous material

$$K \propto \frac{1}{\nu}$$

where ν = kinematic viscosity = μ/ρ = f(temperature). The laboratory or *standard value* of the coefficient of permeability (K_s) is taken as that for pure water at a standard temperature of 20°C. The value of K_t the coefficient of permeability at any temperature t can be converted to K_s by the relation

$$K_s = K_t(\nu_t/\nu_s) \tag{9.7}$$

where v_s and v_t represent the kinematic viscosity values at 20°C and t°C respectively.

The coefficient of permeability is often considered in two components, one reflecting the properties of the medium only and the other incorporating the fluid properties. Thus, referring to Eq. (9.6),a term K_0 is defined as

$$K = K_0 \frac{\gamma}{\mu} = K_0 \frac{g}{v} \tag{9.8}$$

where $K_0 = Cd_m^2$. The parameter K_0 is called specific or *intrinsic permeability* which is a function of the medium only. Note that K_0 has dimensions of $[L^2]$. It is expressed in units of cm^2 or m^2 or in *darcys* where 1 darcy = 9.87×10^{-13} m^2. Where more than one fluid is involved in porous media flow or when there is considerable temperature variation, the coefficient K_0 is useful. However, in groundwater flow problems, the temperature variations are rather small and as such the coefficient of permeability K is more convenient to use. The common units of K are m/day or cm/s. The conversion factor for these two are

1 m/day = 0.0011574 cm/s

or 1 cm/s = 864.0 m/day

Some typical values of coefficient of permeability of some porous media are given in Table 9.2.

Table 9.2 ***Representative Values of the Permeability Coefficient***

No.	Material	K (cm/s)	K_0 (darcys)
A. Granular material			
1.	Clean gravel	1 – 100	10^3–10^5
2.	Clean coarse sand	0.010 – 1.00	10–10^3
3.	Mixed sand	0.005 – 0.01	5–10
4.	Fine sand	0.001– 0.05	1–50
5.	Silty sand	$1 \times 10^{-4} - 2 \times 10^{-3}$	0.1–2
6.	Silt	$1 \times 10^{-5} - 5 \times 10^{-4}$	0.01–0.5
7.	Clay	$< 10^{-6}$	$< 10^{-3}$
B. Consolidated material			
1.	Sandstone	$10^{-6} - 10^{-3}$	$10^{-3} - 1.0$
2.	Carbonate rock with secondary porosity	$10^{-5} - 10^{-3}$	$10^{-2} - 1.0$
3.	Shale	10^{-10}	10^{-7}
4.	Fractured and weathered rock (aquifers)	$10^{-6} - 10^{-3}$	$10^{-3} - 1.0$
At 20° C, for water, v = 0.01 cm^2/s and substituting in Eq. (9.8) K_0 [darcys] = 10^3 K [cm/s] at 20°C			

Consider an aquifer of unit width and thickness B, (i.e. depth of a fully saturated zone). The discharge through this aquifer under a unit hydraulic gradient is

$$T = KB \tag{9.9}$$

This discharge is termed *transmissibility,* T and has the dimensions of $[L^2/T]$. Its units are m^2/s or litres per day/metre width (l pd/m). Typical values of T lie in the range

$1 \times 10^6\ l$ pd/m to $1 \times 10^4\ l$ pd/m. A well with a value of $T = 1 \times 10^5\ l$ pd/m is considered satisfactory for irrigation purposes.

The coefficient of permeability is determined in the laboratory by a *permeameter*. For coarse-grained soils a *constant-head permeameter* is used. In this, the discharge of water percolating under a constant head difference (ΔH) through a sample of porous material of cross-sectional area A and length L is determined. The coefficient of permeability at the temperature of the experiment is found as

$$K = \frac{Q}{A} \frac{1}{(\Delta H/L)}$$

For fine-grained soils, a *falling-head permeameter* is used. Details of permeameters and their use is available in any good textbook in Soil Mechanics, e.g. Ref. 8. It should be noted that laboratory samples are disturbed samples and a permeameter cannot simulate the field conditions exactly. Hence, considerable care in the preparation of the samples and in conducting the tests are needed to obtain meaningful results.

Under field conditions, permeability of an aquifer is determined by conducting pumping tests in a well. One of the many tests available for this purpose consists of pumping out water from a well at a uniform rate till steady state is reached. Knowing the steady-state drawdown and the discharge-rate, transmissibility can be calculated. Information about the thickness of the saturation zone leads one to calculate the permeability. Injection of a tracer, such as a dye and finding its velocity of travel is another way of determining the permeability under field conditions.

9.4.5 Stratification

Sometimes the aquifers may be stratified, with different permeabilities in each strata. Two kinds of flow situations are possible in such a case.

(i) When the flow is parallel to the stratification as in Fig. 9.5(a) equivalent permeability K_e of the entire aquifer of thickness $B = \sum_1^n B_i$ is

$$K_e = \frac{\sum_1^n K_i B_i}{\sum_1^n B_i} \tag{9.10}$$

The transmissibility of the formation is

$$T = K_e \Sigma B_i = \sum_1^n K_i B_i$$

(ii) When the flow is normal to the stratification as in Fig. 9.5(b), the equivalent permeability K_e of the aquifer of length

$$L = \sum_1^n L_i \text{ is}$$

$$K_e = \frac{\sum_1^n L_i}{\sum_1^n (L_i/K_i)} \tag{9.11}$$

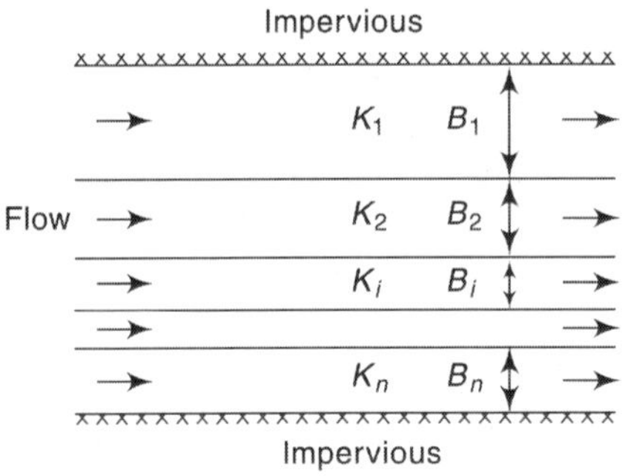

Fig. 9.5(a) ***Flow Parallel to Stratification***

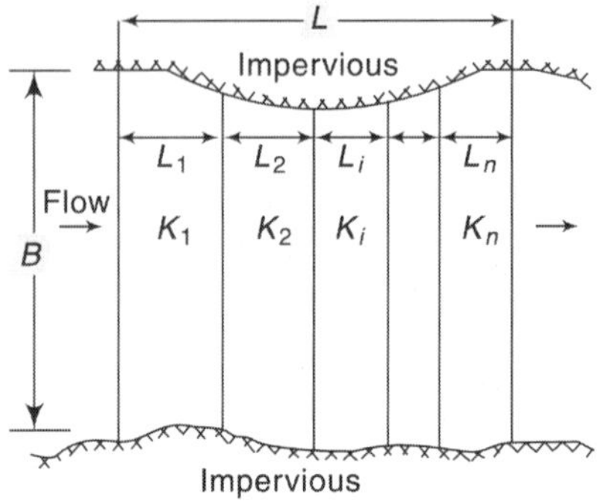

Fig. 9.5(b) ***Flow Normal to Stratification***

(Note that in this case L is the length of seepage and the thickness B of the aquifer does not come into picture in calculating the equivalent permeability)

The transmissibility of the aquifer is $T = K_e \cdot B$

Example 9.1 At a certain point in an unconfined aquifer of 3 km^2 area, the water table was at an elevation of 102.00 m. Due to natural recharge in a wet season, its level rose to 103.20 m. A volume of 1.5 Mm^3 of water was then pumped out of the aquifer causing the water table to reach a level of 101.20 m. Assuming the water table in the entire aquifer to respond in a similar way, estimate (a) the specific yield of the aquifer and (b) the volume of recharge during the wet season.

Solution

(a) Volume pumped out = area × drop in water table × specified yield S_y

$$1.5 \times 10^6 = 3 \times 10^6 \times (103.20 - 101.20) \times S_y$$

$$S_y = 0.25$$

(b) Recharge volume = $0.25 \times (103.20 - 102.00) \times 3 \times 10^6 = 0.9$ Mm^3

Example 9.2 A field test for permeability consists in observing the time required for a tracer to travel between two observation wells. A tracer was found to take 10 h to travel between two wells 50 m apart when the difference in the water-surface elevation in them was 0.5 m. The mean particle size of the aquifer was 2 mm and the porosity of the medium 0.3. If ν = 0.01 cm^2/s estimate (a) the coefficient of permeability and intrinsic permeability of the aquifer, and (b) the Reynolds number of the flow.

Solution

(a) The tracer records the actual velocity of water

$$V_a = \frac{50 \times 100}{10 \times 60 \times 60} = 0.139 \text{ cm/s}$$

Discharge velocity $V = n\, V_a = 0.3 \times 0.139 = 0.0417$ cm/s

$$\text{Hydraulic gradient } i = \frac{0.50}{50} = 1 \times 10^{-2}$$

$$\text{Coefficient of permeability } K = \frac{4.17\times10^{-2}}{1\times10^{-2}} = 4.17 \text{ m/s}$$

$$\text{Intrinsic permeability, } K_0 = \frac{K\nu}{g} = \frac{4.17\times0.01}{981} = 4.25\times10^{-5}\text{ cm}^2$$

Since 9.87×10^{-9} cm^2 = 1 darcy

$$K_0 = 4307 \text{ darcys}$$

(b) Reynolds number $Re = \dfrac{Vd_a}{\nu}$

Taking d_a = mean particle size = 2 mm

$$Re = \frac{0.0417\times2}{10}\times\frac{1}{0.01} = 0.834.$$

Example 9.3 Three wells *A*, *B* and *C* tap the same horizontal aquifer. The distances *AB* = 1200 m and *BC* = 1000 m. The well *B* is exactly south of well *A* and the well *C* lies to the west of well *B*. The following are the ground surface elevation and depth of water below the ground surface in the three wells.

Well	Surface Elevation (metres above datum)	Depth of water table (m)
A	200.00	11.00
B	197.00	7.00
C	202.00	14.00

Determine the direction of groundwater flow in the aquifer in the area *ABC* of the wells.

Solution

Let H = elevation of water table.

H_A = 200.00 – 11.00 = 189.00

H_B = 197.00 – 7.00 = 190.00

H_C = 202.00 – 14.00 = 188.00

Let BA = north direction, designated as Y direction.

The west direction will be called X direction.

The layout of the wells is shown in Fig. 9.6

Along BA: $-\Delta H_y = H_B - H_A$ = 190.00 – 189.00 = 1.00 m

$$i_y = -\frac{\Delta H_y}{L_{AB}} = \frac{1.00}{1200} = 1/1200.$$

$$V_y = K\,.\,i_y = K/1200 \text{ m/s}$$

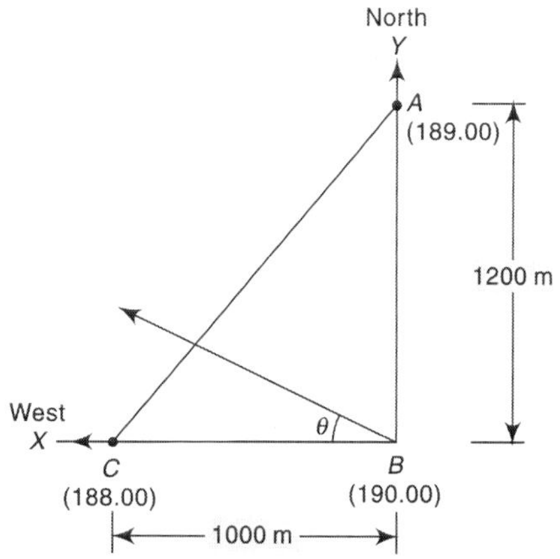

Fig. 9.6 ***Layout of Wells—Example 9.3***

where K = coefficient of permeability.

Along BC, (X direction):

$$-\Delta H_x = H_B - H_C = 190.00 - 188.00 = 2.00 \text{ m}$$

$$i_x = -\frac{\Delta H_x}{L_{BC}} = \frac{2.00}{1000} = 1/500.$$

$$V_x = K \cdot i_x = K/500 \text{ m/s}$$

$$V = \left(V_x^2 + V_y^2\right)^{1/2} = \frac{K}{100}\left[\frac{1}{25} + \frac{1}{144}\right]^{1/2} = 2.167 \times 10^{-3} \text{ K m/s}$$

$$\tan\theta = \frac{V_y}{V_x} = \frac{K}{1200} \times \frac{500}{K} = \frac{1}{2.4}$$

$$\theta = 22.62° = 22° \, 37' \, 11.5''$$

where θ = inclination of V to X-axis (west direction). The groundwater flow will be in a direction which makes 22.62° with line BC and 67.38° with BA. Thus, the direction of groundwater flow is N 67° 27′ 48.5″ W.

9.5 *Geologic Formations as Aquifers*

The identification of a geologic formation as a potential aquifer for groundwater development is a specialised job requiring the services of a trained hydrogeologist. In this section only a few general observations are made and for details the reader is referred to a standard treatise on hydrogeology such as Ref. 4.

The geologic formations of importance for possible use as an aquifer can be broadly classified as (i) unconsolidated deposits, and (ii) consolidated rocks. Unconsolidated deposits of sand and gravel form the most important aquifers. They occur as fluvial alluvial deposits, abandoned channel sediments, coastal alluvium and as lake and glacial deposits. The yield is generally good and may be of the order of 50–100 m^3/h. In India, the Gangetic alluvium and the coastal alluvium in the states of Tamil Nadu and Andhra Pradesh are examples of good aquifers of this kind.

Among consolidated rocks, those with primary porosity such as sandstones are generally good aquifers. Weathering of rocks and occurrence of secondary openings such as joints and fractures enhance the yield. Normally, the yield from these aquifers is less than that of alluvial deposits and typically may have a value of 20–50 m^3/h. Sandstones of Kathiawar and Kutch areas of Gujarat and of Lathi region of Rajasthan are some examples.

Limestones contain numerous secondary openings in the form of cavities formed by the solution action of flowing subsurface water. Often these form highly productive aquifers. In Jodhpur district of Rajasthan, cavernous limestones of the Vindhyan system are providing very valuable groundwater for use in this arid zone.

The volcanic rock basalt has permeable zones in the form of vesicles, joints and fractures. Basaltic aquifers are reported to occur in confined as well as under unconfined conditions. In the Satpura range some aquifers of this kind give yields of about 20 m^3/h.

Igneous and metamorphic rocks with considerable weathered and fractured horizons offer good potentialities as aquifers. Since weathered and fractured horizons are restricted in their thickness, these aquifers have limited thickness. Also, the average permeability of these rocks decreases with depth. The yield is fairly low, being of the order of 5–10 m^3/h. Aquifers of this kind are found in the hard rock areas of Karnataka, Tamil Nadu, Andhra Pradesh and Bihar.

9.6 *Compressibility of Aquifers*

In confined aquifers, the total pressure at any point due to overburden is borne by the combined action of the pore pressure and intergranular pressure. The compressibility of the aquifer and also that of the pore water causes a readjustment of these pressures whenever there is a change in storage and thus have an important bearing on the storage characteristics of the aquifer. In this section, a relation is developed between a defined storage coefficient and the various compressibility parameters.

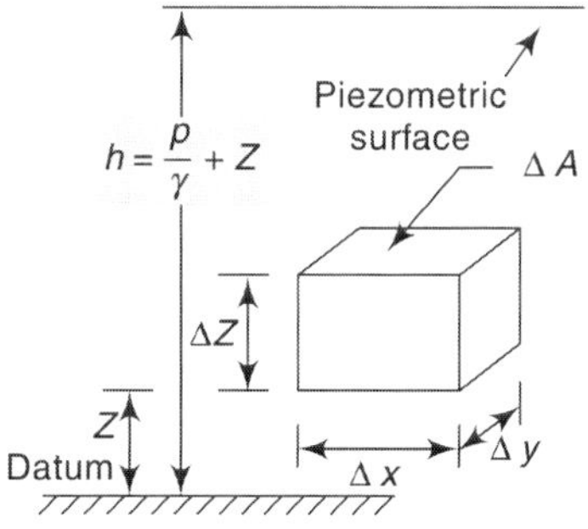

Fig. 9.7 ***Volume Element of a Compressible Aquifer***

Consider an elemental volume $\Delta V = (\Delta x \Delta y)\,\Delta Z = \Delta A\,\Delta Z$ of a compressible aquifer as shown in Fig. 9.7. A cartesian coordinate system with the Z-axis pointing vertically upwards is adopted. Further the following three compressible aquifer assumptions are made:

- The elemental volume is constrained in lateral directions and undergoes change of length in the z-direction only, i.e. ΔA is constant.
- The pore water is compressible
- The solid grains of the aquifer are incompressible but the pore structure is compressible.

By defining the reciprocal of the bulk modulus of elasticity of water as *compressibility of water* β, it is written as

$$\beta = -\frac{(d\,\Delta V_w)}{\Delta V_w}/dp \tag{9.12}$$

where ΔV_w = volume of water in the chosen element of aquifer, and p = pressure. By conservation of mass

$\rho \cdot \Delta V_w$ = constant, where ρ = density of water.

Thus $\rho\, d(\Delta V_w) + \Delta V_w\, d\rho = 0$

Substituting this relationship in Eq. (9.12),

$$\beta = d\rho/(\rho\, dp) \tag{9.13}$$

or

$$d\rho = \rho\beta dp \tag{9.13a}$$

Similarly, by considering the reciprocal of the bulk modulus of elasticity of the pore-space skeleton as the *compressibility of the pores,* α, it is expressed as

$$\alpha = -\frac{d(\Delta V)/\Delta V}{d\sigma_z} \tag{9.14}$$

in which σ_z = intergranular pressure. Since $\Delta V = \Delta A \cdot \Delta Z$ with ΔA = constant,

$$\alpha = -\frac{d(\Delta Z)/\Delta Z}{d\sigma_z} \tag{9.15}$$

The total overburden pressure $w = p + \sigma_z$ = constant.

Thus $dp = -d\,\sigma_z$, which when substituted in Eq. (9.15) gives

$$d\,(\Delta Z) = \alpha\,(\Delta Z)\,dp \tag{9.16}$$

As the volume of solids $\Delta\,V_s$ in the elemental volume is constant,

$$\Delta\,V_s = (1 - n)\,\Delta A \cdot \Delta Z = \text{constant}$$

$$d(\Delta\,V_s) = (1 - n)\,d\,(\Delta Z) - \Delta Z \cdot dn = 0$$

where n = porosity of the aquifer. Using this relationship in Eq. (9.16),

$$dn = \alpha\,(1 - n)\,dp \tag{9.17}$$

Now, the mass of water in the element of volume ΔV is

$$\Delta M = \rho n\,\Delta A\,\Delta Z$$

or $$d(\Delta M) = \Delta\,V\left[n\,d\rho + \rho\,dn + \rho n\frac{d(\Delta Z)}{\Delta Z}\right]$$

i.e. $$\frac{d\,(\Delta M)}{\rho\,\Delta V} = n\frac{d\rho}{\rho} + dn + n\frac{d(\Delta Z)}{\Delta Z}$$

Substituting from Eqs. (9.13), (9.17) and (9.15) for the terms in the right-hand side respectively

$$\frac{d(\Delta M)}{\rho \Delta V} = n\beta\,dp + \alpha(1 - n)\,dp + n\,\alpha\,dp = (n\beta + \alpha)\,dp$$

$$= \gamma(n\beta + \alpha)\,dh = S_s dh \tag{9.18}$$

where $S_s = \gamma(n\,\beta + \alpha)$ and h = piezometric head $= z + \dfrac{\rho}{\gamma}$ and $\gamma = \rho g$ = weight of unit volume of water.

The term S_s is called *specific storage.* It has the dimensions of $[L^{-1}]$ and represents the volume of water released from storage from a unit volume of aquifer due to a unit decrease in the piezometric head. The numerical value of S_s is very small, being of the order of $1 \times 10^{-4}\ m^{-1}$.

By integration of Eq. (9.18) for a confined aquifer of thickness B, a dimensionless *storage coefficient* S can be expressed as

$$S = \gamma(n\beta + \alpha)\,B \tag{9.19}$$

The storage coefficient S (also known as *Storativity)* represents the volume of water released by a column of a confined aquifer of unit cross-sectional area under a unit decrease in the piezometric head. The storage coefficient S and the transmissibility coefficient T are known as the *formation constants* of an aquifer and play a very important role in the unsteady flow through the porous media. Typical values of S in confined aquifers lie in the range 5×10^{-5} to 5×10^{-3}. Values of α for some formation material and β for various temperatures are given in Tables 9.3 and 9.4 respectively.

Table 9.3 *Range of α for Some Formation Materials*

Material	Bulk modulus of elasticity, E_s (N/cm^2)	Compressibility $\alpha = 1/E_s$ (cm^2/N)
Loose clay	$10^2 - 5 \times 10^2$	$10^{-2} - 2 \times 10^{-3}$
Stiff clay	$10^3 - 10^4$	$10^{-3} - 10^{-4}$
Loose sand	$10^3 - 2 \times 10^3$	$10^{-3} - 5 \times 10^{-4}$
Dense sand	$5 \times 10^3 - 8 \times 10^3$	$2 \times 10^{-4} - 1.25 \times 10^{-4}$
Dense sandy gravel	$10^4 - 2 \times 10^4$	$10^{-4} - 5 \times 10^{-5}$
Fissured and jointed rock	$1.5 \times 10^4 - 3 \times 10^5$	$6.7 \times 10^5 - 3.3 \times 10^{-6}$

Table 9.4 *Values of β for Water at Various Temperatures*

Temperature (°C)	Bulk modulus of elasticity, E_w (N/cm^2)	Compressibility $\beta = 1/E_w$ (cm^2/N)
0	2.04×10^5	4.90×10^{-6}
10	2.11×10^5	4.74×10^{-6}
15	2.14×10^5	4.67×10^{-6}
20	2.20×10^5	4.55×10^{-6}
25	2.22×10^5	4.50×10^{-6}
30	2.23×10^5	4.48×10^{-6}
35	2.24×10^5	4.46×10^{-6}

For an unconfined aquifer, the coefficient of storage is given by

$$S = S_y + \gamma(\alpha + n\beta) B_s \qquad (9.20)$$

where B_s = saturated thickness of the aquifer. However, the second term on the right-hand side is so small relative to S_y that for practical purposes S is considered equal to S_y, i.e. the coefficient of storage is assumed to have the same value as the specific yield for unconfined aquifers.

The elasticity of the aquifer is reflected dramatically in the response of the water levels in the wells drilled in confined aquifers to changes in the atmospheric pressure. Increase in the atmospheric pressure causes an increase in the loading of the aquifer. The change in the pressure is balanced by a partial compression of the water and partial compression of the pore skeleton. An increase in the atmospheric pressure causes a decrease in the water level in the well. Converse is the case with the decrease in pressure. The ratio of the water level change to pressure head change is called *Barometric Efficiency* (*BE*) and is given in terms of the compressibility parameters as

$$BE = -\left(\frac{n\beta}{\alpha + n\beta}\right) \tag{9.21}$$

The negative sign indicates the opposite nature of the changes in pressure head and water level. Using Eq. (9.19) in Eq. (9.25), $BE = -(\gamma \beta nB/S)$ and this affords a means of finding S. The barometric efficiency can be expected to be in the range 10–75%. It is apparent that unconfined aquifers have practically no barometric efficiency.

A few other examples of compressibility effects causing water level changes in artesian wells include (i) tidal action in coastal aquifers, (ii) earthquake or underground explosions, and (iii) passing of heavy railway trains.

9.7 *Equation of Motion*

9.7.1 Confined Groundwater Flow

If the velocities of flow in the cartesian coordinate directions x, y, z of the aquifer element, ΔV, are u, v and w respectively, the equation of continuity for the fluid flow is

$$\frac{\partial(\Delta M)}{\partial t} = -\left[\frac{\partial(\rho u)}{\partial x} + \frac{\partial(\rho v)}{\partial y} + \frac{\partial(\rho w)}{\partial z}\right] \cdot \Delta V \tag{9.22}$$

where $\Delta V = \Delta X \cdot \Delta Y \cdot \Delta Z$

From Eq. (9.18), considering the differentials with respect to time

$$\frac{\partial(\Delta M)}{\partial t} \cdot \frac{1}{\Delta V} = S_s \rho \frac{dh}{dt} \tag{9.18a}$$

Further, the aquifer is assumed to be isotropic with permeability coefficient K, so that the Darcy's equation for x, y and z directions can be written as

$$u = -K\frac{\partial h}{\partial x}, v = -K\frac{\partial h}{\partial y} \text{ and } w = -K\frac{\partial h}{\partial z} \tag{9.23}$$

Using Eqs. (9.23) and (9.13) and noting that $h = \dfrac{p}{\gamma} + z$, the various terms of the right hand side of Eq. (9.22) are written as

$$\frac{\partial(\rho u)}{\partial x} = \rho\frac{\partial u}{\partial x} + u\frac{\partial \rho}{\partial x} = -K\rho\frac{\partial^2 h}{\partial x^2} - K\rho^2 \beta g\left(\frac{\partial h}{\partial x}\right)^2 \tag{9.23-a}$$

$$\frac{\partial(\rho v)}{\partial y} = \rho\frac{\partial v}{\partial y} + v\frac{\partial \rho}{\partial y} = -K\rho\frac{\partial^2 h}{\partial y^2} - K\rho^2 \beta g\left(\frac{\partial h}{\partial y}\right)^2 \tag{9.23-b}$$

$$\frac{\partial(\rho w)}{\partial z} = \rho\frac{\partial w}{\partial z} + w\frac{\partial \rho}{\partial z} = -K\frac{\partial^2 h}{\partial z^2} - K\rho^2 \beta g\left[\left(\frac{\partial h}{\partial z}\right)^2 - \frac{\partial h}{\partial z}\right] \tag{9.23-c}$$

Assembling these, Eq. (9.22) can be written as

$$K\rho\left[\frac{\partial^2 h}{\partial x^2} + \frac{\partial^2 h}{\partial y^2} + \frac{\partial^2 h}{\partial z^2}\right] + K\rho^2 \beta g\left[\left(\frac{\partial h}{\partial x}\right)^2 + \left(\frac{\partial h}{\partial y}\right)^2 + \left(\frac{\partial h}{\partial z}\right)^2 - \frac{\partial h}{\partial z}\right] = \rho S_s \frac{\partial h}{\partial t} \tag{9.24}$$

The second term on the left-hand side is neglected as very small, especially for $\partial h/\partial x \ll 1$, and Eq. (9.24) is rearranged to yield

$$\frac{\partial^2 h}{\partial x^2} + \frac{\partial^2 h}{\partial y^2} + \frac{\partial^2 h}{\partial z^2} = \frac{S_s}{K}\frac{\partial h}{\partial t} \tag{9.25}$$

Defining $S_s B = S$, $KB = T$, and $\nabla^2 h = \left(\frac{\partial^2 h}{\partial x^2} + \frac{\partial^2 h}{\partial y^2} + \frac{\partial^2 h}{\partial z^2}\right)$, Eq. (9.25) reads as

$$\nabla^2 h = \frac{S}{T}\frac{\partial h}{\partial t} \tag{9.26}$$

This is the basic differential equation governing unsteady groundwater flow in a homogeneous, isotropic, *confined aquifer.* This form of the equation is known as *diffusion equation.*

If the flow is steady, the $\partial h/\partial t$ term does not exist, leading to

$$\nabla^2 h = 0 \tag{9.27}$$

This equation is known as *Laplace equation* and is the fundamental equation of all potential flow problems. Being linear, the method of superposition is applicable in its solutions.

Equation (9.26) or (9.27) can be solved for suitable boundary conditions by analytical, numerical or analog methods to yield solutions to a variety of groundwater flow problems. The details of solution of the basic differential equation of groundwater are available in standard literature (Refs. 1, 3, 5, 6 and 7).

As an application of the Laplace equation (Eq. 9.27) a simple situation of steady state one-dimensional confined porous media flow is given below.

9.7.2 Confined Groundwater Flow between Two Water Bodies

Figure 9.8 shows a very wide confined aquifer of depth B connecting to water bodies. A section of the aquifer of unit width is considered. The piezometric head at the upstream end is h_0 and at a distance x from the upstream end the head is h.
As the flow is in x direction only, Eq. (9.27) becomes

$$\frac{\partial^2 h}{\partial x^2} = 0 \tag{9.28}$$

On integrating twice, $h = C_1 x + C_2$
On substitution of th boundary condition, $h = h_0$ at $x = 0$

$$h = C_1 x + h_0$$

Also at $x = L$, $h = h_1$ and hence $C_1 = -\left(\frac{h_0 - h_1}{L}\right)$

Thus

$$h = h_0 - \left(\frac{h_0 - h_1}{L}\right)x \tag{9.29}$$

This is the equation of the hydraulic grade line, which is shown to vary linearly from h_0 to h_1.

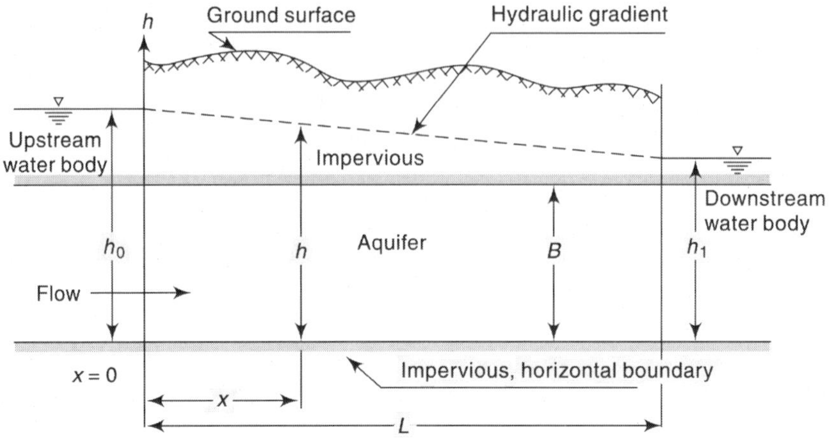

Fig. 9.8 ***Confined Groundwater Flow between Two Water Bodies***

By Darcy's law, the discharge per unit width of the aquifer is

$$q = -KB\frac{\partial h}{\partial x} = -KB\,(-(h_0 - h_1)/L)$$

$$q = \frac{(h_0 - h_1)}{L} K\,B \tag{9.30}$$

9.7.3 Unconfined Flow by Dupit's Assumption

While Eq. (9.26) is specifically for confined aquifers, Eq. (9.27) which is the Laplace equation in h is applicable to steady flow of both confined and unconfined aquifers. However, in unconfined aquifers the free surface of the water table, known as *phreatic surface,* has the boundary condition of constant pressure equal to atmospheric pressure. Also, in a section the line representing the water table, is also a streamline. These boundary conditions cause considerable difficulties in analytical solutions of steady unconfined flow problems by using the Laplace equation in h.

A simplified approach based on the assumptions suggested by Dupit (1863) which gives reasonably good results is described below. The basic assumptions of Dupit are the following:

- The curvature of the free surface is very small so that the streamlines can be assumed to be horizontal at all sections.
- The hydraulic grade line is equal to the free surface slope and does not vary with depth.

Consider an elementary prism of fluid bounded by the water table shown in Fig. 9.9(a).

Let V_x = gross velocity of groundwater entering the element in x direction

V_y = gross velocity of groundwater entering the element in y direction

Assume a horizontal impervious base and no vertical inflow from top due to recharge. By Dupit's assumptions, $\partial V_x/\partial z = 0$ and $\partial V_y/\partial z = 0$. Considering the X direction:

The mass flux entering the element $M_{x1} = \rho\, V_x\, h\, \Delta\, y$

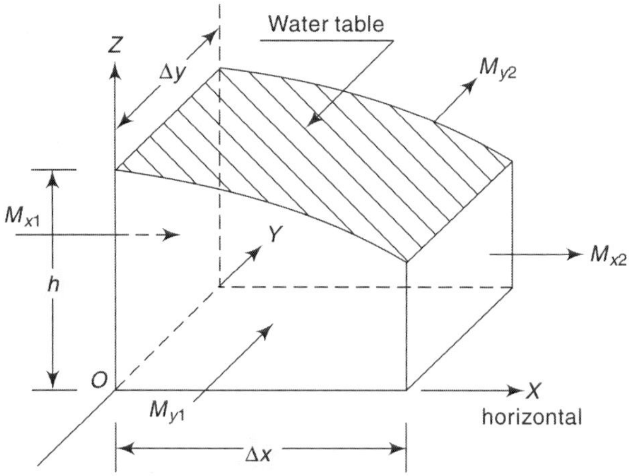

Fig. 9.9(a) ***Definition Sketch—Unconfined Ground-water Flow without Recharge***

The mass flux leaving the element $M_{x2} = \rho V_x h \,\Delta y + \dfrac{\partial}{\partial x}(\rho V_x \, h \,\Delta y)\,\Delta x$

The net mass efflux from the element in x direction, by considering the flow entering the element as postive and outflow as negative, is

$$M_{x1} - M_{x2} = \Delta M_x = -\frac{\partial}{\partial x}(\rho V_x h \,\Delta y)\,\Delta x$$

Similarly the net mass efflux in y direction

$$M_{y1} - M_{y2} = \Delta M_y = -\frac{\partial}{\partial y}(\rho V_y h \,\Delta x)\,\Delta y$$

Further, there is neither inflow or outflow in the Z direction. Thus for steady, incompressible flow, by continuity

$$\Delta M_x + \Delta M_y = 0 \tag{9.31}$$

Substituting for ΔM_x and ΔM_y and simplifying

$$\frac{\partial}{\partial x}(V_x h) + \frac{\partial}{\partial y}(V_y h) = 0 \tag{9.32}$$

By Darcy's law, $V_x = -K\dfrac{\partial h}{\partial x}$ and $V_y = -K\dfrac{\partial h}{\partial y}$

Hence, Eq. (9.32) becomes

$$\frac{\partial}{\partial x}\left(-K h\frac{\partial h}{\partial x}\right) + \frac{\partial}{\partial y}\left(-K h\frac{\partial h}{\partial y}\right) = 0$$

or

$$\frac{\partial^2 h^2}{\partial x^2} + \frac{\partial^2 h^2}{\partial y^2} = 0$$

i.e.

$$\nabla^2 h^2 = 0 \tag{9.33}$$

Thus, the steady unconfined groundwater flow with Dupit's assumptions is governed by Laplace equation in h^2.

1. Unconfined Flow with Recharge

If there is a recharge, i.e. infiltration of water from the top ground surface into the aquifer, at a rate of R (m^3/s per m^2 of horizontal area) as in Fig. 9.9(b), the continuity equation Eq. (9.31) is to be modified to take into account the recharge. Consider the element of an unconfined aquifer as in Fig. 9.9(b) situated on a horizontal impervious bed. Here, in addition to ΔM_x and ΔM_y there will be a net inflow into the element in the Z direction given by

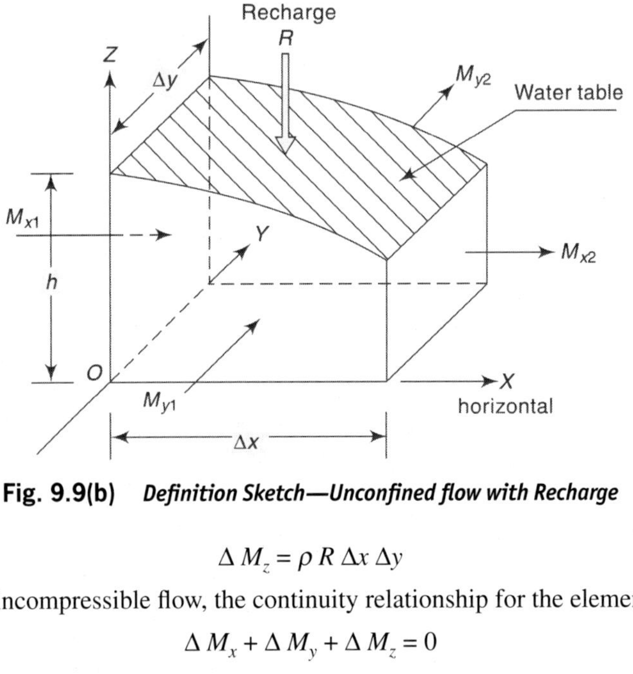

Fig. 9.9(b) ***Definition Sketch—Unconfined flow with Recharge***

$$\Delta M_z = \rho R \Delta x \Delta y$$

For steady, incompressible flow, the continuity relationship for the element is

$$\Delta M_x + \Delta M_y + \Delta M_z = 0$$

i.e. $$-\frac{\partial}{\partial x}(\rho V_x h \Delta x \Delta y) - \frac{\partial}{\partial y}(\rho Vy h \Delta x \Delta y) + \rho R \Delta x \Delta y = 0$$

Substituting $V_x = -K\dfrac{\partial h}{\partial x}$ and $V_y = -K\dfrac{\partial h}{\partial y}$ and simplifying

$$\frac{\partial^2 h^2}{\partial x^2} + \frac{\partial^2 h^2}{\partial y^2} = -\frac{2R}{K} \tag{9.34}$$

Equation (9.34) is the basic differential equation under Dupit's assumption for unconfined groundwater flow with recharge. Note that Eq. (9.33) is a special case of Eq. (9.34) with $R = 0$.

Use of Eq. (9.34) finds considerable practical application in finding the water table profiles in unconfined aquifers. A few examples are (i) an unconfined aquifer separating two water bodies such as a canal and a river, (ii) various recharge situations, (iii) drainage problems, and (iv) infiltration galleries. To illustrate the use of Eq. (9.34), a situation of steady flow in an unconfined aquifer bounded by two water bodies and subjected to recharge from top is given below.

2. One-Dimensional Dupit's Flow with Recharge

(a) The General Case Consider an unconfined aquifer on a horizontal impervious base situated between two water bodies with a difference in surface elevation, as shown in Fig. 9.10. Further, there is a recharge at a constant rate of R m^3/s per unit horizontal area due to infiltration from the top of the aquifer. The aquifer is of infinite length and hence one-dimensional method of analysis is adopted. A unit width of aquifer is considered for analysis.

From Eq. (9.34),
$$\frac{\partial^2 h^2}{\partial x^2} = -\frac{2R}{K} \tag{9.35}$$

On integration with respect to x twice,
$$h^2 = -\frac{R}{K}x^2 + C_1 x + C_2 \tag{9.36}$$

where C_1 and C_2 are constants of integration

The boundary conditions are

(i) at $x = 0$, $h = h_0$ hence, $C_2 = h_0^2$

(ii) at $x = L$, $h = h_1$ hence, $h_1^2 - h_0^2 = -\frac{R}{K}L^2 + C_1 L$

or
$$C_1 = -\frac{\left(h_0^2 - h_1^2 - \frac{RL^2}{K}\right)}{L}$$

Thus, Eq. (9.36) becomes
$$h^2 = -\frac{Rx^2}{K} - \frac{\left(h_0^2 - h_1^2 - \frac{RL^2}{K}\right)}{L}x + h_0^2 \tag{9.37}$$

The water table is thus an ellipse represented by Eq. (9.37). The value of h will in general rise above h_0, reaches a maximum at $x = a$ and falls back to h_1 at $x = L$ as shown in Fig. 9.10. The value of a is obtained by equating $\frac{\partial h}{\partial x} = 0$, and is given by
$$a = \frac{L}{2} - \frac{K}{R}\left(\frac{h_0^2 - h_1^2}{2L}\right) \tag{9.38}$$

The location $x = a$ is called the *water divide.* Figure 9.10 shows the flow to the left of the divide will be to the upstream water body and the flow to the right of the divide will be to the downstream water body.

The discharge per unit with of aquifer at any location x is
$$q_x = -Kh\frac{dh}{dx} = -K\left[-\frac{Rx}{K} - \frac{\left(h_0^2 - h_1^2 - \frac{RL^2}{K}\right)}{2L}\right]$$

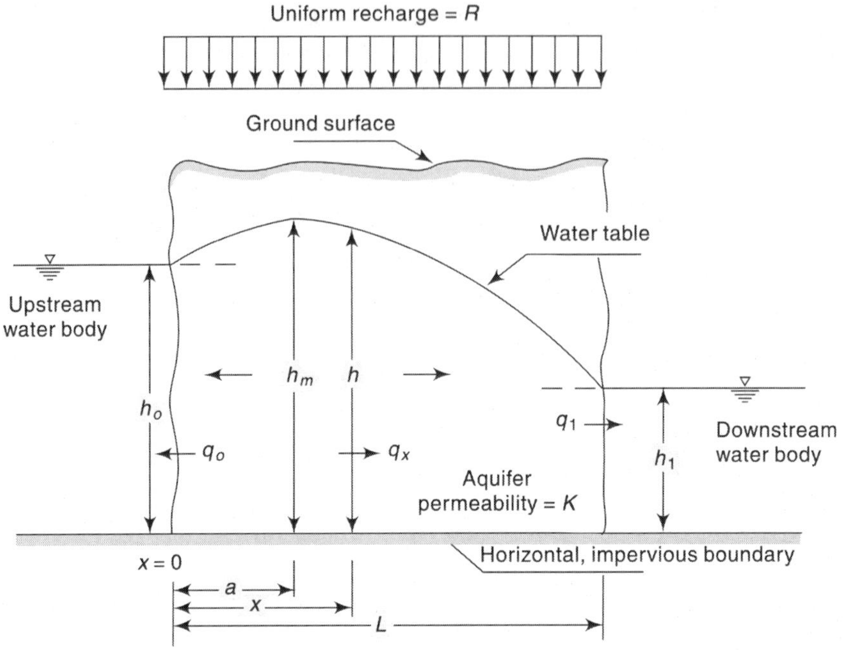

Fig. 9.10 ***One-Dimensional Dupit Flow with Recharge***

$$q_x = R\left(x - \frac{L}{2}\right) + \frac{K}{2L}(h_0^2 - h_1^2) \tag{9.39}$$

It is obvious the discharge q_x varies with x. At the upstream water body, $x = 0$ and

Discharge
$$q_0 = q_{x=0} = -\frac{RL}{2} + \frac{K}{2L}(h_0^2 - h_1^2) \tag{9.40}$$

At the downstream water body, $x = L$ and

$$q_1 = q_{x=L} = \frac{RL}{2} + \frac{K}{2L}(h_0^2 - h_1^2) = RL + q_0 \tag{9.40a}$$

(b) Flow Without Recharge When there is no recharge, $R = 0$ and the flow simplifies to that of steady one-dimensional flow in an unconfined aquifer as in Fig. 9.11.

By putting $R = 0$ in Eq. (9.37), he equation of the water table is given by

$$(h^2 - h_0^2) = \left(\frac{h_1^2 - h_0^2}{L}\right)x \tag{9.41}$$

This represents a parabola (known as *Dupit's parabola*) joining h_0 and h_1 on either side of the aqufer.

Differentiating Eq. (9.41) with respect to x

$$2\,h\frac{dh}{dx} = \frac{(h_1^2 - h_0^2)}{L}$$

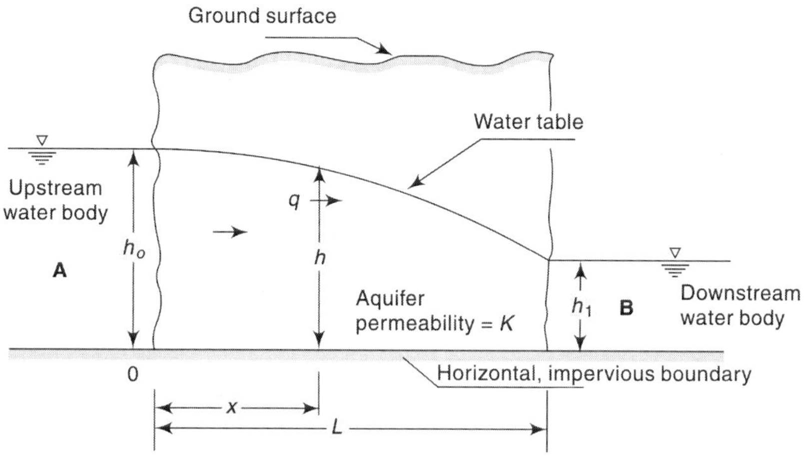

Fig. 9.11 ***One-Dimensional Unconfined Flow Without Recharge***

The discharge q per unit width of the aquifer is

$$q = -K\,h\,\frac{dh}{dx} = \frac{(h_0^2 - h_1^2)}{2L}K \tag{9.42}$$

(c) Tile Drain Problem The provision of a drains system is one of the most widely used method of draining waterlogged areas, the object being to reduce the level of the water table. Figure 9.12 shows a set of porous tile drains maintaining a constant recharge rate of R at the top ground surface.

An approximate expression to the water-table profile can be obtained by Eq. (9.37) by neglecting the depth of water in the drains, i.e. $h_0 = h_1 = 0$. The water table profile will then be

$$h^2 = \frac{R}{K}(L - x)\,x \tag{9.43}$$

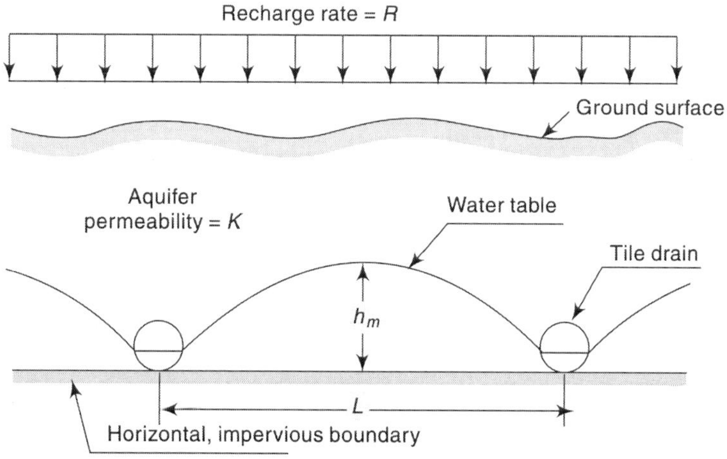

Fig. 9.12 ***Tile Drains under a Constant Recharge Rate***

The maximum height of he water table occurs at $a = L/2$ and is of magnitude

$$h_m = \frac{L}{2}\sqrt{R/K} \tag{9.44}$$

Considering a set of drains, since the flow is steady, the discharge entering a drain per unit length of the drain is

$$q = 2\left(R\frac{L}{2}\right) = RL \tag{9.45}$$

Example 9.4 Two parallel rivers *A* and *B* are separated by a land mass as shown in Fig. 9.13. Estimate the seepage discharge from River *A* to River *B* per unit length of the rivers.

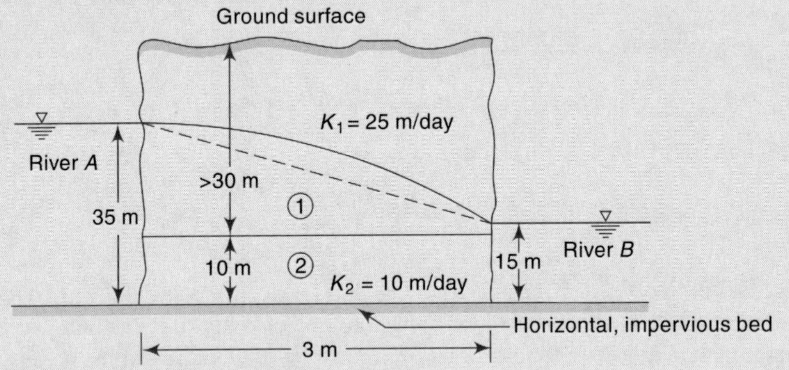

Fig. 9.13 ***Schematic Layout of Example 9.4***

Solution

The aquifer system is considered as a composite of aquifers 1 and 2 with a horizontal impervious boundary at the interface. This leads to the assumptions:

(a) aquifer 2 is a confined aquifer with $K_2 = 10$ m/day

(b) aquifer 1 is an unconfined aquifer with $K_1 = 25$ m/day

Consider a unit width of the aquifers.

For the confined aquifer 2:

From Eq. (9.30) $q_2 = \dfrac{(h_0 - h_1)}{L} KB$

Here $h_0 = 35.0$ m, $h_1 = 15$ m,

$L = 3000$ m, $K_2 = 10$ m/day and $B = 10$ m

$$q_2 = \frac{(35-15)}{3000} \times 10 \times 10 = 0.667 \text{ m}^3\text{/day per metre width}$$

For the unconfined aquifer 1:

From Eq. (9.42), $q_1 = \dfrac{(h_0^2 - h_1^2)}{2L} K$

Here $h_0 = (35 - 10) = 25$ m, $h_1 = (15 - 10) = 5$ m

$L = 3000$ m, $K_1 = 25$ m/day

$$q_2 = \frac{(25)^2 - (5)^2}{2 \times 3000} \times 25 = 2.5 \text{ m}^3\text{/day per metre width}$$

Total discharge from river A to river $B = q = q_1 + q_2$

$$= 0.667 + 2.500 = 3.167 \text{ m}^3\text{/day per unit length of the rivers}$$

Example 9.5 An unconfined aquifer (K = 5 m/day) situated on the top of a horizontal impervious layer connects two parallel water bodies M and N which are 1200 m apart. The water- surface elevations of M and N, measured above the horizontal impervious bed, are 10.00 m and 8.00 m. If a uniform recharge at the rate of 0.002 m^3/day per m^2 of horizontal area occurs on the ground surface, estimate

(i) the water table profile,
(ii) the location and elevation of the water table divide,
(iii) the seepage discharges into the lakes, and
(iv) the recharge rate at which the water table divide coincides with the upstream edge of the aquifer and the total seepage flow per unit width of the aquifer at this recharge rate.

Solution

Consider unit width of the aquifer

Referring to Fig. 9.14

h_0 = 10.0 m, h_1 = 8.0 m,

R = 0.002 m^3/day/m^2, L = 1200 m and K = 5 m/day.

(i) The water-table profile:

By Eq. (9.37), $$h^2 = -\frac{Rx^2}{K} - \frac{\left(h_0^2 - h_1^2 - \frac{RL^2}{K}\right)}{L}x + h_0^2$$

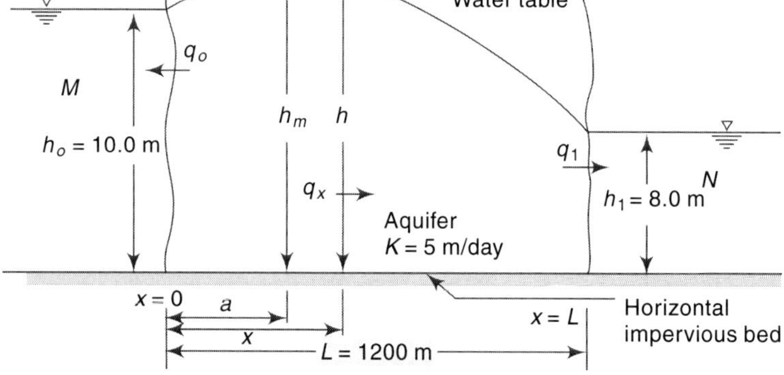

Fig. 9.14 ***Schematic Layout—Example 9.5***

$$= -\left(\frac{0.002}{5}\right)x^2 - \frac{1}{1200}\left[(10)^2 - (8)^2 - \frac{0.002 \times (1200)^2}{5}\right]x + 10^2$$

$$h^2 = -0.0004\,x^2 + 0.45x + 100$$

(ii) Location of water table divide:

From Eq. (9.38), $a = \frac{L}{2} - \frac{K}{R}\left(\frac{h_0^2 - h_1^2}{2L}\right)$

$$a = \frac{1200}{2} - \left(\frac{5.0}{0.002}\right)\left(\frac{(10)^2 - (8)^2}{2 \times 1200}\right) = 562.5 \text{ m}$$

At $x = a = 562.5$ m, $h = h_m$ = height of water-table divide

$$h_m^2 = -0.0004\,(562.5)^2 + 0.45\,(562.5) + 100 = 226.56$$

and $h_m = \sqrt{226.56} = 15.05$ m

(iii) Discharge per unit width of the aquifer:

From Eq. (9.39), $q_x = R\left(x - \frac{L}{2}\right) + \frac{K}{2L}(h_0^2 - h_1^2)$

At $x = 0$, $q_0 = -R\frac{L}{2} + \frac{K}{2L}(h_0^2 - h_1^2)$

$$= \frac{-(0.002 \times 1200)}{2} + \frac{5}{2 \times 1200}[(10)^2 - (8)^2] = -1.20 + 0.075$$

$$q_0 = -1.125 \text{ m}^3\text{/day per metre width}$$

The negative sign indicates that the discharge is in $(-x)$ direction, i.e. into the water body *M*.

At $x = L$, $q_1 = q_L$ and from Eq. (9.40a) $q_L = RL + q_0$

Hence, q_1 = discharge into water body *N*

$= 0.002 \times 1200 + (-1.125) = 1.275$ m^3/day/m width.

(vi) When the distance of the water table divide $a = 0$:

From Eq. (9.36), $a = \frac{L}{2} - \frac{K}{R}\left(\frac{h_0^2 - h_1^2}{2L}\right) = 0$

$$\frac{L}{2} = \frac{K}{R}\left(\frac{h_0^2 - h_1^2}{2L}\right)$$

$$R = \frac{K}{L^2}(h_0^2 - h_1^2) = \frac{5.0}{(1200)^2}[(10)^2 - (8)^2]$$

$= 1.25 \times 10^{-4}$ m^3/day/m^2

Since $a = 0$, $q_0 = 0$ and by Eq. (9.40a)

$$q_1 = q_L = RL$$

$= 1.25 \times 10^{-4} \times 1200 = 0.15$ m^3/day/m width.

9.8 *Wells*

Wells form the most important mode of groundwater extraction from an aquifer. While wells are used in a number of different applications, they find extensive use in water supply and irrigation engineering practice.

Consider the water in an unconfined aquifer being pumped at a constant rate from a well. Prior to the pumping, the water level in the well indicates the static water table. A lowering of this water level takes place on pumping. If the aquifer is homogeneous and isotropic and the water table initially horizontal, due to the radial flow into the well through the aquifer the water-table assumes a conical shape called *cone of depression.* The drop in the water table elevation at any point from its previous static level is called *drawdown.* The areal extent of the cone of depression is called *area of influence* and its radial extent *radius of influence* (Fig. 9.15). At constant rate of pumping, the drawdown curve develops gradually with time due to the withdrawal of water from storage. This phase is called an *unsteady flow* as the water table elevation at a given location near the well changes with time. On prolonged pumping, an equilibrium state is reached between the rate of pumping and the rate of inflow of groundwater from the outer edges of the zone of influence. The drawdown surface attains a constant position with respect to time when the well is known to operate under *steady-flow* conditions. As soon as the pumping is stopped, the depleted storage in the cone of depression is made good by groundwater inflow into the zone of influence. There is a gradual accumulation of storage till the original (static) level is reached. This stage is called *recuperation* or *recovery* and is an unsteady phenomenn. Recuperation time depends upon the aquifer characteristics.

Changes similar to the above take place to a pumping well in a confined aquifer also but with the difference that it is the piezometric surface instead of the water table that undergoes drawdown with the development of the cone of depression. In confined aquifers with considerable piezometric head, the recovery into the well takes place at a very rapid rate.

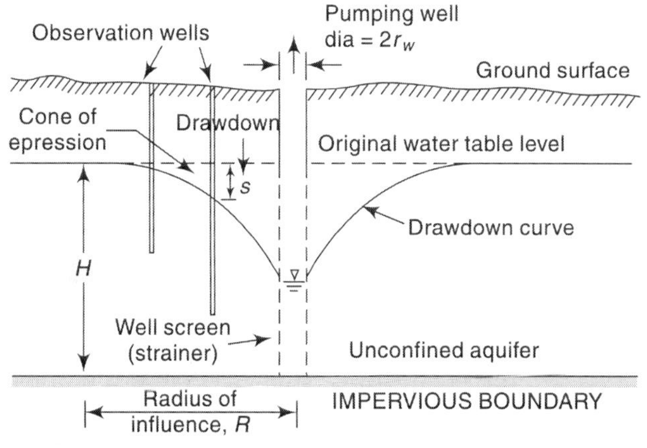

Fig. 9.15 ***Well Operating in an Unconfined Aquifer, (definition sketch)***

9.9 Steady Flow into a Well

Steady-state groundwater problems are relatively simpler. Expressions for the steady state radial flow into a well under both confined and unconfined aquifer conditions are presented below.

9.9.1 Confined Flow

Figure 9.16 shows a well completely penetrating a horizontal confined aquifer of thickness B. Consider the well to be discharging a steady flow, Q. The original piezometric head (static head) was H and the drawdown due to pumping is indicated in Fig. 9.16. The piezometric head at the pumping well is h_w and the drawdown s_w.

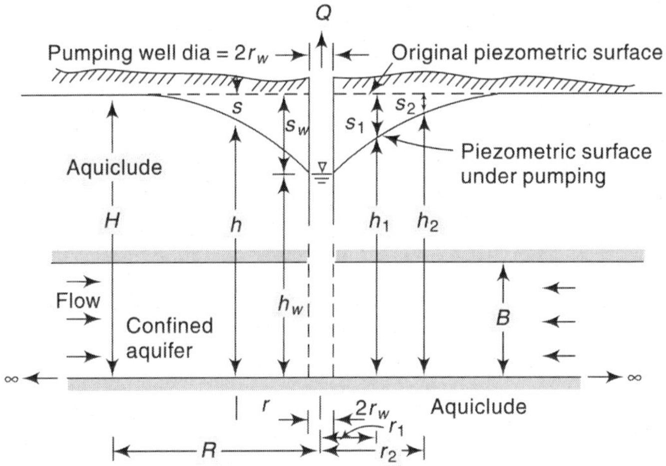

Fig. 9.16 ***Well Operating in a Confined Aquifer***

At a radial distance r from the well, if h is the piezometric head, the velocity of flow by Darcy's law is

$$V_r = K\frac{dh}{dr}$$

The cylindrical surface through which this velocity occurs is $2\pi r B$. Hence by equating the discharge entering this surface to the well discharge,

$$Q = (2\pi r B)\left(K\frac{dh}{dr}\right)$$

$$\frac{Q}{2\pi KB}\frac{dr}{r} = dh$$

Integrating between limits r_1 and r_2 with the corresponding piezometric heads being h_1 and h_2 respectively,

$$\frac{Q}{2\pi KB}\ln\frac{r_2}{r_1} = (h_2 - h_1)$$

or
$$Q = \frac{2\pi KB(h_2 - h_1)}{\ln\dfrac{r_2}{r_1}} \tag{9.46}$$

This is the equilibrium equation for the steady flow in a confined aquifer. This equation is popularly known as *Thiem's equation.*

If the drawdown s_1 and s_2 at the observation wells are known, then by noting that

$$S_1 = H - h_1,\ s_2 = H - h_2 \text{ and } KB = T$$

Equation (9.46) will read as

$$Q = \frac{2\pi T(s_1 - s_2)}{\ln \dfrac{r_2}{r_1}} \tag{9.47}$$

Further, at the edge of the zone of influence, $s = 0$, $r_2 = R$ and $h_2 = H$; at the well wall $r_1 = r_w$, $h_1 = h_w$ and $s_1 = s_w$. Equation (9.47) would then be

$$Q = \frac{2\pi T s_w}{\ln R/r_w} \tag{9.48}$$

Equation (9.47) or (9.48) can be used to estimate T, and hence K, from pumping tests. For the use of the equilibrium equation, Eq. (9.46) or its alternative forms, it is necessary that the assumption of complete penetration of the well into the aquifer and steady state of flow are satisfied.

Example 9.6 A 30 cm diameter well completely penetrates a confined aquifer of permeability 45 m/day. The length of the strainer is 20 m. Under steady state of pumping, the drawdown at the well was found to be 3.0 m and the radius of influence was 300 m. Calculate the discharge.

Solution

In this problem, referring to Fig. 9.16,

$$r_w = 0.15 \text{ m} \qquad R = 300 \text{ m} \qquad s_w = 3.0 \text{ m} \qquad B = 20 \text{ m}$$

$$K = 45/(60 \times 60 \times 24) = 5.208 \times 10^{-4} \text{ m/s}$$

$$T = KB = 10.416 \times 10^{-3} \text{ m}^2/\text{s}$$

By Eq. (9.48)

$$Q = \frac{2\pi T s_w}{\ln R/r_w} = \frac{2\pi \times 10.416 \times 10^{-3} \times 3}{\ln \dfrac{300}{0.15}} = 0.02583 \text{ m}^3/\text{s} = 25.83 \text{ lps}$$

$$= 1550 \text{ lpm}$$

Example 9.7 For the well in the previous example, calculate the discharge (a) if the well diameter is 45 cm and all other data remain the same as in Example 9.6, (b) if the drawdown is increased to 4.5 m and all other data remain unchanged a in Example 9.6.

Solution

(a) $Q = \dfrac{2\pi T s_w}{\ln R/r_w}$

As T and s_w are constants, $\dfrac{Q_1}{Q_2} = \dfrac{\ln R/r_{w_2}}{\ln R/r_{w_1}}$

Putting $R = 300$ m $Q_1 = 1550$ lpm, $r_{w1} = 0.15$ m and $r_{w2} = 0.225$ m.

$$Q_2 = 1550 \frac{\ln 300/0.15}{\ln 300/0.225} = 1637 \text{ lpm}$$

[Note that the discharge has increased by about 6% for 50% increase in the well diameter.]

(b) $$Q = \frac{2\pi T s_w}{\ln R/r_w}$$

$Q \propto s_w$ for constant T, R and r_w.

Thus, $$\frac{Q_1}{Q_2} = \frac{s_{w_1}}{s_{w_2}}$$

$$Q_2 = 1550 \times \frac{4.5}{3.0} = 2325 \text{ lpm}$$

[Note that the discharge increases linearly with the drawdown when other factors remain constant.]

Example 9.8 The discharge from a fully penetrating well operating under steady state in a confined aquifer of 30 m thickness is 2100 litres/minute. The drawdowns observed at two observation wells located at 15 m and 150 m from the well are 3.2 m and 0.28 respectively. Determine the transmissibility and the permeability of the aquifer.

Solution

Here, $s_1 = 3.2$ m and $r_1 = 15$ m

$s_2 = 0.28$ m and $r_2 = 150$ m

$$Q = \frac{2100}{1000 \times 60} = 0.035 \text{ m}^3/\text{s}$$

By Eq. (9.47), $$Q = \frac{2\pi T (s_1 - s_2)}{\ln \frac{r_2}{r_1}}$$

$$0.035 = \frac{2 \times \pi \times T}{\ln \frac{150}{15}} (3.2 - 0.28) = 7.96795 T$$

$$T = \frac{0.035}{7.96795} = 0.00439 \text{ m}^2/\text{s}$$

Transmissibility $T = 0.00439 \times 3600 \times 24 = 379.5$ m^2/day

Permeability $K = \frac{T}{B} = \frac{379.5}{30} = 12.65$ m/day

9.9.2 Unconfined Flow

Consider a steady flow from a well completely penetrating an unconfined aquifer. In this case, because of the presence of a curved free surface, the streamlines are not strictly radial straight lines. While a streamline at the free surface will be curved, the one at the bottom of the aquifer will be a horizontal line, both converging to the well. To obtain a simple solution *Dupit's assumptions* as indicated in Sec. 9.6 are made. In the present case these are the following:

- For small inclinations of the free surface, the streamlines can be assumed to be horizontal and the equipotentials are thus vertical.
- The hydraulic gradient is equal to the slope of the free surface and does not vary with depth. This assumption is satisfactory in most of the flow regions except in the immediate neighbourhood of the well.

Consider the well of radius r_w penetrating completely an extensive unconfined horizontal aquifer as shown in Fig. 9.17. The well is pumping a discharge Q. At any radial distance r, the velocity of radial flow into the well is

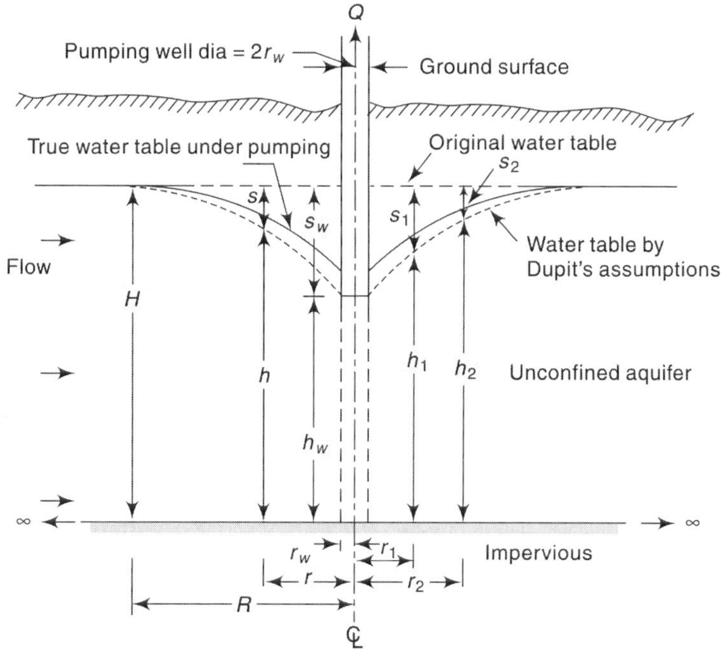

Fig. 9.17 ***Radial Flow to a Well in an Unconfined Aquifer***

$$V_r = K\frac{dh}{dr}$$

where h is the height of the water table above the aquifer bed at that location. For steady flow, by continuity

$$Q = (2\,\pi\,r\,h)V_r = 2\,\pi\,r\,K\,h\,\frac{dh}{dr}$$

or
$$\frac{Q}{2\pi K}\frac{dr}{r} = h\,dh$$

Integrating between limits r_1 and r_2 where the water-table depths are h_1 and h_2 respectively and on rearranging

$$Q = \frac{\pi K (h_2^2 - h_1^2)}{\ln \frac{r_2}{r_1}} \tag{9.49}$$

This is the equilibrium equation for a well in an unconfined aquifer. As at the edge of the zone of influence of radius R, H =saturated thickness of the aquifer, Eq. (9.49) can be written as

$$Q = \frac{\pi K (H^2 - h_w^2)}{\ln \frac{R}{r_w}} \tag{9.50}$$

where h_w = depth of water in the pumping well of radius r_w.

Equations (9.49) and (9.50) can be used to estimate satisfactorily the discharge and permeability of the aquifer by using field data. Calculations of the water-table profile by Eq. (9.49), however, will not be accurate near the well because of Dupit's assumptions. The water-table surface calculated by Eq. (9.49) which involved Dupit's assumption will be lower than the actual surface. The departure will be appreciable in the immediate neighbourhood of the well (Fig. 9.17). In general, values of R in the range 300 to 500 m can be assumed depending on the type of aquifer and operating conditions of a well. As the logarithm of R is used in the calculation of discharge, a small error in R will not seriously affect the estimation of Q. It should be noted that it takes a relatively long time of pumping to achieve a steady state in a well in an unconfined aquifer. The recovery after the cessation of pumping is also slow compared to the response of an artesian well which is relatively fast.

Approximate Equations

If the drawdown at the pumping well $s_w = (H - h_w)$ is small relative to H, then

$$H^2 - h_w^2 = (H + h_w)(H - h_w) \approx 2\,H\,s_w$$

Noting that $T = KH$, Eq. (9.50) can be written as

$$Q = \frac{2\pi T s_w}{\ln \frac{R}{r_w}} \tag{9.50a}$$

which is the same as Eq. (9.48). Similarly, Eq. (9.49) can be written in terms of $s_1 = (H - h_1)$ and $s_2 = (H - h_2)$ as

$$Q = \frac{2\pi T (s_1 - s_2)}{\ln \frac{r_2}{r_1}} \tag{9.49a}$$

Equations (949a) and (9.50a) are approximate equations to be used only when Eq. (9.49) or (9.50) cannot be used for lack of data. Equation (9.50a) over estimates the discharge by $[1/2\,(H/s_w - 1)]$ % when compared to Eq. (9.50).

Example 9.9 A 30 cm well completely penetrates an unconfined aquifer of saturated depth 40 m. After a long period of pumping at a steady rate of 1500 lpm, the drawdown in two observation wells 25 and 75 m from the pumping well were found to be 3.5 m and 2.0 m respectively. Determine the transmissibility of the aquifer. What is the drawdown at the pumping well?

Solution

(a) $$Q = \frac{1500 \times 10^{-3}}{60} = 0.025 \text{ m}^3/\text{s}$$

$h_2 = 40.0 - 2.0 = 38.0$ $\qquad r_2 = 75$ m

$h_1 = 40.0 - 3.5 = 36.5$ m $\qquad r_1 = 25$ m

From Eq. (9.49),

$$Q = \frac{\pi K (h_2^2 - h_1^2)}{\ln \frac{r_2}{r_1}}$$

$$0.025 = \frac{\pi K [(38)^2 - (36.5)^2]}{\ln \frac{75}{25}}$$

$K = 7.823 \times 10^{-5}$ m/s

$T = KH = 7.823 \times 10^{-5} \times 40 = 3.13 \times 10^{-3}$ m^2/s

(b) At the pumping well, $r_w = 0.15$ m

$$Q = \frac{\pi K (H_1^2 - h_w^2)}{\ln \frac{r_1}{r_w}}$$

$$0.025 = \frac{\pi \times 7.823 \times 10^{-5} [(36.5)^2 - h_w^2]}{\ln \frac{25}{0.15}}$$

$h_w^2 = 811.84$ and $h_w = 28.49$ m

Drawdown at the well, $s_w = 11.51$ m

9.10 *Open Wells*

Open wells (also known as *dug wells*) are extensively used for drinking-water supply in rural communities and in small farming operations. They are best suited for shallow and low-yielding aquifers. In hard rocks, the cross sections are circular or rectangular in shape. They are generally sunk to a depth of about 10 m and are lined wherever loose over burden is encountered. The flow into the well is through joints, fissures and such other openings and is usually at the bottom/lower portions of the well. In unconsolidated formations (e.g. alluvial soils) the wells are usually dug to a depth of about 10 m below water table, circular in cross section and lined. The water entry into these wells is from the bottom. These wells tap water in unconfined aquifers.

When the water in an open well is pumped out, the water level inside the well is lowered. The difference in the water-table elevation and the water level inside the well is known as *depression head*. The flow discharge into the well (Q) is proportional to the depression head (H), and is expressed as

$$Q = K_0 H \tag{9.51}$$

where the proportionality constant K_0 depends on the characteristic of the aquifer and the area of the well. Also, since K_0 represents discharge per unit drawdown it is called *specific capacity* of the well. There is a *critical depression head* for a well beyond which any higher depression head would cause dislodging of soil particles by the high flow velocities. The discharge corresponding to the critical head is called as *critical or maximum yield*. Allowing a factor of safety (normally 2.5 to 3.0) a *working head* is specified and the corresponding yield from the well is known as *safe yield*.

9.10.1 Recuperation Test

The specific capacity K_0 of a well is determined from the recuperation test described below.

Let the well be pumped at a constant rate Q till a drawdown H_1 is obtained. The pump is now stopped and the well is allowed to recuperate. The water depth in the well is measured at various time intervals t starting from the stopping of the well. Referring to Fig. 9.18,

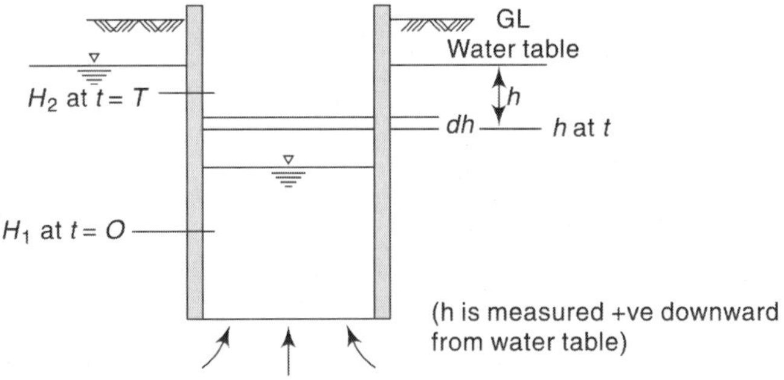

Fig. 9.18 ***Recuperation Test for Open well***

H_1 = drawdown at the start of recuperation, $t = 0$

H_2 = drawdown at a time, $t = T_r$

h = drawdown at any time t

Δh = decrease in drawdown on time Δt

At any time t, the flow into the well $Q = K_0 h$

In a time interval Δt causing a small change Δh in the water level,

$$Q \cdot \Delta t = K_0 h \cdot \Delta t = -A \cdot \Delta h$$

where A is the area of the well. In differential from

$$dt = -\frac{A}{K_0}\frac{dh}{h}$$

Integrating for a time interval T_r

$$\int_0^{T_r} dt = -\frac{A}{K_0}\int_{H_1}^{H_2}\frac{dh}{h}$$

$$T_r = \frac{A}{K_0}\ln\frac{H_1}{H_2} \quad (9.52)$$

or

$$\frac{K_0}{A} = \frac{1}{T_r}\ln\frac{H_1}{H_2} \quad (9.52a)$$

The term $\frac{K_0}{A} = K_s$ represents *specific capacity per unit well area* of the aquifer and is essentially a property of the aquifer. Knowing H_1, H_2 and the recuperation time T_r for reaching H_2 from H_1, and the specific capacity per unit well area is calculated by Eq. (9.52a).

Usually, the K_s of an aquifer, determined by recuperation tests on one or more wells, is used in designing further dug wells in that aquifer. However, when such information is not available the following approximate values of K_s, given by Marriot, are often used.

Type of sub-soil	Value of K_s in units of h^{-1}
Clay	0.25
Fine sand	0.50
Coarse sand	1.00

The yield Q from an open well under a depression head H is obtained as

$$Q = K_s AH \quad (9.5a)$$

For dug wells with masonry sidewalls, it is usual to assume the flow is entirely from the bottom and as such A in Eq. (9.51a) represents the bottom area of the well.

Example 9.10 During the recuperation test of a 4.0 m open well a recuperation of the depression head from 2.5 m to 1.25 m was found to take place in 90 minutes. Determine the (i) specific capacity per unit well area, and (ii) yield of the well for a safe drawdown of 2.5 m (iii) What would be the yield from a well of 5.0 m diameter for a drawdown of 2.25 m?

Solution

$$A = \frac{\pi}{4}\times(4.0)^2 = 12.566\ \text{m}^2$$

From Eq. (9.52a), $\frac{K_0}{A} = \frac{1}{T_r}\ln\frac{H_1}{H_2}$

Here $T_r = 90$ min $= 1.50$ h, $H_1 = 2.5$ m, and $H_2 = 1.25$ m

(i) $K_s = \frac{K_0}{A} = \frac{1}{1.5}\ln\frac{2.5}{1.25} = 0.462\ h^{-1}$

(ii) $Q = K_s \cdot A \cdot H = 0.462 \times 12.566 \times 2.5 = 14.52 \text{ m}^3\text{/h}$

(iii) $A_2 = \frac{\pi}{4} \times (5.0)^2 = 19.635$

$Q = K_s \times A_2 \times H_2 = 0.462 \times 19.635 \times 2.25 = 20.415 \text{ m}^3\text{/h}$

Example 9.11 Design the bottom diameter of a dug well in fine sand region to obtain 10 litres/s of yield under a depression head of 2.5 m. The value of K_s = Specific capacity per unit area may be taken as 0.6 h^{-1}.

Solution

Given : $Q = 0.01 \text{ m}^3\text{/s}$

$= (0.01 \times 60 \times 60) = 36 \text{ m}^3\text{/h}$; $H = 2.5$ m and $K_s = 0.6 \text{ h}^{-1}$

From Eq. (9.51a): $Q = K_s AH$

Hence, $A = \frac{Q}{K_s H} = \frac{36}{0.6 \times 2.5} = 24 \text{ m}^2$

Diameter at bottom of the well $= D = \sqrt{\frac{4A}{\pi}} = \sqrt{\frac{4 \times 24}{\pi}} = 5.53 \text{ m}$

9.11 *Unsteady Flow in a Confined Aquifer*

When a well in a confined aquifer starts discharging, the water from the aquifer is released resulting in the formation of a cone of depression of the piezometric surface. This cone gradually expands with time till an equilibrium is attained. The flow configuration from the start of pumping till the attainment of equilibrium is in unsteady regime and is described by Eq. (9.26).

In polar coordinates, Eq. (9.26), to represent the radial flow into a well, takes the form

$$\frac{\partial^2 h}{\partial r^2} + \frac{1}{r}\frac{\partial h}{\partial r} = \frac{S}{T}\frac{\partial h}{\partial t} \tag{9.53}$$

Making the same assumptions as used in the derivation of the equilibrium formula (Eq. 9.46), Theis (1935) obtained the solution of this equation as

$$s = (H - h) = \frac{Q}{4\pi T}\int_u^\infty \frac{e^{-u}}{u}\,du \tag{9.54}$$

where $s = H - h$ = drawdown at a point distance r from the pumping well, H = initial constant piezometric head, Q = constant rate of discharge, T = transmissibility of the aquifer, u = a parameter = $r^2 S/4Tt$, S = storage coefficient and t = time from start of pumping. The integral on he right-hand side is called the *well function, W* (u), and is given by

$$W(u) = \int_u^\infty \frac{e^{-u}}{u}\,du = -0.577216 - \ln u + u - \frac{u^2}{2.2!} + \frac{u^3}{3.3!} \cdots \tag{9.55}$$

The table of $W(u)$ are available in literature (e.g. Refs. 1, 9 and 10). Values of $W(u)$ can be easily calculated by the series (Eq. 9.55) to the required number of significant digits which rarely exceed 4. For small values of $u(u \le 0.01)$, only the first two terms of the series are adequate. Eq. (9.54) is popularly known as *Theis equation.*

The solution of Eq. (9.54) to find the drawdown s for a given S, T, r, t and Q can be obtained in a straightforward manner. However, the estimation of the aquifer constants S and T from the drawdown *vs* time data of a pumping well, which involve trial-and-error procedures, can be done either by a digital computer or by semigraphical methods such as the use of *Type curve*[1, 8, 9] or by *Chow's method* described in literature[1].

For small values of u ($u \le 0.01$), Jacob (1946, 1950) showed that the calculations can be considerably simplified by considering only the first two terms of the series of $W(u)$, (Eq. 9.55). This assumption leads Eq. (9.54) to be expressed as

$$S = \frac{Q}{4\pi T}\left[-0.5772 - \ln\frac{r^2 S}{4Tt}\right]$$

i.e.

$$s = \frac{Q}{4\pi T}\ln\left[\frac{2.25\,Tt}{r^2 S}\right] \tag{9.56}$$

If s_1 and s_2 are drawdowns at times t_1 and t_2,

$$(s_2 - s_1) = \frac{Q}{4\pi T}\ln\frac{t_2}{t_1} \tag{9.57}$$

If the drawdown s is plotted against time t on a semi-log paper, the plot will be a straight line for large values of time. The slope of this line enables the storage coefficient S to be determined. From Eq. (9.56), when $s = 0$, $t = t_0$ and

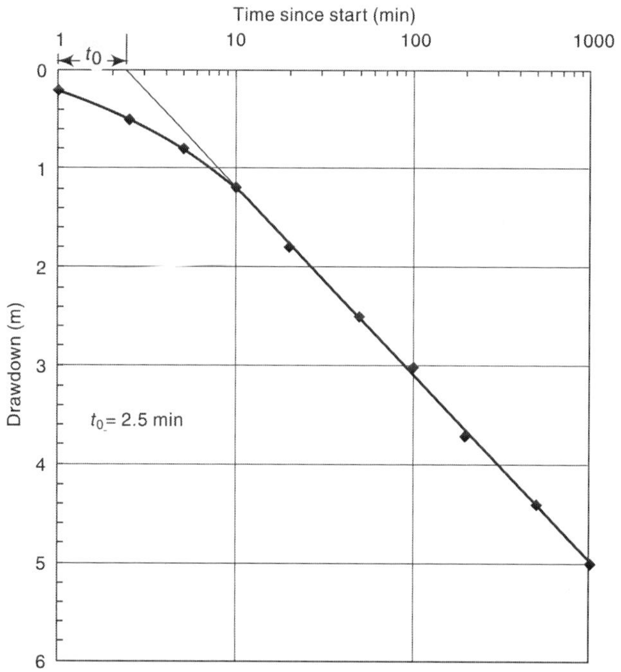

Fig. 9.19 ***Time–Drawdown Plot—Example 9.12***

$$\frac{2.25T\,t_0}{r^2 S} = 1$$

or
$$S = \frac{2.25T\,t_0}{r^2} \tag{9.58}$$

In practice t_0 = time corresponding to *zero* drawdown is obtained by extrapolating the straight-line portion of the semi-log curve of s *vs* t (Fig. 9.19). It is important to remember that the above approximate method proposed by Jacob assumes u to be very small. Example 9.11 describes the Jacob method of estimating aquifer parameters S and T from field data.

9.11.1 Drawdown Test

Equations (9.56 and 9.57) relating drawdown s with time t and aquifer properties is used to evaluate formation constants S and T through pumping test. The method is known as drawdown test.

Procedure

An observation well at a distance r from the production well is selected. The pumping is started and the discharge is maintained at a constant value (Q) throughout the test. Values of the drawdown s are read at the observation well at various times, t. The time intervals between successive readings could progressively increase to cut down on the number of observations. The pumping is continued till nearly steady-state conditions are reached. This may take about 12 to 36 hours depending on the aquifer characteristics. The best values of S and T are obtained from Eqs. (9.56) and (9.57) through semi-log plot of s against time t.

Example 9.12 A 30 cm well penetrating a confined aquifer is pumped at a rate of a 1200 lpm. The drawdown at an observation well at a radial distance of 30 m is as follows:

Time from start (min)	1.0	2.5	5	10	20	50	100	200	500	1000
Drawdown (m)	0.2	0.5	0.8	1.2	1.8	2.5	3.0	3.7	4.4	5.0

Calculate the aquifer parameters S and T.

Solution

The drawdown is plotted against time on a semilog plot (Fig. 9.19). It is seen that for $t > 10$ min. the drawdown values describe a straight line. A best-fitting straight line is drawn for data points with $t > 10$ min. From this line,

when $s = 0,\ t = t_0 = 2.5 \text{ min} = 150\ s$

$$s_1 = 3.1 \text{ m at } t_1 = 100 \text{ min}$$

$$s_2 = 5.0 \text{ m at } t_2 = 1000 \text{ min}$$

Also, $Q = 1200 \text{ lpm} = 0.02 \text{ m}^2/\text{s}$

From Eq. (9.57),

$$s_2 - s_1 = \frac{Q}{4\pi T} \ln \frac{t_2}{t_1}$$

$$(5.0 - 3.1) = \frac{0.02}{4 \times \pi \times T} \ln \frac{1000}{100}$$

$$T = \frac{0.02}{4\pi \times 1.9} \ln 10 = 1.929 \times 10^{-3} \text{ m}^3\text{/s/m} = 1.67 \times 10^5 \text{ lpd/m}$$

From Eq. (9.58),

$$S = \frac{2.25\,T\,t_0}{r^2} = \frac{2.25 \times 1.929 \times 10^{-3} \times 150}{(30)^2}$$

i.e. $S = 7.23 \times 10^{-4}$

Example 9.13 A well is located in a 25 m confined aquifer of permeability 30 m/day and storage coefficient 0.005. If the well is being pumped at the rate of 1750 lpm, calculate the drawdown at a distance of (a) 100 m, and (b) 50 m from the well after 20 h of pumping.

Solution

(a) $$T = K\,B = \frac{30}{86400} \times 25 = 8.68 \times 10^{-3} \text{ m}^2\text{/s}$$

$$u = \frac{r^2 S}{4\,T\,t} = \frac{(100)^2 \times (0.005)}{4 \times (8.68 \times 10^{-3}) \times (20 \times 60 \times 60)} = 0.02$$

Using the *Theis method* and calculating $W(u)$ to four significant digits,

$$W(u) = -0.5772 - \ln(0.02) + (0.02) - \frac{(0.02)^2}{2.2!} + \frac{(0.02)^3}{3.3!}$$

$$= -0.5772 + 3.9120 + 0.02 - 0.0001 + 4.4 \times 10^{-7} = 3.3547$$

$$S_{100} = \frac{Q}{4\pi T} W(u)$$

$$= \left(\frac{1.750}{60}\right) \times \frac{1}{4\pi(8.68 \times 10^{-3})} \times 3.3547 = 0.897 \text{ m}$$

(b) $$r = 50 \text{ m},\; u = \frac{(50)^2 \times (0.005)}{4 \times (8.68 \times 10^{-3}) \times (20 \times 60 \times 60)} = 0.005$$

$$W(u) = -0.5772 - \ln 0.005 + 0.005 = 4.726$$

$$S_{50} = \left(\frac{1.750}{60}\right) \times \frac{1}{4\pi(8.68 \times 10^{-3})} \times 4.726 = 1.264 \text{ m}$$

(Note that for small values of u, i.e. $u < 0.01$, $W(u) \approx -0.5772 - \ln u$).

9.11.2 Recovery of Piezometric Head

Consider a well pumped at constant rate of Q. Let s_1 be the draw-down at an observation well near the well in time t_1. If the pumping is stopped at the instant when the time is t_1, the groundwater flow into the cone of depression will continue at the same rate Q.

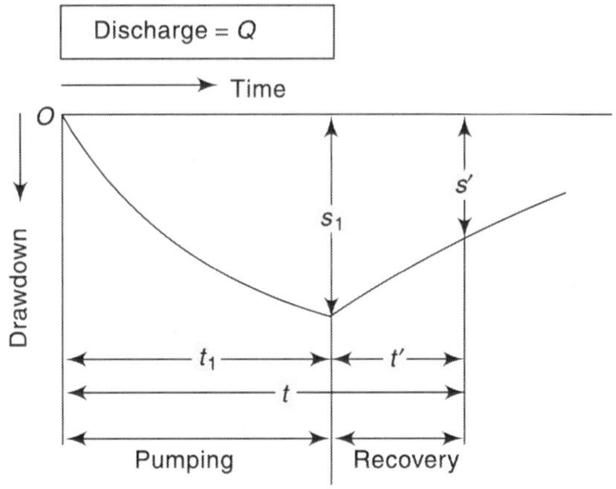

Fig. 9.20 ***Variation of Piezometric Head in Pumping and Recovery Head***

Since there is not withdrawal now, the water level in the observation well will begin to rise and the drawdown will begin to decrease. This is known as the *recovery of the cone of depression*. The variation of the water level with time during pumping and in the recovery phase is shown in Fig. 9.20.

The drawdown at the observation well at any time t' after the cessation of the pumping is kown as *residual drawdown* and can be calculated as

$$s' = \frac{Q}{4\pi T}[W(u) - W(u')] \tag{9.59}$$

where $u = \dfrac{r^2 S}{4\pi T t}$ and $u' = \dfrac{r^2 S}{4\pi T t'}$

$t = t_1 + t'$ = time from start of pump and t' = time since stoppage of pumping (start of recovery).

For small values of r and large values of t', Eq. (9.59) can be approximated as

$$s' = \frac{Q}{4\pi T}\cdot \ln\frac{t}{t'} = \frac{2.302}{4\pi}\cdot\frac{Q}{T}\cdot\log\frac{t}{t'} \tag{9.60}$$

The plot of residual drawdown s' vs (t/t') on semi-log paper represents a straight line with its slope as $\left[\dfrac{2.302Q}{4\pi T}\right]$.

9.11.3 Recovery Test

The relationship of the residual recovery given by Eq. (9.60) is used as a method of assessing the transmissibility T of the aquifer. The procedure is known as *Recovery test*. In this test, the pump is run at constant discharge rate for a sufficiently long time t_1, and then stopped. The value of t_1 depends on the type of aquifer and aquifer characteristics and may range from 12 to 24 hours. The recovery of water level s' in an observation well situated at a distance r from the production well is noted down at various times

(t'). In view of the logarithmic nature of the variation of residual drawdown with the time ratio (t/t'), the time intervals between successive readings could progressively increase. When observation wells are not available, the recovery water levels can be observed in the production well itself and this is a positive advantage of this test.

The value of *transmissibility* T is calculated from plot of s' against (t/t') on semi-log axes.

It is to be noted that the recovery test data does not enable the determination of the storage coefficient S.

Example 9.14 Recovery test on a well in a confined aquifer yielded the following data: Pumping was at a uniform rate of 1200 m^2/day and was stopped after 210 minutes of pumping. Recovery data was as shown below:

Time since stoppage of pump (min)	2	5	10	20	40	90	150	210
Residual drawdown (m)	0.70	0.55	0.45	0.30	0.25	0.19	0.15	0.10

Estimate the transmissibility of the aquifer.

Solution

Here since $t_1 = 210$ min, $t = t_1 + t' = 210 + t'$

The time ratio t/t' is calculated (as shown in the table below) and a semi-log plot of s' *vs* t/t' is plotted (Fig. 9.21).

t'	2	5	10	20	40	90	150	210
t	212	215	220	240	250	300	360	420
t/t'	106	43	22	11.5	6.25	3.33	2.40	2.0
s'	0.70	0.55	0.45	0.30	0.25	0.19	0.15	0.10

A best fitting straight line through the plotted points is given by the equation $s' = 0.1461 \ln (t/t') - 0.0027$

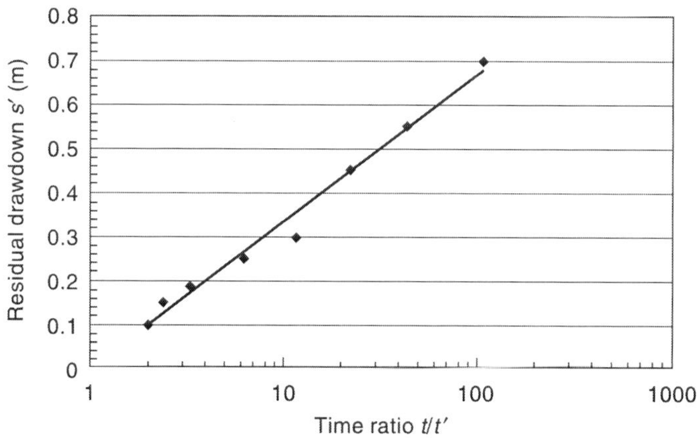

Fig. 9.21 ***Plot of residual drawdown against time ratio (t/t′)—Example 9.12***

By Eq. (9.60), Slope of best fit line = 0.1461 = $\frac{Q}{4\pi T}$

$$T = \frac{1200}{0.1461 \times 4\pi} = 654 \text{ m}^2\text{/day}$$

9.12 *Well Loss*

In a pumping artesian well, the total draw-down at the well s_w, can be considered to be made up of three parts:

1. Head drop required to cause laminar porous media flow, called *formation loss*, s_{wL} (Fig. 9.22);
2. Drop of piezometric head required to sustain turbulent flow in the region nearest to the well where the Reynolds number may be larger than unity, s_{wt}; and
3. Head loss through the well screen and casing, s_{wc}.

Of these three,

$$s_{wL} \propto Q \qquad \text{and} \qquad (s_{wt} \text{ and } s_{wc}) \propto Q^2$$

thus
$$s_w = C_1 Q + C_2 Q^2 \tag{9.61}$$

where C_1 and C_2 are constants for the given well (Fig. 9.21). While the first term C_1Q is the formation loss, the second terms C_2Q^2 is termed *well loss.*

The magnitude of a well loss has an important bearing on the pump efficiency. Abnormally high value of well loss indicates clogging of well screens, etc. and requires immediate remedial action. The coefficients C_1 and C_2 are determined by pump test data of drawdown for various discharges.

Well efficiency

The well loss is a measure of the efficiency of the well that is defined as

$$\text{Well Efficiency} = \eta_w = \frac{\text{Formation loss}}{\text{Total drawdown measured inside the well}}$$

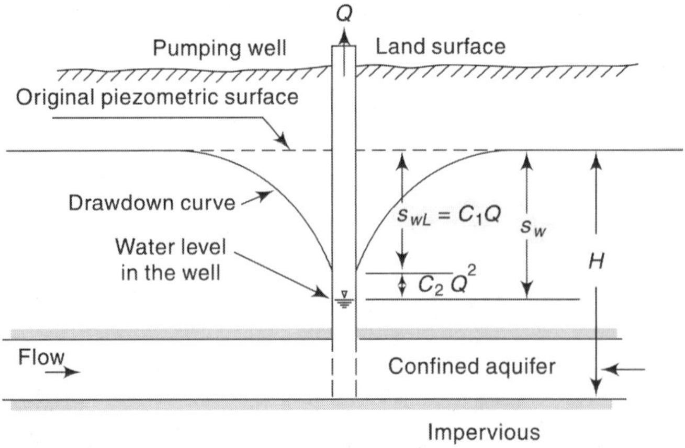

Fig. 9.22 ***Definition Sketch for Well Loss***

$$\eta_w = \frac{S_{wL}}{(S_{wL} + S_{wc} + S_{wt})} = \frac{C_1 Q}{(C_1 Q + C_2 Q^2)} = 1 - \frac{C_2 Q^2}{(C_1 Q + C_2 Q^2)} \quad (9.62)$$

9.13 *Specific Capacity*

The discharge per unit drawdown at the well (Q/s_w) is known as *specific capacity* of a well and is a measure of the performance of the well. For a well in a confined aquifer under equilibrium conditions and neglecting well losses, by Eq. (9.48),

$$\frac{Q}{s_w} = \frac{2\pi}{\ln R / r_w} T \text{ i.e. } \frac{Q}{s_w} \propto T$$

However, for common case of a well discharge in at a constant rate Q under unsteady drawdown conditions, the specific capacity is given by

$$\frac{Q}{s_w} = \frac{1}{\frac{1}{4\pi T}\ln\frac{2.25\,T\,t}{r_w^2 \cdot S} + C_2 Q} \quad (9.63)$$

where t = time after the start of pumping. The term C_2Q is to account for well loss. It can be seen that the specific capacity depends upon T, s, t, r_w and Q. Further, for a given well it is not a constant, but decreases with increases in Q and t.

9.14 *Sea-Water Intrusion*

In coastal areas, there is normally a hydraulic gradient in the water table towards the sea and the fresh groundwater drains into the sea. Since the sea water is heavier than the fresh water, in unconfined groundwater conditions, a salt-water wedge forms at the exit of the fresh water. Figure 9.23 shows an idealised exit of fresh groundwater in to the sea. Assuming the movement of groundwater to be horizontal and the interface of the salt water and freshwater to be sharp and without any diffusion, an expression for the position of the interface relative to the water table is derived as below.

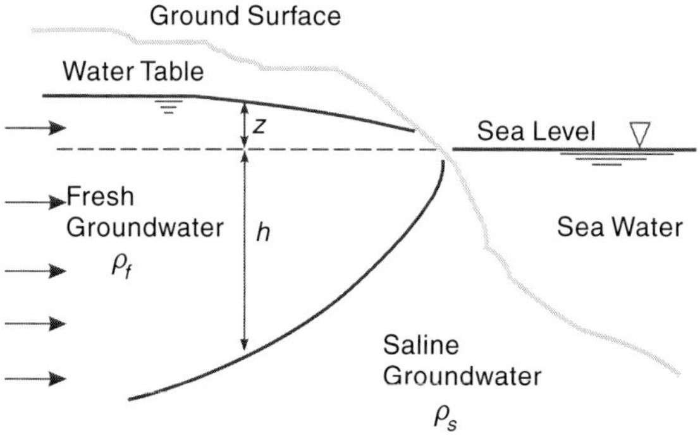

Fig. 9.23 ***Interface Between Fresh and Saline Groundwater in a Coastal Unconfined Aquifer***

The interface represents a surface that is in equilibrium with the hydrostatic pressure forces acting on either side of it. At any vertical section, the pressure force on the top of the interface due to a column of freshwater of unit cross-sectional area is balanced by the pressure force acting on unit area at the bottom of the interface. Thus, from Fig. 9.23,

$$\rho_f (h + z) = \rho_s h$$

where ρ_f and ρ_s are densities of fresh water and salt water respectively. Thus, the depth of interface below sea level, h, is given by

$$h = \left(\frac{\rho_f}{\rho_s - \rho_f} \right) z \qquad (9.64)$$

Taking the commonly used values of ρ_s = 1025 kgf/m^3 and ρ_f = 1000 kgf/m^3, Eq. (9.64) reduces to

$$h = 40z \qquad (9.64\text{-a})$$

Thus, for every metre of the water table above sea level, the fresh water will extend 40 m below the sea level. It also means that in coastal areas, if due to bad groundwater management, the water table drops by 1 metre, it means the salt water at the interface would have risen by 40 metres. Conversely, if due to recharge and good groundwater management the water table rises by 1 m, it implies that the salt water interface has been pushed down by 40 m.

The wedge-shaped freshwater body in an unconfined aquifer near the sea is called *Ghyben–Herzberg Lens,* after two scientists who first derived the relationship of Eq. (9.64) independently. Eq. (9.64) is derived for ideal conditions and is of indicative in nature. In practice, the interface is not a sharp one as assumed but is a diffusion layer where the concentration of salt gradually changes. Tidal conditions and disturbances of discharge conditions cause the interface to fluctuate. In practical situations, the locations of the interface has to be obtained by solving the relevant dynamic groundwater flow equations numerically by using proper boundary conditions.

9.15 *Recharge*

Addition of surface water to zone of saturation is known as *recharge*. Recharge taking place naturally as a part of hydrologic cycle is called *natural recharge* while the process of increasing infiltration of surface water to groundwater systems by altering natural conditions is known as *artificial recharge*.

9.15.1 Natural Recharge

The amount of precipitation that infiltrates into the soil and reaches the zone of saturation is an important component of natural recharge. Seepage from irrigated lands is another important component of recharge of groundwater. In this process, the infiltration phase is natural while the supply of water to the irrigated lands is through artificial means and as such it is sometimes called *incidental recharge*. Other means of natural recharge are seepage from reservoirs; rivers, streams and canals; and other water bodies. Estimation of recharge rates of aquifers is an important component of groundwater resource estimation and in proper utilisation of groundwater.

9.15.2 Artificial Recharge

The process of artificially enhancing the amount of water recharging the aquifer in a given location is known as artificial recharge. In the face of present-day large demands for groundwater, artificial recharge is resorted to

- Conserve runoff
- Improve quantity of available groundwater
- Reduce or correct saltwater intrusion

Various recharging methods commonly adopted are

- Spreading [Flooding, Basin, Ditch, Pit & Channel]
- Through injection wells
- Induced recharge from surface water bodies
- Subsurface dykes
- Percolation tanks, Check dams, Nala bunds and other watershed treatment methods

9.15.3 Estimation of Recharge

Groundwater Resource Estimation Committee[11] (*GEC*-97) recommends two approaches to assessment of recharge to groundwater. These are (i) Groundwater level fluctuation method, and (ii) Rainfall infiltration factor method.

1. Groundwater Level Fluctuation and Specific Yield Method

In this method, the groundwater level fluctuations over a period (usually a monsoon season) is used along with the specific yield to calculate the increase in storage in the water balance equation. Thus, for a given area of extent A (usually a watershed), for a water level fluctuation of h during a monsoon season,

$$h\, S_y A = R_G - D_G - B + I_s + I \qquad (9.65)$$

where
S_y = specific yield
R_G = gross recharge due to rainfall and other sources
D_G = gross water draft
B = base flow into the stream from the area
I_s = recharge from the stream into the ground water body
I = net ground water flow into the area across the boundary (i.e. inflow – outflow)

Writing $R = R_G - B + I + I_s$

Equation (9.65) would be $R = h\, S_y A - D_G$ (9.66)

where R = possible recharge, which is gross recharge minus the natural recharge of the area, and would consist of other recharge factors as

$$R = R_{rf} + R_{gw} + R_{wi} + R_t \qquad (9.67)$$

where
R_{rf} = recharge from rainfall
R_{gw} = recharge from irrigation in the area (includes both surface and ground water sources)
R_{wi} = recharge from water conservation structures
R_t = recharge from tanks and ponds

Computations, using Eq. (9.65) through (9.67) are usually based on the monsoon season rainfall and corresponding groundwater fluctuation covering a span of 30 to 50 years to obtain normal monsoon recharge due to rainfall. The recharge in non-monsoon months is taken as zero if rainfall in non-monsoon months is less than 10% of normal annual rainfall. The computation for calculating the total annual recharge is carried out for both monsoon months and non-monsoon months and the total annual recharge is obtained as a sum of these two.

The specific yield for various hydrogeologic conditions in the country is estimated through norms given in Table 9.5.

2. Rainfall Infiltration Method

In areas where groundwater level monitoring is not adequate in space and time, rainfall infiltration method may be used. The recharge from rainfall in monsoon season is taken as a percentage of normal monsoon rainfall in the area.

Thus $$R_{rf} = f A P_{nm} \tag{9.68}$$

where R_{rf} = recharge from rainfall in monsoon season

f = rainfall infiltration factor

P_{nm} = normal rainfall in monsoon season

A = area of computation for recharge

Table 9.5 ***Norms for Specific Yield (S_y in percentage)***

No.	Description of the area	Recommended value	Minimum value	Maximum value
1	**Alluvial areas**			
	Sandy alluvium	16	12	20
	Silty alluvium	10	8	12
	Clayey alluvium	6	4	8
2	**Hard Rock Areas**			
	Weathered granite, gneiss and schist			
	• with low clay content	3.0	2.0	4.0
	• with significant clay content	1.5	1.0	2.0
	Weathered or vesicular, jointed basalt	2.0	1.0	3.0
	Laterite	2.5	2.0	3.0
	Sandstone	3.0	1.0	5.0
	Quartzite	1.5	1.0	2.0
	Limestone	2.0	1.0	3.0
	Karstified limestone	8.0	5.0	15.0
	Phyllites, Shales	1.5	1.0	2.0
	Massive, poorly fractured rock	0.3	0.2	0.5

The norms for the rainfall factor *f* for various hydrogeological situations in the country are given in the following table.

Table 9.6 ***Norms for Selection of Rainfall Factor f***

No.	Area	Value of *f* in percentage		
		Recommended value	Minimum value	Maximum value
1	**Alluvial areas**			
	• Indo-Gangetic and inland areas	22	20	25
	• East coast	16	20	18
	• West coast	10	8	12
2	**Hard Rock Areas**			
	• Weathered granite, gneiss and schist with low clay content	11	10	12
	• Weathered granite, gneiss and schist with significant clay content	8	5	9
	• Granulite facies like charnokite etc.	5	4	6
	• Vesicular and jointed basalt	8	5	9
	• Weathered basalt	5	4	6
	• Laterite	13	12	14
	• Semi-consolidated sand stone	7	6	8
	• Consolidated sand stone, quartzite, limestone (except cavernous limestone)	7	6	8
	• Phyllites, Shales	12	10	14
	• Massive poorly fractured rocks	6	5	7

The same factors are used for non-monsoon months also with the condition that the recharge is taken as zero if the normal rainfall in non-monsoon season is less than 10% of normal annual rainfall.

Recharge from sources other than rainfall are also estimated by using appropriate factors (e.g., Tables 9.7, 9.8 and 9.9). The total recharge is obtained as the sum of recharge from rainfall and recharge from other sources.

3. Recharge due to Seepage from Canals

When actual specific values are not available, the following norms given in Table 9.7 may be adopted.

4. Recharge from Irrigation

The recharge due to flow from irrigation may be estimated, based on the source of irrigation (groundwater or surface water), the type of crop and the depth of water table below ground level through the use of norms given in Table 9.8.

Table 9.7 ***Recharge due to Seepage from Canals***

1	Unlined canals in sandy soils with some silt content	• 1.8 to 2.5 cumec/million sq.m of wetted area *or* • 15–20 ha. m/day/million sq. m of wetted area *or*
2	Unlined canals in normal soils with some silt content	• 3 to 3.5 cumec/million sq.m of wetted area *or* • 25–30 ha. m/day/million sq. m of wetted area
3	Lines canals and canals in hard rock areas	• 20% of above values for unlined canals

Table 9.8 ***Recharge from Irrigation***

Source of Irrigation	Type of Crop	Recharge as percentage application		
		Water table below Ground level		
		< 10 m	10–25 m	> 25 m
Groundwater	Non-paddy	25	15	5
Surface water	Non-paddy	30	20	10
Groundwater	Paddy	45	35	20
Surface water	Paddy	50	40	25

5. Recharge from Water-Harvesting Structures

The following norms (Table 9.9) are commonly followed in the estimation of recharge from water-harvesting structures:

Table 9.9 ***Recharge factors for Tanks and other Water Harvesting Structures***

S. No.	Structure	Recharge Factor
1	Recharge from Storage Tanks/ Ponds	1.4 mm/day for the period in which the tank has water, based on the average area of water spread. If the data on average water spread is not available, 60% of the maximum water spread area may be used.
2	Recharge from Percolation Tanks	50% of gross storage, considering the number of fillings. Half this value of recharge is assumed to be occurring during monsoon season.
3	Recharge due to Check Dams and Nala Bunds.	50% of gross storage. Half the value of recharge is assumed to be occurring during monsoon season.

Detailed procedure for the estimation of Groundwater resources of an area under Indian conditions available in Ref. 11.

9.16 Groundwater Resource

The quantum of groundwater available in a basin is dependent on the inflows and discharges at various points. The interrelationship between inflows, outflows and accumulation is expressed by the water-budget equation

$$\Sigma I \Delta t - \Sigma Q \Delta t = \Delta S \tag{9.69}$$

where $\Sigma I \Delta t$ represents all forms of recharge and includes contribution by precipitation; infiltration from lakes, streams and canals; and artificial recharge, if any, in the basin.

$\Sigma Q \Delta t$ represents the net discharge of groundwater from the basin and includes pumping, surface outflows, seepage into lakes and rivers and evapotrans-piration. ΔS indicates the change in the groundwater storage in the basin over a time Δt.

Considering a sufficiently long time interval, Δt of the order of a year, the capability of the groundwater storage to yield the desired demand and its consequences can be estimated. It is obvious that too large a withdrawal than what can be replenished naturally leads ultimately to the permanent lowering of the groundwater table. This in turn leads to problems such as drying up of open wells and surface storages like swamps and ponds and change in the characteristics of vegetation that can be supported by the basin. Similarly, too much of recharge and scanty withdrawal or drainage leads to waterlogging and consequent decrease in the productivity of lands.

The maximum rate at which the withdrawal of groundwater in a basin can be carried without producing undesirable results is termed *safe yield*. This is a general term whose implication depends on the desired objective. The "undesirable" results include (i) permanent lowering of the groundwater table or piezometric head, (ii) maximum drawdown exceeding a preset limit leading to inefficient operation of wells, and (iii) salt-water encroachment in a coastal aquifer. Depending upon what undesirable effect is to be avoided, a safe yield for a basin can be identified.

The permanent withdrawal of groundwater from storage is known as *mining* as it connotes a depletion of a resource in a manner similar to the exploitation of mineral resource.

The total groundwater resource of a region can be visualised as being made up of two components: *Dynamic resource* and *Static resource*. The dynamic resource represents the safe yield, which is essentially the annual recharge less the unavoidable natural discharge. The static resource is the groundwater storage available in the pores of the aquifer and its exploitation by mining leads to permanent depletion. Generally, the static resource is many times larger than the dynamic resource. However, mining is resorted to only in case of emergencies such as droughts, etc. and in exceptional cases of planned water resources development. In essence, it is a resource to be used in emergency. As such, the utilisable groundwater resource of a region is the safe yield at a given state of development. It is often said, in a general sense, that the water resource is a replenishable resource. So far as the groundwater resource is concerned, only the dynamic component is replenishable and the static component is non-replenishable.

The annual utilisable groundwater resource of a region is computed by using the water budget method. The total annual recharge is made up of

- rainfall recharge
- seepage from canals
- deep percolation from irrigated areas

- inflow from influent streams, etc.
- recharge from tanks, lakes, submerged lands, and
- artificial recharge schemes, if any

The groundwater losses from aquifers occur due to

- outflow to rivers
- transpiration by trees and other vegetation
- evaporation from the water table

The difference between the enumerated annual recharges and the losses as above is the annual groundwater resource which is available for irrigation, domestic and industrial uses. It should be noted that with the growth in the expansion of canal irrigation the groundwater resources also grow. Further, increase in the recharge of surface waters through artificial methods would also enhance the groundwater resources.

The National water policy (2002) stipulates:

- Exploitation of the groundwater resources should be so regulated as not to exceed the recharging possibilities, as also to ensure social equity. Groundwater recharge projects should be developed and implemented for augmenting the available supplies.
- There should be periodical reassessment on a scientific basis of the groundwater resources taking into consideration the quality of the water available and economic viability.

9.16.1 Categories of Groundwater Development

The groundwater development in areas are categorised as safe, semi-critical and critical based on the stage of groundwater development and long-term trend of pre and post-monsoon groundwater levels.

Category	% of groundwater development	Long term decline of pre & post-monsoon groundwater levels
Safe	< 70%	Not significant
Semi-critical	70% to 90%	Significant
Critical	90% to 100%	Significant
Over-exploited	> 100%	Significant

9.16.2 Groundwater Resources of India

The National Commission on Agriculture (1976) estimated the groundwater resources at 350 km^3, of which 260 km^3 was available for irrigation. The groundwater over exploitation committee (1979) estimated the groundwater potential as 467.9 km^3. The Groundwater estimation committee (1984) suggested a suitable methodology for estimation of groundwater. Using these norms, *CGWB* (1955) has estimated the total replenishable groundwater potential of the country (dynamic) at 431.89 km^3. The statewise and basinwise estimates of dynamic groundwater (fresh) resource made by *CGWB* (1995) are given in Tables 9.10 and 9.11.

Table 9.10 ***State-wise Dynamic Fresh Groundwater Resource***

S. No.	States	Total Replenishable Groundwater Resource from Normal Natural Recharge	Total Replenishable Groundwater Resource due to Recharge Augmentation from Canal Irrigation	Total Replenishable Groundwater Resource
		Km^3 per year	Km^3 per year	Km^3 per year
1	Andhra Pradesh	20.03	15.26	35.29
2	Arunachal Pradesh	1.44	0.00	1.44
3	Assam	24.23	0.49	24.72
4	Bihar	28.31	5.21	33.52
5	Goa	0.18	0.03	0.21
6	Gujarat	16.38	4.00	20.38
7	Haryana	4.73	3.80	8.53
8	Himachal Pradesh	0.29	0.28	.037
9	Jammu and Kashmir	2.43	2.00	4.43
10	Karnataka	14.18	2.01	16.19
11	Kerala	6.63	1.27	7.90
12	Madhya Pradesh	45.29	5.60	50.89
13	Maharashtra	33.40	4.47	37.87
14	Manipur	3.15	0.00	3.15
15	Meghalaya	0.54	0.00	0.54
16	Mizoram		Not Assessed	
17	Nagaland	0.72	0.00	0.72
18	Orissa	16.49	3.52	20.01
19	Punjab	9.47	9.19	18.66
20	Rajasthan	10.98	1.72	12.70
21	Sikkim		Not Assessed	
22	Tamil Nadu	18.91	7.48	26.39
23	Tripura	0.57	0.10	0.67
24	Uttar Pradesh	63.43	20.39	83.82
25	West Bengal	20.30	2.79	23.09
26	Union Territories	0.35	0.05	0.40
	Total	**342.43**	**89.46**	**431.89**

[*Source: Ref. 12*]

Table 9.11 *Basinwise Dynamic Fresh Groundwater Resource*[12] *(Unit: km^3/year)*

S. No.	River Basin	Total Replenishable Groundwater Resource from Normal Natural Recharge	Total Replenishable Groundwater Resource due to Recharge Aug-mentation from Canal Irrigation	Total Repleni-shable Ground-water Resource
1	Indus	14.29	12.21	26.50
2	Ganga-Brahmaputra-Meghna Basin			
	2a Ganga sub-basin	136.47	35.10	171.57
	2b Brahmaputra sub-basin and	25.72	0.83	26.55
	2c Meghna (Barak) sub-basin	8.52	0.00	8.52
3	Subarnarekha	1.68	0.12	1.80
4	Brahmani-Baitarani	3.35	0.70	4.05
5	Mahanadi	13.64	2.86	16.50
6	Godavari	33.48	7.12	40.60
7	Krishna	19.88	6.52	26.40
8	Pennar	4.04	0.89	4.93
9	Cauvery	8.79	3.51	12.30
10	Tapi	6.67	1.6	8.27
11	Narmada	9.38	1.42	10.80
12	Mahi	3.50	0.50	4.00
13	Sabarmati	2.90	0.30	3.20
14	West flowing rivers of Kutchch and Saurashtra	9.10	2.10	11.20
15	West flowing rivers south of Tapi	17.70	2.15	15.55
16	West flowing rivers between Mahanadi and Godavari }			
17	East flowing rivers between Godavari and Krishna }	[12.82]	[5.98]	[18.80]
18	East flowing rivers between Krishna and Pennar }			
19	East flowing rivers between Pennar and Cauvery }	[18.20]	[5.55]	[12.65]
20	East flowing rivers south of Cauvery }			
21	Area North of Ladakh no draining into India		Not Assessed	
22	Rivers draining into Bangladesh		Not Assessed	
23	Rivers draining into Myanmar		Not Assessed	
24	Drainage areas of Andaman, Nicobar and Lakshadweep islands		Not Assessed	
	Total	**342.43**	**89.46**	**431.89**

It is seen from table 9.10 and 9.11 though the rainfall is the principal source of recharge, the contribution of canal seepage and irrigation return flows has also been significant in some states. The contributions are more than 40% in states of Punjab, Haryana, J&K and Andhra Pradesh.

Groundwater extraction by individuals, organisations and local bodies has become a common phenomenon. Consequently, there is considerable development and utilisation of available groundwater sources. Further, the level of extraction has reached critical or over-exploitation level in several pockets in many states. Table 9.12 gives a list of states with considerable exploitation of groundwater resources.

Table 9.12 ***States with more than 40% Groundwater Development***[12]

State	Percent area over-exploited	Level of Groundwater development
Punjab	52	94
Haryana	42	84
Rajasthan	19	51
Tamil Nadu	14	61
Gujarat	6	42

9.16.3 Utilisable Groundwater Resources

As seen from tables 9.10 or 9.11, the sum total of potential for natural recharge from rainfall and due to recharge augmentation from canal irrigation system in the country 431.9 km^3/year. The utilisable groundwater potential is calculated as 396 km^3/year as below:

1. Total replenishable groundwater potential = 432 km^3/year
2. Provision for domestic drinking water and other uses @ 15% of item 1 = 71 km^3/year
3. Utilisable groundwater resource for irrigation @ 90% of (item 1 – item 2) = 325 km^3/year
4. Total utilisable dynamic groundwater resource (Sum of items 2 and 3) = 396 km^3/year

9.17 *Groundwater Monitoring Network in India*

The Central Groundwater Board monitors the groundwater levels from a network of about 15000 stations (mostly dug wells selected from existing dug wells evenly distributed throughout in the country). Dug wells are being replaced by piezometers for water-level monitoring. Measurements of water levels are taken at these stations four times in a year in the months of January, April/May, August and November. The groundwater samples are also collected during April/May measurements for chemicals analyses every year.

The data so generated are used to prepare maps of groundwater level depths, water level contours and changes in water levels during different periods and years. Deeper groundwater level of over 50 metres is observed in Peidmont aquifer in Bhabar belt in foot hills of Himalayas. In Western Rajasthan ground water levels have depths ranging from to 100 metres. In peninsular region, water levels range from 5 to 20 metres below land surface.

REFERENCES

1. Bauwer, H., *Groundwater Hydrology.* McGraw-Hill' Kogakusha, Tokyo, 1978.
2. Bear, J., *Hydraulics of Groundwater,* Dover Pub., 2007.
3. Central Water Commission, *Water Resources of India, CWC* Pub. No. 30/88, April 1988, CWC, New Delhi, India.
4. Davis, S.N. and R.J. M. DeWiest, *Hydrogeology,* John Wiley and Sons; New York, 1966.
5. Halek, V. and J. Svec, *Groundwater Hydraulics,* Elsevier, Amsterdam, 1979.
6. Marino, M.A. and J.N. Luthin, *Seepage and Groundwater,* Elsevier; Amsterdam, 1982.
7. Polubarinova Kochina, P.Y., *Theory of Groundwater Movement,* English Trans. by R.J.M. Dewiest, Princeton Univ. Press, Princeton, 1962.
8. Taylor, D.W., *Fundamentals of Soil Mechanics,* John Wiley and Sons, New York, 1967.
9. Todd, D.K., *Groundwater Hydrology,* John Wiley and Sons, New York, 2006,
10. Walton, W.C., *Groundwater Resource Evaluation,* McGraw-Hill, Kogakusha, Tokyo, 1970.
11. Ministry of Water Resources, *GOI, Report of the Groundwater Resources Estimation Committee*, Groundwater Resource Estimation Methodology, New Delhi, June 1997.
12. Min. of Water Resources, *GOI, Report of the National Commission for Integrated Water Resources Development*, Vol. 1, New Delhi, Sept. 1999.
13. Karanth, K.R., *Hydrogeology*, Tata McGraw-Hill Pub. Co., New Delhi, India, 1989.
14. Karanth, K.R., *Ground Water Assessment, Development and Management*, Tata McGraw-Hill Pub. Co., New Delhi, India, 1992.

REVISION QUESTIONS

9.1 Explain briefly the following terms as used in groundwater flow studies:
(a) Specific yield (b) Storage coefficient
(c) Specific capacity (d) Barometric efficiency

9.2 Distinguish between
(a) Aquifer and aquitard (b) Unconfined aquifer and a leaky aquifer
(c) Influent and effluent streams (d) Water table and piezometric surface
(e) Specific capacity of a well and the specific yield of an aquifer

9.3 Explain the following:
(a) Perched water table (b) Intrinsic permeability
(c) Bulk pore velocity (d) Well loss (e) Recharge

9.4 Discuss the geological formations in India which have potential as aquifers.

9.5 Explain the behaviour of water level in wells in confined aquifers due to changes in the atmospheric pressure.

9.6 Develop the equation relating the steady-state discharge from a well in an unconfined aquifer and depths of water table at two known positions from the well. State clearly all the assumptions involved in your derivation.

9.7 What are Dupit's assumptions? Starting from an elementary prism of fluid bounded by a water table, show that for the steady one-dimensional unconfined groundwater flow with a recharge rate R, the basic differential equation is

$$\frac{\partial^2 h}{\partial x^2} = -\frac{2R}{K}$$

where K = permeability of the porous medium.

9.8 Sketch a typical infiltration gallery. Calculate the discharge per unit length of the infiltration gallery by making suitable assumptions. State clearly the assumptions made.

9.9 Derive the basic differential equation of unsteady groundwater flow in a confined aquifer. State clearly the assumptions involved.

9.10 Describe a procedure by using Jacob's method to calculate the aquifer parameters of a confined aquifer by using the well pumping test data.

9.11 Describe the recovery test to estimate the transmissibility of a confined aquifer.

9.12 The aquifer properties S and T of a confined aquifer in which a well is driven are known. Explain a procedure to calculate the drawdown at a location away from the well at any instant of time after the pump has started.

9.13 Explain briefly the formation of sea-water interface in a coastal unconfined aquifer.

9.14 Derive the equation for Ghyben-Herzberg Lens at the saline water interface in a coastal aquifer.

9.15 Explain briefly

(a) Safe yield of an aquifer (b) Mining of water

(c) Recharge estimation (d) Groundwater estimation

9.16 Discuss the principle of recuperation test of an open well.

9.17 What are the commonly used methods to assess the recharge of groundwater in an area? Explain briefly any one of the methods.

9.18 Describe the groundwater resources of india and its utilization.

PROBLEMS

9.1 In a laboratory test of an aquifer material, a fully saturated sample of 5 litres volume was taken and its initial weight of 105 N was recorded. When allowed to drain completely it recorded a weight of 97 N. The sample was then crushed, dried and then weighed. A weight of 93 N was recorded at this stage. Calculate the specific yield and relative density of the solids. (Assume unit weight of water $\gamma = 9.79\ \text{kN/m}^3$)

9.2 A confined aquifer is 25 m thick and 2 km wide. Two observation wells located 2 km apart in the direction of flow indicate heads of 45 and 39.5 m. If the coefficient of permeability of the aquifer is 30 m/day, calculate (a) the total daily flow through the aquifer, and (b) the piezometric head at an observation well located 300 m from the upstream well.

9.3 In a field test, a time of 6 h was required for a tracer to travel between two observation wells 42 m apart. If the difference in water-table elevations in these wells were 0.85 m and the porosity of the aquifer is 20% calculate the coefficient of permeability of the aquifer.

9.4 A confined aquifer has a thickness of 30 m and a porosity of 32%. If the bulk modulus of elasticity of water and the formation material are 2.2×10^5 and 7800 N/cm^2 respectively, calculate (a) the storage coefficient, and (b) the barometric efficiency of the aquifer.

9.5 An extensive aquifer is known to have a groundwater flow in N 30° E direction. Three wells A, B and C are drilled to tap this aquifer. The well B is to East of A and the well C is to North of A. The following are the data regarding these wells:

Distance (m)	Well	Ground surface elevation (m above datum)	Water table elevation (m above datum)
	A	160.00	157.00
AB = 800 m	*B*	159.00	156.50
AC = 2000 m	*C*	158.00	?

Estimate the elevation of water table at well *C* when the wells are not pumping.

9.6 A confined stratified aquifer has a total thickness of 12 m and is made up of three layers. The bottom layer has a coefficient of permeability of 30 m/day and a thickness of 5.0 m. The middle and top layers have permeability of 20 m/day and 45 m/day respectively and are of equal thickness. Calculate the transmissivity of the confined aquifer and the equivalent permeability, if the flow is along the stratification.

9.7 A pipe of 1.2 m diameter was provided in a reservoir to act as an outlet. Due to disuse, it was buried and completely clogged up for some length by sediment. Measurements indicated the presence of fine sand (K_1 = 10 m/day) deposit for a length of 100 m, at the upstream end and of coarse sand (K_2 = 50 m/day) at the downstream end for a length of 50 m. In between these two layers the presence of silty sand (K_3 = 0.10 m/day) for some length is identified. For a head difference of 20 m on either side of the clogged length, the seepage discharge is found to be 0.8 m^3/day. Estimate the length of the pipe filled up by silty sand.

9.8 A confined horizontal aquifer of 15 m thickness and permeability *K* = 20 m/day, connects two reservoirs *M* and *N* situated 1.5 km apart. The elevations of the water surface in reservoirs *M* and *N* measured from the top of the aquifer, are 30.00 m and 10.00 m respectively. If the reservoir *M* is polluted by a contaminant suddenly, how long will it take the contaminant to reach the reservoir *N*? Assume the porosity of the aquifer *n* = 0.30.

9.9 An Infiltration gallery taps an unconfined aquifer (*K* = 50 m/day) situated over a horizontal impervious bed (Fig. 9.24). For the flow conditions shown, estimate the discharge collected per unit length of the gallery.

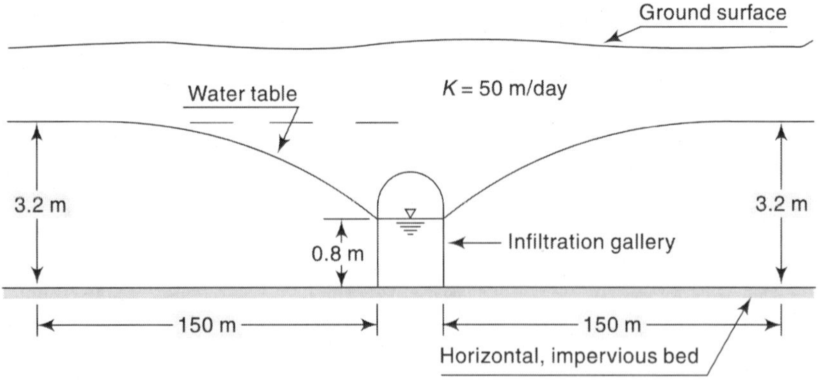

Fig. 9.24 ***Schematic Layout of Problem 9.9***

9.10 An unconfined aquifer has infiltration of irrigation water at a uniform rate *R* at the ground surface. Two open ditches, as show in Fig. 9.25, keep the water table in equilibrium. Show that the spacing *L* of the drains is related as

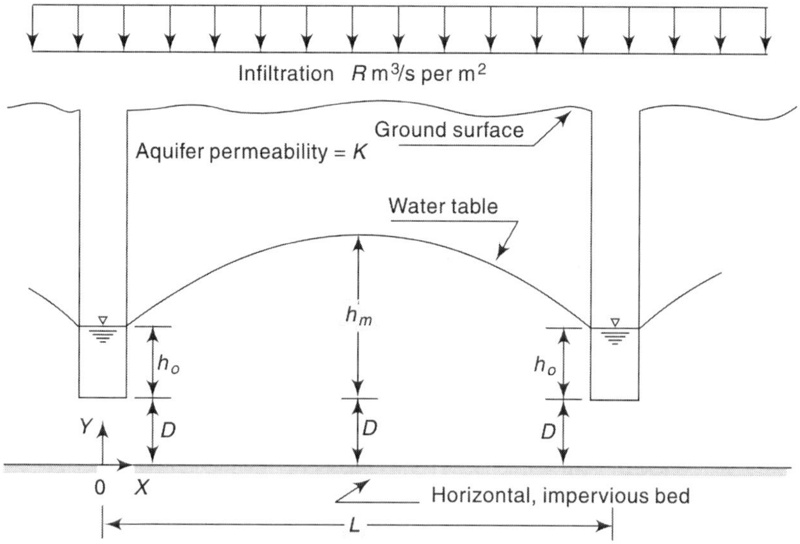

Fig. 9.25 ***Seepage to Open Ditches—Problem 9.10***

$$L^2 = \frac{4K}{R}[h_m^2 - h_0^2 + 2D(h_m - h_0)]$$

where K = coefficient of permeability of the aquifer. Neglect the width of drains. [**Note:** This relation is known as *Hooghoudt's* equation for either open ditch or sub-surface drains. When $h_0 = 0$ and $D = 0$, this equation reduces to Eq. (9.44).]

9.11 A canal and a stream run parallel to each other at a separation distance of 400 m. Both of them completely penetrate an unconfined aquifer (K = 3.0 m/day) located above a horizontal impervious bed. The aquifer forms the separation land mass between the two water bodies. The water surface elevations in the canal and the stream are 5.0 m and 3.0 m, the datum being the top of the horizontal impervious layer. Estimate

(a) the uniform infiltration rate that will create a water table divide at a distance of 100 m from the canal.

(b) the elevation of the water table divide, and

(c) the seepage discharges into the two water bodies.

9.12 Two rivers A and B run parallel to each other and fully penetrate the unconfined aquifer situated on a horizontal impervious base. The rivers are 4.0 km apart and the aquifer has a permeability of 1.5 m/day. In an year, the average water surface elevations of the rivers A and B, measured above the horizontal impermeable bed, are 12.00 m and 9.00 m respectively. If the region between the rivers received an annual net infiltration of 20 cm in that year, estimate

(a) the location of the groundwater table divide, and

(b) the average daily groundwater discharge into the rivers A and B from the aquifer between them.

9.13 A 30 cm well completely penetrates an artesian aquifer. The length of the strainer is 25 m. Determine the discharge from the well when the drawdown at the pumping well is 4.0 m. The coefficient of permeability of the aquifer is 45 m/day. Assume the radius of influence of the well as 350 m.

9.14 A 20 cm dia tubewell taps an artesian aquifer. Find the yield for a drawdown of 3.0 m at the well. The length of the strainer is 30 m and the coefficient of permeability of the aquifer is 35 m/day. Assume the radius of influence as 300 m.

If all other conditions remain same, find the percentage change in yield under the following cases:

(a) The diameter of the well is 40 cm (b) the drawdown is 6.0 m

(c) the permeability is 17.5 m/day

9.15 The discharge from a fully penetrating well operating under steady state in a confined aquifer of 35 m thickness is 2000 lpm. Values of drawdown at two observation wells 12 and 120 m away from the well are 3.0 and 0.30 m respectively. Determine the permeability of the aquifer.

9.16 A confined aquifer of thickness B has a fully penetrating well of radius of r_0, pumping a discharge Q at a steady rate. An observation well M is located at a distance R from the pumping well. Show that the travel time for water to travel from well M to the pumping well is

$$t = \frac{\pi B \eta}{Q}(R^2 - r_0^2) \text{ where } \eta = \text{Porosity of the aquifer.}$$

9.17 A 45 cm well penetrates an unconfined aquifer of saturated thickness 30 m completely. Under a steady pumping rate for a long time the drawdowns at two observation wells 15 and 30 m from the well are 5.0 and 4.2 m respectively. If the permeability of the aquifer is 20 m/day, determine the discharge and the drawdown at the pumping well.

9.18 A 30 cm well fully penetrates an unconfined aquifer of 25 m saturated depth. When a discharge of 2100 lpm was being pumped for a long time, observation wells at radial distances of 30 and 90 m indicated drawdown of 5 and 4 m respectively. Estimate the coefficient of permeability and transmissibility of the aquifer. What is the drawdown at the pumping well?

9.19 For conducting tests on a 30 cm diameter well in an unconfined aquifer two observation wells A and B are bored at distances 25 m and 40 m respectively from the centre of the pumping well. When water is pumped at a rate of 10 litres/s the water depth in the pumping well is 10.0 m above the horizontal impervious layer up to which the well is driven. The median grain size of the aquifer is 2.5 mm and the permeability is known to be 0.1 cm/s. Calculate (a) the depth of water above the impervious layer in the observation wells A and B; (b) the Reynolds number of flow at the pumping well and observation wells A and B. [Assume kinematic viscosity of water = 0.01 cm^2/s].

9.20 A 45 cm well in an unconfined aquifer of saturated thickness of 45 m yields 600 lpm under a drawdown of 3.0 m at the pumping well, (a) What will be the discharge under a drawdown of 6.0 m? (b) What will be the discharge in a 30 cm well under a drawdown of 3.0 m? Assume the radius of influence to remain constant at 500 m in both cases.

9.21 For conducting permeability tests in a well penetrating an unconfined aquifer, two observation wells A and B are located at distances 15 and 30 m respectively from the centre of the well. When the well is pumped at a rate of 5 lps, it is observed that the elevations of the water table above the impervious layer, up to which the well extends are 12.0 and 12.5 m respectively at A and B. Calculate the permeability of the aquifer in m/day.

9.22 Calculate the discharge in m^3/day from a tubewell under the following conditions:

Diameter of the well	=	45 cm
Drawdown at the well	=	12 m
Length of strainer	=	30 m
Radius of influence of the well	=	200 m
Coefficient of permeability	=	0.01 cm/s
Aquifer	=	unconfined

9.23 A fully penerating well of 30 cm diameter in an unconfined aquifer of 50 m saturated thickness was found to give the following drawdown-discharge relations under equilibrium condition.

Drawdown at the pumping well (m)	Discharge (lpm)
3.0	600
11.7	1800

If the radius of influence of the well can be assumed to be proportional to the discharge through the well, estimate the flow rate when the drawdown at the well is 6.0 m.

9.24 A 45 cm well in an unconfined aquifer was pumped at a constant rate of 1500 lpm. At the equilibrium stage the following drawdown values at two observation wells were noted:

Observation	Radial distance from pumping well (m)	Drawdown (m)
A	10	5.0
B	30	2.0

The saturated thickness of the aquifer is 45 m. Assuming the radius of influence to be proportional to the discharge in the pumping well, calculate:
(a) Drawdown at the pumping well; (b) transmissibility of the aquifer;
(c) drawdown at the pumping well for a discharge of 2000 lpm; and
(d) radius of influence for discharges of 1500 and 2000 lpm.

9.25 A 4.5 m diameter open well has a discharge of 30.0 m^3/h with a drawdown of 2.0 m. Estimate the (i) specific capacity per unit well area of the aquifer, and (ii) discharge from a 5.0 m open well in this aquifer under a depression head of 2.5 m.

9.26 In a recuperation test of a 3.0 m diameter open well the water level changed from 114.60 m to 115.70 m elevation in 120 minutes. If the water table elevation is 117.00 m, diameter (i) the specific capacity per unit well area of the aquifer, and (ii) discharge in the well under a safe drawdown of 2.75 m.

9.27 During a recuperation test, the water in an open well as depressed by pumping by 2.5 m it recuperated 1.8 m in 80 minutes. Calculate the yield from a well of 4.0 m diamater under a depression heat of 3.0 m.

9.28 The drawdown time data recorded at an observation well situated at a distance of 50 m from the pumping well is given below:

Time (min)	1.5	3	4.5	6	10	20	40	100
Drawdown (m)	0.15	0.6	1.0	1.4	2.4	3.7	5.1	6.9

If the well discharge is 1800 lpm, calculate the transmissibility and storage coefficients of the aquifer.

9.29 Estimate the discharge of a well pumping water from a confined aquifer of thickness 20 m with the following data:
Distance of observation well from the pumping well = 100 m
Drawdown at the observation well after 4 hours of pumping = 1.5 m
Drawdown at the observation well after 16 hours of pumping = 2.0 m
Storage coefficient, $S = 0.0003$

9.30 A fully penetrating well in a confined aquifer is being pumped at a constant rate of 2000 Lpm. The aquifer is known to have a storage coefficient of 0.005 and transmissibility of 480 m^2/day. Find the drawdown at a distance of 3.0 m from the production well after (i) one hour, and (ii) 8 hours after pumping.

9.31 A fully penetrating well in a confined aquifer is pumped at the rate of 60 m^3/h from an aquifer of storage coefficient and transmissibility 4×10^{-4} and 15 m^2/h respectively. Estimate the drawdown at a distance of 100 m after 8 hours of pumping.

9.32 A fully penetrating confined aquifer is pumped at a constant rate of 100 m^3/h. At an observation well located at 100 m from the pumping well, the drawdown was observed to be 0.65 m and 0.80 m after one and two hours of pumping respectively. Estimate the formation constants of the aquifer.

9.33 A well in a confined aquifer was pumping a discharge of 40 $m^3/hour$ at uniform rate. The pump was stopped after 300 minutes of running and the recovery of drawdown was measured. The recovery data is shown below. Estimate the transmissibility of the aquifer.

Time since stopping of the pump (min)	1	3	5	10	20	40	90	150	200
Residual drawdown (m)	1.35	1.18	1.11	1.05	0.98	0.88	0.80	0.70	0.68

OBJECTIVE QUESTIONS

9.1 A geological formation which is essentially impermeable for flow of water even though it may contain water in its pores is called
(a) aquifer (b) aquifuge (c) aquitard (d) aquiclude

9.2 An aquifer confined at the bottom but not at the top is called
(a) semiconfined aquifer (b) unconfined aquifer
(c) confined aquifer (d) perched aquifer

9.3 A stream that provides water to the water table is termed
(a) affluent (b) influent (c) ephemeral (d) effluent

9.4 The surface joining the static water levels in several wells penertating a confined aquifer represents
(a) water-table surface (b) capillary fringe
(c) piezometric surface of the aquifer (d) cone of depression

9.5 Flowing artesian wells are expected in areas where
(a) the water table is very close to the land surface (b) the aquifer is confined
(c) the elevation of the piezometric head line is above the elevation of the ground surface
(d) the rainfall is intense

9.6 Water present in artesian aquifers is usually
(a) at sub-atmospheric pressure (b) at atmospheric pressure
(c) at 0.5 times the atmospheric pressure (d) above atmospheric pressure

9.7 The volume of water that can be extracted by force of gravity from a unit volume of aquifer material is called
(a) specific retention (b) specific yield
(c) specific storage (d) specific capacity

9.8 Which of the pairs of terms used in groundwater hydrology are not synonymous?
(a) Permeability and hydraulic conductivity (b) Storage coefficient and storativity
(c) Actual velocity of flow and discharge velocity
(d) Water table aquifer and unconfined aquifer

9.9 The permeability of a soil sample at the standard temperature of 20°C was 0.01 cm/s. The permeability of the same material at a flow temperature of 10°C is in cm/s
(a) < 0.01 (b) > 0.01
(c) $= 0.01$ (d) depends upon the porous material

9.10 A soil has a coefficient of permeability of 0.51 cm/s. If the kinematic viscosity of water is 0.009 cm^2/s, the intrinsic permeability in darcys is about
(a) 5.3×10^4 (b) 474 (c) 4.7×10^7 (d) 4000

9.11 Darcy's law is valid in a porous media flow if the Reynolds number is less than unity. This Reynolds number is defined as
(a) (discharge velocity × maximum grain size)/μ
(b) (actual velocity × average grain size)/ν
(c) (discharge velocity × average grain size)/ν
(d) (discharge velocity × pore size)/ν

9.12 Two observation wells penetrating into a confined aquifer are located 1.5 km apart in the direction of flow. Heads of 45 m and 20 m are indicated at these two observation wells. If the coefficient of permeability of the aquifer is 30 m/day and the porosity is 0.25, the time of travel of an inert tracer from one well to another is about
(a) 417 days (b) 500 days (c) 750 days (d) 3000 days

9.13 A sand sample was found to have a porosity of 40%. For an aquifer of this material, the specific yield is
(a) = 40% (b) > 40% (c) < 40% (d) dependent on the clay fraction

9.14 An unconfined aquifer of porosity 35%, permeability 35 m/day and specific yield of 0.15 has an area of 100 km^2. The water table falls by 0.20 m during a drought. The volume of water lost from storage in Mm^3 is
(a) 7.0 (b) 3.0 (c) 4.0 (d) 18.0

9.15 The unit of intrinsic permeability is
(a) cm/day (b) m/day (c) darcy/day (d) cm^2

9.16 The dimensions of the storage coefficient S are
(a) L^3 (b) LT^{-1} (c) L^3/T (d) dimensionless

9.17 The dimensions of the coefficient of transmissibility T are
(a) L^2/T (b) L^3T^2 (c) L/T^2 (d) dimensionless

9.18 The coefficient of permeability of a sample of aquifer material is found to be 5 m/day in a laboratory test conducted with water at 10°C. If the kinematic viscosity of water at various temperatures is as below:

Temp in °C	10	20	30
ν (m^2/s)	1.30×10^6	1.00×10^6	0.80×10^6

The standard value of the coefficient of permeability of the material, in m/day, is about
(a) 4.0 (b) 5.0 (c) 6.5 (d) 9.0

9.19 In an aquifer extending 150 ha, the water table was 20 m below ground level. Over a period of time the water table dropped to 23 m below the ground level. If the porosity of the aquifer is 0.40 and the specific retention is 0.15, what is the change in groundwater storage of the aquifer?
(a) 67.5 ha. m (b) 112.5 ha. m (c) 180.0 ha. m (d) 450 ha. m

9.20 A groundwater basin consists of 10 km^2 area of plains. The maximum groundwater table fluctuation has been observed to be 1.5 m. Consider the specific yield of the basin as 10% and specific retention as 20%. What is the available groundwater storage in million cubic metres?
(a) 1.0 (b) 1.5 (c) 2.5 (d) 3.0

9.21 A well of 20 cm diameter fully penetrates a confined aquifer. After a long period of pumping at a constant rate of 2720 litres/minute, the observation of drawdown taken at 10 and 100 m distances from the centre of the well are found to be 3 m and 0.5 m respectively. The transmissivity of the aquifer is
(a) 675 m^2/day (b) 575 m^2/day (c) 525 m^2/day (d) 250 m^2/day

9.22 The surface joining the static levels in several non-pumping wells penetrating a continuous confined aquifer-represents
(a) water-table surface
(b) capillary fringe
(c) physical top surface of the aquifer
(d) piezometric surface of the aquifer

9.23 What would be the volume stored in a saturated column of aquifer with a porosity of 0.35, cross-sectional area of 1 square metre and of 3.0 metres depth?
(a) 3.0 m^3 (b) 2.0 m^3 (c) 1.05 m^3 (d) 0.105 m^3

9.24 A stratified unconfined aquifer has three horizontal layers as below:

Layer	Coefficient of permeability (m/day)	Depth (m)
1	6	2.0
2	16	4.0
3	24	3.0

The effective vertical coefficient of permeability of this aquifer, in m/day, is about
(a) 13 (b) 15 (c) 24 (d) 16

9.25 An aquifer confined at top and bottom by impervious layers is stratified into three layers as follows:

Layer	Thickness (m)	Permeability (m/day)
Top layer	3.0	30
Middle layer	2.0	10
Bottom layer	5.0	20

The transmissibility of the aquifer in m^2/day is
(a) 6000 (b) 18.2 (c) 20 (d) 210

9.26 The specific storage is
(a) storage coefficient/aquifer depth (b) specific yield per unit area
(c) specific capacity per unit depth of aquifer (d) porosity-specific detention

9.27 When there is an increase in the atmospheric pressure, the water level in a well penetrating a confined aquifer
(a) decreases (b) increases
(c) does not undergo any change
(d) decreases or increases depending on the elevation of the ground.

9.28 The specific capacity of a well is the
(a) volume of water that can be extracted by the force of gravity from unit volume of aquifer
(b) discharge per unit drawdown at the well
(c) drawdown per unit discharge of the well
(d) rate of flow through a unit width and entire thickness of the aquifer

9.29 In one-dimensional flow in an unconfined aquifer between two water bodies, when there is a recharge, the water table profile is
(a) a parabola (b) part of an ellipse
(c) a straight line (d) an arc of a circle

9.30 In one-dimensional flow in a confined aquifer between two water bodies, the piezometric head line is
(a) a straight line (b) a part of an ellipse
(c) a parabola (d) an arc of a circle

9.31 For one-dimensional flow without recharge in an unconfined aquifer between two water bodies the steady water table profile is
(a) a straight line
(b) a parabola
(c) an ellipse
(d) an arc of a circle

9.32 The discharge per unit drawdown at a well is known as
(a) specific yield
(b) specific storage
(c) safe yield
(d) specific capacity

9.33 The specific capacity of a well in confined aquifer under equilibrium conditions and within the working limits of drawdown
(a) can be taken as constant
(b) decreases as the drawdown increases
(c) increases as the drawdown increases
(d) increases or decreases depending upon the size of the well

CHAPTER 10

EROSION AND RESERVOIR SEDIMENTATION

10.1 *INTRODUCTION*

Erosion, transportation and deposition of sediment in a watershed are natural processes which are intimately connected with the hydrologic processes. Soil and water conservation in watershed and reservoir sedimentation are important parameters affecting the success and economy of many water-resources development activities in a basin. This chapter briefly deals with erosion, sediment yield and reservoir sedimentation aspects of the erosion phenomenon. This chapter is only a brief introduction to the topic. For details excellent treatises are available and Refs 2 and 5 contain some valuable source material on this topic.

Erosion is the wearing away of land. Natural agents such as water, wind and gravity are eroding the land surface since geologic times. Out of many erosion-causing agents, the role of water in detachment, transportation and deposition is indeed very significant. Since recent past, human activities like agricultural practice, mining, building activities, railway and road construction are contributing significantly to erosion of land surface. Water storage structures like reservoirs, tanks and ponds act as receptacles for deposition of eroded material.

10.2 *Erosion Processes*

10.2.1 Processes

Erosion takes place in the entire watershed including the channels. During a rainfall event, when raindrops impact on a soil surface, the kinetic energy of the drops breaks the soil aggregates and detaches the particles in the impact area. The detached particles are transported by surface run off. Depending upon the flow conditions, topography and geometry of the channel, etc. there may be some deposition of the eroded material enroute. Erosion takes place in various modes, which can be classified as follows:

1. Inter-Rill Erosion

In this, the detached particles due to raindrop impact are transported over small distances in surface flow of shallow depth without formation of elementary channels called *rills*. The mode of transport is essentially *sheet flow* and the inter-rill erosion from this mode is known as *sheet erosion.* Sheet erosion removes a thin covering of soil from large areas, often from the entire fields, more or less, uniformly during every rain which produces a run-off. The existence of sheet erosion is reflected in the muddy colour of the run-off from the fields.

2. Rill Erosion

Rills are elementary channels which form during the surface runoff event due to the concentration of flow. These are temporary features and facilitate channelling of overland flow. The flow in rills cuts the surface, detaches and transports the sediment in surface runoff.

3. Gulley Erosion

Gullies are formed due to confluence of many rills and formation of a major rill. When a major rill becomes deeper and steeper a gulley is formed. Gullies are capable of transporting larger amounts of sediment. The sediment removed due to formation, enlargement and deepening of gullies is known as *gulley erosion*. Enlarged gullies become permanent topographic features. Gullies are the most visible evidence of the destruction of soil. The gullies tend to deepen and widen with every heavy rainfall. Further, they cut up large fields into small fragments and, in course of time, make them unfit for agricultural operations.

4. Channel Erosion

Channels are permanent topographic features formed due to confluence of gullies. Channel erosion includes stream bed and bank erosion and flood plain scour. Channel erosion is significantly larger than sheet erosion.

10.2.2 Factors Affecting Erosion

The quantity of sediment that is produced by erosion in a watershed depends upon a host of factors related to climate, soil, topography soil cover and human activities in that watershed. The major effects of these parameters are summarised in Table 10.1.

Table 10.1 ***Factors Affecting Erosion***

Factor	Parameter	Effect
Climate	Rainfall intensity	Splash erosion
	Duration of rainfall	Flow erosion
	Temperature	Weathering action
Soil Characteristics	Soil Mass characteristics (Granulation, Porosity, Moisture content)	Infiltration and Runoff rates and hence erosion rate.
	Grain size and shape	Erosion rate and transportation mode.
Topography	Slope (Orientation, Degree and Length)	Steeper slope: Higher energy of flow, higher erosion and transportation rates.

(Contd.)

(Contd.)

Soil Cover	Vegetation/plant cover	Retardation of flow and erosion rates, Protection from splash erosion.
Land Use (Human Activities)	Agricultural practice, Mining, Roads, Building construction, etc.	Increased erosion rates
	Reservoirs	Sedimentation

10.2.3 Gross Erosion, Sediment Yield and Delivery Ratio

Gross erosion is the sum of all erosions in watershed. Total sediment outflow from a watershed at a reference section in a selected time interval is known as *sediment yield*. Not all the sediment produced due to erosion in a watershed is transported out of it as there will be considerable temporary depositions in various phases and locations. As such, the sediment yield is always less than the gross erosion. The ratio of sediment yield to gross erosion is known as *sediment delivery ratio*.

The sediment yield of a watershed varies with the size of the contributing area. For purposes of comparing the sediment production rate of different areas, it is customary to convert the sediment yield data to the yield per unit of drainage area to obtain *sediment–production rate* of the catchment, which is usually expressed in units of tonnes/km^2/year (or in ha-m/km^2/year).

10.3 *Estimation of Sheet Erosion*

Estimation of sheet erosion is of utmost importance in soil and water-conservation practice and management of watershed. Considering different weightages to the various factors affecting the erosion process, several methods have been proposed to estimate the sheet erosion rate in a watershed. Two popular methods are described below.

10.3.1 Universal Soil Loss Equation (*USLE*)

The universal soil loss equation is the most widely used tool for estimation of soil loss from agricultural watersheds for planning erosion control practices. The *USLE* is an erosion prediction model for estimating long term averages of soil erosion from sheet and rill erosion modes from a specified land under specified conditions. The equation is written as

$$A = RKLSCP \tag{10.1}$$

where A = the soil loss per unit area in unit time. Usually the units of A are metric tonnes/ha/year
R = rainfall erosivity factor
K = soil erodibility factor
L = slope length factor
S = slope-steepness factor
C = cover management factor
P = Support practice factor (Ratio of soil loss with a support practice like contouring, strip-cropping or terracing to that with straight row farming up and down the slope)

The various factors of the *USLE* equation are as below:

1. Rainfall Erosivity Factor (*R*)

The factor R is the number of rainfall erosion index units (EI_{30}) in a given period at the study location. The rainfall erosion index unit (EI_{30}) of a storm is defined as

$$EI_{30} = \frac{KE \times I_{30}}{100} \tag{10.2}$$

where KE = kinetic energy of the storm. The KE in metric tones/ha-cm is expressed as

$KE = 210.3 + 89 \log I$

where I = rainfall intensity in cm/h

I_{30} = maximum 30 minutes rainfall intensity of the storm.

The study period can be a week, month, season or year. The storm EI_{30} values for that length of period is summed up.

Annual EI_{30} values are usually computed from data available at various meteorological stations and lines of equal EI_{30} lines (known as *Iso-erodent* lines) are drawn for the region covered by the data stations for ready use in *USLE*. Iso-erodent maps of Karnataka and Tamil Nadu are available in Ref. (9). In most parts of Karnataka, annual EI_{30} values range from 250 to 500, except in Western Ghats where they range from 500 to 1500. In Tamil Nadu annual EI_{30} values range from 300 to 700.

2. Soil Erodibility Factor (*K*)

The factor K relates the rate at which different soils erode due to soil properties. These are usually determined at special experimental runoff plots or by use of empirical erodibility equations which relate several soil properties to factor K. Table 10.2 shows some computed values of K at several research stations in the country (Ref. 5).

Table 10.2 ***Values of K at Several Stations***

Station	Soil	Values of K
Agra	Loamy sand, alluvial	0.07
Dehradun	Dhulkot silt, Loam	0.15
Hyderabad	Red chalka sandy loam	0.08
Kharagpur	Soils from lateritic rock	0.04
Kota	Kota-clay loam	0.11
Ootakamund	Laterite	0.04
Rehmankhera	Loam, alluvial	0.17
Vasad	Sandy Loam, alluvial	0.06

[*Source:* Ref. 5]

3. Topographic Factor (*LS*)

The two factors L and S are usually combined into one factor LS called *topographic factor* and is given by

$$LS = \left(\frac{\lambda}{22.13}\right)^m [65.41 \sin^2 \theta + 4.56 \sin \theta + 0.065] \tag{10.3}$$

where λ = field slope length in metres

m = exponent factor varying from 0.2 to 0.5
θ = angle of slope

4. Crop-Management Factor (*C*)

This factor reflects the combined effect of various crop-management practices. Values of factor *C* for regions surrounding some stations are given in Table 10.3.

Table 10.3 ***Values of Factor C***

Station	Crop	Soil Loss (Tonnes/ha per year)	Value of *C*
Agra	Cultivated fallow	3.80	1.0
	Bajra	2.34	0.61
	Dichanhium annualtu	0.53	0.13
Dehradun	Cultivated fallow	33.42	1.0
	Cymbopogon grass	4.51	0.13
	Strawberry	8.89	0.27
Hyderabad	Cultivated fallow	5.00	1.0
	Bajra	2.00	0.40

[*Source:* Ref. 5]

5. Supporting Cultivation Practice (*P*)

This factor is the ratio of soil loss with a support practice to that with straight row farming up and down the slope. Table 10.4 gives the factor *P* for some support practices.

Table 10.4 ***Value of Factor P***

Station	Practice	Factor *P*
Dehradun	Contour cultivation of maize	0.74
	Up and down cultivation	1.00
	Contour farming	0.68
	Terracing and bunding in agricultural watershed	0.03
Kanpur	Up and down cultivation of Jowar	1.00
	Contour utilization of Jowar	0.39
Ootacamund	Potato up and down	1.00
	Potato on contour	0.51

[*Source:* Ref. 5]

6. Use of *USLE*

USLE is an erosion prediction model and its successful application depends on the ability to predict its various factors with reasonable degree of accuracy. Based on considerably large experimental data base relating to various factors of *USLE* available in *USA*, this equation is being used extensively in that country to provide reliable estimates of erosion in a variety of situations related to in small agricultural watersheds. It should be noted that to estimate sediment yield of a watershed using *USLE*, information on sediment delivery ratio of the watershed would be needed.

Based on 21 observed points and 64 estimated erosion values of soil loss obtained by use of *USLE* at points spread over different regions of the country, Gurmeet Singh *et al* (6) have prepared the Iso-erosion rate map of India. Soil-erosion rates have been classified by them into 6 categories and the area of the country under different classes of erosion are found to be as shown in Table 10.5.

Table 10.5 ***Distribution of Various Erosion Classes in India***

Range (Tonnes/ha/year)	Erosion Class	Area (km^2)
0– 5	Slight	801,350
5–10	Moderate	1,405,640
10–20	High	805,030
20–40	Very high	160,050
40–80	Severe	83,300
> 80	Very severe	31,895

[*Source:* Ref. 6]

10.3.2 Modified Universal Soil Loss Equation (*MUSLE*)

The USLE was modified by Williams in 1975 to *MUSLE* by replacing the rainfall energy factor with a runoff factor. The *MUSLE* is expressed as

$$Y = 11.8(Q \times q_p)^{0.56}\ K(LS)CP \tag{10.4}$$

where Y = the sediment yield from an individual storm (in metric Tonnes)
Q = the storm runoff volume in m^3
q_p = the peak rate of runoff in m^3/s

and other factors K, (LS), C and P retain the same meaning as in *USLE* (Eq. 10.1).

In this equation, Q and q_p are obtained by appropriate runoff models (chapters 5 and 6). In this model Q is considered to represent detachment process and q_p the sediment transport. It should be noted that *MUSLE* is a sediment yield model and does not need separate estimation of sediment delivery ratio. Also, it is applicable to individual storms. It is believed that *MUSLE* increases sediment yield prediction accuracy. From the modelling point of view, *MUSLE* has the advantage that daily, monthly and annual sediment yields of a watershed can be modelled by combining appropriate hydrologic models with *MUSLE*.

10.4 *Channel Erosion*

The channel erosion comprises erosion in bed, sides and also flood plain of the stream. A channel flowing in a watershed transports the runoff that is produced in the catchment and also the erosion products, out of the watershed. The total sediment load that is transported out the catchment by a stream is classified into components depending upon their origin as:

1. Wash load
2. Bed material load
   (i) Bed load (ii) Suspended load

1. Wash Load

It is sediment originating from the land surface of the watershed and is transported to the stream channel by means of splash, sheet, rill and gully erosion. Wash load is generally composed of fine-grained soils of very small fall velocity.

2. Bed Material Load

The sediment load composed of grain sizes originating in the channel bed and sides of the stream channel.

(a) Bed Load It is the relatively coarse bed material load that is moved at the bed surface through sliding, rolling, and saltation.

(b) Suspended Load The relatively finer bed material that is kept in suspension in the flow through turbulence eddies and transported in suspension mode by the flowing water is called suspended load. The suspended load particles move considerably long distances before settling on the bed and sides.

In a general sense, bed load forms a small part of total load (usually < 25%) and wash load forms comparatively very small part of the total load. The mechanics of bed material transport in channels, viz. bed load and suspended load, have been studied in extensive detail and treatises on the subject are available (for example Ref. 4).

10.4.1 Measurement

While a large number of devices are available for measuring bed load for experimental/special investigations, no practical device for routine field measurement of bed load is currently in use. For planning and design purposes the bed load of a stream is usually estimated either by use of a bed load equation such as Einstein bed load equation[4] or is taken as a certain percentage of the measured suspended load. Table 10.6 gives some recommended values for use in preliminary planning purposes.

Table 10.6 ***Approximations for Bed Load***

Concentration of suspended load (ppm)	Type of material forming the stream channel	Texture of suspended material	Percent of measured suspended load that could be taken as Bed load
Less than 1000	Sand	Similar to bed material	25 to 150
Less than 1000	Gravel, rock or consolidated clay	Small amount of sand	5 to 12
1000 to 7500	Sand	Similar to bed material	10 to 35
1000 to 7500	Gravel, rock or consolidated clay	25 percent sand or less	5 to 12
Over 7500	Sand	Similar to bed material	5 to 15
Over 7500	Gravel, rock or consolidated clay	25 percent sand or less	2 to 8

[*Source:* Ref. 7]

The suspended load of a stream is measured by sampling the stream flow. Specially designed samplers that do not alter the flow configuration in front of the sampler are available. The sediment from the collected sample of sediment laden water is removed by filtering and its dry weight is determined. It is usual to express suspended load as parts per million (ppm) on weight basis as

$$C_s = \left[\frac{\text{Weight of sediment in sample}}{\text{Weight of (sediment + water) of the sample}}\right] \times 10^6$$

Thus the sediment transport rate in a stream of discharge Q m^3/s is

$$Q_s = (Q \times C_s \times 60 \times 60 \times 24)/10^6 = 0.0864\ QC_s \text{ tonnes/day}$$

Routine observations of suspended load are being done at many stream-gauging stations in the country. At these stations in addition to stream flow discharge Q the suspended sediment concentration and hence the suspended sediment load Q_s is also noted. The relation between Q_s (tones/day) and stream discharge Q (m^3/s) is usually represented in a logarithmic plot (Fig. 10.1) known as *sediment rating curve.* The relationship between Q_s and Q can be represented as

$$Q_s = KQ^n \tag{10.5}$$

where the exponent n has a range of 1.0 to 3.0 and its usual value is round 2.0

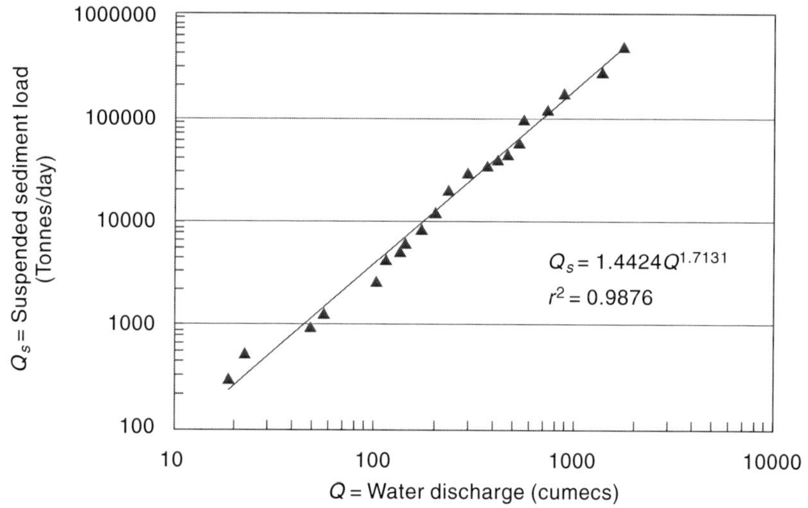

Fig. 10.1 ***Sediment rating curve (Schematic)***

The sediment rating curve in conjunction with the stream flow hydrograph can be used to estimate the suspended sediment load transport in the stream in a specified time interval. However, it should be remembered that due to inherent inaccuracies, the sediment load obtained by using sediment rating curve is less accurate than that obtained by observed flow hydrograph and sediment concentration data at the gauging station. A method of estimating the annual sediment yield of a watershed by using the sediment rating curve is described in Sec. 10.6.

10.5 *Movement of Sediment from Watersheds*

As indicated earlier, not all sediment produced in an erosion process in the watershed is transported out of the catchment in a real time basis. Due to loss of momentum of the conveying mechanism, considerable deposition occurs mostly in areas of the catchment with low slope, high roughness or very low velocities due to large expansion of flow area. The ratio of sediment yield to the gross erosion in the watershed, called *Sediment Delivery Ratio* (*SDR*), is an important parameter in quantitative estimation of sediment yield. The values of average annual *SDR* vary in a wide range as this parameter depends on a host of parameters. Out of the many parameters the significant ones are: (i) the size of the watershed, (ii) the channel density, and (iii) the relief length ratio.

The size of the watershed plays an important role in controlling the deposition opportunities for the eroded sediment, the larger the area larger is the opportunity for the sediment to be deposited in the catchment and hence lower *SDR*. The relief ratio of the watershed is a measure of the average slope of the watershed and as such higher values of relief ratio can be associated with larger transportation rates and hence larger values of *SDR*. Similarly the *SDR* is generally higher for well-defined channel network of higher density as the transportation of erosion products out of the catchment is highly facilitated by such channel networks. The variation of *SDR* with catchment area and relief can be expressed as

$$SDR = KA^{-m} (R/L)^n \tag{10.6}$$

where A = watershed area

R = watershed relief (elevation difference)

L = watershed length

K, m and n are positive coefficients.

Values of K, m and n applicable to a homogeneous region can be estimated by using observed data from experimental watersheds.

10.6 *Sediment Yield from Watersheds*

Estimation of sediment yield from a watershed is of utmost importance in the soil and water-conservation practice in the watershed and in planning, design and operation of reservoirs. While the procedure for estimation of sediment yield is generally problem specific in view of many practical constraints relating to availability of quality data, a few commonly used procedures are described below.

(1) Flow duration curve and sediment rating curve procedure
(2) Reservoir sedimentation surveys
(3) Estimation of watershed erosion and sediment delivery ratio

10.6.1 Flow Duration Curve and Sediment Rating Curve Procedure

This procedure uses the sediment rating curve of a stream to operate on the flow duration curve of the stream at the same location to obtain weighted expected daily suspended sediment transport load. This, when multiplied by 365 days, gives the estimated annual suspended sediment load yield at the site.

The flow duration curve of the stream is developed by using available gauged daily discharge data records of sufficiently long length (Sec. 5.7, Chap. 5). The result is expressed as a graph or a table of exceedance probabilities (in percentages) of levels of daily discharge Q. The sediment rating curve of the steam at the gauging site is prepared by using gauged suspended sediment load Q_s (tonnes/day) and the corresponding daily discharge Q (m^3/s), as described in Sec. (10.4). The sediment rating curve is operated on the flow–duration curve as mentioned below:

- The flow duration curve is considered divided into a large number of sections (say 10–20 sections). The sections need not be uniform and it is desirable to provide more sections at low exceedance probabilities.
- Note the ranges of various intervals and also the mid-point value of each interval, (p_i).
- The discharge value Q_i (cumecs) corresponding to each mid-point value of exceedence, p_i is noted.
- Using the sediment rating curve, the value of suspended sediment load Q_{si} (tonnes/day) corresponding to each Q_i is calculated.
- Expected daily suspended sediment discharge at each probability level p_i is
$$EQ_{si} = p_i Q_{si}.$$
- Total weighted expected daily suspended sediment load $Q_{sd} = \Sigma p_i Q_{si}$
- Expected total annual suspended sediment yield $Y_{ss} = 365 \times Q_{sd}$
- The bed load (Y_b) is calculated either as a percentage of annual suspended load yield (Y_{ss}) as in Table 10.4 or by using appropriate bed load equation.
- Total annual sediment yield at the station $Y = Y_b + Y_{ss}$

Example 10.1 given below explains this procedure.

Example 10.1 The salient co-ordinates of the flow duration curve of a stream at a gauging station is given below:

Exceedance Probability range (%)	Mean Daily Discharge at mid-point of the interval (Cumecs)	Exceedance Probability range (%)	Mean Daily Discharge at mid-point of the interval (Cumces)
0.1—1.0	250	40—50	65
1.0—5.0	200	50—65	50
5.0—10.0	160	65—80	40
10.0—15.0	135	80—90	25
15—20	120	90—95	15
20—30	100	95—99.0	10
30—40	80		

The gauging station is at the outlet of a watershed of 3000 sq. km area. The sediment rating curve at the station is given by the relation $Q_s = 0.80\, Q^{1.84}$ where Q_s is the suspended sediment load in tonnes/day and Q = discharge in m^3/s. Estimate (i) the total sediment yield of the watershed by assuming the bed load to be 5% of the suspended load, and (ii) the annual average concentration of suspended load.

Solution

Plot the flow duration curve on a logarithmic plot for using it to determine the discharge corresponding to any chosen exceedance probability through interpolation, if necessary (Fig. 10.2).

Thirteen sections of the flow duration curve are selected and through appropriate interpolations their mid-values of the intervals are read from Fig. 10.2 and values entered in Table 10.7.

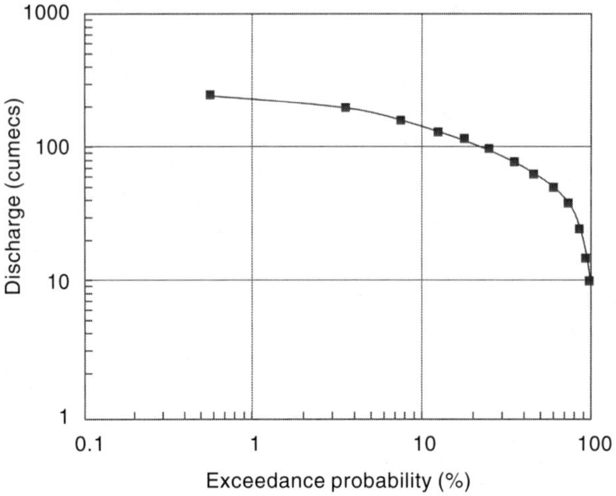

Fig. 10.2 ***Flow Duration Curve—Example 10.1***

The computations are effected in Table 10.7. In this table

(i) Column 1 shows the 13 sections into which the exceedance probability is divided.
(ii) Column 2 is the interval of probability range of Col. 1.
(iii) Column 3 is the mid-point of interval of Col. 1.
(iv) Column 4 is the mean daily discharge corresponding to mid point value (Col. 3) and is obtained from flow duration curve (Fig. 10.2).
(v) Suspended sediment discharge in tonnes/day is calculated for each value of Co1. 4 by using the equation $Q_s = 0.80\ Q^{1.84}$.
(vi) Column 6 is the Expected daily sediment load of known level of exceedance probability and is obtained as a product of Col. 2 and Col. 5 and divided by 100 to account for percentage values.
(vii) Co1umn 7 is the corresponding water flow in units of cumec. day: = [Col. 2 × Column 4/100]

The total expected mean daily suspended sediment load is 2904 tonnes, obtained as the sum of Co1umn 6 values.

Table 10.7 ***Computation of Suspended Sediment Load—Example 10.1***

1	2	3	4	5	6	7
Exceedance Frequency range (%)	Interval (%)	Midpoint (%)	Mean Daily Discharge (Cumec)	Sediment Discharge (tonnes/day)	Expected Sediment Load per day (Tonnes)	Volume of Water flow (cumec. day)
					[2] × [5]/100	[2] × [4]/100
0.1–1.0	0.9	0.55	250	20668	186.0	2.25
1.0–5.0	4	3.0	200	13708	548.3	8.00
5.0–10.0	5	7.5	160	9092	454.6	8.00
10.0–15.0	5	12.5	135	6651	332.6	6.75
15–20	5	17.5	120	5355	267.8	6.00
20–30	10	25	100	3829	382.9	10.00

(Contd.)

(Contd.)

30–40	10	35	80	2540	254.0	8.00
40–50	10	45	65	1733	173.3	6.50
50–65	15	57.5	50	1070	160.4	7.50
65–80	15	72.5	40	709	106.4	6.00
80–90	10	85	25	299	29.9	2.50
90–95	5	92.5	15	117	5.8	0.75
95–99.0	4	97.0	10	55	2.2	0.40
				Total	**2904**	**72.65**

Over one year, suspended sediment load yield is = 365 × 2904 = 1,059,960 tonnes.
Thus suspended load yield = 1.06 million tonnes/year.
Bed load (at 5% of suspended load yield) = 0.05 × 1,059,960 = 52998 Tonnes
Total Sediment Yield = 1,112,958 tonnes = say 1.113 million tonnes/year
For a catchment area of 3000 sq.km, Sediment yield rate = 1,113,000/3000
= 371 tonnes/sq. km per year.

Water Yield: Mean daily yield = 72.65 cumec. day
Annual yield = 365 × 72.65 = 26517.25 cumec. days
= 26517.25 × (60 × 60 × 24)/10^6 = 2291 million m^3.

Considering annual values

$$\text{Average cncentration of Suspended Load} = \frac{1.113\times10^6}{(2291\times10^6)+(1.113\times10^6)}\times 10^6$$

= 485.6

= say **486** parts per million (ppm) by weight.

10.6.2 Reservoir Sedimentation Surveys

Reservoir sedimentation surveys are conducted to get reliable data relating to various aspects of sedimentation such as (i) rate of sedimentation, (ii) sediment densities, (iii) depositional pattern, and (iv) loss of storage capacity of the reservoir at various elevations. The conventional reservoir survey is a hydrologic survey using echo depth recorder along pre-established range lines on the water spread of the reservoir. Depth measurements are taken from a good quality boat with adequate safety provisions by positioning the boat at a desired point on a given range line. The basic measurements are depth of water at the location of the boat and the fixing of the position of the boat with respect to appropriate reference co-ordinates. The depth measurement is done usually by using an echo depth recorder of appropriate accuracy. The position fixing of the boat is through standard land survey techniques through use of *Theodolites*. Nowadays, use of *Total Station* units are very common as it considerably reduces the observational time and computational effort. In addition to the depth and position observations, sediment samples are taken at a number of locations in the reservoir to assess the composition and density of the sediment deposits.

Use of *Differential Global Positioning System* (*DGPS*) along with an echo recorder in the boat enables faster data acquisitioning with better accuracy. Details of the *DGPS* methodology adopted by *CWC* are given in Ref. (1).

The data from a reservoir survey is analysed to produce contour map of the bed of the reservoir. Some of the end products of the analysis of reservoir survey data are

- Area–Elevation–Capacity curve of the reservoir
- Description of deposition pattern (Qualitative and quantitative)
- L-section of the delta deposition
- Sediment density at key locations
- Average sediment yield from the catchment during the interval between two successive surveys

Periodic reservoir surveys are essential to efficient management of major reservoirs.

Using remote-sensed images of the reservoir taken at frequent intervals in monsoon and non-monsoon periods, the areal extent of the water spread at various elevations can be established very accurately. Further, using the observed reservoir levels corresponding to the date and time of the images, the area elevation capacity curve of the reservoir covering a substantial portion of the reservoir can be established by back office computations only. It may be noted that the reservoir levels are routinely observed at dam site in all reservoirs. This procedure of using remote sensed images is extremely useful in monitoring reservoir sedimentation and efficacy of catchment soil conservation practice.

10.6.3 Estimation of Watershed Erosion and Sediment Delivery Ratio

For very small watersheds having predominantly agricultural land use, the sediment yield can be calculated by using *USLE* or *MUSLE* with appropriate factors applicable to the site. If *USLE* is used the sediment delivery ratio (*SDR*) will have to be estimated. When regional relations for *SDR* are not available, *SDR* applicable to the site will have to be established by empirical equations calibrated by using local information and observed values. The sediment yield is obtained by multiplying gross watershed erosion by *SDR*.

10.6.4 Empirical Equations

Many empirical equations and procedures have been developed for estimating sediment yield at the outlet of a watershed. A few of these in common use in India or developed by use of Indian data are given below:

1. Khosla's Equation (1953) (Ref. 8)

The annual sediment yield on volume basis is related to catchment area as

Annual sedimet yield rate (on volume basis)

$$q_{sv} = \frac{0.00323}{A^{0.28}} \text{ Mm}^3/\text{km}^2/\text{year} \tag{10.7a}$$

or Volume of sediment yield per year from the catchment is

$$Q_{sv} = 0.00323\, A^{0.72} \text{ Mm}^3/\text{year} \tag{10.7b}$$

where A = area of catchment in km^2.

This equation is in common use in many parts of the country to estimate the annual sediment volume inflow into a reservoir. The observed data of Khosla had an upper average limit of 3.6 ha.m/100 sq.km and the absolute maximum limit of observed data was 4.3 ha.m/100 sq.km. While this equation has been used in many of the reservoirs in the country up to about 1970, the observed data of actual sedimentation of many reservoirs indicate that the Eq. 10.7(a) underestimates the sedimentation rate.

2. Joglekar's Equation (1960): (Ref. 8)

Based on data from reservoirs from India and abroad, Joglekar expressed the annual sediment yield rate as

$$q_{sv} = \frac{0.00597}{A^{0.24}} \quad \text{Mm}^3/\text{km}^2/\text{year} \tag{10.8a}$$

or volume of sediment yield per year from a catchment area is

$$Q_{sv} = 0.00597 A^{0.76} \quad \text{Mm}^3/\text{year} \tag{10.8b}$$

In these equations A = area of catchment in km^2.

3. Dhruv Narayan et al's Equation (1983): (Ref. 3)

In this, annual sediment rate is related to annual runoff as

$$Q_s = 5.5 + 11.1\, Q \tag{10.9}$$

where Q_s = annual sediment yield rate in tonnes/year from the watershed
Q = annual runoff volume in M.ha.m

4. Garde and Kothyari's Procedure

A detailed procedure for sediment yield estimate for Indian catchments has been developed by Garde and Kothyari[3] (1987). Reference 3 contains a map of India with iso-erosion lines (in tonnes/km^2/year) developed by using this procedure.

10.7 *Trap Efficiency*

10.7.1 Deposition Process

When a river enters a reservoir, it suffers a massive enlargement of cross section of flow and consequently a large reduction of flow velocity results. The heavy sediment particles are deposited at the mouth of the reservoir in the form of a delta deposit. The sands and gravels are deposited first and the finer particles are deposited farther downstream. The sediment deposits could be classified as *top set beds*, *foreset beds* and *bottomset beds*. The topset beds are composed of coarse sediments of large particle size and foreset beds are of coarse sandy particles. Bottomset beds are of fine particles. Generally topset beds have flat slopes approximately at half the slope of the original channel bed. The foreset slopes are steeper and are about 5 to 7 times steeper than the topset slopes. Delta deposit forms at the mouth of the river entering the reservoir and may cause rising of the backwater profile in the channel upstream of the reservoir. Profile of a typical reservoir delta is shown in Fig. 10.3.

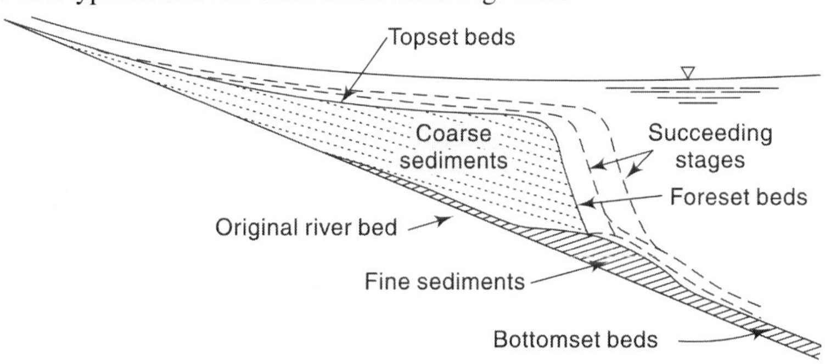

Fig. 10.3 ***Schematic Representation of Reservoir Delta***

Very fine particles of clays and colloids remain suspended and are transported to the remaining parts of the reservoir. The reservoir acts as a sedimentation tank and the suspended particles settle down gradually in course of time depending upon their settling velocities and reservoir operation. In the process of overflow and reservoir withdrawals, some suspended sediment passes out of the reservoir to downstream locations.

Sometimes river water containing high levels of fine to very fine sediment behaves like a high-density fluid and flows at the bottom of the reservoir as a gravity current with its own identity. Such flows termed *density currents* or *turbidity currents,* generally move very slowly causing a layer of high density sediment matter suspension at the reservoir bed. The bottom layers of this gradually gets compacted over a long length of time.

10.7.2 Trap Efficiency

Out of the total quantity of sediments brought to the reservoir through the channel system of the catchment, a major portion of the sediment is deposited in the reservoir and the balance is moved downstream by overflow and reservoir withdrawals. The amount of sediment trapped in the reservoir is of importance in the long term planning and operation of the reservoir. The ability of a reservoir to trap and retain incoming sediment is known as *trap efficiency* and is usually expressed as a percent of sediment yield of the catchment retained in the reservoir.

Trap efficiency of a reservoir depends on a host of parameters the important ones being (a) sediment characteristics, (b) detention–storage time, (c) nature of outlets, and (d) reservoir operation. As such, the trap efficiency becomes reservoir specific. However, for planning purposes the correlation of trap efficiency with capacity—Inflow ratio (C/I) of the reservoir, developed by Brune[2] is commonly used. Figure 10.4 shows the median and envelope curves for normal ponded reservoir relating the trap efficiency (η_t) with *C*/*I* ratio. Brune developed these curves on the basis of observed data from 40 reservoirs covering *C*/*I* values ranging from 0.0016 to 2.00.

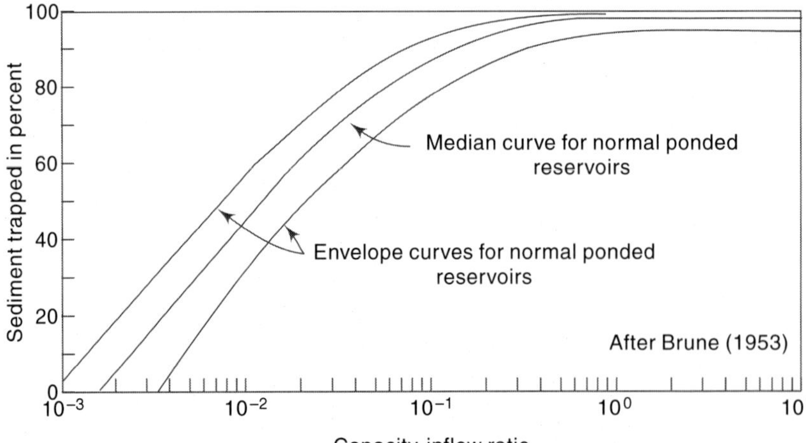

Fig. 10.4 ***Brune's Curve of Trap Efficiency of a Reservoir***

Figure 10.4 shows that for the median curve $\eta_t \approx 100\%$ for $C/I > 0.70$. For ranges smaller than 0.70, the trap efficiency can be expressed as

$$\eta_t = K \ln(C/I) + M \tag{10.10}$$

where η_t = trap efficiency in percent

C/I = capacity –inflow ratio where C = capacity of reservoir at FRL in Mm^3

and I = annual inflow into reservoir in Mm^3

K and M are coefficients dependent on C/I.

C/I	K	M
0.002 to 0.03	25.025	158.61
0.03 to 0.10	14.193	119.30
0.10 to 0.70	6.064	101.48

10.8 *Density of Sediment Deposits*

Sediment load whether computed by equations or by direct observations are usually expressed in terms of its dry weight basis. However, to estimate the volume occupied by a given weight of sediment, it is necessary to know the unit weight of deposited sediment. The unit weight (also known as *specific weight*) is expressed as the ratio of dry weight of the sediment (say tonnes) in unit volume (say in m^3) of the sediment deposit in the reservoir. The unit weight or sediment deposits varies over a wide range depending upon the composition, reservoir operation and consolidation undergone by the deposit over time. Typical values lie in the range of 0.3 to 2.0 tonnes/m^3, with an average of around 1.3 tonnes/m^3. Since sediment deposited in a reservoir gets compacted during time period through a consolidation process, the unit weight increases with time logarithmically. A commonly used formula for rough estimation of unit weight is due to *Koelzer* and *Lara* and is given by

$$W_T = \frac{P_{sa}}{100}(W_1 + B_1 \log T) + \frac{P_{si}}{100}(W_2 + B_2 \log T) + \frac{P_{cl}}{100}(W_3 + B_3 \log T) \tag{10.11}$$

in which W_T = unit weight of deposit of age T years

p_{sa}, p_{si} and p_{cl} = percentage of sand, silt and clay, respectively, on weight basis present in the sediment deposit.

W_1, W_2 and W_3 = unit weight (dry) of sand, silt and clay, respectively, at the end of the first year.

B_1, B_2 and B_3 = constants relating to compacting characteristics of the sediment components.

T = age of sediment in years.

Typical values of the parameters given by Koelzer and Lara are given in Table 10.8.

Table 10.8 *Values of Coefficients W and B in Eq. 10.11*

Reservoir operation	Sand		Silt		Clay	
	W_1 (kgf/m³)	B_1	W_2 (kgf/m³)	B_2	W_3 (kgf/m³)	B_3
Sediment always submerged or nearly submerged	1490	0	1040	91.3	480	256.3
Normally a moderate reservoir drawdown	1490	0	1185	43.3	737	171.4
Normally considerable reservoir drawdown	1490	0	1265	16.0	961	96.1
Reservoir normally empty	1490	0	1315	0	1250	0

[*Source:* Ref. 8] [**Note:** 1 kgf = 9.81 N, 1 Tonne = 1000 kgf = 9.81 kN]

Using Eq. 10.11, the average unit weight of deposit W_{av} during a period of T years is obtained as

$$W_{av} = W_{T1} + 0.4343 B_w \left[\left\{ \left(\frac{T}{T-1} \right) \ln T \right\} - 1 \right] \qquad (10.12)$$

where W_{T1} = initial unit weight in tonnes/m³

and B_w = weighted value of B in Eq. (10.11) in decimal, weightages being fraction of sand, silt and clay in the sample

$= (p_{sa} \cdot B_1 + p_{si} \cdot B_2 + p_{cl} \cdot B_3)/100$

The average unit weight value W_{av} is used in estimating the time period required to reduce the capacity by a defined fraction due to sedimentation. The value of W_1, W_2 and W_3 as well as B_1, B_2 and B_3 used by different agencies differ over a wide margin. As such the values of Table 10.7 are only indicative values. Reference (7) contains valuable data pertaining to unit weights of sediments of many Indian reservoirs.

In estimating the time required for a certain capacity of a reservoir to be filled up by sediment, a trial and error and step-wise procedure is adopted. Example 10.4 illustrates the method.

Example 10.2 A reservoir has a capacity of 15 Mm³ at full reservoir level. The catchment area is 480 km². The average annual runoff of the catchment is 250 mm and the sediment yield is 12.5 MN/km². The sediment can be assumed to have an average insitu specific weight of 13.5 kN/m³ after deposition. Estimate the time required for the reservoir capacity to be reduced to 5.0 Mm³. Use uniform steps of capacity interval of 2 Mm³. The variation of the trap efficiency is given by the relation:

$$\eta_t = 100 \times \left[1 - \frac{1}{(1 + 65(C/I))} \right]^2$$

where η_t = percentage trap efficiency, C = capacity of the reservoir, and I = annual inflow into the reservoir.

Solution

Annual inflow volume $= \dfrac{250 \times 480}{1000} = 120\,\text{Mm}^3$

Capacity at start = C = 15.0 Mm^3

Annual sediment yield = 12.5 × 480 = 6000 MN

Volume of annual sediment yield = $\dfrac{6000\times10^6}{13.5\times10^3} = 0.444$ Mm^3

If η_t is the percentage trap efficiency, Volume of sediment trapped per year

$$= \frac{\eta_t \times 0.444}{100} = 0.00444\,\eta_t \text{ Mm}^3$$

Time, in years, to fill a volume of ΔV (Mm^3) = $\Delta t = \dfrac{\Delta V}{0.00444\,\eta_t} = 225\left(\dfrac{\Delta V}{\eta_t}\right)$

Using uniform steps of ΔV = 2 Mm^3, the calculations are performed in tabular form as shown below. In this, I = annual inflow = 120 Mm^3. The percentage trap efficiency is calculated as $\eta_t = 100\times\left[1-\dfrac{1}{(1+65(C/I)}\right]^2$.

Table for calculation of Δt for various capacities

Capacity C (Mm^3)	Average capacity over the step	C/I	η_t (Percentage)	Δt (Years)
1	2	3 = (Col.1/120)	4 By given equation	5 = (225 ×2)/(Col.4)
15				
13	14	0.117	78.056	5.77
11	12	0.100	75.111	5.99
9	10	0.083	71.260	6.31
7	8	0.067	66.016	6.82
			Total =	**32.58 Years**

Time required for the reservoir to have its capacity reduced to 5 Mm^3 ≈ 33 years.

Example 10.3 Assuming the relative density of a sand particle as 2.6 and unit weight (dry) of a cubic metre of sediment as 980 kgf, estimate the weight of 1 m^3 of deposited sediment in the reservoir bed.

Solution

In 1 m^3 of sediment

Volume of solids = $(1 - p)$ m^3 where p = porosity.

Volume of water = p m^3

Weight of solids = 980 = $(1 - p)$ × 2.60 × 1000

Hence $p = 0.623$

Weight of 1 m^3 of sediment deposit = (1 – 0.623) × 2.60 × 1000 + 0.623 × 1000

= 980 + 623 = 1603 kgf.

Example 10.4 Estimate the unit weight of a reservoir sediment in the first year of its deposition if the sediment contains 20% sand, 35% silt and 45% clay by weight. Estimate the volume occupied by 1000 tonnes of sediment in the first year and in 50th year. The reservoir can be assumed to have normally a moderate drawdown. Assume the reservoir operation is such that the sediment is always submerged.

Solution

The unit weight (dry) of reservoir sediment deposit is given by Eq. (10.11) as

$$W_T = \frac{P_{sa}}{100}(W_1 + B_1 \log T) + \frac{P_{si}}{100}(W_2 + B_2 \log T) + \frac{P_{cl}}{100}(W_3 + B_3 \log T)$$

Here $p_{sa} = 20, p_{si} = 35$ and $p_{cl} = 45$

Referring to Table 10.7, values of coefficients B_1, B_2 and B_3 and W_1, W_2 and W_3 are found and introduced into Eq. (10.11) to obtain

$$W_T = \frac{20}{100}(1490) + \frac{35}{100}(1185 + 43.3 \log T) + \frac{45}{100}(737 + 171.4 \log T)$$

$$W_T = 1044.4 + 92.285 \log T$$

When $T = 1$ year, $W_{T1} = 1044.4$ kgf/m^3

When $T = 50$ year, $W_{T50} = 1201.2$ kgf/m^3

volume of 1000 tonnes of sediment in first year $V_1 = \dfrac{1000}{1044.4/1000} = 957.5$ m^3

in 50th year $V_{50} = \dfrac{1000}{1201.2/1000} = 832.5$ m^3.

Example 10.5 A reservoir has a capacity of 130 Mm^3 at its full reservoir level. The average water inflow and average sediment inflow into the reservoir are estimated as 200 Mm^3/year and 2.00 M tonnes/year respectively. The sediment inflow was found to have a composition of 30% sand, 30% silt and 40% clay. Estimate the time in years required for the capacity of the reservoir to be reduced to 50% of its initial capacity. Assume the sediment is always submerged.

Solution

Initial reservoir capacity = 130 Mm^3

Final reservoir capacity = $0.5 \times 130 = 65$ Mm^3

Amount of sediment deposit = 65 M tonnes

First trial of average unit weight:

Assume a unit weight of 1.0 t/m^3.

Volume of total sediment deposit = 65 Mm^3

Assuming a C/I ratio > 0.70 throughout and trap efficiency $\eta_t = 100\%$

Approximate time required to fill 50% of initial capacity = 65/2.0 = 32.5 years

Assume $T = 33$ years to calculate W_{av}

Second Trial: Using Table 10.7,

Initial unit weight $W_1 = (1490 \times 0.30) + (1040 \times 0.30) + (480 \times 0.40) = 951$ kgf/m^3

$= 0.951$ tonnes/m^3

$B_w = (p_{sa} B_1 + p_{si} B_2 + p_{cl} B_3) = [0 + (0.3 \times 91.3) + (0.4 \times 256.3)] = 129.9$ kgf/m^3

By Eq. 10.12

$$W_{av} = 951 + 0.4343 \times (129.9)\left[\left\{\frac{33}{32}\ln 33\right\} - 1\right] = 1096 \text{ kgf/m}^3 = 1.096 \text{ tonnes/Mm}^3$$

The calculation for estimating the time to fill 65 Mm^3 of capacity by sediment is performed in tabular form (Table 10.9).

Table 10.9 ***Computation of Time for filling Part Capacity by Sediment—Example 10.5***

1	2	3	4	5	6	7	8	9
					Second Trial		Third Trial	
Capacity C (in Mm^3)	ΔC	Capacity-Inflow ratio (C/I)	Trap Efficiency	Average Trap Efficiency η_t	Volume of Sediment Deposited per year	Time to fill ΔC (in years)	Volume of Sediment Deposited per year	Time to fill ΔC (in years)
130		0.650	98.87					
115	15	0.575	98.12	98.50	1.797	8.35	1.78	8.41
105	10	0.525	97.57	97.85	1.786	5.60	1.77	5.64
95	10	0.475	96.97	97.27	1.775	5.63	1.76	5.67
85	10	0.425	96.29	96.63	1.763	5.67	1.75	5.71
75	10	0.375	95.53	95.91	1.750	5.71	1.74	5.76
65	10	0.325	94.66	95.10	1.735	5.76	1.72	5.80
					Total	36.73	Total	37.00

In Table 10.9:

Col. 1 gives the capacity of the reservoir. The decrease in the capacity due to sedimentation is considered in 6 steps. (Note that the steps need not be equal.)

Col. 2 is the decrease in capacity between two successive steps.

Col. 3 is the Capacity – Inflow ratio at the beginning of the step. Col. 3 = (Col. 1)/200

Col. 4 is the trap efficiency corresponding to C/I value of Col. 3 calculated by using Eq. (10.10). Thus Col. 4 = 6.064 × ln (Col. 3) + 101.48

Col. 5 is the average trap efficiency in a step

Col. 6 is the volume of sediment deposited in the reservoir per year during the event represented by the step. Col. 6 = 2.00 × ((Col. 5)/100)/1.096

Col. 7 is the time in years required to fill the capacity represented by the step.
Col. 7 = Col. 2/Col. 6.

The sum of the time periods represented by Col. 7 is the time required to fill the capacity by 65 Mm^3. Thus the total time required to reduce the reservoir capacity by 50% is T_{50} = 36.73 years = say **37 years**.

Since the value of T_{50} differs from the assumed value of 33 years by more than 5% further trials to refine the value of W_{av} are necessary.

Third trial: Take T = 37 years

$$W_{av} = 951 + 0.4343 \times (129.9) \left[\left\{\frac{37}{36} \ln 37\right\} - 1\right] = 1104 \text{ kgf/m}^3 = 1.104 \text{ tonnes/Mm}^3$$

Using this value of W_{av} = 1.104 tonnes/Mm^3, the second Trial of Table 10.8 is refined as shown in Cols. 8 and 9 of Table 10.8.

The total time required for filling 50% of capacity is indicated as 37 years which is the same as assumed at the beginning of this trial. Thus, $\boldsymbol{T_{50}}$ **= 37 years** is the desired period.

10.9 *Distribution of Sediment in the Reservoir*

10.9.1 Classification of Reservoirs

When a sediment-laden stream enters a reservoir, the sediment is deposited not only at the head of the reservoir as delta deposit but also all along the internal surface of the reservoir. The area as well as the volume distribution of accumulated sediment at various levels is an important factor in the design of reservoirs. Further, the distribution and the rate of growth of deposits decide the location of various outlets as well as the operation strategy of the reservoir.

The deposition pattern of sediment depends on a host of factors which include the slope, geometry of the reservoir, particle size distribution of sediment and the operation pattern of the reservoir. On the basis of extensive field data of reservoirs in *USA*, Borland and Miller classified the reservoir into four standard types as mentioned below:

Classification Number	Reservoir Type	Parameter *m*
IV	Gorge	1.0–1.5
III	Hilly region	1.5–2.5
II	Flood plain, Foot-hill region	2.5–3.5
I	Lake	>3.5

The parameter *m* is the reciprocal of the slope of best-fitting line obtained by plotting reservoir elevation above bed as ordinate and reservoir capacity at that elevation as abscissa on a log-log paper [Fig. 10.5(a)]. (Note that linear scales are used to measure the slope). The typical distribution of sediment in these four types of reservoirs is shown in Fig. 10.5(b).

It is seen that Type I reservoirs have considerably lower percentage of silt at lower depths when compared to Type IV reservoirs. Conversely, Type IV and III reservoirs have very low percentage of sediment at top portions when compared to Type I and II reservoirs.

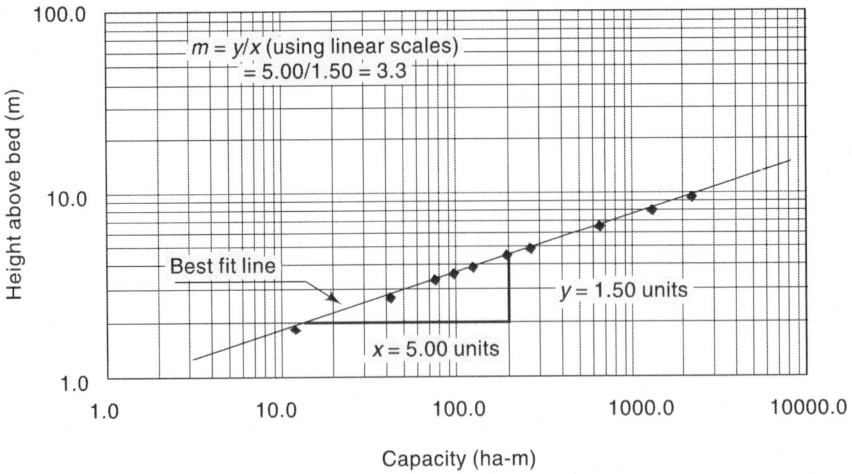

Fig. 10.5(a) ***Determination of Parameter m***

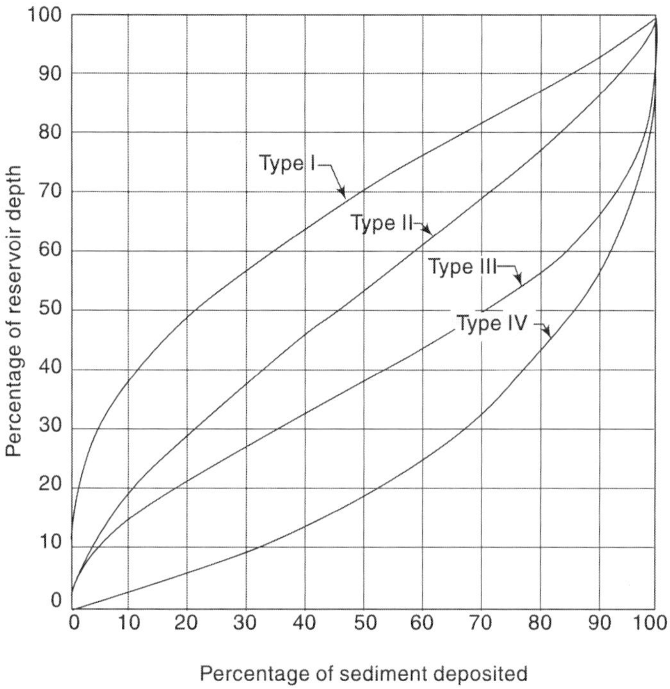

Fig. 10.5(b) ***Typical Distribution of Sediment in various Reservoir Types***

10.9.2 Methods of Predicting Sediment Distribution

Two methods, both suggested by Borland and Miller, known as (i) *empirical area reduction method*, and (ii) *area increment method* are in common use. These are described below.

1. Empirical Area Reduction Method

(a) Data Needed

1. Elevation–Area–Capacity relationship for the reservoir at any time T_i after the construction of the reservoir. Usually, at $T_i = 0$, i.e. at the start of the reservoir project this data will be available.
2. Estimated volume of sediment V_s to be deposited in a period $(T_f - T_i) = \Delta T$ years. Usually for design purposes $\Delta T = 50$ years or 100 years is selected.

(b) Distribution Pattern Based on the observed reservoir sediment distribution data, Borland and Miller have expressed the distribution of sediment area at any level h above the bed as

$$A_p = C\ p^{m_1}\ (1-p)^{n_1} \qquad (10.13)$$

where A_p = a dimensionless relative area = $\dfrac{\text{Area at elevation } h \text{ above the bed}}{\text{Area at initial zero elevation}}$

p = relative depth = h/H

where h = height above the reservoir bed to any given elevation in the reservoir, and

H = Difference in the elevations of FRL and original bed of the reservoir
= depth or reservoir at Full reservoir level (FRL). (Obviously, p has a range of 0 to 1.0.)

C, m_1 and n_1 = coefficients dependent on the type classification of the reservoir as given below (Ref. 7):

Reservoir Type	C	m_1	n_1
I	5.074	1.85	0.36
II	2.487	0.57	0.41
III	16.967	1.15	2.32
IV	1.486	–0.25	1.34

(c) Procedure A trial-and-error procedure is adopted. The reservoir surface can be taken to be somewhat conical in shape. When a sediment volume V_s is deposited gradually over a time ΔT, some part of the conical portion of the reservoir completely fills up with sediment (say up to a height of h_o above the vertex) and in the remaining portion the deposition will be on the surface and the cross sectional area at any elevation will be diminished, Fig. 10.6. Let the volume of sediment filled in the conical portion to a depth of $h_o = V_{so}$.

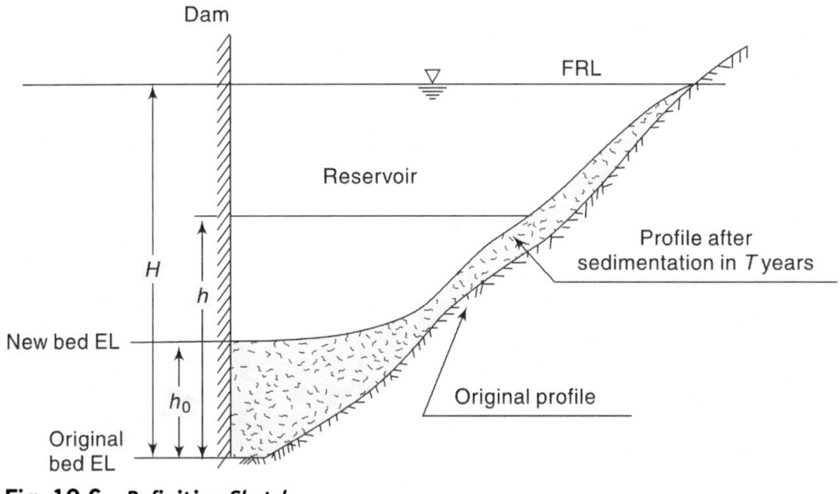

Fig. 10.6 ***Definition Sketch***

1. The elevation h_o, relative to the original bed of the reservoir, up to which the sediment completely fills up the reservoir, is assumed. The top of this filled up portion is taken as the new datum, i.e., the new zero elevation. The area A_o at this depth is determined. The value of p at this level = $p_o = h_o/H$.
2. The new total depth of the reservoir = $H - h_o$
3. Volume of sediment to be distributed = $V_s - V_{so}$
4. The type classification of the reservoir is determined
5. Values of A_p are determined for various values of p (= h_o/H) by using Eq. (10.13)
6. At $p = p_o$, $A_p = A_{po}$

7. Find $K = A_o/A_{po}$
8. Using this value of K, the sediment area A_s at any height h above, the new datum is determined as $A_s = A_p K$
9. ΔV_s = Volume or sediment deposited between two consecutive heights h_1 and h_2 above the new datum is determined either as

   $\Delta V_s = (A + A_2)\, \Delta h/2$ ...(average end area method) or as

   $\Delta V_s = (A_1 + A_2 + \sqrt{A_1 A_2}\,)\dfrac{\Delta h}{3}$ ...(weighted area method).

   where $\Delta h = (h_2 - h_1)$
10. Accumulated sediment at various elevations starting from the original bed elevation are now determined.
11. The total volume of deposited sediment up to the top of the reservoir, obtained at step No. 10, must be equal to the given value of V_s. If the value found in step 10 differs from V_s considerably, say more than 2%, then the entire procedure is repeated by assuming a new value of h_o, i.e. new zero elevation.

Example 10.5 given below illustrates the use of this procedure.

Example 10.6 For a reservoir the capacity—area—elevation data is given below. Estimated total accumulation of sediment in the reservoir in 50 years of its operation is 100 MCM. Original bed elevation is El. 535.00 m and the spillway crest is at 546.50 m. Determine the distribution of 100 MCM of sediment in the reservoir by the empirical area reduction method. The reservoir can be taken as of Type II. [Note: For the first trial assume the level up to which the reservoir is fully covered by sediment at the end of 50 years as 539.40 m].

Elevation (m)	535.00	536.50	538.0	539.0	539.40	540.00
Original Water spread area (sq. km)	2.0	4.51	6.89	8.71	11.52	15.74
Original Reservoir Capacity (MCM)	0	5.18	13.13	20.92	24.96	33.10
Elevation (m)	541.00	542.00	543.00	544.00	545.00	546.50
Original Water spread area (sq km)	26.88	38.02	47.84	57.66	67.09	81.15
Original Reservoir Capacity (MCM)	54.35	86.76	244.76	355.97	744.76	355.97

Solution

Table 10.10 shows the computations.

First, the given elevation-area data is enlarged through linear interpolation to cover the elevation at an average interval of 0.50 m. The incremental area between two area elements A_1 and A_2 separated by a height Δh is calculated as

$$\Delta V = \frac{(A_1 + A_2)}{2} \times \Delta h$$

The accumulated volume starting from the original bed El. 535.00 m is calculated to get original reservoir capacity-elevation data.

Table 10.10 *Empirical Area Reduction Method of Determining Sediment Deposition—Example 10.6*

Sediment Deposition Computations by Empirical Area Reduction Method
Period of Sedimentation: 50 years
Total Sedimentation accumulation in the period: 100 MCM
Spillway crest: 546.50 m | **Original River bed = 535.00** | **Depth up to spillway crest = 11.50 m**
First Trial (K_1 = 9.761) | **Final Trial (K_2 = 9.646)** | **Final Values**

1	2	3	4	5	6	7	8	9	10	11	12	13	14	15	16
Sl. No.	Elevation (m)	Original Water spread Area (sq. km)	Height above original bed (m)	Original Reservoir Capacity (MCM)	Relative Depth (p)	Function *A* (Type II)	Sediment Area (KA_p) (Sq. km)	Incremental sediment volume (MCM)	Accumulated Sediment Volume (MCM)	Revised Sediment Area (KA_p) (sq. km)	Revised Incremental Sediment Volume (MCM)	Revised Accumulated Sediment Volume (MCM)	Reservoir Area at the end of 50 years (Sq. km) (MCM)	Reservoir Capacity at the end of 50 years (MCM)	Available reservoir elevation after at the end of 50 years
1	535.0	2.00	0.00	0.00	0	0									
2	535.5	2.77	0.50	1.19	0.043	0.409									
3	536.0	3.55	1.00	2.76	0.087	0.595									
4	536.5	4.35	1.50	4.73	0.130	0.735									
5	536.6	4.51	1.60	5.18	0.139	0.760									
6	537.0	5.16	2.00	7.11	0.174	0.848									
7	537.5	6.02	2.50	9.90	0.217	0.942									
8	538.0	6.89	3.00	13.13	0.261	1.021									
9	538.5	7.80	3.50	16.80	0.304	1.088									
10	539.0	8.71	4.00	20.92	0.348	1.143									
11	**539.4**	11.52	4.40	24.96	0.383	1.180	11.52	0.00	24.96	11.52	0.00	24.96	0.00	0.00	539.4
12	539.5	12.22	4.50	26.13	0.391	1.189	11.60	1.16	26.11	11.46	1.15	26.11	0.76	0.03	539.5
13	540.0	15.74	5.00	33.10	0.435	1.224	11.95	5.89	32.00	11.81	5.82	31.92	3.93	1.18	540.0
14	540.5	21.31	5.50	42.33	0.478	1.251	12.21	6.04	38.04	12.07	5.97	37.89	9.24	4.44	540.5
15	541.0	26.88	6.00	54.35	0.522	1.268	12.38	6.15	44.19	12.24	6.08	43.97	14.64	10.38	541.0

Table 10.10 (*Contd.*) Example 10.6

1	2	3	4	5	6	7	8	9	10	11	12	13	14	15	16
16	541.5	32.45	6.50	69.16	0.565	1.277	12.46	6.21	50.40	12.32	6.14	50.11	20.13	19.05	541.5
17	542.0	38.02	7.00	86.76	0.609	1.276	12.45	6.23	56.63	12.30	6.16	56.26	25.72	30.50	542.0
18	542.5	42.93	7.50	106.99	0.652	1.264	12.34	6.20	62.83	12.19	6.12	62.39	30.74	44.60	542.5
19	543.0	47.84	8.00	129.67	0.696	1.242	12.12	6.12	68.94	11.98	6.04	68.43	35.86	61.24	543.0
20	543.5	52.75	8.50	154.80	0.739	1.207	11.78	5.97	74.92	11.64	5.90	74.33	41.11	80.47	543.5
21	544.0	57.66	9.00	182.40	0.783	1.157	11.29	5.77	80.68	11.16	5.70	80.03	46.50	102.36	544.0
22	544.5	62.37	9.50	212.40	0.826	1.089	10.63	5.48	86.16	10.50	5.42	85.45	51.87	126.95	544.5
23	545.0	67.09	10.00	244.76	0.870	0.996	9.72	5.09	91.25	9.61	5.03	90.48	57.48	154.28	545.0
24	545.5	71.81	10.50	279.47	0.913	0.868	8.47	4.55	95.80	8.37	4.49	94.97	63.44	184.50	545.5
25	546.0	76.53	11.00	316.55	0.957	0.670	6.54	3.75	99.55	6.47	3.71	98.68	70.06	217.87	546.0
26	546.5	81.15	11.50	355.97	1.000	0.000	0.00	1.64	101.19	0.00	1.62	100.30	81.15	255.67	546.5

Capacity after 50 years
= 256 (MCM)

$K_1 = 9.761$
$K_2 = 9.646$

El.539.40 m is the determined new zero elevation after 50 years

In Table 10.9:

Col. 2 = elevation in metres

Col. 3 = original water surface area at given elevation, sq.km

Col. 4 = height above original bed elevation = [Elevation of the item – 535.00] = h

Col. 5 = original reservoir capacity in MCM. This list contains given data as well as newly generated data

Col. 6 = relative depth $p = h/H$ = Col. 4 /11.50

Col. 7 = function A_p calculated by Eq. 10.13 as $A_p = 2.487\, p^{0.57}\, (1 - p)^{0.41}$.

First Trial: A trial-and-error procedure is adopted. For the first trial, the level up to which the reservoir is fully covered by sediment at the end of 50 years is taken as 539.40 m, as per the suggestion in the problem. This would be the new datum for the bed of the reservoir at the end of 50 years. At this level:

From Col. 7, $A_1 = 1.180$, 3, From Col. 3, Original reservoir area = $A_o = 11.52$ km^2.

Coefficient K (for the first trial) = $K_1 = 11.52/1.180 = 9.761$

Col. 8 = Represents $K A_p = 9.761 \times$ Col. 7 for all elevations higher than 539.40 m. This column represents the area covered by sedimentation at a particular level and hence called *Sediment area.*

Col. 9 = Incremental sediment volume between two successive elevation calculated as (average sediment area × incremental depth).

Col. 10 = Accumulated sediment volume starting from 24.96 MCM at El.539.40 m.

Note that the value 24.96 represents the original volume of the reservoir at the elevation 539.40 m and this volume is taken as completely filled up by sediment at the end of 50 years.

The last value in Col. 10 is obtained as 101.19 MCM whereas the given sediment data indicates a deposition of 100 MCM in 50 years. This indicates a need for second trial. The calculations are very near the logical final values and since the difference is slightly more than 0.5% only minor corrections are needed. Tweaking the coefficient K does this. Thus for second trial K_2 = Adjusted value of K_1 on a pro-rata basis of total accumulated sediment volume = $\dfrac{K_1 \times 100}{101.19} = 9.646$. Values of this second trial (with $K_2 = 9.646$) are shown in Cols. 11, 12 and 13.

Second Trial:

Col. 11 = Revised sediment area with $K_2 = 9.646$.

Col. 12 = Revised incremental sediment volume with $K_2 = 9.646$

Col. 13 = Revised accumulated sediment volume with $K_2 = 9.646$

Note that this second trial improves the result considerably and the accumulated sediment volume at El. 546.50 m is 100.30 MCM. The difference between this and the given value of 100.0 MCM is negligible being less than 0.5% and thus this distribution could be taken as final values.

Col. 14 = Reservoir area distribution at the end of 50 years

Col. 15 = Reservoir volume distribution at the end of 50 years

Col. 16 = Available reservoir elevations at the end of 50 years

Note that the new bed level of the reservoir at the end of 50 years is 539.40 m.

The distribution of the area and capacity with elevation, at the beginning and at the end of 50 years is shown in Fig. 10.7 (a and b).

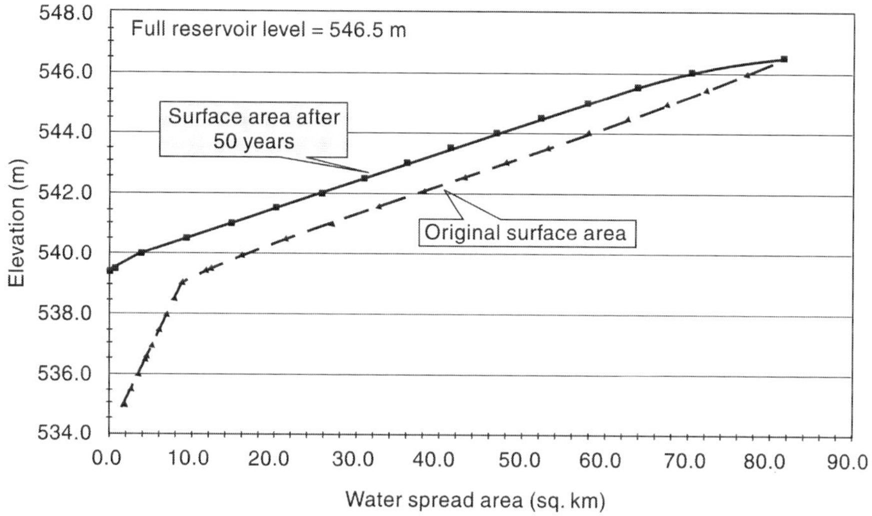

Fig. 10.7(a) ***Reservoir Elevation–Area Curves—Example 10.6***

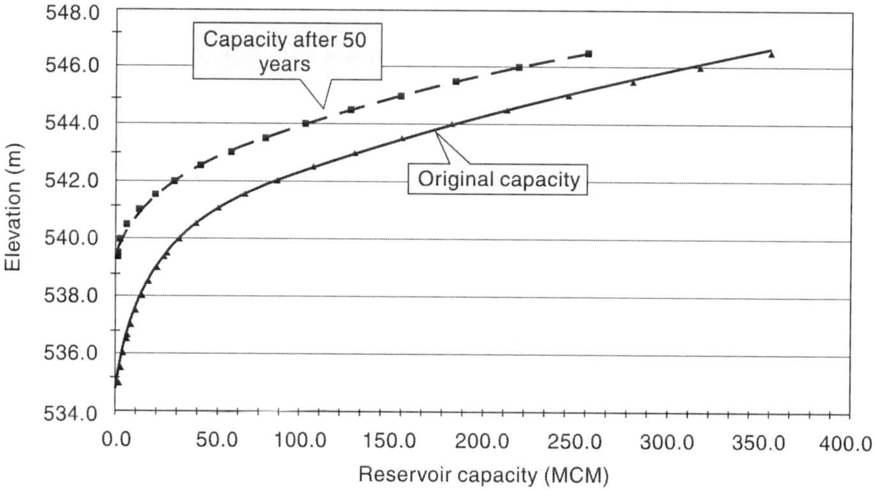

Fig. 10.7(b) ***Reservoir Elevation–Capacity Curves—Example 10.6***

2. Area Increment Method

The basic assumption of this method is that the volume of sediment deposition per unit height in the reservoir is constant. This is same as assuming that the area-elevation curve after sedimentation is parallel to the curve before sedimentation. Thus, the sediment area is constant at all elevations and is equal to the sediment area at the new zero elevation, h_0. The distribution of sediment is given by

$$V_s = A_o(H - h_0) + V_0 \tag{10.14}$$

where V_s = sediment volume to be distributed in the reservoir

A_o = original reservoir area at the new zero level

h_o = depth at which reservoir is completely filled up = elevation of new zero elevation with respect to old bed elevation as datum

V_o = sediment volume between old zero and new zero bed level

H = Difference in the elevations of *FRL* and original bed of the reservoir

= depth of reservoir at Full reservoir level (*FRL*), (original value).

The procedure or distributing the given total sediment volume V_s is done by a trial and error procedure as detailed below:

(i) Assume h_o and find corresponding V_o and A_o

(ii) Check whether V_s given by Eq. (10.14) agree with the given value. If not repeat with a new value of h_o

(iii) Sediment area at any level is A_o. Establish water surface area at any level as (original area – A_o).

(iv) Over a depth Δh. Incremental sediment volume = $A_o \cdot \Delta h$.

Hence reservoir capacity after sedimentation at any level h' above new zero

$$= (\text{Original volume} - V_o - A_o h')$$

Example 10.7 illustrates the procedure

Example 10.7 The reservoir of Example 10.6 is expected to have 50 MCM of sediment accumulated in first 25 years of its operation. Determine the distribution of this 25 MCM of sediment by Area Increment method.

Solution

Given data: $V_s = 50 \text{ Mm}^3$

$H = 546.50 - 535.00 = 11.50$ m

First Trial:

Assume that the new zero elevation is 536.50 m giving $h_o = 536.5 - 535.00 = 1.50$ m

$A_o = 4.35 \text{ km}^2$ (from given data. Col. 3 in Table 10.11)

V_o = volume corresponding to El. 536.50 = 4.73 Mm^3 (from Col. 6 of Table 10.11)

By Eq. 10.14, $V_s = V_o + A_o (H - h_o) = 4.73 + 4.35(11.5 - 1.5) = 48.235 \text{ Mm}^3$

While this value is close to the given value of $V_s = 50 \text{ Mm}^3$, a new trial to get better agreement is needed.

Second Trial:

For the second trial, assume new zero elevation as 535.60 m.

$h_o = 536.6 - 535.00 = 1.60$ m

$A_o = 4.51 \text{ km}^2$

V_o = volume corresponding to El. 536.60 = 5.18 Mm^3 (From Col. 6 of Table 10.11)

By Eq. (10.14), $V_s = V_o + A_o(H - h_o) = 5.18 + 4.51\ (11.5 - 1.6) = 49.83 \text{ Mm}^3$

This V_s value is nearly equal to the given value of 50.00 Mm^3 and differs by only 0.34%. Since the difference is less than the permissible 1% value, no new trial is required. The assumed zero elevation of 536.60 is considered to represent the bed elevation at the end of 25 years satisfactorily. Using this value, the distribution of sediment at different elevations is worked out as shown in Table 10.11.

In Table 10.11:

Cols. 2 and 3 are given data. Col. 4 is incremental height in m

Table 10.11 ***Area Increment Method of determining Sediment Deposition—Example 10.7***

Period of Sedimentation: 25 years

Total Sediment accumulation in the period : 50 MCM   Spillway crest: 546.50 m

Sl. No.	Elevation (m)	Water spread Area (sq. km)	Incremental height (m)	Incremental volume (MCM)	Reservoir Capacity (MCM)	Sediment Area = A_o (sq. km)	Incremental Sediment Volume (MCM)	Cumulative Sediment Volume (MCM)	Reservoir area at the end of 25 years (sq. km)	Reservoir Capacity at the end of 25 years (MCM)
1	2	3	4	5	6	7	8	9	10	11
1	535.0	2.00	0.00	0.00	0.00	0	0	0	0	0.0
2	535.5	2.77	0.50	1.19	1.19	0	0	1.19	0	0.0
3	536.0	3.55	0.50	1.58	2.76	0	0	2.76	0	0.0
4	536.5	4.35	0.50	1.97	4.73	0	0	4.73	0	0.0
5	**536.6**	4.51	0.10	0.44	5.18	4.51	0.00	5.18	0	0.0
6	537.0	5.16	0.40	1.93	7.11	4.51	1.804	6.98	0.65	0.1
7	537.5	6.02	0.50	2.79	9.90	4.51	2.255	9.24	1.51	0.7
8	538.0	6.89	0.50	3.23	13.13	4.51	2.255	11.49	2.38	1.6
9	538.5	7.80	0.50	3.67	16.80	4.51	2.255	13.75	3.29	3.0
10	539.0	8.71	0.50	4.13	20.92	4.51	2.255	16.00	4.20	4.9
11	539.5	12.22	0.50	5.21	26.13	4.51	2.255	18.26	7.71	7.9
12	540.0	15.74	0.50	6.97	33.10	4.51	2.255	20.51	11.23	12.6
13	540.5	21.31	0.50	9.23	42.33	4.51	2.255	22.77	16.80	19.6
14	541.0	26.88	0.50	12.02	54.35	4.51	2.255	25.02	22.37	29.3

Table 10.11 (*Contd.*) Example 10.7

1	2	3	4	5	6	7	8	9	10	11
15	541.5	32.45	0.50	14.81	69.16	4.51	2.255	27.28	27.94	41.9
16	542.0	38.02	0.50	17.60	86.76	4.51	2.255	29.53	33.51	57.2
17	542.5	42.93	0.50	20.23	106.99	4.51	2.255	31.79	38.42	75.2
18	543.0	47.84	0.50	22.68	129.67	4.51	2.255	34.04	43.33	95.6
19	543.5	52.75	0.50	25.14	154.80	4.51	2.255	36.30	48.24	118.5
20	544.0	57.66	0.50	27.59	182.40	4.51	2.255	38.55	53.15	143.8
21	544.5	62.37	0.50	30.00	212.40	4.51	2.255	40.81	57.86	171.6
22	545.0	67.09	0.50	32.36	244.76	4.51	2.255	43.06	62.58	201.7
23	545.5	71.81	0.50	34.72	279.47	4.51	2.255	45.32	67.30	234.2
24	546.0	76.53	0.50	37.08	316.55	4.51	2.255	47.57	72.02	269.0
25	546.5	81.15	0.50	39.41	355.97	4.51	2.255	**49.83**	**76.64**	**306.1**

El. 536.60 m is the determined new bed elevation.

Total sediment load up to El. 546.50 m at the end of 25 years = 49.83 = say **50 MCM**

Col. 5 is the incremental reservoir capacity between two elevations calculated as (average area × incremental height)

Col. 6 is cumulative value of Col. 5

Col. 7 is sediment area at chosen h_o and is constant over the full depth of the reservoir measured above the chosen new zero elevation

Col. 8 is incremental sediment volume calculated as equal to [A_o × (incremental height)].

Col. 9 is cumulative sediment volume. Note that the values are the same as in Col.6 up to El. 536.60.

Col. 10 is estimated reservoir area at the end of 25 years. This area starts from new zero elevation of 536.60 m and is equal to (Col. 3–Col. 7).

Col. 11 is estimated reservoir volume at the end of 25 years. This volume starts with a zero value at new zero elevation of 536.60 m and is equal to (Col. 6–Col. 9).

10.10 *Life of a Reservoir*

A reservoir is designed to serve one or more specific purposes. Sedimentation causes progressive reduction in the capacity of the reservoir and this may impact on the desired performances of the reservoir at some point in time. With this in view, various definitions of specific terms related to the general term *life of a reservoir* are defined.

1. Useful Life

Period through which deposited sediment does not impact on the intended purposes of the reservoir. Useful life is considered to be over when an additional reservoir is to be built (or water is to imported from another source) to meet the original intended demand.

2. Economic Life

A point of time since the commissioning of the project at which the physical depreciation due to sedimentation, in conjunction with other economic and physical factors cause the operation of the reservoir, to meet intended demands, economically inefficient.

3. Design Life

A fixed period (usually 50 years or 100 years) adopted by the designers as estimate of minimum assured useful life.

The present-day practice in India is to adopt 100 years as the design life of the reservoir. The *CWC* practice in this connection is as follows:

- Volumes of sediment estimated to be deposited in the reservoir in 50 years (V_{50}) and in 100 years (V_{100}) of operation are worked out.
- Distribution of V_{100} is performed and the zero level after sedimentation is established.
- Minimum drawdown level (*MDDL*) is fixed a little above the zero level corresponding to V_{100} found in the second step above.
- Distribution of sediment V_{50} is performed to develop area–elevation–capacity curves. This set of curves corresponding to 50 years of sedimentation is used in working-table studies, reservoir performance simulation studies, etc.

10.11 *Reservoir Sedimentation Control*

Sedimentation of reservoirs causes great economic loss primarily due to reduction of storage capacity of the reservoirs. Other impacts of sedimentation such as increased high flood levels due to flattening of the bed slope in the river upstream of the reservoir leading to frequent inundation and waterlogging in the up-reservoir areas are serious in many instances. As such, monitoring and control of sedimentation forms a prime item in the management of any major reservoir project. Considering the basic natural process of erosion and transportation inherent in the phenomenon, it is obvious that the reservoir sedimentation can never be stopped but with good effort can be retarded considerably. The basic methods available for control can be listed as

- Reduction in sediment yield from the catchment
- Reduction in the rate of accumulation of sediment in the reservoir
- Physical removal of already deposited sediment

10.11.1 Reduction in Sediment Yield

Various control measures that can be adopted to reduce erosion and transportation of eroded products in the catchment are dealt under the specialised interdisciplinary practice known as *Soil Conservation technology*. After a thorough study of the catchment area, soil and water-conservation practices best suited for each sub watershed of the catchment have to be established by the specialists in the area of soil and water conservation. In a general sense, the soil conservation practices involve components such as

- Terraces, strip cropping and contour bunding to retard overland flow and hence reduction in sheet erosion
- Check dams, ravine reclamation structures etc. to reduce sediment inflow into the stream
- Vegetal covers, grassed waterways and afforestation to reduce runoff rates and hence to reduce erosion

In view of the interlocking and interdependency feature of various aspects of soil, water, biomass production and livelihood of people living in the watersheds, an integrated approach should be adopted in watershed management. Integrated operation of soil and water conservation aspects in a watershed can be represented as in Fig. 10.8.

10.11.2 Reduction of Accumulation of Sediment

Accumulation of the sediment that flows into the reservoir can be reduced if arrangements are made for venting out the sediment through structural arrangements in the dam and appropriate reservoir operation. Some of the measures that can be adopted towards this are

- Provision of scouring sluices at lower elevations in the dam to flush out high concentration sediments and density currents
- Appropriate operation of gated overflow outlets and other sluices in the dam in such a manner to allow passage of freshets with high concentration of sediments to the downstream of the dam and to catch only relatively clear water flows for storage in the reservoir.

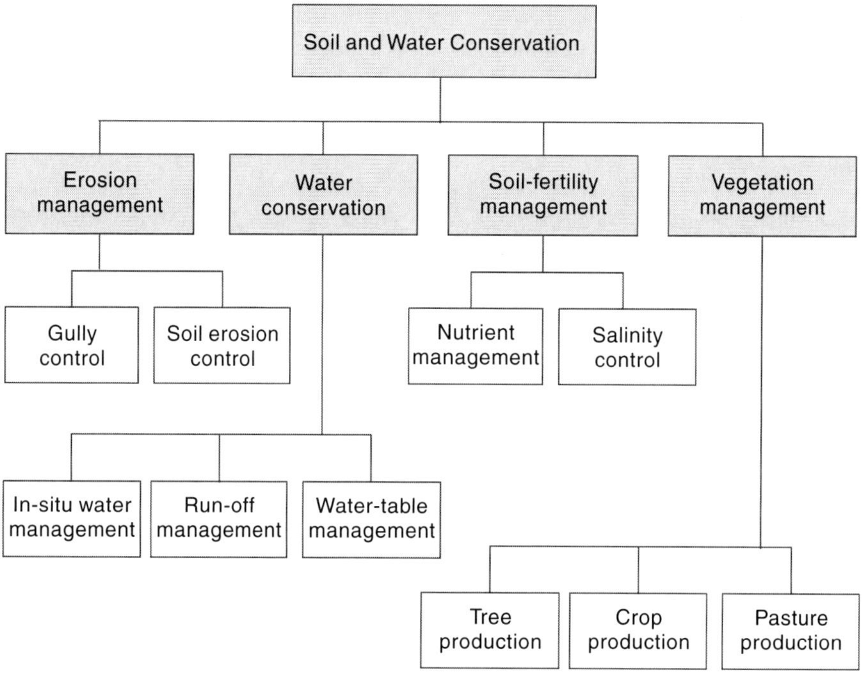

Fig. 10.8 ***Integrated Soil and Water Conservation***

10.11.3 Physical Removal of Deposited Sediments

Deposited sediments can be removed by hydraulic or mechanical means. However, for large reservoirs the disposal of removed sediment does pose environmental problems and also the entire operation may not be economically feasible. However, for many small reservoirs, sediment removal, popularly known as *desilting,* can be a feasible proposition. As a typical example, desilting of irrigation tanks of south India can be cited. Several thousands of tanks in South India, particularly in Andhra Pradesh, Karnataka and Tamil Nadu are in existence since several decades and are serving as sources of minor irrigation. Many of these tanks have been successfully desilted, in recent past, and their capacity restored to their original values. Acute scarcity of water, community participation and use of tank silt as soil amendment in both irrigated command area and in the up-catchment rain-fed agricultural lands have made these ventures economically viable.

10.12 *Erosion and Reservoir Sedimentation Problems in India*

10.12.1 Erosion Problem

The India, practically all the regions are subjected to fairly serious erosion problems due to several reasons. It is estimated that out of 305.9 M ha of reported area in the country about 145 M ha is in need of soil conservation measures. Table 10.12 gives details of soil-conservation problem areas in India under different land cover/use. It is seen that major part of agricultural land suffers from erosion problem and consequent loss of productivity, nutrient and soil resource. The distribution of soil-erosion problem areas (as of 1985) statewise is shown in Table 10.13(a).

Table 10.12 *Soil Conservation Problem Areas in India*

Particulars	Total Area (M ha)	Soil-Conservation problem area (M ha)
Forest	61.170	20
Culturable wasteland	17.362	15
Permanent pastures and other grazing land	14.809	14
Land under miscellaneous tree crops and groves	4.218	1
Fallow lands:		
(i) Fallow lands other than current fallows	9.168	8
(ii) Current fallows	11.132	7
Total for Fallow lands	15	20.5
Net area under cultivation	137.9	80
Other land uses, not available for agriculture, forest, etc.	50.188	—
Grand Total	305.947*	145

*305.947 million hectares is the reported area for land-utilisation statistics out of a geographical area of 328.809 million hectares.

Table 10.13(a) *State-wise Distribution of Soil Erosion Problem Areas (as of 1985)*

Sl. No.	State	Extent of Problem area due to Soil Erosion (M ha)	Sl. No.	State	Extent of Problem area due to Soil Erosion (M ha)
1.	Andhra Pradesh	11.502	14.	Nagaland	0.405
2.	Assam	2.217	15.	Orissa	4.578
3.	Bihar	4.260	16.	Punjab	1.007
4.	Gujarat	9.946	17.	Rajasthan	19.902
5.	Haryana	1.591	18.	Sikkim	0.303
6	Himachal Pradesh	1.914	19.	Tamil Nadu	3.640

Contd.

Table 10.13(a) *Contd.*

7.	Jammu & Kashmir	0.883	20.	Tripura	0.167
8.	Kamataka	10.989	21.	Uttar Pradesh	7.110
9.	Kerala	1.757	22.	West Bengal	1.033
10.	Madhya Pradesh	19.610	23.	Arunachal Pradesh	2.444
11.	Maharashtra	19.181	24.	Goa	0.200
12.	Manipur	0.374	25.	Mizoram	0.421
13.	Meghalaya	0.837	26.	Union Territories	0.349

(Source: Ref. 5)

For purposes of developing appropriate technologies for soil conservation, Central Soil and Water Conservation Institute has considered the erosion problem areas of the country under three categories as hilly regions, ravine regions and semi-arid, black and red soil regions. The characteristic features of these regions are as in Table 10.13(b).

Table 10.13(b) ***Region-wise Soil Erosion Problems***

Region	Area	Extent (M ha)	Sub-areas	Details
Hilly Region	Western, North-Western & Central Himalaya	31.13	J & K, UP (hill Districts) and HP	Area is prone to erosion haz-ards due to weak geology, deforesta-tion and hill road construction.
	Eastern Himalayan Region	17.70	Assam. Eastern states, Sikkm and West Bengal	Shifting cultivation has caused denudation and degradation of land. Heavy runoff and massive soil erosion.
	Western Ghats	7.74	T.N., Kerala, Karnataka and Maharashtra	Heavy rainfall in the range 1300 to 6000 mm. Prone to severe erosion due to deforestation, faulty land use and overgrasing.
Ravine Region	Arid to Semi arid regions	3.67	Parts of UP., M.P., Bihar, Rajasthan, Gujarat, Punjab, W.B., I.N. and Maharashtra	Faulty land use, overgrasing and loss of natural land covering are the chief causes of gully and ravine formation and soil erosion.
Semi-arid Black and Red Soil Region	Arid to Semi-arid tropical region	97.00	Rest of the country in semi-arid tropical region not covered in the above	About 500 to 2000 mm of rainfall and 80 to 90% of rainfall received in 30 to 70 hours. Sheet erosion caused by high intensity rainfall.

(Source: Ref. 5)

Table 10.14 *Rate of Silting in Some Reservoirs in India*

Sl. No.	Name of Reservoir	River	Storage Capacity (Mm^3)	Catchment Area (km^2)	Year of Impounding	Sedimentation in ha-m/100 km^2/year		Period (years)	Yearly Average Loss in capacity (%)
						Designed rate	Observed rate (Average for period)		
1	Sriramasagar	Godavari	3171.9	91751	1970	3.57	6.19	14	1.79
2	Nizamsagar	Manjira	841.2	21694	1930	2.38	4.89	45	1.26
3	Panchet Hill	Damodar	1581.0	10878	1956	6.67	5.89	29	0.40
4	Maithon	Barakae	1348.8	6294	1955	9.05	10.25	24	0.48
5	Ukai	Tapi	8510.0	62224	1972	1.49	7.16	12	0.53
6	Kadana	Mahi	1543.0	255520	1977	1.30	3.92	7	0.65
7	Pongh	Beas	8579.0	12562	1974	NA	27.85	12	0.41
8	Tungabhadra	Tungabhadra	3751.2	28180	1953	4.29	6.48	32	0.49
9	Bhadar	Bhadar	239.2	2435	1963	7.60	11.61	11	1.18
10	Gandhi Sagar	Chambal	7740.0	23025	1960	3.57	8.96	16	0.29
11	Girna	Girna and Panzam	608.8	4729	1965	0.56	7.49	14	0.58
12	Shivaji Sagar (Koyna)	Koyna	2987.8	891	1961	6.67	7.71	10	0.02
13	Hirakud	Mahanadi	8105.0	83395	1957	2.50	6.62	27	0.61
14	Bhakra	Satluj	9869.0	56980	1958	4.29	5.57	29	0.32
15	Matatila	Betwa	11.3	20720	1956	1.33	6.00	28	1.10
16	Ramganga	Ramganga	2449.6	3134	1975	4.25	22.94	10	1.10
17	Ichari	Tons	11.6	4913	1972	NA	1.33	6	0.65
18	Dhukwan	Betwa	106.5	21340	1907	0.43	0.30	73	0.61
19	Mayurakshi	Mayurakshi	607.7	1860	1955	3.75	16.83	15	0.52

Source: Compendium on Silting of Reservoirs in India, *CWC*, 1991

10.12.2 Reservoir Sedimentation

A large number of reservoirs have been built in India since 1950. Sedimentation in the reservoirs has been found to be fairly high particularly in those reservoirs sited in arid and semi-arid tropical erodible regions. Reservoir surveys conducted during 1958 to 1986 indicated that all of the surveyed reservoirs were found to be silting at a rate faster than what was anticipated. In a majority of the reservoirs about 50% of sediment is deposited in the upper 20–30% of the depth indicating deposition in the head reaches of the reservoir. Reference 7 reports the details of surveys on a large number of reservoirs. A summary of sedimentation surveys on 19 reservoirs covering different regions of the country is given in Table 10.14. It is seen from this table that the reservoirs are losing annually a capacity of about 0.75% of its original value. Further, the range of loss of reservoir capacity is 1.79 to 0.02% and in majority of cases the actual rate of sedimentation is many times more than the designed rates.

As per Compendium of Reservoirs by *CWC* (2001) , the live and gross storages created by the country is about 177 and 217 km^3 respectively. Based on sediment data on 144 reservoirs, the weighted average annual loss in gross storage due to siltation is 0.44%. Thus, the likely annual loss in the gross storage of 217 km^3 is 0.95 km^3. Another set of 42 reservoirs indicated an annual loss in live storage of 0.31%. Thus, the likely loss in the total live storage is 0.55 km^3. Considering an average density of 1.137 tonnes/m^3, the weight of the total annual sediment deposits in all the reservoirs in India is about 1080 million tonnes. This underscores the magnitude of annual loss of reservoir capacity and also the problem of sediment erosion in the catchments of reservoirs.

REFERENCES

1. Agarwal, K.K. and Idicula, K.K., *Reservoir Sedimentation Surveys Using Geographical Positioning Systems,* The Asian *GPS* Conf. New Delhi, Oct. 2000, pp 47–50.
2. Chow, V.T. (Ed), *Hand Book of Applied Hydrology,* Section 17, McGraw-Hill Book Co., New York, 1964, pp 17-1 to 17–67.
3. Garde, R.J. and Kothyari, D.C., *Sediment Yield Eistimations, Jl. of Irrigation and Power, CBIP*, India, Vol. 44, No. 3, 1987.
4. Garde, R.J., and Rangaraju, K.G., *Mechanics of Sediment Transportation and Alluvial Stream Problems,* 2nd Ed. Wiley Eastern Ltd., New Delhi, 1977.
5. Gurmel Singh et al., *Manual of Soil & Water Resources Conservation Practices,* Oxford & *IBH*, New Delhi, 1990.
6. Gurmel Singh et al., Soil Erosion Rates in India, *Jl. of Soil and Water Conservation*, Vol. 47, No. 1, 1992, pp. 97–99.
7. Murthy, B.N., *Life of Reservoir, CBIP* Tech. Report No. 19, New Delhi, March 1977.
8. Mutreja, KN., *Applied Hydrology, Tata McGraw-Hill*, New Delhi, 1986.
9. Ram Babu et al., *Distribution of Erosion Index and Iso-Erodent Maps of Tamil Nadu and Karnataka States of India,* Int. conf. on Managing Natural Resources for Sustainable Agricultural Production in the 21st Century, Vol. 2, New Delhi, 2000, pp. 23–24.
10. Central Water Commission, Govt of India, "*Compendium of Silting of Reservoirs*", New Delhi, 1991.
11. Gurmeet Singh et al., *Manual of Soil and Water Conservation Practices*, Oxford & *IBH* Pub. Co., New Delhi., India, 1990.
12. Yang Xiaoquing, *Manual on Sediment Management and Measurement, WMO*-No. 948, Operational Hydrology Report No. 47, *WMO* Secretariat, Geneva, Switzerland, 2003.

REVISION QUESTIONS

10.1 Describe different forms of land erosion by water.

10.2 Describe the flow-duration and sediment rating curve procedure of estimating the sediment yield of a watershed.

10.3 Explain briefly the Universal Soil Loss Equation (*USLE*).

10.4 What is Modified Universal Soil Loss Equation (*MUSLE*)? What is its chief advantage over *USLE*?

10.5 Briefly explain:
(a) Erosion index
(b) Sediment delivery ratio (*SDR*)
(c) Bed load
(d) Suspended load
(e) Reservoir delta

10.6 What is meant by Trap Efficiency of a reservoir? What factors influence its value?

10.7 Describe a commonly used method of estimating the trap efficiency of a reservoir.

10.8 Describe the procedure of conducting a reservoir survey.

10.9 List the factors affecting the density of sediment deposited in a reservoir. What is the commonly used method of estimating the average density of sediment deposited over a period of T years in a reservoir?

10.10 How are reservoirs classified for purposes of estimating the deposition pattern?

10.11 Explain the empirical–area–reduction method of determining the sediment distribution in a reservoir.

10.12 Explain the area–increment method of determining the sediment distribution in a reservoir.

10.13 Explain a procedure to estimate the time taken for a reservoir to lose $x\%$ of its initial volume.

10.14 List different methods available for reservoir sediment control.

10.15 Write a brief note on a procedures to be adopted towards reduction of sediment yield of a catchment.

PROBLEMS

10.1 For a catchment area of 1500 km^2, estimate the sediment yield in ha-m/100 sq. km/year by using (i) Khosla's formula, and (ii) Joglekar's formula.

10.2 If the dry unit weight of a sediment deposit in a reservoir is 850 kgf/m^3, estimate the (a) porosity and (b) weight of 1 m^3 of sediment deposit in the reservoir. [Assume relative density of sediment particles as 2.65.]

10.3 Estimate the (dry) unit weight of a reservoir sediment deposit having 25% sand, 35% silt and 40% clay, at the end of (a) one year and (b) 25 years. [Assume that normally the reservoir undergoes considerable drawdown.]

10.4 Reservoir sediment deposition survey of a reservoir indicated the following composition of sediment in a sample: Sand = 25%, Silt = 21% and Clay = 54%. The sample can be taken as a 10- year old deposit. If the dry unit weight was 1650 kgf/m^3, determine the accuracy of estimate of unit weight of the sample by the use of Koezler equation. The reservoir operation is of 2nd kind viz, normally moderate drawdown is expected in its operation.

10.5 A reservoir, sediment is estimated through use of Koezler equation to have average unit weight of 1100 kgf/m^3 at the end of 35 years and 1120 kgf/m^3 at the end of 50 years. Estimate the average unit weight at the end of first year of deposit and at the end of 100 years.

10.6 In a reservoir, the average weight of deposited sediment was found to be (i) 500 kgf/m^3 over a period of first ten years, and (ii) 600 kgf/m^3 over a period of first twenty years. Estimate the average unit weight over a period of first 50 years of the reservoir's life.

10.7 A reservoir with a capacity of 50 Mm^3 is proposed at a location in a river having the following properties:
Catchment area = 200 km^2
Average Annual water yield at the site = 35 cm
Average Annual Sediment yield at the site = 150,000 Mm^3
Estimate the time required for the loss of 30% of initial capacity of the reservoir due to sedimentation. [Assume three equal steps of capacity loss.]
The reservoir is expected to be a normal ponded reservoir.

10.8 A reservoir has an initial capacity of 90000 ha-m and the annual sediment load in the stream is estimated as 600 ha-m. If the average annual inflow into the reservoir is 400000 ha-m, estimate the time in years for the reservoir to lose 50% of initial capacity. In the relevant range the trap efficiency η_t can be assumed to be given by

$\eta_t = 6.064\, Ln(C/I) = 101.48$

Use five steps.

10.9 A reservoir has a capacity of 8 Mm^3 at full reservoir level. The catchment area is 500 km^2. The average annual runoff of the catchment is 300 mm and the sediment yield is 15.0 MN/km^2. The sediment can be assumed to have an average *insitu* specific weight of 12.5 kN/m^3 after deposition. Estimate the time required for the reservoir capacity to be reduced to 25% of its initial volume. Use uniform steps of capacity interval of 1.0 Mm^3. The following relation gives the variation of the trap efficiency:

$$\eta_t = 100 \times \left[1 - \frac{1}{1 + 100\,(C/I)}\right]^{1.5}$$

where η_t = percentage trap efficiency, C = capacity of the reservoir, I = annual inflow in to the reservoir.

10.10 A proposed reservoir has a catchment of 2660 km^2. It has a capacity of 360 Mm^3 and the annual yield of the catchment is estimated as 40 cm. Assuming the average composition of sediment as 20% sand, 35% silt and 45% clay, estimate the probable life of the reservoir to a point where 40% of the reservoir capacity is lost by sedimentation. The sediment yield is estimated independently as 360 tonnes/km^2/year. [Assume the reservoir to have normally a moderate reservoir drawdown. Take five capacity steps for the life calculation.]

10.11 Coordinates of suspended load rating curve and flow duration curve of a river at a gauging site is given below. Plot the respective curves and using them estimate the (a) total sediment yield at the gauging station and (b) concentration of suspended load in ppm. [Assume bed load as 10% of suspended load.]

Flow Duration Curve		Suspended Sediment Rating Curve	
Percent times flow equalled or exceeded	**Average daily discharge**	**Water discharge (m^3/s)**	**Suspended load (tonnes/day)**
0.5–1.0	2550	2550	355,000
1.5–5.0	1275	1250	200,000
5.0–15.0	735	750	62,500
15.0–35.0	450	450	22,500
35.0–55.0	200	350	17,500

Contd.

55.0–75.0	110	225	10,000
75.0–95.0	50	125	4000
95.0–99.5	20	85	2000
		50	500
		25	50

10.12 A reservoir has a capacity of 180 Mm^3 at its full reservoir level. The average water inflow and average sediment inflow into the reservoir are estimated as 400 Mm^3/year and 3.00 M tonnes/year respectively. The sediment inflow was found to have a composition of 20% sand, 30% silt and 50% clay. Estimate the time in years required for the capacity of the reservoir to be reduced to 35% of its initial capacity. [Assume the sediment is always submerged.]

10.13 The area of a reservoir at different elevations as obtained by survey is shown in the following table. Estimate the capacity of the reservoir by using the weighted area method. Plot the capacity–elevation above bed curve on log-log axes and estimate the value of parameter *m* and the reservoir type.

Elevation (m)	Area (ha)	Capacity in (ha.m)
560.52	0	0
562.35	19.42	11.8
563.27	47.75	41.8
563.88	62.32	75.3
564.18	79.72	96.5
564.49	96.72	123.8
565.10	137.59	194.8
565.55	191.41	268.7
566.93	366.64	647.2
568.45	513.95	1313.3
569.97	679.06	2217.1

10.14 Original Reservoir the capacity–area–elevation data of Bhakra reservoir, India, is given below. Estimated total accumulation of sediment in the reservoir in 25 years of its operation is 92250 ha-m. Original bed elevation is E1.350.52 m and the spillway crest is at 512.06 m. Determine the distribution of 92,250 ha-m of sediment in the reservoir by the empirical area reduction method. The reservoir can be taken as of Type II. Assume the level up to which the reservoir is fully covered by sediment at the end of 25 years as 365.76 m. [Use average end area method for computing incremental volume.]

Elevation (m)	Original area (ha)	Original capacity	Elevation (m)	Original area (ha)	Original capacity (ha.m)
350.52	0	0	426.72	4452	148830
365.76	364	2460	441.96	5382	222630
381.00	1295	13530	457.20	6799	317340
396.24	2428	43050	472.44	8620	436650
400.00	2752	55350	487.68	11048	587940
411.48	3561	87330	502.92	13760	772440
			512.06	15378	910200

10.15 For a reservoir the original area–elevation–capacity relation is as given below. Over a period of 10 years, this reservoir expects a total sediment inflow of 10,000 ha-m. Determine the distribution of 10,000 ha-.m of sediment in this reservoir by the area-increment method. [An accuracy of >99.5% in gross sediment volume is expected.]

Elevation (m)	Original area (ha)	Original capacity (ha.m)	Elevation (m)	Original area (ha)	Original capacity ha.m)
97.53	0	0	112.77	1659	7258
100.58	43.7	98	115.82	2241	13242
103.63	168.4	471	118.88	3083	21300
105.76	411.2	1177	121.92	4346	32550
106.68	464.2	1509	124.97	6206	42710
109.73	720.7	3321			

OBJECTIVE QUESTIONS

10.1 In a reservoir the capacity is 20 M cm and the annual inflow is estimated as 25 M cm. The trap efficiency of this reservoir under normal operating conditions is about
(a) 10% (b) 45% (c) 75% (d) 95%

10.2 In a reservoir, the sediment deposit is found to be made up of only sand and this deposit is always found to be submerged. The unit weight of this sediment deposit at any time *T* years after the commencement of operation of the reservoir is about
(a) 1500 kgf/m^3
(b) 1500 + B Ln T kgf/m^3 where B is a positive non-zero coefficient
(c) 750 kgf/m^3
(d) 750 + B Ln T kgf/m^3 where B is a positive non-zero coefficient

10.3 Borland and Miller's classification of reservoirs for distribution of sediments in the reservoir is based on a parameter m. The reservoir is classified as
(a) Type I if m is in the range 2.5 to 3.5 (b) Type II if m is in the range 1.0 to 1.5
(c) Type III if m is in the range 1.5 to 2.5 (d) Type IV if m is greater than 3.5

10.4 A reservoir had an original capacity of 720 ha-m. The drainage area of the reservoir is 100 sq.km and has a sediment delivery rate of 0.10 ha-m/sq.km. If the reservoir has a trap efficiency of 80% the annual percentage loss of original capacity is
(a) 1.39% (b) 1.11% (c) 1.74% (d) 0.28%

10.5 The sediment delivery ratio (SDR) of a watershed is related to watershed area (A), relief (R) and watershed length (L) as
(a) $SDR = KA^m\,(R/L)^n$
(b) $SDR = KA^{-m}\,(R/L)^{-n}$
(c) $SDR = KA^{-m}\,(R/L)^n$
(d) $SDR = KA^m\,(R/L)^{-n}$
where K, m and n are positive coefficients.

10.6 If erosion in a watershed is estimated as 30 tonnes/ha/year, his watershed is in erosion class designated as
(a) severe (b) very high (c) high (d) moderate

10.7 The suspended sediment concentration C_s in ppm is determine from a sample of suspended sediment mixture as

(a) $C_s = \dfrac{[\text{Weight of sediment in sample}]}{[\text{Weight of water in sample}]} \times 10^6$

(b) $C_s = \dfrac{[\text{Weight of sediment in sample}]}{[\text{Weight of (sediment + water) in sample}]} \times 10^6$

(c) $C_s = \dfrac{[\text{Volume of sediment in sample}]}{[\text{Volume of (sediment + water) in sample}]} \times 10^6$

(d) $C_s = \dfrac{[\text{Weight of (sediment + water) in sample}]}{[\text{Volume of (sediment + water) in sample}]} \times 10^6$

10.8 The current *CWC* practice in design of reservoirs adopts minimum drawdown level (*MDDL*) based on the bed elevation that will be reached in the reservoir after *N* years of sedimentation, where *N* is equal to

(a) 25 years (b) 50 years (c) 100 years (d) 500 years

10.9 The present *CWC* practice in design of reservoirs adopts area – capacity – elevation curves expected after *M* years of sedimentation for working table studies and checking for the performance of the project. In this *M* is equal to

(a) 25 years (b) 50 years (c) 100 years (d) 500 years

10.10 A new reservoir has a capacity of 12 Mm^3 and its catchment area is 400 km^2. The annual sediment yield from the catchment is 0.1 ha.m/km^2 and the trap efficiency can be assumed to be constant at 0.90. The number of years it takes for the reservoir to loose 50% of its initial capacity is about

(a) 167 years (b) 33 years (c) 17 years (d) 6 years

10.11 The relation between suspended sediment transport Q_s and stream flow Q is often represented by an equation of the form

$$Q_s = KQ^n$$

where the exponent n usually lies in the range

(a) 0.1 to 0.3 (b) 0.4 to 0.6 (c) 0.6 and 1.0 (d) 1.0 and 3.0

APPENDIX

A

MULTIPLE CHOICE OBJECTIVE QUESTIONS

A.1 Match List–1 with List–2 and select the correct answer using the codes given below the lists:

List–1		List–2	
A	Conservation of reservoirs	1	Uncontrolled outlets
B	Retarding basins	2	Flood fighting
C	Flood plains	3	Temporary storage of flood waters
D	Flood walls	4	Controlled outlets

Codes:

	A	*B*	*C*	*D*
(a)	1	4	3	2
(b)	4	1	2	3
(c)	4	1	3	2
(d)	1	4	2	3

A.2 The following four hydrological features have to be estimated or taken as inputs before one can compute the flood hydrograph at any catchment outlet:

1. Unit hydrograph
2. Rainfall hydrograph
3. Infiltration Index
4. Base flow

The correct order in which they have to be employed in the computation is

(a) 1, 2, 3, 4 (b) 2, 1, 4, 3 (c) 2, 3, 1, 4 (d) 4, 1, 3, 2

A.3 The following steps are involved in arriving at a unit hydrograph:

1. Estimating the surface runoff in terms of depth
2. Estimating the surface runoff in volume
3. Separation of base flow
4. Dividing the surface runoff ordinates by depth of runoff

The correct sequence of these steps is

(a) 3, 2, 1, 4 (b) 2, 3, 4, 1 (c) 3, 1, 2, 4 (d) 4, 3, 2, 1

A.4 Match List–1 with List–2 and select the correct answer using the codes given below the lists:

List–1		List–2	
A	Specific yield	1	Volume of water retained per volume of aquifer
B	Specific capacity	2	Volume of water drained by gravity per unit volume of aquifer
C	Specific retention	3	Difference of porosity and specific storage
D	Specific storage	4	Well yield per unit drawdown
		5	Volume of water released from unit volume of aquifer for unit decline in piezometric head

Codes:

	A	*B*	*C*	*D*
(a)	2	4	1	5
(b)	4	2	3	1
(c)	2	3	1	4
(d)	4	2	3	5

A.5 Consider the following statements:

1. A 100-year flood discharge is greater than a 50-year flood discharge
2. 90% dependable flow is greater than a 50% dependable flow
3. Evaporation from salt-water surface is less than that from freshwater surface

Which of these statements are true?

(a) 1 and 2 (b) 2 and 3 (c) 1 and 3 (d) 1, 2 and 3

A.6 Match List–1 with List–2 and select the correct answer using the codes given below the lists:

List–1 (Name of Scientist)		List–2 (Contribution to field of Hydrology)	
A	Dalton	1	Unit hydrograph
B	Snyder	2	Evaporation
C	Blaney-Criddle	3	Empirical flood formula
D	Sherman	4	Synthetic unit hydrograph
		5	Consumptive use equation

Codes:

	A	*B*	*C*	*D*
(a)	2	4	5	1
(b)	1	4	3	2
(c)	2	3	5	1
(d)	1	3	4	5

A.7 Which of the following principles relate to a unit hydrograph?

1. The hydrograph of direct runoff due to effective rainfall of equal duration have the same base
2. Effective rainfall is not uniformly distribution within the duration
3. Effective duration is uniformly distributed throughout the whole area of drainage basin
4. Hydrograph of direct runoff from a basin due to a given period of effective rainfall reflects the combination of all the physical characteristics of the basin.

Select the correct answer using the code given below:

(a) 1, 2 and 3 (b) 1, 2 and 3 (c) 2, 3 and 4 (d) 1, 3 and 4

A.8 Match List–1 with List–2 and select the correct answer using the codes given below the lists:

List–1 (Equation)		List–2 (Applicability/Principle of Equation)	
A	Anemometer	1	Humidity
B	Rain simulator	2	Evapotranspiration
C	Lysimeter	3	Infiltration
D	Hygrometer	4	Wind speed

Codes:

	A	*B*	*C*	*D*
(a)	4	3	1	2
(b)	4	3	2	1
(c)	3	4	1	2
(d)	3	4	2	1

A.9 Match List–1 with List–2 and select the correct answer using the codes given below the lists:

List–1 (Equation)		List–2 (Applicability/Principle of Equation)	
A	Thiem's equation	1	Based on energy conservation principle
B	Dupit's assumption	2	Based on mass conservation principle
C	Bernoulli's equation	3	Applicable to steady flow towards a well in a confined aquifer
D	Continuity equation	4	Applicable to steady flow in a confined aquifer

Codes:

	A	*B*	*C*	*D*
(a)	4	3	1	2
(b)	3	4	2	1
(c)	4	3	2	1
(d)	3	4	1	2

A.10 Consider the following statement:

1. An ephemeral stream is one that has a base-flow contribution.
2. Flow characteristics of a stream depend upon rainfall and catchment characteristics and also the climatic factors which influence evapotranspiration.
3. Sequent peak algorithm is used for estimation of runoff from rainfall.

The correct statements are

(a) 1, 2 and 3 (b) 1 and 3 (c) 2 and 3 (d) Only 2

A.11 Match List–1 with List–2 and select the correct answer using the codes given below the lists:

List–1		List–2	
A	φ index	1	Used for measurement of evapotranspiration for given vegetation
B	Lysimeter	2	Used for flow measurement of a steam
C	Dilution technique	3	Average rainfall above which the rainfall is equal to runoff volume
D	*USLE*	4	Relates to soil loss estimation from a catchment

Codes:

	A	*B*	*C*	*D*
(a)	4	2	1	4
(b)	3	1	2	4
(c)	3	2	1	4
(d)	4	1	2	3

A.12 Match List–1 with List–2 and select the correct answer using the codes given below the lists:

List–1 (Well Hydraulics Parameter)		List–2 (Definition)	
A	Specific yield	1	Discharge per unit drawdown
B	Safe yield	2	Same as specific retention
C	Specific capacity	3	Measure of water that can be removed by pumping
D	Field capacity	4	Limit of withdrawal from a well without depletion of the aquifer
		5	Water bearing capacity of aquifer

Codes:

	A	*B*	*C*	*D*
(a)	4	3	2	5
(b)	3	4	1	2
(c)	3	4	2	5
(d)	4	3	1	2

A.13 Match List–1 with List–2 and select the correct answer using the codes given below the lists:

List–1		List–2	
A	Thiessen polygon	1	Average depth of rainfall over an area
B	Mass curve	2	Relationship of rainfall intensity and time
C	Hyetograph	3	Relationship of accumulated rainfall and time
D	*DAD* curve	4	Relationship of river runoff and time
		5	Always a falling curve

Codes:

	A	*B*	*C*	*D*
(a)	1	5	3	2
(b)	1	3	2	5
(c)	4	3	2	5
(d)	4	5	3	2

A.14 Match List–1 with List–2 and select the correct answer using the codes given below the lists:

List–1 (Type of Precipitation)		List–2 (Principal cause)	
A	Convective	1	Atmospheric disturbance
B	Cyclonic	2	Mountain barrier
C	Frontal	3	Pressure difference
D	Orographic	4	Temperature difference
		5	Warm and cold air masses

Codes:

	A	*B*	*C*	*D*
(a)	4	3	5	2
(b)	1	4	5	2
(c)	1	4	2	5
(d)	4	3	2	5

A.15 Consider the following statements:

1. Time area histogram method aims at developing an *IUH*
2. Isochrone is a line joining equal rainfall on a map
3. Linear reservoir is having straight boundaries
4. Linear channel is a fictitious channel in which an inflow hydrograph passes through with only translation and no attenuation.

The correct statements are

(a) 1, 2 and 3 (b) 1 and 4 (c) 2, 3 and 4 (d) 1, 2, 3 and 4

A.16 Match List–1 with List–2 and select the correct answer using the codes given below the lists:

List–1		List–2	
A	Rising limb of a hydrograph	1	Depends on intensity of rainfall
B	Falling limb of a hydrograph	2	Function of total channel length
C	Peak rate of flow	3	Function of catchment slope
D	Drainage density	4	Function of storage characteristics

Codes:

	A	*B*	*C*	*D*
(a)	1	4	3	2
(b)	3	2	1	4
(c)	1	2	3	4
(d)	3	4	1	2

A.17 Match List–1 with List–2 and select the correct answer using the codes given below the lists:

List–1		List–2	
A	Unit hydrograph	1	Design flood
B	Synthetic unit hydrograph	2	Permeability
C	Darcy law	3	Ungauged basin
D	Rational method	4	1 cm of runoff

Codes:

	A	*B*	*C*	*D*
(a)	2	3	4	1
(b)	4	3	2	1
(c)	2	1	4	3
(d)	4	1	2	3

A.18 Match List–1 with List–2 and select the correct answer using the codes given below the lists:

List–1		List–2	
A	φ Index	1	Dependable flow
B	Slope area method	2	Reservoir regulation
C	Flow-duration curve	3	Steady stream discharge
D	Dilution technique	4	Runoff volume
		5	Unsteady stream discharge

Codes:

	A	*B*	*C*	*D*
(a)	3	5	1	4
(b)	3	1	2	4
(c)	4	1	2	3
(d)	4	5	1	3

A.19 Consider the following statements:
In case of flood rounting in a river by using Muskingum method, the coefficient *x* represents:

1. A dimensionless constant indicting relative importance of inflow and outflow in determining storage.
2. A storage constant having the dimension of time
3. A coefficient that has a value in the range 0.2 to 0.3 in natural channels
4. A coefficient that represents equal influence of inflow and outflow when its value is 0.5.

The correct statements are

(a) 1, 2, 3 and 4 (b) 1, 3 and 4 only
(c) 1, and 3 only (d) 2, 3 and 4 only

A.20 Match List–1 with List–2 and select the correct answer using the codes given below the lists:

List–1		List–2	
A	Rainfall intensity	1	Isohyets
B	Rainfall excess	2	Cumulative rainfall
C	Rainfall averaging	3	Hyetograph
D	Mass curve	4	Direct runoff hydrograph

Codes:

	A	*B*	*C*	*D*
(a)	1	3	2	4
(b)	1	3	4	2
(c)	3	4	1	2
(d)	3	4	2	1

A.21 Match List–1 with List–2 and select the correct answer using the codes given below the lists:

List–1		List–2	
A	Thiessen method	1	Unsteady flow in a well
B	Thies' solution	2	Steady flow in a confined aquifer
C	Thiem's equation	3	Flow through porous media
D	Laplace equation	4	Average Precipitation

Codes:

	A	*B*	*C*	*D*
(a)	4	2	3	1
(b)	1	4	2	3
(c)	1	4	3	2
(d)	4	1	2	3

A.22 Match List–1 with List–2 and select the correct answer using the codes given below the lists:

List–1		List–2	
A	Hydraulic conductivity	1	Psychrometer
B	Infiltration capacity	2	Phytometer
C	Transpiration	3	Permeameter
D	Relative humidity	4	Rainfall simulator

Codes:

	A	*B*	*C*	*D*
(a)	3	4	2	1
(b)	3	4	1	2
(c)	4	3	2	1
(d)	3	2	4	1

A.23 Match the List 1 and 2

List–1		List–2	
A	Aquiclude	1	Contains water and also transmits water
B	Aquifer	2	Contains water but can transmit water at very low
C	Aquifuge	3	Contains water but cannot transmit it
D	Aquitard	4	Neither porous not permeable to water

Codes:

	A	B	C	D
(a)	1	3	4	2
(b)	2	4	1	2
(c)	3	4	1	2
(d)	2	1	4	3

A.24 Match the List 1 and 2

	List–1		List–2
A	Specific yield	1	Discharge per unit drawdown
B	Specific Capacity	2	Volume of water that can be extracted by gravity force from a unit volume of aquifer
C	Specific storage	3	Volume of water that is held back in a unit of aquifer after it is drained by gravity
D	Specific retention	4	Volume of water that is released from storage from a unit volume of aquifer due to unit decrease in piezometric head

Codes:

	A	B	C	D
(a)	1	2	3	4
(b)	3	2	4	1
(c)	2	1	4	3
(d)	3	1	4	2

A.25 Match the List 1 and 2

	List–1 (Parameter)		List–2 (Dimensions)
A	Coefficient of permeability, K	1	$M^0L^0T^0$
B	Intrinsic permeability, K_0	2	LT^{-1}
C	Transmissibility, T	3	L^2
D	Storage coefficient, S	4	L^2T^{-1}

Codes:

	A	B	C	D
(a)	2	3	4	1
(b)	3	2	4	1
(c)	2	4	3	1
(d)	1	3	4	2

A.26 Consider different zones of vertical distribution of subsurface water shown in the figure given below. Match List-II that contain the nomenclature of water in different zones with List-I contain in the list of zones.

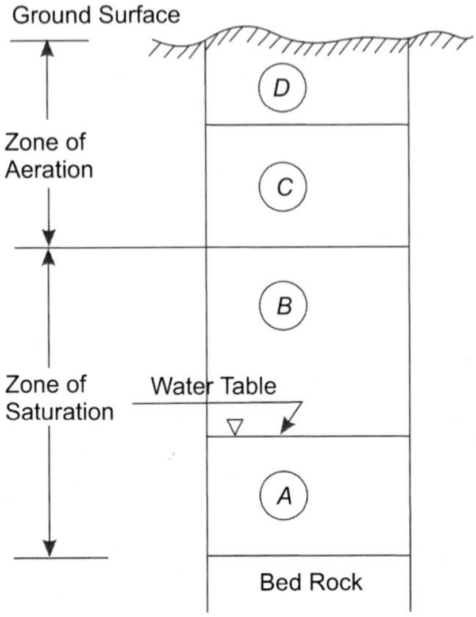

Fig. A.1 ***Question A. 26***

	List–I (Zones in Fig. A.1)		List–II (Nomenclature)
A	Zone A	1	Soil water
B	Zone B	2	Groundwater
C	Zone C	3	Pelicular water
D	Zone D	4	Capillary water

Codes:

	A	*B*	*C*	*D*
(a)	4	3	2	1
(b)	2	3	4	1
(c)	2	4	1	3
(d)	2	4	3	1

A.27 Match the two list

	List–1		List–2
A	Hydrograph due to a continuous effective rainfall at a constant rate for infinite period	1	DRH
B	Hydrograph due to an *ERH*	2	S-curve
C	Unit hydrograph with ordinates expressed as percent of total direct runoff	3	IHU
D	Unit hydrograph whose duration of effective rainfall is infinitesimally small	4	Distribution graph

Codes:

	A	B	C	D
(a)	2	1	4	3
(b)	1	2	4	3
(c)	2	3	4	1
(d)	1	4	3	2

A.28 Match the following two lists:

	List–1		List–2
A	Rainfall simulator	1	Increased peak in flood hydrograph
B	Interception loss	2	Study of infiltration characteristics
C	Urbanization	3	Not significant in maximum flood
D	Soil conservation measures	4	Reduction in peak in small and medium floods

Codes:

	A	B	C	D
(a)	2	3	1	4
(b)	2	1	3	4
(c)	4	1	3	2
(d)	2	3	4	1

A.29 Figure A.2 shows a 6-h unit hydrograph of a catchment. Consider the following statements with respect to this figure.

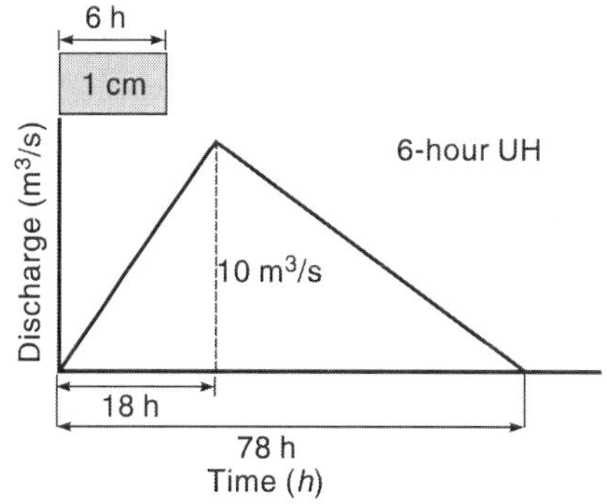

Fig. A.2 ***Question A. 29***

1	The area of the catchment is 1404 km^2
2	The *DRH* produced from this catchment due to a rainfall excess of 5 cm in 6 hours will have a peak of 50 m^3/s.
3	The 12-h unit hydrograph of this catchment will have a peak of 5 m^3/s
4	An excess rainfall of 6 cm in 12 hours occurring on this catchment will produce a *DRH* having a base of 84 h.

Codes: The correct statements are

(a) 2 and 4 only (b) 1 and 3 only (c) 1, 2 and 4 (d) 2, 3 and 4

A.30 A *DRH* of a catchment of 216 km² area is shown in Fig. A.3. With respect to this *DRH* consider the following statements:

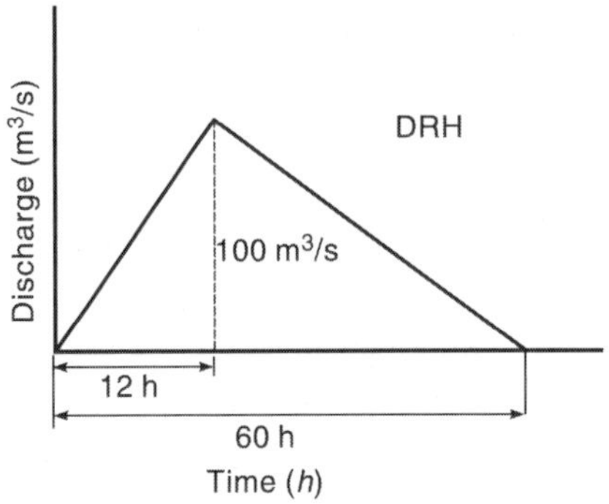

Fig. A.3 ***Question A. 30***

1	The effective rainfall producing the *DRH* is 5 cm
2	The ordinate of a 6-h unit hydrograph of the catchment is 10 m³/s
3	The peak of a *DRH* due to excess rainfall of 10 cm in 6 hours is 200 m³/s
4	The base of a *DRH* of excess due to 10 cm of excess rainfall in 6 hours is 66 hours.

Codes: The correct statements are

(a) 1, 2 and 3 (b) 3 and 4 only (c) 1 and 3 only (d) 2, 3 and 4

A.31 Figure A.4 shows a typical flood hydrograph. Match the labeled items using lists I and II

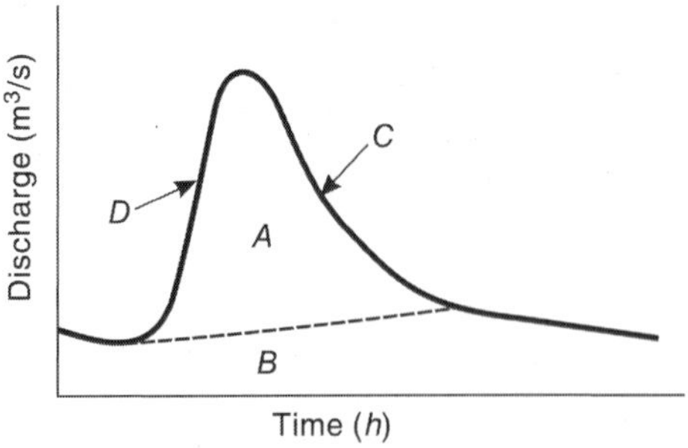

Fig. A.4 ***Question A. 31***

	List I (Items in Fig. A.4)		List II (Labeled items are known as)
A	Item-A	1	Recession Limb
B	Item-B	2	Rising Limb
C	Item-C	3	Direct runoff
D	Item-D	4	Base flow

Codes:

	A	B	C	D
(a)	3	4	2	1
(b)	3	4	1	2
(c)	4	3	1	2
(d)	4	3	2	1

A.32 List I contain parameters and List II contains methods/instruments. Match the items.

	List–1 (Parameters)		List–2 (Methods/Instruments)
A	Stream flow velocity	1	Anemometer
B	Evapotranspiration rate	2	Penman's method
C	Infiltration rate	3	Harton's method
D	Wind velocity	4	Current meter

Codes:

	A	B	C	D
(a)	1	4	2	3
(b)	4	2	1	3
(c)	1	3	2	4
(d)	4	2	3	1

A.33 Consider the following statements:

1. Over the oceans there is more evaporation than precipitation
2. On land it is more precipitation tan evapotranspiration.

Which of these two statements are correct?

(a) Both 1 and 2 (b) Neither 1 nor 2
(c) 1 only (d) 2 only

A.34 A watershed was transformed from rural to urban over a period of time. If two storm hydrographs, occurring under identical conditions, but belonging to before and after urbanisation phase are compared, the hydrograph of after urbanisation phase would show:

(1) a decrease in the volume of runoff
(2) an increase in the volume of runoff
(3) decrease in the time base
(4) decrease in the time to peak
(5) Increase in the peak discharge

The correct answers are

(a) 2, 4 and 5 (b) 1, 3 and 5 (c) 1 and 3 only (d) 2 and 5 only

A.35 Consider the following zones in a soil profile:

1. Saturated zone 2. Capillary zone
3. Intermediate zone 4. Soil-water zone

Which of these does **not** relate to the zone of aeration in the soil profile?

(a) 1 and 2 (b) 2 and 3 (c) 4 only (d) 1 only

Answers to Multiple Choice Objective Questions of Appendix-*A*

Appendix-A	0	1	2	3	4	5	6	7	8	9
A. 00	–	c	c	a	a	c	d	d	b	d
A. 10	c	b	b	b	a	b	d	b	d	c
A. 20	c	d	a	b	c	a	d	a	a	a
A. 30	c	b	d	a	a	d				

APPENDIX

B

ABBREVIATIONS

AET	Actual Evapotranspiration
AI	Aridity Index
AMC	Antecedent Moisture Condition
BE	Barometric efficiency
CBIP	Central Board of Irrigation and Power (India)
CFC	Chlorofluorocarbons
CGWB	Central Groundwater Board (India)
CN	Curve Number
CWC	Central Water Commission (India)
DAD	Maximum Depth–Area–Duration
DGPS	Differential Global Positioning System
DRH	Direct Runoff Hydrograph
DVC	Damodar Valley Corporation
EF	Environmental Flows
EFA	Environmental Flow Assessment
EFR	Environmental Flow Requirement
EI_{30}	Rainfall Erosion Index Unit
ENSO	El Nino and Southern Oscillation
ER	Effective (or Excess) Rainfall
ERH	Effective (Excess) Rainfall Hyetograph
EU	European Union
FAO	Food and Agriculture Organisation of United Nations Organisation.
FEM	Finite Element Method
FRL	Full Reservoir Level
GCM	Global Circulation Model

GDP	Gross Domestic Product
GHG	Green House Gases
GIS	Geographical Information System
GOI	Government of India
GPS	Geographical Positioning System
GSI	Geological Survey of India
HFS	High Flow Season
IMD	India Meteorological Department
IPCC	Inter-Governmental Panel on Climate Change
ISMR	Indian Summer Monsoon Rainfall
IUH	Instantaneous Unit Hydrograph
IWRM	Integrated Water Resource Management
KWDT	Krishna Water Dispute Tribunal
LFS	Low Flow Season
MAF	Mean Annual Flow
MAI	Moisture availability index
MCM	Million Cubic Metre
MDDL	Minimum Drawdown Level
MOC	Method of Characteristics
MOEF	Ministry of Environment and Forestry, Government of India
MSL	Mean Sea Level
MUSLE	Modified Universal Soil Loss Equation
NBSS & LUP	National Bureau of Soil Survey and Land Use Planning
NCIWRD	National Commission for Integrated Water Resources Development (1999)
NIH	National Institute of Hydrology (India)
NRSA	National Remote Sensing Agency
PET	Potential Evapotranspiration
PI	Palmer Index
PMF	Probable Maximum Flood
PMP	Probable Maximum Precipitation
RBA	Rashtriya Barh Ayog (National Flood Commission)
RTWH	Roof Top Water Harvesting
SCS	US Soil Conservation Service
SDR	Sediment Delivery Ratio
SO	Southern Oscillation

SPF	Standard Project Flood
SPS	Standard Project Storm
SRK	Standard Runga-Kutta Method
SWM	Stanford Watershed Model
TMC	Thousand Million Cubic Feet
UH	Unit Hydrograph
UK	United Kingdom
UNESCO	United Nations Economic, Social and Cultural Organisation
USA	United States of America
USLE	Universal Soil Loss Equation
WMO	World Meteorological Organisation
WQAA	Water Quality Assessment Authority

APPENDIX

C

CONVERSION FACTORS

C.1 Volume

$$1\ m^3 = 35.31 \text{ cubic feet} = 264 \text{ US gallons} = 220 \text{ Imp. gallons}$$
$$= 1.31 \text{ cubic yards} = 8.11 \times 10^{-4} \text{ acre feet} = 1000 \text{ litres}$$

C.2 Flow Rate (Discharge)

Unit	Cubic metres per second (m^3/s)	Litres per minute (lpm)	Litres per second (lps)
1 cft/s (cusec)	0.02832	1699	28.32
1 Imp. gpm	7.577×10^{-5}	4.546	0.07577
1 US gpm	6.309×10^{-5}	3.785	0.06309
1 Imp. mgd	0.05262	3157	52.62
1 acre ft/day	0.01428	856.6	14.28

C.3 Permeability

1. *Specific permeability, K_0*
   1 darcy = $9.87 \times 10^{-13}\ m^2 = 9.87 \times 10^{-9}\ cm^2$
2. *Coefficient of permeability, K*
   1 lpd/m^2 = 1.1574×10^{-6} cm/s
   1 m/day = 1.1574×10^{-3} cm/s = 20.44 Imp. gpd/ft^2 = 24.53 US gpd/ft^2
   = 0.017 US gpm/ft^2

C.4 Transmissibility

1 m^2/day = 67.05 Imp. gpd/ft = 80.52 US gpd/ft = 0.056 US gpm/ft

C.5 Equivalents of Some Commonly used Units

1 Metre	=	3.28 feet	1 Foot	=	30.48 cm = 0.3048 m
1 Kilometre	=	0.6215 mile	1 Mile	=	1.609 km
1 Hectare	=	2.47 acres	1 Acre	=	0.405 ha
1 sq. km	=	100 ha	1 Sq. Mile	=	259 ha = 640 acres
1 Million Cubic metre (MCM)	=	810.71 Acre ft. = 0.0353 TMC	1 TMC = one thousand million cubic feet	=	28.317 million cubic metres (MCM)
			1 Million acre feet	=	1233.48 million cubic feet
			1 cusec. day	=	86400 cft = 2446.9 m^3
			1 million gallons (imperial)	=	160544 Cubic feet = 4546.09 Cubic metres
			1 kgf	=	9.81 N
			1 Tonne	=	1000 kgf = 9.81 kN

APPENDIX

D

ANSWERS TO OBJECTIVE QUESTIONS

Chapter	0	1	2	3	4	5	6	7	8	9
1.00		c	d	a	b	c	d	c	a	c
2.00		d	b	c	d	c	a	b	b	c
2.10	c	d	a	d	c	b	b	d	b	b
2.20	b	a	a	b	c	a	b			
3.00		c	b	d	c	c	d	b	d	c
3.10	c	b	b	b	c	d	b	d	b	c
3.20	c									
4.00		d	d	b	d	c	b	a	c	d
4.10	b	a	c	b	a	b	c	b	c	c
5.00		a	a	b	a	c	c	b	c	b
5.10	d	c	b	a	b	c	b	c	a	
6.00		c	a	b	a	a	c	b	b	b
6.10	b	c	b	d	c	a	c	d	a	d
6.20	a	d	b	d	b	a	b	c	b	
7.00		b	c	a	d	a	c	c	b	b
7.10	d	a	a	c	c	c	a	c	a	c
7.20	c	b	d	d						
8.00		a	b	a	d	c	b	d	b	d
8.10	c	d	b	a	b	a	b	b	c	d
8.20	b	d	a							
9.00		d	b	b	c	c	d	b	c	a
9.10	b	c	c	b	b	d	d	a	c	b
9.20	b	b	d	c	a	d	a	a	b	b
9.30	a	b	d	b						
10.00		d	a	c	b	c	b	b	c	b
10.10	c	d								

APPENDIX E

Answers to Problems

Chapter 1

1.1 $Q = 57.87\ m^3/s$

1.2 (i) 0.61 (ii) Increase in abstraction = 18.492 Mm^3

1.3 $Q = 6.191\ m^3/s$

1.4 $S_2 = 19.388$ ha.m

1.5 $\overline{P} = 1105$ mm, $\overline{E} = 532.4$ mm, $r_a = 0.485$, $r_b = 0.472$, $r_c = 0.522$, $r_d = 0.538$, $r_{total} = 0.518$

1.6 (i) $T_r = 8.2$ days (ii) $T_r = 4800$ years (iii) $T_r = 28{,}500$ years

Chapter 2

2.1 (a) 6.53%, (b) $\varepsilon = 6.53\%$, Additional gauges = 5

2.2 12.86 cm

2.3 (a) 1955 (b) Correction ratio = 0.805, mean $P_A = 143.9$ cm

2.4

Time since start of the storm (minutes)	30	60	90	120	150	180	210
Intensity of rainfall in the interval (cm/h)	3.50	4.50	12.0	9.0	5.0	3.0	1.5
Cumulative rainfall since start (cm)	1.75	4.00	10.00	14.50	17.00	18.50	19.25

Average intensity = 5.5 cm/h

2.5 36.06 mm

2.6 Years 1964, 1971, 1972, 1976 and 1980 were drought years.

2.7 (i) 114.8 mm, (ii) 121.2 mm

2.8 7.41 cm

2.9 135 cm

2.10 (a) 10.6 cm (b) 10.80 cm (c) 11.07 cm

2.11 112.03 cm

2.12 (i) Average depth = 20.1 mm (ii) Depth at storm centre = 21.5 mm.

2.13 (a)

Time (min)	0	10	20	30	40	50	60	70	80	90
Intensity (mm/h)		114	132	42	120	138	198	168	48	36

(b)

Duration (min)	10	20	30	40	50	60	70	80	90
Max. Depth (mm)	198	183	168	156	134.4	133	130.3	1200	110.7

2.14

Duration (min)	10	20	30	40	50	60	70	80	90
Max. Depth (mm)	16	25	31	40	47	55	60	64	67

2.15

Maximum Intensity	75.0	62.1	49.8	40.5	37.0	33.0	30.1	27.2	24.7
Duration in Min	10	20	30	40	50	60	70	80	90
Maximum Depth (mm)	12.5	20.7	24.9	27.0	30.9	33.0	35.1	36.2	37.0

2.16 (a) 132.50 cm (b) 143.0 cm
2.17 (a) 118.0 cm (b) $T = 3.5$ years (c) $p_{66.7} = 88.0$ cm; $p_{75} = 84.5$ cm
2.19 (a) 0.167 (b) 0.0153 (c) 0.183
2.20 (a) 0.605 (b) 0.01
2.21 10 years
2.22 (a) 0.155 (b) 0.00179 (c) 0.0845

Chapter 3

3.1 10.7 mm/day **3.2** 24.5 mm **3.3** Decrease, 48 Mm^3
3.4 175 mm/month **3.5** (a) 27.1 cm (b) 32.28 cm
3.6 23.4 cm **3.7** 46.8 cm **3.8** 11.25 cm/month
3.9 3.9 mm/day for Day 2 and Day 7; 3.6 mm/day on Day 9
3.10 $f_p = 1.0 + 10.45e^{-3.1t}$
3.11 $K_h = 4.1235\ h^{-1}, f_c = 3.21$ cm/h, $f_o = 41.278$ cm/h
3.12 Kostiakov: $F_p = 6.733t^{0.7393}$

Green – Ampt: $f_p = 9.0239\left(\frac{1}{F_p}\right) + 3.8375$

Philip: $f_p = 2.9735t^{-0.5} + 2.0461$

3.13 $K_h = 3.06\ h^{-1}, f_c = 1.50$ cm/h, $f_o = 27.935$ cm/h
3.14 Kostiakov: $F_p = 4.245\ t^{0.7841}$

Philip: $f_p = 1.911\ t^{-0.5} + 1.485$

Green – Ampt: $f_p = 3.975\left(\frac{1}{F_p}\right) + 2.305$

3.15 (a) (i) $f_p = 0.7598\ t^{-1.185}$ (ii) $F_p = 7.4827 \ln t + 34.781$
(b) $K_h = 1.91\ h^{-1}$, $f_c = 0.10$ cm/h, $f_o = 10.244$ cm/h
3.16 $f_{av} = 1.02$ cm/h
3.17 (i) At $t = 2.0$ h $f_p = 1.7$ cm/h
(ii) At $t = 3.0$ h $f_p = 1.51$ cm/h
3.18 φ-index = 0.42 cm/h **3.19** φ-index = 0.657 cm/h, $t_e = 3.5$ hours
3.20 W-index = 2.52 mm/h **3.21** R = 2.50 cm **3.22** R = 2.24 cm

Chapter 4

4.1 6.426 m^3/s

4.2 (i) $\frac{\bar{v}}{v_{0.6}} = \frac{1}{(0.4)^m\,(m+1)}$ (ii) (a) $\frac{\bar{v}}{v_{0.6}} = 1.001$ (b) $\frac{\bar{v}}{v_{0.6}} = 1.00036$

4.3 11.895 m^3/s **4.4** 3458.9 m^3/s **4.5** 10.0 m^3/s **4.6** 500 m^3/s

4.7 142.8 m^3/s **4.8** 11 km **4.9** 44.25 m^3/s **4.10** 30.18 m^3/s

4.11 (1) $Q = 159.44\,(G - a)^{1.371}$, (2) 0.968 (3) 2525 m^3/s, 5368 m^3/s

4.12 426.9 m^3/s **4.13** $a = 18.60$ m

4.14 $(G - 20.5) = 0.1641\,Q^{0.4648}$, Stage = 26.842 m **4.15** 164.4 m^3/s

Chapter 5

5.1 (a)121 cm (b) 62.7%

5.2 $R = 0.4828\,P - 0.2535$; 5.06 Mm^3 **5.3** $R = 0.6163\,P - 21.513$; 40.12 cm

5.4 10143 Mm^3; 12842 Mm^3 **5.5** 0.144

5.6

Month	July	August	September	October
Monthly Yield (m^3)	41580	430290	177120	751410
Total seasonal yield (Mm^3)	1.44			

5.7 (a) 59550 m^3 (b) 180800 m^3

5.8 $CN_I = 51.4$, $CN_{II} = 70.7$, $CN_{III} = 85.0$

5.9 346080 m^3 **5.10** 620500 m^3 **5.11** 90984 m^3

5.12 140.15 mm **5.13** (a) 122.4 mm, (b) 105.2 mm; 16%

5.14 $Q_{75} = 14$ m^3/s **5.15** 9545 cumec.day

5.16 (a) 5700 cumec.day; (b) 82 m^3/s **5.17** 365 Mm^3

5.18 389.12 Mm^3 **5.19** 2280 cumec.day **5.20** 389.12 Mm^3

5.21 16.74 Mm^3; 1.41 Mm^3/day; Nil **5.22** 27 Mm^3 less water

5.23

Month	Jun	Jul	Aug	Sep	Oct	Nov	Dec	Jan	Feb	Mar	Apr	May
EFR (m^3/s)	10.5	10.5	12.8	10.5	3.5	3.5	3.5	3.5	3.5	3.5	10.5	10.5

Chapter 6

6.1 $K_{rh} = 0.886$, $K_{rs} = 0.2217$, $S_{t7} = 9.12$ cumec.day

6.2 (a)$K_{rb} = 0.966$ (b) $Q_3 = 28.28$ m^3/s **6.4** 2.47 Mm^3

Time (h)	1	2	3	4
Average *ER* (min)	1.44	25.50	3.94	0

6.5 (Given ordinates/4.32)

6.6 Base = 66 hour; $q_p = 15.91$ m^3/s at 10 hours from start

6.7

Time (h)	0	6	12	18	24	30	36	42	48	54	60	66	72
6-h *UH* ord. (m^3/s)	0	5.0	35.0	64.0	72.0	62.0	45.8	32.8	20.8	11.6	5.6	1.6	0

6.8 ER = 5.76 cm

Time (days)	0	1	2	3	4	5	6	7	8
1-day Dist. graph (%)	0	10.50	31.50	26.25	15.50	9.25	5.25	2.00	0

6.9 Ordinates of 8-h *UH* = (Given ordinates/4.5) **6.10** 70 m^3/s

6.11 Volume of Direct Surface runoff = 5.78 Mm^3
Peak runoff rate = 1376 m^3/s

6.12

Time (h)	0	3	6	9	12	18	24	30
Q(m^3/s)	30	300	480	1410	2060	4450	6010	6010
Time (h)	36	42	48	54	60	66	72	78
Q(m^3/s)	5080	3996	2866	1866	1060	500	170	30

6.13

Time (h)	0	6	12	18	24	30	36	42
Q(m^3/s)	10	30	90	220	280	220	166	126
Time (h)	48	54	60	66	72			
Q(m^3/s)	92	62	40	20	10			

6.14

Time (h)	0	6	12	18	24	30	36	42
12-h *UH* ord. (m^3/s)	0	10	40	105	135	105	78	58
Time (h)	48	54	60	66	72			
12-h *UH* ord. (m^3/s)	41	26	15	5	0			

6.15

Time (h)	0	2	4	6	8	10	12	14
S-curve ord. (m^3/s)	0	25	125	285	475	645	755	825
12-h *UH* ord. (m^3/s)	0	12.5	62.5	130	175	180	140	90
Time (h)	16	18	20	22	24	26		
S-curve ord. (m^3/s)	855	875	881	881	881	881		
12-h *UH* ord. (m^3/s)	50	25	13	3	0	0		

6.16 160 m^3/s **6.17** Area = 7.92 km^2

Time (h)	0	1	2	3	4	5	6	7
S-curve ord. (m^3/s)	0	5	13	18	21	22	22	22
2-h *UH* ord. (m^3/s)	0	2.5	6.5	6.5	4.0	2.0	0.5	0

6.18

Time (h)	0	6	12	18	24	30	36	42	48	54
6-h *UH* ord. (m^3/s)	0	20	54	98	124	148	152	154	138	122
Time (h)	60	66	72	78	84	90	96	102	108	114
6-h *UH* ord. (m^3/s)	106	92	76	66	50	42	28	22	12	10
Time (h)	120	126	132	138						
6-h *UH* ord. (m^3/s)	6	3	1	0						

6.19

Time (h)	0	3	6	9	12	15	18	21	24	27	30	33
9-h *UH* ord. (m^3/s)	0	4	29	73	129	174	191	183	158	135	115	99
Time (h)	36	39	42	45	48	51	54	57	60	63	66	
9-h *UH* ord. (m^3/s)	83	69	54	43	33	25	18	12	6	2	0	

6.20

Time (h)	0	6	12	18	24	30	36	42
Q (m^3/s)	20	80	200	550	620	890	698	530
Time (h)	48	54	60	66	72	78		
Q (m^3/s)	380	280	178	100	60	20		

6.21 $A = 1296\ km^2$

Time (h)	0	6	12	18	24	30	36	42
Q (m^3/s)	25	75	225	375	525	600	525	450
Time (h)	48	54	60	66	72	78		
Q (m^3/s)	375	300	225	150	75	25		

6.22 80.5 m^3/s

6.23

Time (h)	0	2	6	10	14	18	22	26	30	34
DRH ord. (m^3/s)	0	4.3	19.4	44.4	39.6	28.1	14.1	4.9	1.4	0

6.24

Time (Units of 6 h)	0	2	6	10	14	18	22	26	30	34
Dist. Graph ord. (%)	0	4.3	19.4	44.4	39.6	28.1	14.1	4.9	1.4	0

6.25 $Q_p = 42.09\ m^3/s$, $Q_s = 231.5\ m^3/s$

6.26

Time (h)	0	3	6	9	12	15	18	21	24	27
3-h *UH* ord. (m^3/s)	0	60	120	90	50	30	20	10	5	0

6.27 $t_R = 6.0\ h$; $t'_p = 27.75\ h$; $Q_p = 126\ m^3/s$; $W_{50} = 79\ h$; $W_{75} = 45\ h$; $T_b = 156\ h$

6.28 $t'_p = 9\ h$; $Q_p = 86.5\ m^3/s$; $W_{50} = 30.7\ h$, $W_{75} = 17.5\ h$; $T_b = 52\ h$

6.29 $Q_p = 126\ m^3/s$, $T_p = 30.75$ h and to be used with Table 6.12 in the Text.

6.30

Time (h)	0	1	2	3	4	5	6	7	8
Ordinate of 4-h *UH* (cm/h)	0	1	2	3	4	3	2	1	0
Ordinate of S_4-curve (cm/h)	0	1	2	3	4	4	4	4	4
Ordinate of 3-h *UH* (cm/h)	0	1.33	2.67	4.0	4.0	2.67	1.33	0	

6.31

Time (h)	0	1	2	3	4	5	6	7	8
Ordinate of S_2-curve (cm/h)	0.5	0.5	0.5	0.5	0.5	0.5	0.5	0.5	0.5
Ordinate of 4-h *UH* (cm/h)	0.25	0.25	0.25	0.25					

6.32

Time (h)	0	1	2	3	4	5	>5
Ordinate of *DRH* (m^3/s)	300	300	300	300	300	300	0

6.33 $Q_p = 20.48$ m^3/s, $T_p = 1.025$ h, $T_b = 2.75$ h
DRH: A triangle with peak of 81.92 m^3/s occurring at 1.025 h from start. Base = 2.75 h

6.34

Time (h)	Ord. of 2-h *UH* (m^3/s)	Time (h)	Ord. of 2-h *UH* (m^3/s)
0	0.00	8	17.50
2	2.50	10	19.29
4	7.50	12	17.8
6	12.50	14	16.4
16	15.00	28	6.43
18	13.57	30	5.00
20	12.14	32	3.57
22	10.71	34	2.14
24	69.29	36	0.71
26	37.86	38	0.00

6.35 Area = 257.8 km^2

Time (h)	Ord. of 3-h *UH* (m^3/s)	Time (h)	Ord. of 3-h *UH* (m^3/s)
0	0.0	12	42.3
1	1.8	13	36.0
2	9.8	14	30.0
3	26.0	15	24.3
4	46.0	16	19.0
5	62.3	17	14.5
6	70.7	18	10.8
7	71.8	19	7.5
8	68.5	20	4.5
9	62.8	21	2.0
10	56.0	22	0.5
11	49.0	23	0.0

Chapter 7

7.1 4.0 m^3/s **7.2** 10.0 m^3/s **7.3** 19.65 m^3/s **7.4** $Q_p = 55.08$ m^3/s

7.5 $q_m = 311.88$ m^3/s occurs at the end of 5 hours after the commencement of the storm

7.6 $Q_p = 2.08$ m^3/s **7.7** $Q_p = 6.42$ m^3/s

7.8 See Fig. AnsP–7.8 **7.9** $Q_p = 0.345$ m^3/s

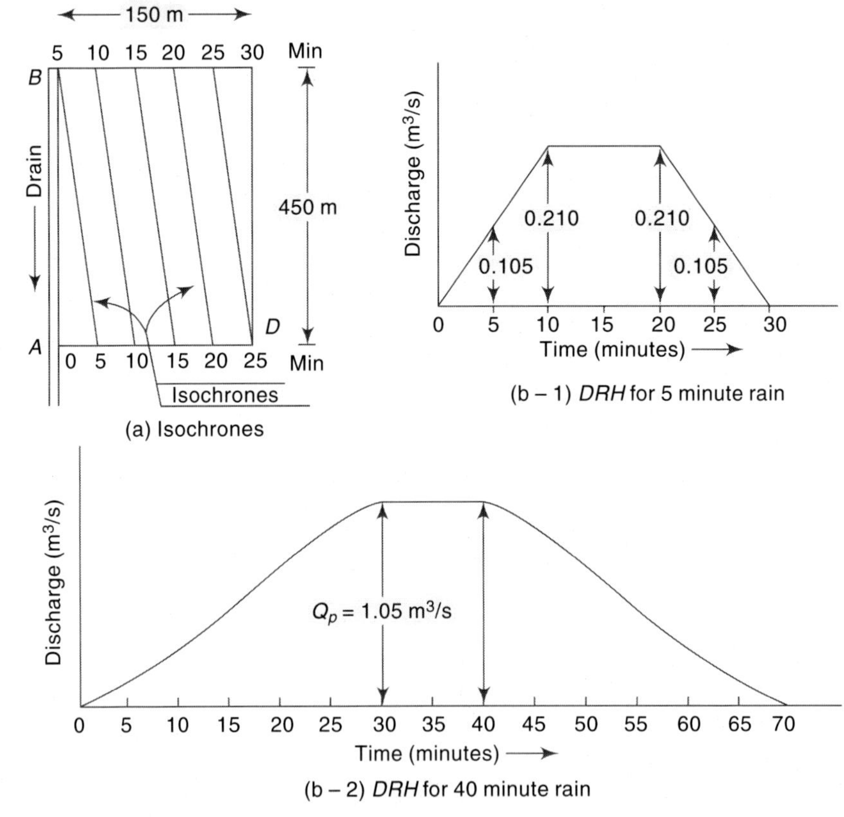

Fig. AnsP–7.8 ***Answers to Problem 7.8***

7.10 (a) 0.025 (b) 0.397 (c) 0.975

7.11 (a) 0.41 (b) 0.358 (c) 0.02 (d) 0.000833 (e) 0.636

7.12

N	35	45	55	60	65	5	>5
$K(T, N)$	5.6421	5.5221	5.4420	5.4100	5.3832	300	0

7.13

Q_T m^3/s for T =	100	200	1000
Gumbel	7916	8784	10806
Log Pearson-III	5060	5428	6254
Log Normal	5286	5738	6794

7.14 100 years **7.15** 85 years **7.16** $Q_{1000} = 1347$ m^3/s

7.17

Q_T m^3/s for T =	50	100	200	1000
Log Pearson-III	794	928	1077	1496
Log Normal	696	771	846	1025

7.18 22950 m^3/s **7.19** (i) 7896 m^3/s, (ii) 8103 m^3/s

7.20 Ganga: $Q_{100} = 16359 \pm 2554$ m^3/s $Q_{1000} = 22023 \pm 3744$ m^3/s

Yamuna: $Q_{100} = 17298 \pm 3885$ m^3/s $Q_{1000} = 23935 \pm 5721$ m^3/s

7.21 567 m^3/s **7.22** (a) $\bar{x} = 385$ m^3/s; $\sigma_{n-1} = 223$ m^3/s (b) 1525 m^3/s

7.23

Time (h)	0	12	24	36	48	60	72	84
Q (m^3/s)	50	104	482	1669	3139	3699	3358	2603
Time (h)	96	108	120	132	144	156	168	
Q (m^3/s)	1928	1268	753	393	183	51	50	

7.24 Design Storm

Time (h)	0	6	12	18	24
Design strom rainfall excess (cm)	0	2.0	4.4	6.4	2.4

Flood Hydrograph

Time (h)	0	6	12	18	24	30	36
Q (m^3/s)	20.0	45.0	125	285	475	620	666
Time (h)	42	48	54	60	66	72	78
Q (m^3/s)	568	416	264	132	46	20	20

7.25 $T = 390$ years $\bar{R} = 12\%$ **7.26** $R_e = 0.603$

7.27 (a) $T_a = 10$ years (b) $T_b = 190$ years **7.29** 4908 m^3/s

7.30 $T = 247$ years; $x_t = C_{hf} = 26700$ m^3/s; $C_{af} = 34{,}710$ m^3/s; Safety margin = 8010 m^3/s

Chapter 8

8.1

Time (h)	0	3	6	9	12	15	18	21	24	27
Q (m^3/s)	0	1	10	27.6	38.29	41.88	40.26	35.35	29.30	23.27
Elevation (300.0 +) m	0.0	0.30	0.45	0.95	1.17	1.24	1.21	1.11	0.99	0.87

8.2

Time (h)	0	3	6	9	12	15	18	21	24	27
Q (m^3/s)	60.84	39.41	38.25	45.86	50.92	49.53	43.81	39.91	32.17	25.08
Elevation (300.0 +) m	1.50	1.19	1.17	1.29	1.36	1.34	1.26	1.20	1.04	0.90

8.3 62.38 m, 55.21 m^3/s, 4.8 m^3/s **8.4** 2 h 25 min

8.5

Time (h)	0	2	4	6	8	10	12	14
Q (m^3/s)	0	6.67	17.78	27.41	28.86	24.38	21.87	18.71
Elev. (200.0 +) (m)	0	0.67	1.78	2.74	2.89	2.44	2.19	1.87
Time (h)	16	18	20	22	24	26		
Q (m^3/s)	15.76	12.75	9.75	6.75	3.75	075		
Elev. (200.0 +) (m)	1.58	1.27	0.98	0.67	0.38	0.07		

8.6

Time (h)	0	2	4	6	8	10	12	14	16	27
Q (m^3/s)	0	4.0	18.4	43.04	61.82	65.09	59.06	49.43	39.66	25.08
Time (h)	18	20	22	24	26	28	30	32		
Q (m^3/s)	29.80	19.88	1193	7.16	4.29	2.58	1.55	0.93		

8.7 $K = 10.0$ h; $x = 0.3$

8.8 21 h, 35 m^3/s

8.9

Time (h)	0	6	12	18	24	30
Q (m^3/s)	35.00	29.94	25.49	28.91	41.14	69.49

8.10

Time (h)	0	4	8	12	16	20	24	28
Q (m^3/s)	8	8	12	21	25.5	25.25	22.63	18.81

8.11

Time (h)	0	3	6	9	12	15	18	21
Q (m^3/s)	10	13.91	24.63	35.48	41.55	44.08	43.99	41.97
Time (h)	24	27	30					
Q (m^3/s)	38.53	34.02	28.73					

8.12

Time (h)	0	1	2	3	4	5	6	7
IUH Ord. (m^3/s)	0	2.78	13.07	17.32	28.10	34.18	33.55	30.75
1-h *UH* ord. (m^3/s)	0	1.39	7.92	15.19	22.71	31.14	33.86	32.15
Time (h)	8	9	10	11	12	13	14	
IUH Ord. (m^3/s)	27.67	24.90	22.41	20.17	18.16	16.34	14.71	so on
I-h *UH* ord. (m^3/s)	29.21	26.29	23.66	21.29	19.16	17.25	15.52	so on

8.13 3 h, 0.6 m^3/s

8.14 $Q = \beta t \left(\frac{\beta}{\alpha} - I_0\right) e^{-\alpha t} - \frac{\beta}{\alpha}$

8.15

Time (h)	$u(t)$ in (cm/h)	$u(t)$ (m^3/s)	Time (h)	$u(t)$ (cm/h)	$u(t)$ (m^3/s)
0	0	0	33	0.01889	26.25
3	0.00211	2.93	36	0.01487	20.67
6	0.01022	14.20	39	0.01147	15.94
9	0.02092	29.08	42	0.00869	12.08
12	0.03007	41.08	45	0.00648	9.01
15	0.03563	49.52	48	0.00477	6.63
18	0.03734	51.90	51	0.00347	4.82
21	0.03596	49.99	54	0.00250	3.47
24	0.03256	45.26	57	0.00178	2.48
27	0.02812	39.09	60	0.00126	1.75
30	0.02340	32.52	63	0.00089	1.23

8.17

Time (h)	*IUH* $u(t)$ (m^3/s)	1-h *UH* (m^3/s)	Time (h)	*IUH* $u(t)$ (m^3/s)	1-h *UH* (m^3/s)
0	0	0	10	24.6	29.78
1	25.4	12.68	11	17.0	20.80
2	69.1	47.25	12	11.5	14.22
3	97.2	83.17	13	7.6	9.55
4	104.3	100.73	14	5.0	6.32
5	96.4	100.32	15	3.2	4.13
6	81.1	88.75	16	2.1	2.67
7	64.0	72.56	17	1.3	1.71
8	48.1	56.07	18	0.8	1.08
9	34.9	41.54	20	0.525	0.68

8.18

Time (h)	$u(t)$ in (m^3/s)	Time (h)	$u(t)$ in m^3/s
0	0.000	22	6.062
2	2.242	24	4.571
4	7.692	26	3.386
6	12.857	28	2.472
8	16.008	30	1.782
10	16.962	32	1.270
12	16.228	34	0.896
14	14.471	36	0.627
16	12.255	38	0.435
18	9.978	40	0.300
20	7.876	42	0.206

8.19 $n = 4.18$, $K = 3.35$ h

Chapter 9

9.1 $S = 16.34\%$, RD of solids = 2.517

9.2 (a) 4125 m^3/day (b) 44.175 m

9.3 1.92 cm/s

9.4 (a) 3.816×10^{-3} (b) –1.12%

9.5 156.459 m

9.6 31.46 m/day

9.7 1.73 m

9.8 4.6 years

9.9 3.2 m^3/day/m length

9.11 (i) $R = 6 \times 10^{-4}$ m^3/day/m width (ii) $h_m = 5.196$ m (iii) $q_o = -0.06$ m^3/day/m width (iv) $q_1 = 0.18$ m^3/day/m width

9.12 $a = 1978.4$ m, $q_a = -1.0842$ m^3/day/m width, $q_b = 1.108$ m^3/day/m width

9.13 2532 lpm

9.14 1716 lpm; (a) 9.5% increase (b) 100% increase (c) 50% inrease

9.15 11.17 m/day

9.17 2556 lpm, 10.54 m

9.18 $K = 25.8$ m/day; $T = 645$ m^2/day, $s_w = 10.78$ m

9.19 $h_A = 10.784$ m, $h_B = 10.853$ m; $(\mathbf{Re})_{PW} = 2.653$, $(\mathbf{Re})_A = 0.0148$, $(\mathbf{Re})_B = 0.00917$

9.20 (a)1159 lpm (b) 570 lpm

9.21 7.78 m/day

9.22 2303 m/day

9.23 1512 lpm

9.24 (a) 17.80 m (b) $T = 136.5$ m^2/day (c) 29.33 m (d) 65.2 m and 86.1 m

9.25 (i)0.943 h^{-1} (ii) 46.3 m^3/h

9.26 (i)0.3066 h^{-1} (ii) 5.96 m^3/h

9.27 36.0 m^3/h

9.28 $T = 105$ m^2/day; $S = 1.976 \times 10^{-4}$

9.29 $Q = 1611$ lpm

9.30 (i) $s = 3.3$ m, (ii) $s = 4.3$ m

9.31 1.234 m

9.32 $S = 4.06 \times 10^{-4}$; $T = 36.87$ m^2/h

9.33 $T = 28.1$ m^2/h

Chapter 10

10.1 (i) 4.17 ha-m/100 sq. km/year (ii) 10.32 ha-m/100 sq. km/year

10.2 $p = 0.679$, $W = 1529$ kgf

10.3 $W_{T1} = 1199.65$ kgf/m^3, $W_{T25} = 1261.2$ kgf/m^3

10.4 7% under prediction

10.5 1160 kgf/m^3

10.6 730 kgf/m^3

10.7 102 years

10.8 83 years

10.9 16 years

10.10 196 years

10.11 (a) 9.75 M. Tonnes/year; (b) 875 ppm

10.12 $T_{35} = 22$ years

10.13 $m = 3.2$

10.14

Sl. No.	Elevation (m)	Final Reservoir Area (ha)	Final Reservoir Capacity (ha.m)
1	350.52	0.0	0
2	365.76	0.0	0
3	381.00	777.3	4351
4	396.24	1807.9	25201
5	400.00	2112.0	35132
6	411.48	2871.5	59481
7	426.72	3720.0	110148
8	441.96	4632.7	172661
9	457.20	6059.1	256023
10	472.44	7921.4	364372
11	487.68	10435.7	505672
12	502.92	13325.1	682193
13	512.06	15378.0	817966

10.15

Elevation (m)	106.68	109.73	112.77	115.82	118.88	121.92	124.97
Reservoir Area after 10 years (ha)	0	256.65	1194.8	1176.8	2618.8	3881.8	5741.8
Capacity after 10 years (ha.m)	0	396	2922	7490	14128	23967	32711

INDEX

A

Actual Evapotranspiration (AET) 85, 94
Adjustment curve 151
Aeration zone 389
Agricultural drought 211
Agricultural drought and drought indices 212
Analytical methods of evaporation estimation 79
Annual rainfall 23
Annual series 305
Antecedent Moisture Condition (AMC) 187
Antecedent rainfall 175
Anticyclones 20
Apparent velocity 395
Application of unit hydrograph 256
Applications in engineering 12
Applied hydrology 1
Approximate methods 359
Approximations for bed load 458
Aquiclude 391
Aquifer 391
Aquifer properties 393
Aquifuge 391
Aquitard 391
Area increment method 479
Area-velocity method 131
Aridity index 212
Arithmetical-mean method 40
Artesian 392
Artesian aquifer 391
Artificial recharge 432, 433
Attenuation 351
Automatic stage recorders 124
Auxiliary gauge 151
Available water 85

B

Backwater effect 151
Barlow's Tables 178
Barometric efficiency 403
Base flow 165
Base flow separation 248
Basin lag 272
Bed load 457
Bed material load 457
Binnie's percentages 178
Blaney–Criddle formula 90
Brune's curve 466
Bubble gauge 125
Bulk pore velocity 395

C

Calculation of storage volume 198
Calibration 128
Calibration equation 128
Capillary fringe 390
Cascade of reservoirs 364
Catchment area 4
Catchment characteristics 169
Cetyl alcohol 84
Channel erosion 453, 457
Channel improvement 375
Channel routing 340
Characteristics of soil 99
Check dams 215
Chemical films 83

Chemical method 135
Chow's method 425
Chronically drought-prone area 211
Circulatory ratio 172
Clark's Method For IUH 360
Class A evaporation pan 75
Classification of infiltration capacities 109
Classification of reservoirs 472
Climatic factors 244
Coefficient of correlation 174
Coefficient of determination 174
Coefficient of permeability 394, 395
Coefficient of skew 317
Colorado sunken pan 76
Compactness coefficient 172
Complete numerical methods 359
Components of a hydrograph 245
Compressibility of aquifers 401
Compressibility of the pores 401
Compressibility of water 401
Conceptual hydrograph development 360
Cone of depression 415
Confidence limits 315
Confidence probability 316
Confined aquifer 391, 405
Confined flow 416
Confined groundwater flow 404, 405
Consistency of record 33
Constant-fall curve 151
Constant-head permeameter 397
Constant-rate injection method 136
Construction of S–curve 266
Continuous simulated models 184
Control section 141
Convective precipitation 20
Conveyance method 153
Convolution integral 284
Crest segment 245
Critical depression head 422
Critical storm 327, 328
Crop-management factor (C) 456
Current meters 126
Curve Number (CN) 186
Cyclone 19

D

DAD analysis 46
DAD curves 46, 47
Dalton's law of evaporation 74
Darcys 396
Darcy's law 394
Data for frequency studies 324
Delayed interflow 164
Delivery ratio 454
Demand and available water resources 223
Demand line 199
Density currents 466
Density of sediment deposits 467
Depletion curve 240
Deposition of sediment 452
Deposition process 465
Depression head 422
Depression storage 96
Depth-Area relation 45
Depth of snowfall 28
Derivation of D-hour unit hydrograph from IUH 284
Derivation of IUH 284
Derivation of T-hour unit hydrograph 267
Derivation of unit hydrograph 259
Design flood 324, 325
Design life 483
Design storm 326
Detention reservoirs 374
Deterministic watershed simulation 184
Dickens formula 303
Differential global positioning system (DGPS) 463
Diffusion equation 405
Dikes 374
Dilution technique 135
Direct runoff 165
Direct runoff hydrograph (DRH) 250
Direct storm runoff 165
Discharge-frequency curve 194
Distribution graph 271
Distribution of sediment in the reservoir 472

Divide 4
Double-mass curve technique 33
Double-ring infiltrometer 101
Downsizing 380
Drainage area 4
Drainage basin 4
Drainage density 170, 242
Drawdown 415
Drawdown test 426
Drizzle 18
Drought management 212
Drought prone area 211
Droughts 210
Droughts in India 216
Drought year 211
Dug wells 421
Duhamel integral 284
Dupit's assumption 406
Dupit's flow with recharge 409
Dupit's parabola 410
Dynamic resource 437

E

Echo-depth recorder 131
Economic life 483
Effective rainfall (ER) 250
Effective rainfall hyetograph (ERH) 250
Effluent streams 392
Electromagnetic method 138
El Nino 62
Elongation ratio 172
Empirical area reduction method 473
Empirical evaporation equations 77
Empirical probability distribution curve 306
Energy balance in a water body 80
Energy-budget method 80
Engineering hydrology 1
Envelope curves 304
Environmental flow assessment (EFA) 226
Environmental flows 225
Ephemeral stream 168, 169
Equation of continuity 341
Equation of motion 341
Equivalent runoff coefficient 299
Erosion 452
Erosion problem 485
Erosion processes 452
Estimation of missing data 31
Estimation of parameters of infiltration models 103
Estimation of sheet erosion 454
Estimation of watershed erosion 464
Evacuation and relocation 377
Evaporation 73
Evaporation stations 77
Evaporimeters 75
Evapotranspiration 73, 84
Evapotranspiration equations 86
Evidences of climatic change 379
Evolution of EF concept 225
Excess rainfall 250
Excess rainfall hyetograph 250
Extrapolation of rating curve 153
Extratropical cyclone 20
Extremes of extremes 322

F

Factors affecting erosion 453
Factors affecting runoff hydrograph 241
Falling-head permeameter 397
Field capacity 98
Float-gauge recorder 124
Flood 296
Flood control 373
Flood control in India 377
Flood embankments 374
Flood forecasting 376
Flood frequency plot 306
Flood frequency studies 305
Flood hydrograph 240
Flooding-type infiltrometer 100
Flood insurance 377
Flood management 373
Flood peak-area relationships 303
Flood plain zoning 376
Flood routing 340
Flood walls 375
Floodwater farming 215

Floodways 375
Flow duration durve 194, 462
Flow-mass curve 197
Flow-measuring structures 140
Flow without recharge 410
Fluid characteristics 99
Flumes 142
Formation constants 403
Formation loss 430
Form factor 172
Forms of subsurface water 389
Fow–mass curve 198
Free flows 141
Freezing rain 18
Frequency analysis 305, 306
Frequency of point rainfall 49
Freshwater lakes 11
Front 19
F-test 324

G

Gaining streams 392
Gauging site 131
Geologic formations 400
Ghyben–Herzberg lens 432
Glacier resources 224
Glaciers and permanent ice caps 11
Glaze 18
Global circulation models 379
Global freshwater resources 10
Global warming 378
Goodrich method 347
Green–Ampt equation 103
Greenhouse effect 378
Gross erosion 454
Groundwater 11
Groundwater flow 164
Groundwater monitoring network in India 441
Groundwater resources of India 438
Groundwater runoff 164
Gulley erosion 453
Gulp or integration method 135
Gumbel probability paper 310, 314
Gumbel's distribution 308
Gumbel's method 308

H

Habitat simulation 227
Hagen-Poiseuille flow 395
Hail 18
Heat storage in water bodies 74
History of hydrology 11
Holistic method 227
Homogeneity 324
Horton's equation 102
Hurricane 19
Hydraulic conductivity 395
Hydraulic method of flood routing 359
Hydraulic rating method 227
Hydrograph 167, 240
Hydrograph analysis 101
Hydrological drought 210, 211
Hydrological index method 227
Hydrologic channel routing 352
Hydrologic cycle 1, 2
Hydrologic equation 4
Hydrologic soil groups 109
Hydrologic storage routing 342
Hydrology 1
Hydrometry 122
Hydrometry stations 155
Hyetograph 37
Hyetometer 25
Hypsometric curve 171, 172

I

Incidental recharge 432
Indian summer monsoon rainfall 62
Infiltration 97, 390
Infiltration capacity 98
Infiltration indices 110
Inflow design flood 326
Influent streams 393
Inglis formula 303
Initial loss 95
Instantaneous unit hydrograph (iuh) 282
Interception 95
Interflow 164
Intermediate zone 390

Intermittent stream 168
Inter-rill erosion 452
Intrinsic permeability 396, 399
ISI standard pan 75
Isochrone 333
Iso-erodent lines 455
Isohyet 41
Isohyetal method 40, 41
Isopleths 93
Isopluvial map 57, 58

J

ϕ–index 110

K

Khosla's equation 464
Khosla's formula 182
Kirpich equation 298
Kostiakov equation 103

L

Lag 351
Lag time 272
Land use 188, 243
Land use/cover 186
La nina 63
Laplace equation 405, 407
Latent heat of vapourisation 116
Leaky aquifer 392
Length 171
Length of reach 136
Levees 374
Level pool routing 342
Life of a reservoir 483
Linear channel 360
Linear reservoir 353, 360
Linear response 253
Linear storage 353
Logarithmic-plot method 154
Log normal distribution 318
Log-Pearson type III distribution 317
Long-base weirs 142
Looping curve 354
Loop rating curve 152
Losing streams 393
Lysimeters 86

M

Manual gauges 123
Mass curve of rainfall 37
Mass-transfer method 81
Maximum depth-duration relationship 54
Maximum intensity-duration-frequency relationship 53, 54
Maximum intensity-duration relationship 53
Maximum yield 422
Mean annual flood 310
Mean grain size 395
Measurement of evapotranspiration 86
Measurement of infiltration 100
Measurement of precipitation 25
Measurement of stage 123
Measurement of velocity 126
Meteorological drought 210, 211
Meteorological radars 28
Method of mid-sections 132
Method of superposition 263
Methods of base-flow separation 249
Methods of predicting sediment distribution 473
Methods to reduce evaporation losses 83
Meyer's formula 78
Micro catchment system 215
Mining 437
Modelling infiltration capacity 102
Modified Pul's method 342
Modified universal soil loss equation (MUSLE) 457
Modular limit 141
Moisture availability index 212
Monsoon 20
Monsoon studies in India 61
Monsoon trough 21
Monthly rainfall of India 64
Moving average 38
Moving-Boat method 133
Moving means 38
MS Excel 357

Muskingum equation 353
Muskingum method of routing 356
Muskingum routing equation 357

N

Nalabunds 215
Nash's conceptual model 364
National water policy 223
Natural flow 165
Natural recharge 432
Natural-syphon type 27
Network density 29
Nonrecording gauges 25
Non-structural methods 373
Normal annual precipitation 31
Normal rainfall 31
Normal ratio method 32

O

Ombrometer 25
Open wells 421
Orographic precipitation 20
Overland flow 164

P

Palmer index 212
Pan coefficient Cp 76
Partial duration series 321
Penman's equation 86
Perched water table 392
Percolation 390
Percolation tanks 216
Perennial rivers 392
Perennial stream 167, 168
Permanent control 146
Permeameter 397
Philip's equation 103
Phreatic surface 406
Piezometric 424
Piezometric surface 424
Plateau gauging 136
Plotting position 50
Plotting-position formulae 51
Pluviometer 25
Point rainfall 38
Porosity 393
Post-monsoon 21
Potential evapotranspiration over India 93
Potential evapotranspiration (PET) 85
Precipitation 17
Prediction of monsoon rainfall 63
Pre-monsoon 21
Preparation of data 31
Prism storage 352, 353
Probable Maximum Flood (PMF) 325
Probable maximum precipitation 60
Procedure for calculation of ϕ-index 110
Prompt interflow 164

Q

Quantitative geomorphology 169
Quick return flow 164

R

Radar measurement of rainfall 27
Radius of influence 415
Rain 18
Rainfall 17
Rainfall data in India 63
Rainfall erosivity factor (R) 455
Rainfall excess 110
Rainfall infiltration method 434
Rainfall–runoff correlation 174
Rainfall simulator 101
Raingauge network 29
Rating curve 145
Rating equation 147
Rational method 296
Recession limb 245
Recharge 391, 432
Recharge due to seepage from canals 436
Recharge from irrigation 435
Recharge from water-harvesting structures 436
Recording gauges 26
Recovery 415
Recovery of piezometric head 427

Recovery of the cone of depression 428
Recovery test 428
Recuperation 415
Recuperation test 422
Recurrence interval 49
Reduced mean 311
Reduced standard deviation 311
Reduced variate 309
Reference crop evapotranspiration 90
Reference evapotranspiration 90
Regional flood-frequency analysis 322
Regulation diagram 203
Relief 171
Reservoir evaporation 82
Reservoir regulation effect 196
Reservoir routing 340
Reservoirs 11
Reservoir sedimentation 452, 489
Reservoir sedimentation control 484
Reservoir sedimentation surveys 463
Residence time 9
Residual drawdown 428
Residual mass curve 206, 207
Return flow 173, 218
Return period 49
Reynolds number 395
Rill erosion 453
Rising limb 245
Risk and reliability 329
Rivers 11
Rod float 130
Rohwer's formula 78
Rooftop water harvesting 214
Routing 362
Runoff 2, 164
Runoff characteristics of streams 168
Runoff coefficient 298
Runoff coefficient (c) 298
Runoff iso-chrone 361
Runoff process 164
Runoff volume 173
Ryves formula 303

S

Safety factor 329
Safety margin 329
Safe yield 422, 437
Saturated zone 389
Science of climate change 379
S-Curve 264
SCS-cn equation for Indian conditions 190
SCS-cn method 185
SCS dimensionless unit hydrograph 276
SCS triangular unit hydrograph 276
Sea-Water Intrusion 431
Secondary gauge 151
Sediment area 478
Sediment delivery ratio 454, 460, 464
Sediment–production rate 454
Sediment rating curve procedure 460
Sediment rating curve (Schematic) 459
Sediment yield 454
Sediment yield from watersheds 460
Seepage 392
Sequent peak algorithm 206
Shape 171
Shape factor 172
Shape of the basin 242
Sheet erosion 453
Sheet flow 453
Shifting control 145, 150
S-hydrograph 265
Simple (tube type) infiltrometer 100
Size 242
Sleet 18
Slope 171, 242
Slope-area method 142
Snow 18
Snow boards 28
Snow course 29
Snowfall 28
Snow gauges 28
Snow stakes 28
Snow tubes 29
Snyder's method 272
Snyder's synthetic-unit hydrograph 272

Soil conservation technology 484
Soil erodibility factor (k) 455
Soil type 186
Soil water zone 390
Soluble salts 74
Sorptivity 103
Sounding weights 130
Sources of data 13
Sources of groundwater 389
Southern oscillation 62
South-west monsoon 20
Specific capacity 431
Specific capacity per unit well area 423
Specific retention 394
Specific storage 402
Specific yield 394
Specific yield method 433
Spillway design flood 325
Spreading of water 216
Springs 392
Staff gauge 123
Stage 122
Stage data 126
Stage-discharge relationship 145
Stage for zero discharge 147
Stage hydrograph 126
Standard current meter method 132
Standard error 30
Standard Fourth-Order Runge-Kutta Method (SRK) 350
Standard Project Flood (SPF) 325
Standard SCS-CN equation 190
Standard value 395
Static resource 437
Station rainfall 38
Station-year method 70
Steady-flow 415
Steady flow into a well 416
Stemflow 96
Stilling wells 124
Storage coefficient 402
Storage reservoirs 373
Storage-time constant 353
Storativity 402
Storm hydrograph 240
Storm runoff 165
Storm seepage 164
Strange's tables 178
Stratification 397
Stream density 170
Stream flow 2, 122
Streamflow measurement 122
Stream order 170
Structural measures 373
St. Venant equations 341
Submerged flow 141
Subsurface storm flow 164
Sudden-injection method 135
Summer monsoon 20
Supporting cultivation practice (p) 456
Surface float 130
Surface of entry 99
Surface runoff 164
Surface-water resources of india 218
Suspended load 457
Symons' gauge 25
Synthetic unit hydrograph 272

T

Telemetering raingauges 27
Tennant's method 227
Theis equation 425
Thiem's equation 417
Thiessen-polygon method 40
Thiessen polygons 41
Thin-plate structures 141
Thornthwaite formula 92
Through flow 164
Tile drain problem 411
Time–area curve 361
Time–area histogram 360
Time invariance 253
Time of concentration 297
Tipping-bucket type 26
Topographic Factor (LS) 455
Total station 463
Towing tank 129
Tracers 136
Transmissibility 396, 402
Transmission zone 97
Transpiration 84

Transportation 452
Trap efficiency 465, 466
Tropical cyclone 19
t-test 324
Turbidity currents 466
Type curve 425
Types of evaporimeters 75
Typhoon 19

U

Ultimate infiltration capacity 102
Ultrasonic method 139
Unconfined aquifer 391
Unconfined flow 418
Unconfined flow with recharge 408
Unit hydrograph 252
Unit hydrograph from a complex storm 262
Unit hydrograph method 252, 304
Unit hydrographs of different durations 263
Universal Soil Loss Equation (USLE) 454
Unsteady-flow effect 152
Unsteady flow in a confined aquifer 424
Urbanisation 243
Useful life 483
Use of USLE 456
US geological survey floating pan 76
USNWS method 68
Utilisable water resources 220

V

Validation 184
Vapour pressure 74
Variable demand 203
Variation of AET 85
Velocity measurement by floats 130
Villemonte formula 141
Virgin flow 165

W

Wading 129
Warning 376
Wash load 457
Water-budget equation 4, 5
Water-budget method 79
Water divide 409
Water equivalent of snow 28
Water harvesting 213
Watershed 4, 169
Watershed management 375
Watershed simulation 183
Water table 392
Water table aquifer 391
Water year 167
Weather systems 18
Wedge storage 352, 353
Weibull formula 50, 305
Weighing-bucket type 26
Weighting factor 353
Well efficiency 430
Well function 424
Well loss 430
Wells 415
Wetlands 11
Wetting zone 97
W- index 112
Winter season 21
Wire gauge 123
Within-field catchment system 215
WMO criteria for hydrometry station density 156
Working head 422
World bank 225
World's fifteen largest rivers 219
World water balance 7

Y

Year books 157
Yield 173

Z

Zone of aeration 390

Made in United States
North Haven, CT
10 November 2026

22096531R00300